Handbook of Ayurvedic Medicinal Plants

L.D. Kapoor, Ph.D., F.I.A.C., F.A.S.P.

*Scientist (retired)
National Botanical Research Institute, Lucknow
Council of Scientific and Industrial Research
India*

CRC Press
Boca Raton   London   New York   Washington, D.C.

Cover image courtesy of
T. Michael Kengla
GrassRoots Productions

Library of Congress Cataloging-in-Publication Data

Kapoor, L.D.
  [CRC handbook of ayurvedic medical plants]
  Handbook of ayurvedic medicinal plants / L.D. Kapoor.
    p. cm.-- (Herbal reference library)
  Previously published under the title: CRC handbook of ayurvedic medicinal plants.
  Includes bibliographical references and index.
  ISBN 0-8493-2929-9 (alk. paper)
    1. Materia medica, Vegetable--India--Handbooks, manuals, etc. 2. Medicine,
Ayurvedic--Handbooks, manuals, etc. I. Title. II. Herbal reference library series.

RS180.I3 K365 2000
615'.321'0954—dc21   00-049360

This book contains information obtained from authentic and highly regarded sources. Reprinted material is quoted with permission, and sources are indicated. A wide variety of references are listed. Reasonable efforts have been made to publish reliable data and information, but the author and the publisher cannot assume responsibility for the validity of all materials or for the consequences of their use.

Neither this book nor any part may be reproduced or transmitted in any form or by any means, electronic or mechanical, including photocopying, microfilming, and recording, or by any information storage or retrieval system, without prior permission in writing from the publisher.

The consent of CRC Press LLC does not extend to copying for general distribution, for promotion, for creating new works, or for resale. Specific permission must be obtained in writing from CRC Press LLC for such copying.

Direct all inquiries to CRC Press LLC, 2000 N.W. Corporate Blvd., Boca Raton, Florida 33431.

Trademark Notice: Product or corporate names may be trademarks or registered trademarks, and are used only for identification and explanation, without intent to infringe.

First pulished by CRC Press LLC © 1990
© 2001 by CRC Press LLC

Herbal Reference Library edition
No claim to original U.S. Government works
International Standard Book Number 0-8493-2929-9
Library of Congress Card Number 00-049360
Printed in the United States of America 1 2 3 4 5 6 7 8 9 0
Printed on acid-free paper

PREFACE

The ancient system of medical treatment is based on the rich experiences of innumerable Vaidyas over thousands of years, having trials on hundreds and thousands of human beings to its credit, to which no modern system of treatment in the world can ordinarily lay claim. This is one possible reason why this system has survived the critics through the ages and is still catering to the health needs of millions all over the world. The system is basically within the economic reach of the common individual because of nature's bounty to mankind in herbal wealth of medicinal value.

The active ingredients of a drug, synthesized in the small unicellar laboratory of the leaf and then transported to reservoirs embedded within roots, fruits and leaves, bark, and flowers under photosynthetic influence, are nature's gift to mankind. This alone speaks well for the inexpensive nature of the treatment afforded by traditional medicine in comparison to what the modern method of treatment through allopathy makes available and that, too, only to the affluent few in human communities.

The Indian materia medica contains over 2000 drugs, the majority of which are of vegetable origin. During the time of the Emperor Ashoka, the Hindu materia medica contained only 700 plants which were used by the Vaidyas. They were mostly cultivated and grown all over the country and their method of collection, manner of storage, and preservation were all well known. Since the number of drugs commonly used in those days was not very large, no elaborate descriptions were given for their identification. The student of medicine used to live with his guru in "gurukulas" (residential schools) and received practical training in the correct identification of drugs and plants and their collection at the proper time.

In the course of time, more and more vegetable herbs were gradually added to the indigenous materia medica, but unfortunately standards of purity and correct identification did not keep pace with this process of expansion.

The identity of a drug plant was based on the doctrine of signatures, viz., "pashanbhed", meaning thereby the plant which can split a stone or grow through rocks. It is also indicative of dissolving or splitting of stones in the bladder or kidney. But there are quite a few plants which have the same diagnostic character as described above but which are different from each other and yet are called "pashanbheds" in different parts of India, such as *Aerva lanata* Juss.; *Bergenia ligulata* (Wall.) Engl., *Coleus amboinicus* Lour., *Homonoia riparia* Lour., *Kalanchoe pinnata* Pers., *Ocimum basilicum* Linn., *Rotula aquatica* Lour., etc. Similarly, a Unani drug plant called "gaozaban" (like the tongue of a cow) may be *Macrotomia benthami* DC.; *Onosma bracteatum* Wall.; or *Anchusa trigosa, Caccinia glauca* Savi, etc. The doctrine of signatures was not often recorded, but the knowledge was handed down from "guru" to disciple or father to son where the Ayurvedic profession was confined to the family.

As time passed, the herbal gardens disappeared due to rapid colonization of rural forest areas. Much confusion has been created in the correct identification of such plants. Many plants are known by the same name, or many names are given to the same plant in different parts of the country. This has made matters more complex.

An attempt has been made in the present volume to describe some 251 plants, giving their different names in different languages in India and abroad. Their habitat, distribution and chemical constituents, and pharmacological or therapeutic uses and dosage as prescribed in the Ayurveda are also indicated. Pharmacognostic details, wherever possible, are also incorporated to confirm identification. A bibliography of references is included.

It is hoped that this information will now help correct the identification of Ayurvedic drug plants within the debris of confusion.

The vast country of India has varying agro-climatic and soil conditions ranging from arid to alpine zones, and every type of plant can be found growing in nature; further, many more exotics can be grown here. With its lush and diverse flora, India is rightly called the vegetable emporium of medicinal plants. The mountain range of the Himalayas has been known to be a repository of rare important medicinal plants since time immemorial. The great sages of Ayurveda fame used to assemble here for discussion and discourses on the theory and practice of Ayurveda and its materia medica.

Drug plants, which form about 90% of the materia medica of traditional medicine, play the pivotal role in the efficacy of the treatment. A wrong or dubious plant used in the formulations or treatment can give a bad name to the system itself.

This project was originally conceived by Dr. D. P. Sharma Ayurved Chakravarti, Pranacharya, who, in consultation with the late Dr. C. Dwarkanath and Prof. R. R. Pathak, worked on the format and sketched out a model of Ayurvedic description, properties, and action/uses for each plant, based on the literary records and personal experiences of leading practicing Ayurveds.

The number of Ayurvedic medicinal plants described in the text is by no means exhaustive. Some of these are more frequently prescribed than others. It would be rewarding if any of these plants gives a lead for more thorough and systematic research for a potential therapeutic drug against such fell diseases as cancer and AIDS, etc. Sometimes lesser-known plants may unfold latent virtues with the help of modern sophisticated technology.

When this manuscript was in preparation, I received the shocking news of Dr. Sharma's premature sad demise. It was tragic to lose such a friend, philosopher, and guide. He was a man of eternal vision and he pursued a missionary zeal to map Ayurveda on the scientific forefront.

It is my proud privilege to dedicate this volume to the memory of the late Dr. D. P. Sharma, who devotedly served the cause of the development of Ayurveda in India and in neighboring countries.

THE AUTHOR

L. D. Kapoor, Ph.D. is a retired scientist from the National Botanical Research Institute, Lucknow, India. He received his B.Sc. from Punjab University, his M.Sc. from Banaras Hindu University, Varanasi, and his Ph.D. from London University. He worked in the Drug Research Laboratory of Jammu and the Kashmir Government and was appointed Botanist-in-Charge, Botany Division. He initiated research in pharmacognosy and in the survey and cultivation of medicinal and aromatic plants, some of which were grown on a semi-commercial scale. He was Editor-in-Chief of the Regional Research Laboratory Bulletin, Jammu. For the first time in the temperate Himalayas, he successfully introduced the culture of ergot on rye, now grown on a commercial scale. His Majesty's Government, Nepal, invited Dr. Kapoor to advise on the establishment of herbal farms for the commercial production of ayurvedic drugs, for pharmaceutical preparations. He was the recipient of an award for his invention of a new distillation apparatus by the President of India.

Dr. Kapoor is a fellow of the Indian Academy of Science, Bangalore, and the American Society of Pharmacognosy. He is co-author of two books and has published more than 200 research papers. He now resides in New Jersey.

ACKNOWLEDGMENTS

I received helpful cooperation and encouragement from a number of scientists, for which I am very grateful. The names of Dr. James A. Duke, Dr. G. M. Hocking, Dr. William C. Steere, Dr. G. V. Satyavati, Prof. P. N. Mehra, Prof. T. S. Sadasivan, Dr. S. R. Tandon, and Dr. S. B. Malla need special mention.

For permission to reproduce plates from their reprinted book *Indian Medicinal Plants,* by Kirtikar and Basu, I am indebted to Messrs. Bishen Singh and Mahinderpal Singh of Dehradun. To Little Brown & Company, Boston, I am grateful for permission to reproduce the drawing of *Rauwolfia serpentina* and its pharmacognostical description.

I am especially indebted to the Director General of the Indian Council for Medical Research (ICMR), New Delhi for permission to reproduce excerpts from *Medicinal Plants of India,* Volumes 1 and 2. I am also grateful to Dr. W. C. Evans for permission to reproduce some excerpts from his book, *Pharmacognosy* (12th Edition, 1983).

For providing the up-to-date botanical nomenclature, I am particularly grateful to the Director of the Royal Botanical Gardens, Kew, England; the President of the Forest Research Institute, Dehradun; Prof. Cecil J. Saldanha, S. J., Director of the Center for Taxonomic Studies, St. Joseph's College, Bangalore; and to Dr. H. O. Saxena, Scientist, Regional Research Laboratory, Bhubneshwar (Orissa).

Special grateful thanks are due to Mr. B. J. Starkoff of CRC Press, Inc., for making publication of this volume possible. To Mr. Promod Kumar Sharma and his colleagues Mr. R. K. Gupta and Mr. Vinay Kumar of Baidyanath Ayurved Bhawan Patna, who made available the recorded material, I am gratefully indebted.

For her inspiration and moral support I am very grateful to my wife Prakash. I am equally thankful to all the members of my family for their helpful cooperation in many ways. I owe my thanks to the staff of the Library of Science and Medicine, Rutgers University, New Jersey for their helpful cooperation.

It would not have been possible to finalize this project without the active cooperation and secretarial help of Mrs. Judy Ann Reilly, to whom I owe my sincere thanks.

L. D. Kapoor

TABLE OF CONTENTS

Introduction 1

Ayurvedic Medicinal Plants 5

Appendices
    Basic Concepts of Ayurveda 347
    Introductory Notes on the Fundamental Principles of Ayurvedic Pharmacology 349

Glossary 353

References 361

List of Figures and Credits 389

Index 391

INTRODUCTION

THE ANTIQUITY OF INDIAN MATERIA MEDICA

The history of medicine in India can be traced to the remote past. The earliest mention of medicinal use of plants is to be found in the Rigveda, which is one of the oldest, if not the oldest, repositories of human knowledge, having been written between 4500 and 1600 B.C. In this work, mention has been made of the Soma plant and its effects on man. In the Atharvaveda, which is a later production, the use of drugs is more varied, although it takes the form, in many instances, of charms, amulets, etc. It is in the Ayurveda, which is considered as an Upaveda (or supplementary hymns designed for the more detailed instruction of mankind), that definite properties of drugs and their uses have been given in some detail. Ayurveda, in fact, is the very foundation stone of the ancient medical science of India.[1] It has eight divisions that deal with different aspects of the science of life and the art of healing: (1) *kaya cikitsa*, or internal medicine; (2) *salya tantra*, or surgery; (3) *salakya tantra*, or the treatment of diseases of the head and neck; (4) *agada tantra*, or toxicology; (5) *bhuta vidya*, or the management of seizures by evil spirits and other mental disorders; (6) *bala tantra*, or pediatrics; (7) *rasayana tantra*, or geriatrics, including rejuvenation therapy; and (8) *vajikarana tantra*, or the science of aphrodisiacs. The age of Ayurveda is fixed by various Western scholars at somewhere about 2500 to 600 B.C. The eight divisions of the Ayurveda were followed by two works written later, i.e., *Susruta*[5] and *Charaka*.[4] About the date of *Susruta*, there is a great deal of uncertainty, but it could not have been written later than 1000 B.C. In this work, surgery is dealt with in detail, but there is a comprehensive chapter on therapeutics. *Charaka*, written about the same period, deals more with medicine, and its seventh chapter is taken up entirely with the consideration of purgatives and emetics. In its twelve chapters there is to be found a remarkable description of materia medica as it was known to ancient Hindus. The simple medicines alone are arranged by this author under forty-five headings. The methods of administration of drugs are fully described and bear a striking resemblance to those in use at the present time; even administration of medicaments by injections for various diseased conditions has not failed to attract notice and attention. From *Susruta* and *Charaka*, various systems dealing with different branches of medicine sprang up. Dr. Wise (1845; quoted in Reference 6) mentions two systems of Hindu surgery and nine systems of medicine, three of materia medica, one of posology, one of pharmacy, and three of metallic preparations alone. From these one can gather the strength and dimensions of the scientific knowledge of ancient India regarding therapeutic agents of both organic and inorganic origin. Even anesthetics in some form or other were not unknown."Bhojaprabandha", a treatise written about 980 A.D., contains a reference to inhalations of medicaments before surgical operations, and an anesthetic called "sammohini" is said to have been used in the time of Buddha.[1]

From this period down to the Mohammedan invasion of India, Hindu medicine flourished. Its progress may briefly be traced through four distinct stages, namely, (1) the Vedic period; (2) the period of original research and classical authors; (3) the period of compilers and also of *tantra* and *siddhas* (chemists-physicians); and (4) the period of decay and recompilation. During the second and third periods, progress was remarkable in every respect, and Ayurveda then attained its highest development. Towards the close of this period, Ayurvedic medicine made its way far beyond the limits of India. The nations of the civilized world of that time eagerly sought to obtain information regarding the healing art from the Hindus of those times; the influence of Hindu medicine permeated far and wide into Egypt, Greece, and Rome, and molded Greek and Roman medicine and through the former, Arabic medicine also. Jacolliot[7] very rightly and pertinently remarked, "We should not forget that India, that immense and luminous center in olden times, was in constant communication with all the peoples of Asia and that all the philosophers and sages of antiquity went there to study the science of life." There are unmistakable evidences in Grecian and Roman medicine of the influence of Hindu medicine. Hellenic civilization came most intimately in contact with Indian civilization through the conquests of Alexander the Great. During this period, Indian medicine was at its zenith, and the knowledge of the Hindu physicians in the domain of drug therapy and toxicology was far in advance of others. They made an intensive study of the properties of every product of the soil and systematically devoted their attention to the study of disease and its treatment with drugs. The skill of these physicians in curing snake bites and other ailments among the soldiers of the Grecian camp bears testimony to this. No wonder then that Grecian medicine imbibed in large measure the knowledge of the healing science and enriched its materia medica from those of the Hindus. There is reason to believe that many Greek philosophers, like Paracelsus, Hippocrates, and Pythagoras, actually visited the East and helped in the transmission of Hindu culture to their own countries. The work of the great physician Dioscorides definitely shows to what extent the ancients were indebted to India and the East for their medicine. Many Indian plants are mentioned in his first work, particularly the aromatic group of drugs for which India has always been famed. The smoking of datura in cases of asthma, the use of nux vomica in paralysis and dyspepsia, and the use of croton as a purgative, can be definitely traced to an origin in ancient India. Even the effects produced by excessive smoking of datura came to their notice.

The Romans also took a great interest in Indian drugs. There is evidence to show that an extensive trade in Indian drugs existed between India and Rome many centuries ago. The country, with enormous variations of climate and with such wonderful ranges of mountains as the Himalayas, was from the earliest times recognized as a rich nursery of the vegetable materia medica. In the days of Pliny, this drug traffic assumed such enormous proportions that he actually complained of the heavy drain of Roman gold to India in buying costly Indian drugs and spices. The following extract from the writing of an English student of Oriental literature will be of interest in this connection. In the course of a lecture, Captain Johnston Saint mentioned the extraordinary advance made both in surgery and medicine in India when Europe was groping for light in her cradle in Greece.[8] Says he, "If then this is what we found in surgery, what may we not find in medicine from India — that vast and fertile country which is a veritable encyclopedia of the vegetable world. The materia medica of the ancient Hindu is a marvel from which both the Greeks and the Romans freely borrowed."

AYURVEDA DURING THE BUDDHIST REGIME

The advent of Buddhism in India brought considerable change in the practice of Ayurveda.[2] Surgery, the performance of which is invariably associated with pain, was treated as a form of *himsa* or violence, and therefore, its practice was banned. To compensate for this loss, and to alleviate the sufferings of ailing humanity, more drugs were added during this period to Ayurvedic materia medica. Prior to this period, metals and minerals were no doubt used for therapeutic purposes. But their use was in a crude form and they were sparingly used. Buddhist scholars added considerably to the metals and minerals in Ayurvedic materia medica, inasmuch as *rasa sastra,* or iatro-chemistry, formed a specialized branch and many authentic texts were composed on this subject. Thus, the loss by discarding the practice of surgery was well compensated by the addition of iatro-chemistry to the materia medica of Ayurveda during this period.

Some of the Buddhist rulers, like Asoka, established several herb gardens, so that people could obtain drugs conveniently for the treatment of their diseases. Buddhist monks were encouraged to learn Ayurveda and practice it, which was considered to be the most convenient and popular method of spreading the teachings of their preceptor. Thus, through Buddhism, Ayurveda spread to Sri Lanka, Nepal, Tibet, Mongolia, the Buriyat Republic of Soviet Russia, China, Korea, Japan and other Southeast Asian countries. Ayurvedic texts, including texts on materia medica, were translated into the languages of these countries and these are still available in translated form, even though some of the originals have been extinct in the country of their origin. Based upon the fundamental principles of Ayurveda, some local herbs, diet, and drinks were included in the traditional medicine of the respective countries. The materia medica of Ayurveda was thus enriched.

India was severely invaded by outsiders such as the Greeks, the Saks, and the Huns. When these people came into contact with the rich tradition of India, especially its medical science, they carried back with them medicines and doctors. Indian doctors who went out with them translated Ayurvedic texts into their languages. The new drugs which were specially used in those places were incorporated into the Ayurvedic pharmacopoeia. Such repeated exchanges of scholars enriched Ayurveda and its materia medica. Through commercial channels, Ayurvedic drugs and spices were exported to these countries and were held in high esteem there.

Several universities were established during this period for imparting theoretical and practical training in different religious and secular subjects. Taxila and Nalanda were two such universities which not only attracted intellectuals from different parts of this country, but also many from abroad. In Taxila, there was a medical faculty with Atreya as its Chairman. In Buddhist literature, there are many interesting stories and ancedotes about the activities of these universities, their scholars, and faculty members. One such anecdote is related to Jivaka, who was three times crowned as the King of Physicians because of his proficiency in the art of healing. He was an expert in pediatrics and brain surgery. It was the practice at that time for candidates desirous of admission to the faculty in the university to appear for a test before the "Dvarapala" or the gate-keeper. Jivaka and several other noble princes had to face these tests before being considered for admission to the medical faculty. They were asked to go to nearby forests and collect as many plants as possible which did not possess any medicinal property. By evening, candidates returned with several plants which they considered to be free from medicinal properties. Jivaka did not return for several days and, when he did, he was empty-handed. On a query from the gate-keeper, Jivaka replied, "I could not find any plant, or for that matter anything which does not have medicinal value." He alone was selected for admission to the medicinal faculty. This was the status of knowledge of the candidates for the medicinal faculty at that time. This was the time when the knowledge of materia medica became highly developed.

THE DECAY OF INDIAN MEDICINE

After the period of the Tantra and Siddhas, the glories of Hindu medicine rapidly waned and declined. During the invasion of India by the Greeks, Scythians, and Mohammedans successively, no original works were written, and Hindu medicine gradually began to decay. During the disturbed times that followed, a good deal of existing Ayurvedic literature was mutilated or lost, and degeneration became discernible everywhere. Various branches of medicine passed into the hands of priests, and drugs and herbs gave way to charm and amulets. The medicine man himself became a member of a subcaste of Brahmins to whom knowledge and learning were chiefly confined. A large section of them began to think that the study and practice of the healing art, especially surgery, led to pollution. To touch the dead body was considered sinful, and, dissection of dead bodies being discontinued, advancement in anatomical and surgical knowledge declined. The Buddhist doctrine of "Ahimsa" also exercised a great influence in that direction. Though surgery declined to a great extent during the Buddhistic period, medicine again made rapid progress. As already stated, a large number of valuable drugs began to be systematically cultivated and investigated in this period. With the decline of Buddhism, degeneration set in all round — in knowledge, learning, and practice of both medicine and surgery — and the process of decay became well advanced about the time of the Mohammedan invasion.

With the advent of the Muslim conquerors, the decline was even more rapid. The invaders brought their own healing system, which was fairly advanced for that period, and as Mohammedan rule became established, the old Hindu or the Ayurvedic system of treatment was rapidly thrown into the background. The Arabic system thus introduced became the system of relief.

Professor Brown,[9] in his lectures before the Royal College of Physicians, showed how greatly Arabic medicine was influenced by Greek learning in the early centuries of the Christian era. Although the chief pursuit of the chemists about this time was the philosophers' stone and the elixir of life, they nevertheless made many real discoveries. How many of these we owe to the Arabs is apparent from such words as "alcohol", "alembic", and others still current at the present time. There is no doubt that it is in the domain of chemistry and materia medica that the Arabs added most to the body of scientific doctrine which they inherited from the Greeks. Leclerc,[11] in his *Histoire de la Medicine Arabe,* points out that even a century earlier than the Arab conquest of Egypt, the process of assimilating Greek medicine had begun. Arabian medicine was also influenced by the Persian Jundi-Shapor school, which flourished in Persia in the 5th century A.D. This is evident from the fact that there is undoubtedly a Persian element, especially in materia medica, in which Arabic nomenclature plainly revealed in many instances Persian origin. About the middle of the 8th century, when the city of Baghdad was newly founded, the great stream of ancient learning began to pour into the Mohammedan world and to reclothe itself in Arabian dress. The Mohammedan system of medicine thus brought with it a rich store of its own materia medica, quite unknown to the country.

THE ADVENT OF ARABIAN AND WESTERN MEDICINE

The Arabian and Mohammedan medicine prevalent during the reign of the Pathan and Moghul dynasties unfortunately did not make much progress after its introduction into the country and with the fall of the Moghuls it rapidly decayed. During the intimate contact between the old Hindu medicine and the Arabian medicine, which lasted for a century, there was a great deal of intermingling, and each utilized the materia medica of the other. The result was that, though both the systems had declined, a rich store of the combined materia medica was left behind. With the advent of Europeans — first Portuguese, then the French, and last the English — the decline was still further marked.

When British rule was established, the Western system was introduced, and it was primarily intended to give relief to those who administered the country. As there was no proper system of medical relief in vogue at that time, the newly introduced Western system found its way among the people and was welcomed by them; the appreciation and the demand for it extended all over the country, especially as its surgical achievements appealed strongly to the people and made a great effect on them. It also brought with it its own materia medica and there was further intermingling and introduction of new medicinal plants into the country.

ABIES SPECTABILIS (D. Don) G. Don *Family:* CONIFERAE
(Syn. *A. WEBBIANA* Lindl.)

Vernacular names — Sanskrit, Talisapatram; Hindi, Talispatra; English, Himalayan silver fir; Bengali, Talispatra; Nepalese, Tulsi; Sinhalese, Talispaturu; Japanese, Himaraymoni; French, Sapin; Unani, Tulsi; Arabian, Shahasfarm; Persian, Rayhan; Tibetan, Ba-iu; German, Edeltanne.

Habitat — This is a lofty tree growing on the higher ranges of the Himalayas.

Parts used — Leaves.

Morphological characteristics — A tall tree, mostly evergreen. Leaves are spirally arranged, very variable in length, flat, about 2 mm broad; petiole very short. Flowers monoecious, perianth 0. Male flowers in deciduous catkins with scale-like stamens bearing 2 to 5 one-celled pollen sacs on the lower surface. Female flowers in cones consisting of scale-like open carpels which are flat and bear 1 to 2 naked ovules. Fruit usually a woody cone. Seeds often winged, testa thin, albumen fleshy.[28]

Ayurvedic description — *Rasa— tikta; Guna — laghu, teekshna; veerya — ushna; vipak — madhura.*

Action and uses — *Kafa vat samak, badana nasak, rochan, dipan, swashar, jawrahar, mutrajanak, bal bardhak.*

Chemical constituents — It contains a crystalline alkaloid known as taxin.

Pharmacological action — Leaves are carminative, stomachic, tonic, astringent, antispasmodic, and expectorant.[3]

Medicinal properties and uses — Decoctions or tinctures in doses of 2 to 4 ml or infusions in doses of 16 to 48 ml of the dried leaves are useful in cases of cough, phthisis, asthma, chronic bronchitis, and catarrh of the bladder, and other pulmonary affections. It is also given in enlarged spleen along with musk.[1]

Doses — Powder — 0.5 to 1 g; infusion — 16 to 48 ml; tincture — 2 to 4 ml.

ABROMA AUGUSTA (Linn.) L. f.　　　　Family: STERCULIACEAE

Vernacular names — Sanskrit, Pinchaskarpas; Hindi, Olat kambal; English, Devil's cotton; Bengali, Olatkambal; Nepalese, Mhashukapay; Sanukpasi, Sinhalese, Peevary; Japanese, Aogiri; Chinese, Taimeo; German, Abrome; Unani, Ulat kambal.

Habitat — *A. augusta* grows wild throughout the hotter parts of India, from Utter Pradesh to Sikkim, Khasia Hills, and Assam. It is also cultivated for its showy, deep-scarlet flowers.

Parts used — Root, root bark, stem, and leaves.

Pharmacognostical characteristics — Root is brownish in color; fractures fibrously; and has indistinct odor. Root bark has a dull brown outer surface which is longitudinally wrinkled along with small lenticels. The inner surface is smooth, dark brown, and longitudinally striated. Bark is not brittle and can be easily separated. Mitra and Prasad[13] studied detailed microscopic characters of young and mature root and root bark as well as the young stem and stem bark and its mature bark. Mucilaginous cavities, starch grains, and rosette crystals of calcium oxalate have been reported in root and stem, as well as the other normal tissues.

Ayurvedic description — *Rasa — katu, tikta; Guna — laghu, rooksha, teekshna, snigdh; Veerya — ushna; Vipaaka — katu.*

Action and uses — *Kafa vat samak, garvasaya balya, artaw janan, badanasthapan, mutral, snahan.*

Chemical constituents — The roots of the plant were reported to contain an alkaloid abromine, $C_{16}H_{13}NO$, mp 283 to 285°C, friedelin, and abromasterol A, mp 125.5°C. Later on isolation of taraxeryl acetate, taraxerol, β-sitosterol, and low-melting neutral compound from the petroleum ether extract of the leaves was reported.[14-16] The identity of abromine was confirmed as betaine. Besides, β-sitosterol and stigmasterol were also isolated from the nonsaponifiable fraction of the petroleum ether extract of root.[17]

Pharmacological action — Leaves and stem are demulcent, root bark is an emmenagogue. It is reported as oxytocic and liable to produce dermatitis.[3]

Medicinal properties and uses — Infusion in cold water of fresh leaves and stem is very efficacious in gonorrhea. Fresh juice of the root bark is more efficacious and is given in dysmenorrhea. A single administration during the menses will regulate the menstrual flow; also acts as a uterine tonic and to facilitate conception in young married women.[1]

Doses — Root bark — 1 to 2 g; decoction of root — 14 to 56 ml; leaf infusion — 15 to 25 ml.

ABRUS PRECATORIUS Linn.   *Family:* LEGUMINOSAE

Vernacular names — Sanskrit, Gunja; Hindi, Gunja; English, Jequirity; Bengali, Kunch; Nepalese, Ratigedi; Sinhalese, Olinda; Japanese, Toazuki; German, Paternostererbsen; Unani, Ghongchi; Arabian, Ain-ud-dik; Persian, Chashm-e-Kharoos; Burmese, Gyingwe; Malaysian, Akar be limbing; Tamil, Kunri.

Habitat — Throughout India from the Himalayas down to southern India and Sri Lanka.

Morphological characteristics — It is a beautiful woody climber, flowers in August to September, and pods ripen by the end of the cold season. The seeds are slightly smaller than ordinary peas. They are usually of bright scarlet color with a black spot at one end. White seeds are also present. The root is woody, tortuous, and much branched. The seeds are reported to be a hot, dry tonic and an aphrodisiac. The seeds are poisonous when ground into a paste and made into needles which are inserted under the skin of animals to kill them. The seeds are also reported to be abortifacient, antiseptic, antitubercular, and antidysenteric.[1,3]

Parts used — Roots, leaves, and seeds.

Ayurvedic description — *Rasa* — *tikta, kashaaya; Guna* — *laghu, rooksha, teekshna; Veerya* — *ushna; Vipaaka* — *katu.*

Action and uses — *Tridoshahar, khustaghan bran ropan, kesya, bajikaran, grvonirodhak, jwaraghan, gravapatak.*

Chemical constituents — Hemagglutinating principle of *A. precatorius* was found to be a heat-labile protein located in the cotyledon. Two new alkaloids, the methyl ester of *N,N*-dimethyl tryptophan methocation and picatorine, were isolated from the seeds. In addition, abrine, hypaphorine, choline, and trigonelline were also isolated from seeds. Root and leaves contain about 10% of glycyrrihizin and abrin and acid resin.[18] Abrasine, precasine, and hypaphosrine alkaloids have been reported from roots and seeds.[10]

Pharmalogical action — Seeds are purgative, emetic, tonic, antiphlogistic, and aphrodisiac.

Medicinal properties — The leaves are demulcent; their juice is given for the cure of sore throat and dry cough. It is also a good blood purifier. A paste of the leaves made with some bland oil is applied over swellings and in rheumatism. The seeds are strongly purgative and emetic. Seed paste is used as rubefacient in sciatica, stiff shoulders, paralysis, and other nervous diseases. It is also useful for skin diseases, ulcers, and inflammations. A 3.0% infusion of the decorticated seeds is a useful lotion for eye diseases, such as granular eyelids and corneal opacity.

Doses — Leave decoction — 56 to 112 ml; root powder — 0.5 to 1 g; seed powder — 1 to 3 gr.

ACACIA CATECHU (Linn. f.) Willd.   *Family:* MIMOSACEAE

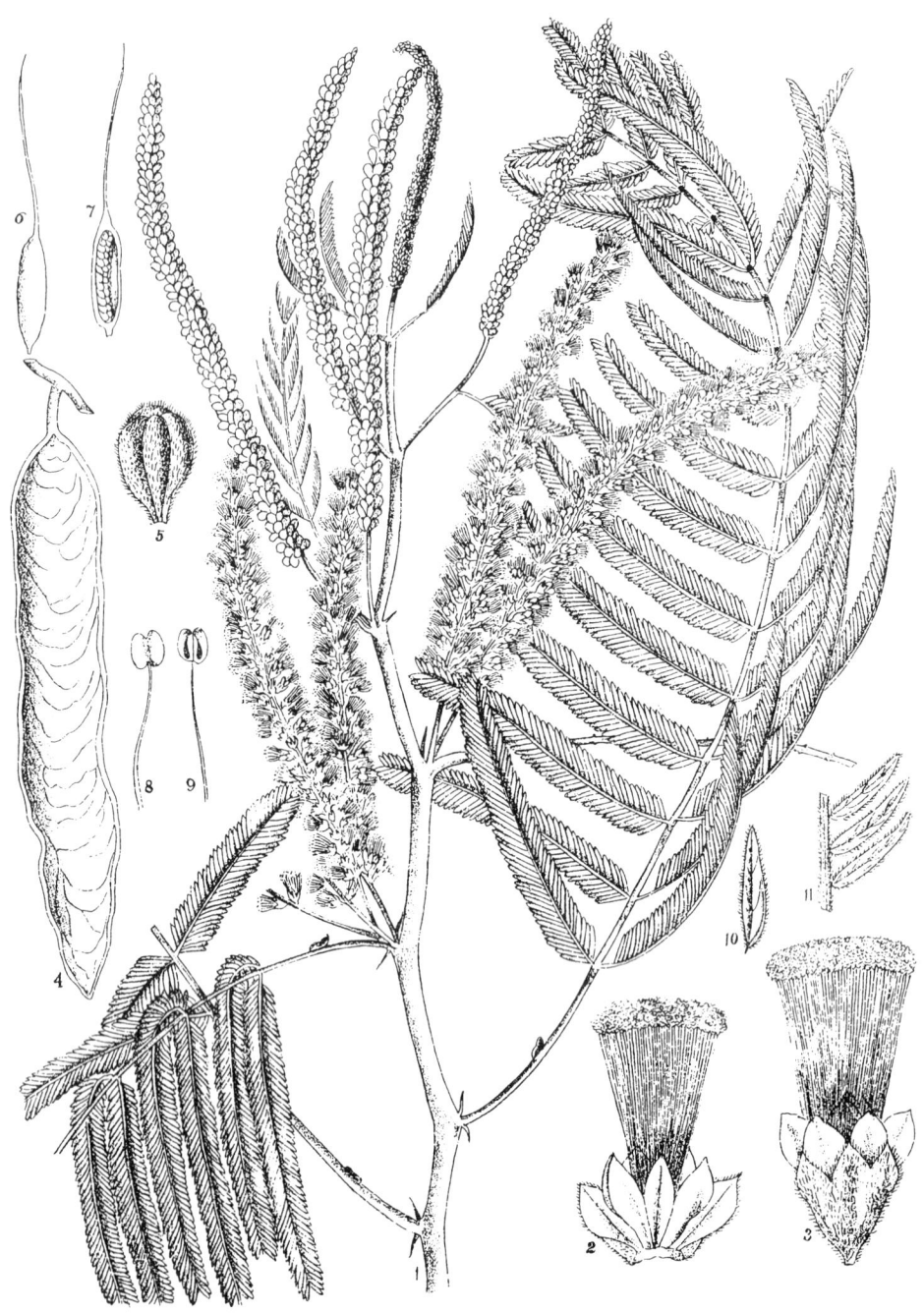

Vernacular names — Sanskrit, Khadira; Hindi, Katha; English, Catechu; Begali, Khayer; Nepalese, Khaya; Sinhalese, Ratkihiri; Japanese, Penguasenyaku; German, Katechubaum; French, Cachoutier; Unani, Katha; Arabian, Kaat; Persian, Kaat; Tibetan, Sen loen; Burmese, Sha; Malaysian, Khadiram; Tamil, Karugali.

Habitat — The tree is widely distributed in India from the Indus eastward to Assam and throughout the peninsula. It is also common in the dry plains and lower hill forests of upper and lower Burma.

Parts used — Extract, bark, wood, flowering tops, and gum.

Pharmacognostical characteristics — *A. catechu* is a moderate-sized tree. Its dark grayish-brown bark is nearly $1/2$ in. in thickness, exfoliating in long, narrow strips, brown or red inside. The most important products obtained from *A. catechu* is catechu. This is obtained by boiling the heartwood with water. As sold in the market, *katha* is found in irregular pieces or small square blocks of grayish color, which after breaking show crystalline fracture.

Ayurvedic description — *Rasa — tikta, kasaya; Guna — lagu, rooksha; Veerya — sheeta; Vipak — katu.*

Action and uses — *Kaf pit samak, raktsodhak, kustaghan, soth har, asthamban, krimighan.*

Chemical constituents — The main constituents of the heartwood are catechin and catechutannic acid; the catechin content varies from 4 to 7%. The catechin of *A. catechu*, also called acacatechin, is a colorless crystalline material insoluble in cold water but soluble in hot water. Wood contains α-, β-, γ- catechin[19] and *l*-epicatechin.[20]

Pharmacologic action — It is a powerful astringent.

Medicinal properties and uses — The wood extract, known as *katha,* is very useful in chronic diarrhea and dysentery, bleeding piles, uterine hemorrhage, leukorrhea, and gleet, in doses of 0.5 to 1.5 g. Catechu is very efficacious for mercurial salivation, bleeding or ulcerated or spongy gums, hypertrophy of the tonsils, and ulceration of the mouth. It may be used as a lozenge. An infusion of catechu is a very useful douche in uterine hemorrhage, leukorrhea, and gonorrhea. Tincture of catechu is particularly useful for bed sores and painful mammary glands. A mixture of catechu and myrh is used as a tonic and as a galactagogue for nursing mothers. This plant is also alleged to have antiseptic and antidysenteric properties.[1]

Doses — Bark and wood in powder form — 1 to 2 g; decoction — 28 to 56 ml; flowers — 1 to 2 g; extract — 0.5 to 1 g.

ACACIA FARNESIANA (L.) Willd. Family: MIMOSACEAE

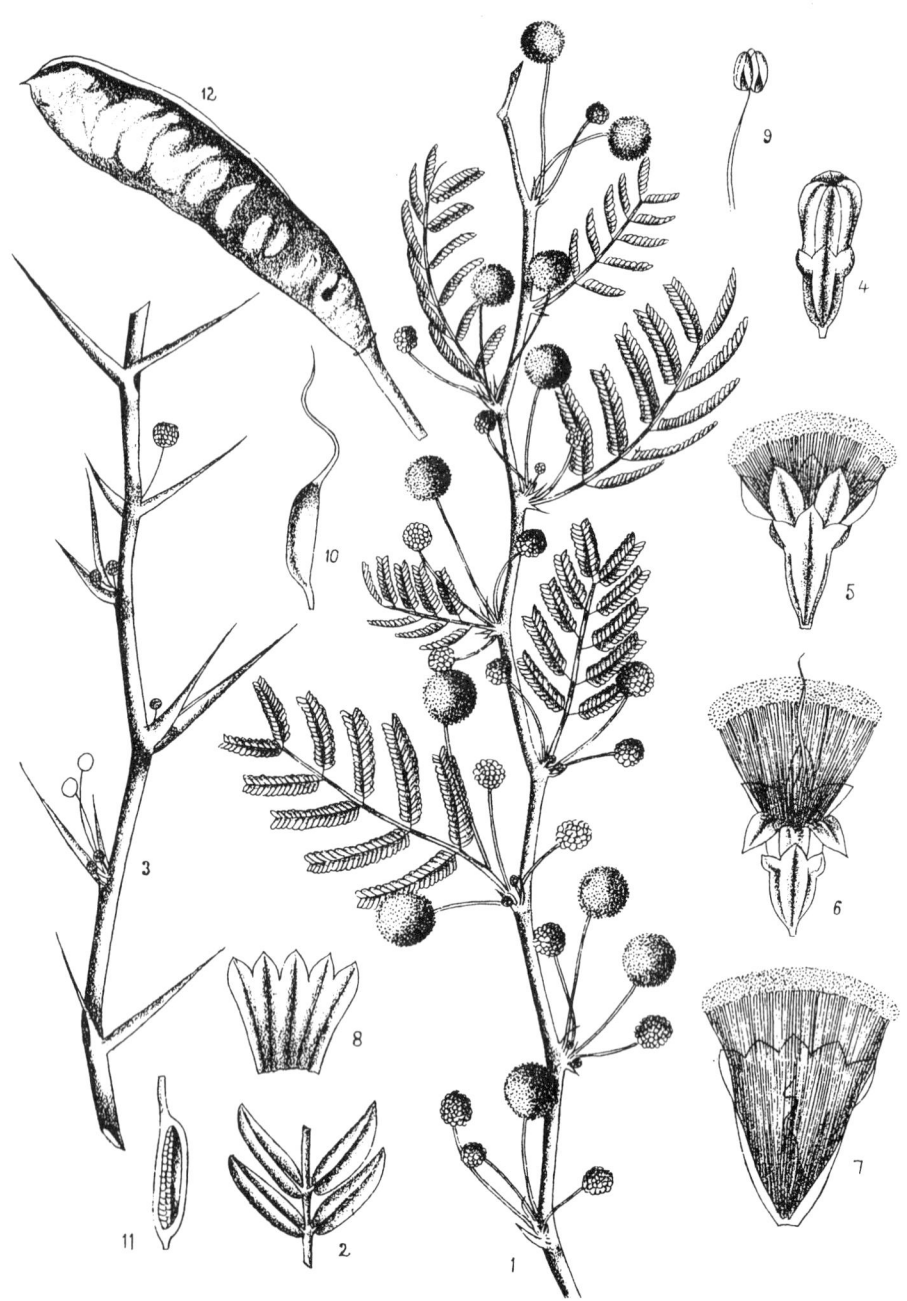

Vernacular names — Sanskrit, Arimaedah; Hindi, Gand babul; English, Cassia flower; Bengali, Guyabebula; Sinhalese, Seenidda; Japanese, Kinko; Chinese, Keota; German, Akazie; Unani, Kikar; Arabian, Ummughilan; Persian, Mughilan; Tamil, Priki aruvel.

Habitat — Found throughout India, often planted in gardens.

Parts used — Bark, leaves, gum, pods, and flowers.

Morphological characteristics — A small erect tree, usually armed; leaves 1 to 1.5 inches long; flowers aromatic; deep orange in color.

Ayurvedic description — *Rasa* — *tikt, kasaya; Guna* — *laghu, rooksha; Veerya* — *sheeta; Vipak* — *katu.*

Action and uses — *Kapha pitta samak, sthamvan, ruchi vardhak, krimighan, sonit sthapan, kus taghan, sothahar.*

Chemical constituents — Oil from flowers contains benzaldehyde; salicylic acid; methyl salicylate; and benzyl alcohol. Pods contain tannin.[23]

Pharmacological action — Astringent, demulcent, and alterative.

Medicinal properties and uses — Decoction of the bark with ginger is an astringent wash for the teeth and is useful in bleeding of the gums. A gum exudes from the bark of the tree which is a good substitute for gum arabic. Tender leaves are bruised and swallowed with a liitle water for gonorrhea. The oil is employed as an adjunct to an aphrodisiac in spermatorrhea. Purified *silajit*, a drug of mineral origin soaked in a decoction of this plant, is reported to increase efficacy.[1]

Doses — Bark decoction — 28 to 56 ml.

ACACIA NILOTICA (Linn.) Delile   *Family:* MIMOSACEAE
  ssp. *INDICA* (Benth.) Brenan
(Syn. *ACACIA ARABICA* [Lam.] Willd.)

Vernacular names — Sanskrit, Babhoola; Hindi, Babhul; English, Babul tree; Bengali, Babla; Nepalese, Babhul; Sinhalese, Babboola; Japanese, Indogom; German, Babool Baum; Unani, Babool; Arabian, Mughilan; Persian, Ummughiion; Burmese, Huanlong kyain; Tamil, Vel.

Habitat — Common all over India in dry and sandy localities, abundant in western peninsula, the Deccan and Coromandel coast.

Parts used — Bark, gum, leaves, seeds, and pods.

Morphological characteristics — A medium-sized evergreen tree with dark brown bark; inner surface smoother than outer; leaves 2.5 to 5 cm long, leaflets 10 to 20 pairs 3 to 6 mm long; membranous heads, flowers yellow, fragrant; pods subindehiscent, persistently gray-brown, 3 to 6 in. long and 0.5 in. broad, contain eight to twelve seeds.

Ayurvedic description — *Rasa* — *kashaaya;* *Guna* — *guru, rooksha; Veerya* — *sheeta; Vipaak* — *katu.*

Action and uses — *Kapha pitta samak, rakta rodhak, bran ropan, sthamvan, sankoshak, krimighan, sanhan, grahi, rakt pitta samak, mutral, sthamvan, daah pra saman.*

Chemical constituents — Gum contains arabic acid combined with calcium, magnesium, and potassium, as well as a small quantity of malic acid, sugar, moisture 14%, ash 3 to 4%. Bark contains a large amount of tannin; pods contain about 22.44% tannin.

Pharmacological action — Astringent, demulcent, aphrodisiac, nutritive, and expectorant. Antiprotozoal and CNS depressant activity of stem bark was reported by Bhakuni et el.[11]

Medicinal properties and uses — A paste of the tender tops made with water and sugar is given as a demulcent in cough. An infusion of the tops is given as a douche in gonorrhea and leukorrhea. The tender leaves are eaten as an astringent for diarrhea and dysentery. The decoction is a useful gargle for relaxed sore throat and spongy gums; it is also used for washing bleeding ulcers and wounds. An ointment made of the burnt leaves with coconut oil is an efficacious remedy for skin diseases. Decoction of the bark is used as a demulcent in prolapse of the anus; it is also prescribed for the treatment of gonorrhea, cystitis, leukorrhea, vaginal discharge, and prolapse of the uterus. The gum is an emollient and demulcent; it is particularly useful in diarrhea, dysentery, and diabetes; it also soothes the inflamed membranes of the pharynx, alimentary canal, and genitourinary organs; it is prescribed for cough and chest complaints. A decoction of the gum is used as a tonic or as an aphrodisiac. The decoction of the bark or infusion is used as a collyrium in ophthalmia and as a wash for ulcers. The gum of this tree is an efficient substitute for true gum acacia.[1] The plant is alleged to have antiseptic and antidysenteric properties.

Doses — Leaves — 2 to 4 g; bark decoction — 28 to 56 ml; seed powder — 1.5 to 3 g; gum — 2 to 6 g.

ACHYRANTHES ASPERA Linn. *Family:* AMARANTHACEAE

Vernacular names — Sanskrit, Apamarga; Hindi, Chirchira Apamarg, Latjira; English, Rough chaff tree; Bengali, Apang; Nepalese, Datiwan; Sinhalese, Gaskara, Karalheba; Japanese, Hinata inokozuchi; Chinese, Niu-hsi; Unani, Chirchita; Arabian, Ankamah; Persian, Khar-vazbun; Tibetan, A-pa-ma-rga; Burmese, Kivalamon; Malaysian, Kada; Tamil, Nayurivi.

Habitat — *A. aspera* is a small herb found as a weed all over India, up to 3000 ft.

Parts used — Herb, leaves, seeds, and root.

Pharmacognostical description — Epidermis of the stem is covered by glandular hairs and rubiaceous-type stomata. The pericycle has lignified fibers arranged in a discontinuous ring. The vascular tissue, having five to six rings of xylem with small strands of phloem, is occupied by pith with two medullary bundles, either separate or fused. The epidermis of the leaves has glandular and covering trichomes and stomata and rosettes of calcium oxalate crystals are present.[21]

Some of the distinguishing features of the stem and root are the stem shows six to ten ridges and collenchyma is present under each ridge; both stem and root show anomalous growth and have four to six successive rings of xylem and phloem. The primary diarch structure is visibly intact in the mature root. The true seed contains a dicotyledonous curved embryo embedded in a hard horny endosperm.[22]

Ayurvedic description — *Rasa — katu, tikta; Guna — laghu, rooksha, teekshana; Veerya — ushna; Vipaaka — katu.*

Action and uses — *Krimghna, paachana, rakta vardhaka, sphota hara.*

Chemical constituents — The whole plant of *A. aspera* has been found to contain traces of basic substances, one of which was identified as betaine and a water-soluble alkaloid—achyranthine.[23,24] From the seeds of the plant two well-defined and pure saponins, which have been called Achyranthes saponins A and B, have been isolated.[25] The seeds yield hentriacontane from petroleum ether extract and 2% saponin from alcoholic extract, also oleanolic acid-oligosaccharide saponin. The root extract gives oleanolic acid from the glycosidic fraction.[178]

Pharmacological action — Astringent, diuretic, alterative, antiperiodic and purgative.[3]

Medicinal properties and uses — The leaves, seeds, and twig of the plant are used in the treatment of renal dropsy and bronchial conditions. It is also very useful in the treatment of leprosy. Decoction of the plant, 56 g boiled with 900 ml of water for 20 min, is a good diuretic found efficacious in renal dropsies and general anasarca; 28 to 56 ml of the mixture is given two or three times daily; leaf juice is also useful in stomachache and bowel complaints, piles, boils, and skin eruption. In large doses it produces abortion or labor pains. A decoction of powdered leaves with honey or sugar candy is useful in the early stage of diarrhea and dysentery.

Doses — Decoction — 28 to 56 ml.

ACONITUM FEROX Wall ex Ser. *Family*: RANUNCULACEAE

Vernacular names — Sanskrit, Vatsanaba; Hindi, Mithzaahar; English, Indian aconite; Bengali, Mithvaish; Nepalese, Bisha; Sinhalese, Vachchavi; Japanese, Hanatorikabuto; Chinese, Wu-t'ou; German, Wilder Sturmhut; Unani, Bichnak; Arabian, Bish; Persian, Zahar; Tamil, Nabi.

Habitat — *A. ferox* grows wild in temperate and subalpine regions of the Himalayas from Kashmir to Sikkim and Nepal.

Parts used — Dried tuberous root.

Pharmacognostical characteristics — Flowers light blue, similar to the pea flower; fruits small and spiny. Root 1 to 3 in. long, externally brown and internally white in color.

Ayurvedic description — *Rasa* — *madhura; Guna* — *rooksha, teekshna, laghu, boyabai, bikashr; Veerya* — *ushna; Vipaak* — *madhura*.

Action and uses — *Vat kapha samak, badana, stapan, sotha har, dipan, pachan, swadak, mutral, jwaraghan.*

Chemical constituents — The tubers contain a crystalline toxic alkaloid called pseudaconitine, similar to aconitins, with a transparent vitreous appearance, soluble in boiling water, slightly soluble in ether, chloroform and alcohol, and a small quantity of aconitine (0.97 to 1.23%), picro-aconine, aconine, benzyl aconine, and homonapelline.[50]

Pharmacological action — Roots are diaphoretic, diuretic, antiperiodic, anodyne, antidiabetic, antiphlogistic, and antipyretic in very small doses. In large doses it acts as a powerful sedative, narcotic, and poison.[1,3]

Medicinal properties and uses — Used in diseases like neuralgia and muscular rheumatism, acute and chronic, itching as in erythema; in nasal catarrh, tonsilitis, sore throat, coryza, acute gout, and other painful affections; and in leprosy it is alterative and is a nervine tonic; in small doses in cases of paralysis, the root in the form of linament or paste is applied externally.

Doses — $1/4$ gr (after purification).

ACONITUM HETEROPHYLLUM Wall   Family: RANUNCULACEAE

Vernacular names — Sanskrit, Ativisha; Hindi, Atis; English, Indian, atees; Bengali, Ataicha; Nepalese, Atiras; Sinhalese, Athividayam; Japanese, Hanatorikabuto; Chinese, Wu-tou; German, Eisenhut; Unani, Atis; Persian, Vijjiturki; Tibetan, Bon-na-dkar-po; Tamil, Athividayam.

Habitat — It is found in subalpine and alpine zones of the Himalayas from the Indus to Kumaon, at an altitude of 7500 to 12,000 ft. in central and western Himalayas.[26]

Parts used — Dried tuberous root.

Pharmacognostical characteristics — A biennial herb. Flowers yellow-green. The root of a heterophyllum is 2 to 8 cm long and 0.4 to 1.5 cm thick. The root can be identified in transverse section by the presence of four to nine xylem bundles near the periphery embedded in secondary phloem tissue. Four to nine small cambium rings are present near the periphery. Sieve islets are conspicuous and present in the medullary region.[27]

Ayurvedic description — *Rasa — jikta, katu; Guna — laghu, rooksha; Veerya — ushna; Vipaka — katu.*

Action and uses — *Paachana, deepana, samgraabe, aamateesraghna, kapha ateesaraghna, pittaatee saraghna, aamavishaghna, krimighna, kaphapita jwaraghna, vishara shophagna.*[26]

Chemical constituents — As reported by Chopra et al.,[1] the roots contain the amorphous nontoxic alkaloid atisine (0.4%) to which Lawson and Tops[29] have assigned the formula, $C_{22}H_{33}O_2N$. Jacobs and Craig[30] have further isolated two crystalline alkaloids, viz., hetratisine $C_{22}H_{33}O_5N$, mp 262 to 267°C (decomp), and hetisine $C_{20}H_{27}O_{35}N$, mp 253 to 256°C (decomp). Recently the alkaloids benzoylheteratisine, heteratisine, heterophyllidine, heterophylline have been isolated from the rhizomes of this plant.

Pharmacological action — Roots of *A. heterophyllum* are bitter, tonic, astringent, febrifuge, stomachic, antiperiodic, and aphrodisiac.

Medicinal properties — Root contains nontoxic amorphous alkaloids and is one of the best bitter tonics for children.[27] The alkaloid "atisine" is employed medicinally in India as an antiperiodic, aphrodisiac, and tonic. In combating debility after fevers it is a very useful tonic and aphrodisiac; very efficacious in diarrhea, dysentery, and acute inflammatory affections. It is also very useful in cough and dyspepsia. The plain powder of the tuberous root mixed with honey is given to children for cough, coryza, fever, and vomiting.

Doses — Powder — 0.3 to 3 gr.

ACORUS CALAMUS Linn.   Family: ARACEAE

Vernacular names — Sanskrit, Vacha; Hindi, Bach; English, Sweet flag; Bengali, Bach; Nepalese, Sapee; Sinhalese, Vadakala; Japanese, Shobu; Chinese, Shui chang; German, Kalmus; Unani, Buch; Arabian, Waj Turki; Persian, Agar turki; Tibetan, Su-dag; Burmese, Linhe; Malaysian, Deringu; Tamil, Vasamber.

Habitat — *A. calamus* is a semiaquatic perennial plant cultivated in damp, marshy places in India and Burma. It is quite common in Manipur and the Naga Hills on the edges of lakes and streams,[26] also growing wild, ascending to 3000 to 6000 ft in the Himalayan ranges.[28]

Parts used — Rhizomes.

Pharmacognostical characteristics — A semiaquatic perennial herb with a creeping and much branched aromatic rhizome which is cylindrical or slightly compressed. Externally it is light brown or pinkish-brown, but white and spongy within. In transverse section the rhizome shows a prominent endodermis and scattered vascular bundles both in the central as well as in cortical regions. The outer surface is bounded by an epidermis or cork beneath which is to be seen a broad cortex separated from a large central cylinder or stele by the epidermis. Under the epidermis there are three to five layers of collenchymatus cells forming the hypodermis. Under the hypodermis there are many layers of parenchymatous cortex; their cortical cells are of various dimensions and generally with intercellular spaces. Thin-walled parenchyma cells are filled with starch grains. They are arranged in chains surrounding large intercellular spaces, each chain containing one or more large spheroidal sacs with a yellowish volatile oil content. A number of concentric bundles with fibers occur scattered in the cortex. The endodermis is a continuous single layer of barrel-shaped cells with a thickened side wall. The vascular bundles of the stelar region form almost complete rings under the endodermis and are also scattered throughout the ground tissue. Here, almost all the bundles are typically leptocentric. Sometimes fibers occur in the phloem region, too. The ground parenchyma is the same as has been described for cortical parenchyma; here, also, large spheroidal secretion sacs containing yellowish volatile oil are found.

Ayurvedic description — *Rasa — tikta, katu; Guna — lagu, tikshna, sar; Veerya — ushna; Vipak — katu.*

Action and uses — *Kapha vat samak, pittawardhak, sothar, badanasak, sangaya asthapak, kas har, swedjenak, lakhan, jawaraghan, vamak, anuloman.*

Chemical constituents — The dry rhizomes contain 1.5 to 3.5% of a yellow aromatic volatile oil. It has a mellow odor resembling that of patchouli. The chief constituent of the oil is asaryl-aldehyde. There are also present a bitter glycoside named acorin and some other substances such as eugenol, ascarone, pinene, and camphene. The Indian oil has much higher asarone content than the European commercial oil. Kelkar and Roe[31] found that in addition to asarone, $C_{12}H_{16}O_3$ (mp 62 to 63°C), the oil contains a small amount of sesquiterpenes and sesquiterpenic alcohols.

Fractionation of the active principle from volatile oil by gas phase chromatography revealed the presence of two components isolated in a pure state, i.e., α-asarone and β-asarone, which were *trans* and *cis* isomers, respectively, of 2,4,5-trimethoxy-l-propenyl benzene.[32] *Acorus* sp. is reported to contain an alkaloid choline.[10]

Pharmacological action — The rhizome is an aromatic, stimulant, bitter, tonic, carminative, antispasmodic, emetic, expectorant, emmenagogue, aphrodisiac, laxative, and diuretic.[3] The alcoholic extract of the plant has been shown to possess sedative and analgesic properties; it causes a moderate depression in the blood pressure and respiration. The water-soluble fraction of the dealcoholized extract produced relaxation of isolated intestine and negative inotropic action on frog's heart.[1] The insecticidal activity of the solvent extracts and steam-distilled volatile principle of rhizome against common houseflies is quite marked. The insecticidal activity of the oil and the extract appears to be due to the presence of the *trans*-isomer of asarone.[33,34]

Medicinal properties and uses — Its infusion is given in diarrhea, dysentery, bronchial and chest affections, and epilepsy. The burned root mixed with some bland oil is applied over the abdomen in flatulent colic; the poultice is also usefully applied to paralyzed limbs and rheumatic swelling.

Doses — Powder — 2 to 7 gr; as emetic — 10 to 15 gr.

ADIANTUM PHILIPPENSE Linn.
(Syn. *ADIANTUM LUNULATUM* Burm. f.)

Family: POLYPODIACEAE

Vernacular names — Sanskrit, Hansaraj; Hindi, Hansapadi; English, Maiden-hair fern; Bengali, Kalijhant; Nepalese, Hanaraj; Japanese, Hongekujaku; Unani, Hansraj; Arabian, Sharuljin; Persian, Parsiyaushan.

Habitat — The fern is found in moist places throughout North India, in the western side, in the plains and lower slopes of the hills in South India.

Parts used — Whole plant.

Morphological characteristics — Rhizome soft and reddish. Leaves small, soft, margin oval, 0.5 to 1.3 in. and curved; lower surface of the leaves contains reproductive organs.

Ayurvedic description — *Rasa* — *madhur, tikta, kasaya; Guna* — *guru, snigdha; Veerya* — *sheeta; Vipaak* — *madhur.*

Action and uses — *Vatta pitta samak, kafa nisark, daaha prasaman, vishaghan, bran ropan, krimighan, grahi, rakt sodhak, rakta pitta samak, kasahar, swas har, mutral, balya.*

Pharmacological action — Febrifuge, diuretic, alterative and anthelmintic.

Medicinal properties and uses — As an expectorant, it is given in cough and respiratory troubles; due to cooling properties it is used as diuretic, febrifuge, and tonic. Rhizomes are used in leprosy, fever, and erysipelas.[50]

Doses — Infusion — 4 to 12 ml; powder — 1 to 3 g.

AEGLE MARMELOS (L.) Corr. ex Roxb.   Family: RUTACEAE

Vernacular names — Sanskrit, Bilva-shriphala; Hindi, Bael; English, Bael fruit; Bengali, Bael; Nepalese, Bhyacha; Sinhalese, Beli; Japanese, Igurusumarumero; Chinese, Mak toum; German, Bela-Fruchte; Unani, Baelgari; Arabian, Safarjal hindi; Persian, Balegari; Tibetan, Kabed; Burmese, Okshit; Tamil, Vilvo, Villuvam.

Habitat — The tree is indigenous to India and is found wild all over the sub-Himalayan tract, in Bengal, in Central and South India, and in Burma. It is also cultivated to a great extent.

Parts used — Fruits (both ripe and unripe), root bark stem, leaves, rind of the ripe fruit, and flowers.

Pharmacognostical characteristics — A deciduous tree, 20 to 25 ft in height and 3 to 4 ft in girth, with straight, sharp, axillary thorns and trifoliate aromatic leaves. The flowers are greenish white. The fruits are globose, 2 to 4 in. in diameter, gray or yellowish, and with smooth, hard, aromatic rind. Seeds are numerous, oblong, and compressed, and the pulp is mucilaginous, thick, orange red in color.

The root bark is 3.5 mm thick, curved, with its surface cream yellow or grayish in color. The surface is rough, irregular, and shallow with ridges along the line of lenticels and ruptured all over. The stem bark is externally gray and internally cream in color. The outer surface is rough and warty. It is 4 to 8 mm thick, firm in texture, and occurs as flat or channeled pieces. The fracture is tough and gritty in the outer region and fibrous in the inner; taste is sweet.

The root in transection shows a pentarch to heptarch stele, the cork cambium arising in the pericycle. The cork is lignified and stratified, the phelloderm is composed of a wide zone of parenchyma cells with strands of stone cells in the mature bark. The medullary group in the inner region is uni- to triseriate, while in the outer region it is bi- to pentaseriate.[38]

Ayurvedic description — *Rasa — kasaya, tikta; Guna — rooksha, lagu; Veerya — ushna; Vipaka — katu.*

Action and uses — *Kapha bat nasak, soth har, badana asthapan, dipan, pachan, grahi, mridurachan, hirdya, rakt stamphan, jaraghan.*

Chemical constituents — The leaves contain a neutral alkaloid, aegelin, and a sterol having mp 144 to 145°C, identified as γ-sitosterol. Further studies on the leaves resulted in the isolation of new alkaloid coumarins, viz., aegelinin (a minor alkaloid) and marmin, a new coumarin from the bark of the trunk.[39-42] The leaves also contain 0.6% of essential oil, mostly composed of *d*-limonene. The fruits contain 4.6% sugar and tannin, 9% in the pulp and 20% in the rind.[43,44] The seeds yield 11.9% bitter fatty oil. Chopra et al.[178] reported the presence of γ-sitosterol in the leaves. Fruits yield essential oil containing *d*-α-phellandrene, alloimperatorin, and β-sitosterol, its carbohydrate. The heartwood contains a ferroquinoline alkaloid, dictamine, a dihydrofuro coumarin, "marmasin", and β-sitosterol.

Pharmacological action — Ripe fruit is sweet, aromatic, cooling, alterative, and nutritive. When taken fresh it possesses laxative properties. Unripe fruit is astringent, digestive and stomachic.

Medicinal properties — A decoction of the leaves is a febrifuge and expectorant; it is particularly used for asthmatic complaints. Decoction of the root bark is used in intermittent fevers, melancholia, and palpitation of the heart. It is best given in subacute or chronic cases of diarrhea and dysentery and in irritation of the alimentary canal.[1,3] Decoction of the root, and root bark, is useful in intermittent fever; also hypochondiasis, melancholia, and palpitation of the heart. Powdered pulp is given in doses of 2 to 4 gr in acute dysentery with griping pain.

Doses — Powder — 2 to 12 gr; infusion — 12 to 20 ml; decoction — 28 to 56 ml.

ALANGIUM SALVIIFOLIUM (Linn. f.) Wang.
(Syn. ALANGIUM LAMARCKII Thwaits)

Family: ALANGIACEAE

Vernacular names — Sanskrit, Ankola; Hindi, Ankola; English, Sage leaves, alangium; Bengali, Akaranta; Nepalese, Ankol; Sinhalese, Rukanguna; Japanese, Nagabaurinoki; Chinese, Choimoi; Unani, Ankol; Tibetan, A-rko-ta; Tamil, Azhunjil.

Habitat — A deciduous shrub or a small tree found in tropical forests throughout India and Burma. Common in tropical forests of South India, Burma, and Sri Lanka.

Parts used — Root, bark, seeds, and leaves

Morphological characteristics — The bark is brown, $1/2$ in. thick; spiny branch; leaves 3 to 6 in. long and 1 to 2 in. broad. Flowers yellowish white. Seeds red containing oil.

Ayurvedic description — *Rasa* — *tikta, katu, kasaya; Guna* — *laghu, snigdha, teekshna, sara; Veerya* — *ushna; Vipaak* — *katu.*

Action and uses — *Kapha-vat samak, pitta sansodhak, badana sthapan, bran ropan, vishaghan, sothahar, shul prsaman, balya, brighana jwaraghan.*

Chemical constituents — The plant yields an alkaloid, alangine-A (0.82%) in the air-dried material.[52] Dasgupta isolated a new crystalline phenolic alkaloid ankorine from the leaves, having a mp of 174 to 176°C and molecular formula $C_{19}H_{29}O_4N$. It has two OCH_3 groups, one phenolic-OH group and possibly one furan ring.[53] Choline was isolated from aqueous solution of leaves.[54] Some steroid and terpenoids have also been isolated from the leaves.[55] A new alkaloid from the leaves, alangimarckine ($C_{29}H_{37}N_3O_3$, (mp 184 to186C), was found to be identical with ankorine.[56]

Root bark showed the presence of acetyl alcohol and nonsaponifiable matter (16.8%), myristic acid (6.73%), palmitic acid (9.36%), oleic acid (85.8%), linoleic acid (3.7%), and resin acids (2.3%). The alcoholic components were myricyl alcohol and sterols, among which stigmasterol, sitosterol and β-sitosterol were detected in nonsaponifiable matter.[57,58] The chemical investigation of root bark and stem bark showed a mixture of alkaloids; one of those was identified as tuburlosine $C_{31}H_{39}N_3O_4$. Pakrashi and Eshakali[59] isolated two other new alkaloids, desmethylpsychotrine and alangicine, from *A. lamarckii*. The seed kernels of this plant showed the presence of betulinic acid, betulinaldehyde, betulin, and lupeol.[60] Willaman and Li[10] have compiled a list of different alkaloids isolated from different parts of *A. lamarckii*: root bark — alangicine, alangimarckine, alkaloid AL 60, cephaeline, demethylpsychotrine, demethyltublosine, emetine, protoemetinol, psychotrine, tubulosine; leaf — ankorine, deoxytublosine, dihydroprotoemetine; stem bark — cephaeline, demethyltublosine, emetine, lamarchinine, psychotrine, tubulosine; root — ankorine, isotubulosine; and a few unnamed alkaloids from flowers, seeds, root bark, and stem bark.

Pharmacological action — The root bark is anthelmintic and purgative.[3,50] In small doses the alkaloid lowers blood pressure temporarily, depresses the heart, and produces irregular respiration. It also increases peristaltic movement of the intestines. The fruit is cooling, nutritive, and tonic. The alcoholic extract of the leaves of *A. salvifolium* showed hypoglycemic activity in albino rats. It also possessed antiprotozoal activity against *Entamoeba histolytica*.[51]

Medicinal properties — Root bark is an antidote for several poisons. Rubbed in rice water it is given with honey in diarrhea. It has a reputation in many diseases like leprosy and syphilitic and other skin diseases. Decoction or infusion of the root is given with "ghee" for dog bites; it is also useful in worms. Its oil is useful for external application in acute rheumatism. In doses of 5 to 6 gr it is used as a diuretic.

Doses — 0.3 to 0.5 gr (as a diuretic and diaphoretic); 3 gr as an emetic; 0.3 gr as a blood purifier.

ALBIZZIA LEBBEK (Willd.) Benth. *Family:* MIMOSACEAE

Vernacular names — Sanskrit, Shirish; Hindi, Siris; English, Siris; Bengali, Siris gach; Nepalese, Shirish; Sinhalese, Huremara; Japanese, Pabane munoki; Chinese, Chres; Unani, Siris; Arabian, Sultan-ul-ashjar; Persian, Darekht-e-zakaria; Tibetan, Si-ri-sa; Burmese, Kakko; Tamil, Vagai.

Habitat — A tree found throughout India up to 4000 ft above sea level in the Himalayas; the plant is also cultivated.

Parts used — Bark, seeds, leaves, and flower.

Pharmacognostical characteristics — The bark is dark brown to greenish black and rough, covering being acrid. In the young stem the epidermis has a wavy cuticle and unicellular trichomes are present. Some cortical cells are lignified. Prismatic crystals of calcium oxalate are present. Rhytidome shows alternate layer of cork and cortex, and the cork has several rows of radially suberized lignified cells arranged. A large number of pitted stone cells containing prismatic crystals of calcium oxalate occur in the primary cortex. Phloem is transversed by medullary rays which are two to four cells wide and at the periphery are tortuous and broader. Starch grains are present in cortical and phloem parenchyma cells but absent in medullary ray cells.[45]

Ayurvedic description — *Rasa* — *madhura, tikta, kasaaya; Guna* — *lagu, rooksha, teekshna; Veerya* — *sheeta; Vipak* — *katu.*

Action and uses — *Vishanaashana, vedana sthapan, kaashara, shwashar, shirovirachan, sthambhan.*[26]

Chemical constituents — Tannins and pseudotannins were found in the stem bark of *A. lebbeck*.[45] From the seeds of *A. lebbeck,* saponin based on echinocystic acid has been obtained, while[46-48] friedelin and γ-sitosterol were identified in the bark.[49] The saponin mixture from the seeds from Uttar Pradesh yielded oleanolic acid and echinocystic acid. The beans yield albigenic acid, an isomer of echinocystic acid.[178]

Pharmacological action — Alterative and insecticidal. The alcoholic extract of the roots was found to possess anticancer activity against sarcoma 180 in mice. The stem bark of the plant had hypoglycemic activity in albino rats. The pods possessed antiprotozoal activity against *Entamoeba histolytica*. It also showed hypoglycemic activity in albino rats and anticancer activity in human epidermal carcinoma of nasopharynx in tissue culture.[51]

Medicinal properties and uses — Bark is used as a remedy for bronchitis, leprosy, paralysis, gum inflammation, and helminthic infection. Whole plant is used in snake bites and scorpion stings. Leaves are useful in night blindness.[1] Leaves are made into a poultice and applied to ulcers.

Doses — Bark powder — 3 to 6 gr; seed powder — 1 to 2 gr; infusion of leaves and flowers — 12 to 24 ml.

ALHAGI PSEUDALHAGI (Beib.) Desv.   *Family:* LEGUMINOSAE
(Syn. *ALHAGI CAMELORUM* Fisch.;
   A. MAURORUM Baker)

Vernacular names — Sanskrit, Jawasa; Hindi, Jawasa; Bengali, Dulallabha; Nepalese, Durlambha; Sinhalese, Duralaba; German, Persischemanna; Unani, Jawasa; Arabian, Haaj; Persian, Khar-e-shutr.

Habitat — The Punjab, the Rajasthan upper Gangetic Plains, the Konkhan.

Parts used — Whole plant.

Pharmacological characteristics — A small herb covered with numerous, hard, $1/2$ to 1-in.-long prickles. Leaves simple, drooping, oblong, leathery, smooth, apex rounded. Flowers 1 to 6, borne on a spine or on a short stalk, small, corolla reddish, much longer than the calyx. Pods up to 1 in. long, curved or straight, greenish gray, and hard.

Ayurvedic description — *Rasa — madhur, tikta, kasaya; Guna — laghu, snigdha; Veerya — sheeta; Vipak — madhur.*

Action and uses — *Vata pitta samak, kapha nisarak, sotha har, badana nasak, raky rodhak, samak, anuloman, kapha nisarak, virisya, balya and brighan.*

Chemical constituents — The plant consists mostly of sugar: melizitose 47.1%, sucrose 26.4%, and invert sugar 11.6%.[61] The sugary secretion "manna" obtained from this plant is said to be collected in Persia and exported to India. There is no record of an Indian plant yielding this manna.

Pharmacological action — Laxative, diuretic, and expectorant.

Medicinal properties and uses — Its fresh juice is given with aromatics for relieving the suppression of urine. Juice of the plant locally used in corneal opacity. The sweet exudation of the leaves and branches is restorative, aphrodisiac, aperient, cholagogue, expectorant, and a blood purifier.

Doses — infusion — 1 to 2 gr; decoction — 48 to 96 ml.

ALLIUM CEPA Linn. *Family:* LILIACEAE

Vernacular names — Sanskrit, Palandu; Hindi, Piyaz; English, Onion; Bengali, Piyaj; Nepalese, Pyaaz; Sinhalese, Ralu lunu; Japanese, Tamanegi; Chinese, Hu T'sung T'song; German, Zwiebel; French, Oignon; Unani, Piyaz; Arabian, Basal; Persian, Piyaz; Burmese, Kesounni; Tamil, Vengayan.

Habitat — Cultivated all over India.

Parts used — Leaves and bulbs.

Morphological characteristics — Leaves linear, hollow, fleshy. Flower heads dense with purple color and bulbils; main stalk erect in umbels forming a glabrous head, white or purplish. Bulbs rounded at top and bottom.

Ayurvedic description — *Rasa* — *madhur, katu; Guna* — *guru, tikshna, snigdha; Veerya* — *ushna; Vipak* — *madhur.*

Action and uses — *Vata samak, pitta kapha wardhak, dipan, anuloman, sothahar, kaphanisaak, mutrajanan, shukrajanan, vajekaran, artawajanan, balya, oza wardhak, kandu ghan, chadan kafanisaak.*

Chemical constituents — Bulbs contain an acrid volatile oil. The outer skin of the bulb contains a yellow coloring matter, "quercetin". Active principle is reported to be glycollic acid.[178] It also contains a small amount of sugar, essential oil, and organic sulfides. Dry brown outer scales and the fleshy scales contain flavonoid and other phenolic compounds. Phenolic acids have been identified in different parts of the plant.[178] Scales contain catechol and protocatechic acid; essential oil 0.05% of whole plant. Chief constituent of crude oil is allyl-propyl disulfide; fresh expressed juice moderately bactericidal.[1]

Pharmacological action — Stimulant, diuretic, expectorant, rubefacient, aphrodisiac, and antiseptic.[1]

Medicinal properties and uses — The bulb is a stimulant, diuretic and expectorant. As an emmenagogue and diuretic it is eaten raw; its juice is given with sugar in piles. Its decoction is given in cough and stranguary; with vinegar, the bulb is given in jaundice, spleenic enlargement, and dyspepsia. Onion is also used in obstruction of the intestines, prolapse of the anus, and as sedative.[3] As an emollient, crushed onion or its juice is applied over skin diseases and insect bites; a poultice of the roasted bulb is used over indolent boils, wounds, broken chilblains, and suppurating ears. Onion juice applied as liniment with mustard oil over painful joints, inflammatory swelling, and skin diseases.

Doses — infusion — 15 to 20 ml; seed powder — 1 to 3 gr.

ALLIUM SATIVUM Linn.

Family: LILIACEAE

Vernacular names — Sanskrit, Rasonam; Hindi, Lasan; English, Garlic; Bengali, Rasun; Nepalese, Lobha; Sinhalese, Sudulunu; Japanese, Ninniku; Chinese, Ta-suam; German, Lauch; French, Ail; Unani, Lasan; Arabian, Foom; Persian, Seer; Burmese, Kesumphin; Malaysian, Dawang; Tamil, Velluli.

Parts used — Bulb and oil.

Habitat — Cultivated all over India.

Pharmacognostical characteristics — It is a hardy perennial herb having bulbs with seed cloves and narrow, flat leaves and bears small, white flowers. The bulb and clove are surrounded by a thin white, or pinkish, sheath.

The leaves are yellowish green with a strong alliaceous odor. The scales consist of two to three layers of rectangular cells containing plenty of rhomboid crystals of calcium oxalate. Several vascular bundles having xylem and phloem arranged alternately are found. The epidermis consists of smaller cubical cells.[26,62]

Ayurvedic description — *Rasa — madhur, lavan, katu, kasaya; Guna — snigdha, teekshna, pichila, guru, sara; Veerya — ushna; Vipak — katu.*

Action and uses — *Kapha vata samak, sotha har, badana, sthapan, dipan, pachan, anuloman, shula pra saman, krimighan, kafanisark, rasayan, amapachan, khutaghana.*

Chemical constituents — The bulb contains moisture (62.8%), protein (6.3%), fat (0.1%), carbohydrates (29.0%), calcium (0.03%), potassium (0.31%), iron (1.3 mg); vitamin C (13 mg/100 g).[63] It is also reported to contain copper. The bulbs, on distillation, yield 0.06 to 1% of an essential oil, containing allylpropyl disulfide, $C_6H_{12}S_2$ (6%), diallyl disulfide $C_6H_{10}S_2$ (6%), and two more sulfur-containing compounds.[64] The genuine active principle of garlic, named alliin, was first isolated by Stoll in Basle and called 5-allyl cysteine sulfoxide, crystallized from water-acetone, and had a mp 163.5°C (decomp).[62] Amino acids are also found in Indian garlic.

Pharmacological action — Stimulant, carminative, emmenagogue, antirheumatic, and alterative.[1,3]

The juice and essential oil of garlic showed significant protective action against fat-induced increase in serum cholesterol and plasma fibrinogen and decrease in fibrinolytic activity as well as coagulation time.[65,66] Garlic and onion were effective in controlling the hyperglycemic effect of glucose feeding as assessed by the maximum percentage fall in blood sugar levels.[67] The plant produces inhibitory effects on Gram-negative bacteria of the typhoid-paratyphoid-enteritis group. The plants possess outstanding germicidal properties, all cultures having been destroyed under their influence.[178]

Medicinal properties — It is very useful in pulmonary phthisis, bronchiectasis gangrene of the lung, and whooping cough. Garlic juice is given in the treatment of laryngeal tuberculosis, lumps, and duodenal ulcers. Inhalation of its fresh juice is useful in pulmonary tuberculosis. It has been found useful in atonic dyspepsia, flatulence, and colic. Externally, the juice is used as a rubefacient in skin diseases and as ear drops in earache. It is also a vermifuge.

Doses — Fresh tubers — 1.5 to 6 gr; oil — 1 to 2 minims.

ALOCASIA MACRORRHIZA (L.) Schott   *Family:* ARACEAE

Vernacular names — Sanskrit, Manka; Hindi, Mankanda; English, Great-leaved caladium, Giant taro; Bengali, Man kocku; German, Taro; Unani, Mankanda.

Habitat — Found all over tropical Asia, wild and cultivated.

Parts used — Rootstock and leaves.

Morphological characteristics — Ground stem swollen and succulent; underground rhizome horizontal. Leaves attached to the petiole at their basal margin, entire, ovate-cordate, up to 3 ft long, pointed apex deflexed, margin wavy; leaf stalks 2 to 3 ft long. Flower stalk 4 to 8 in. long, several, usually paired; bract enclosing the spadix 8 to 12 in. long, yellowish green, occasionally marked with reddish streaks, foul smelling, finely wrinkled.

Ayurvedic description — *Rasa—madhur; Guna—guru, snigdha; Veerya—sheeta; Vipak—madhur.*

Action and uses — *Vatta pitta samak, sotha har, badansthapan, vata samak, sula prasaman, anuloman, rakta rodhak, sotha har, mutral, bala wardhak.*

Chemical constituents — The plant contains acicular crystals of oxalate of lime which may be cause for its acridity. Rhizomes contain glucose, fructose, a phytosterol-like compound, and an alkaloid.[178]

Pharmacological action — Digestive, laxative, diuretic, lactagogue, and leaves are astringent.[1]

Medicinal properties — The juice of the leaves or their decoction is given for colic and constipation, the acrid juice extracted from the slightly roasted leaf stalk is astringent and styptic. The rootstock is a mild laxative and diuretic; it is generally used in piles, constipation, and dropsy in the form of a gruel; the ash of the rootstock is used as an anthelmintic; mixed with honey it is applied in aphthous condition of the mouth. It is also used in scorpion sting.[1]

Doses — Rootstalk — 6 to 12 gr; infusion — 14 to 28 ml.

ALOE BARBADENSIS Mill.
(Syn. *ALOE VERA* Linn.)

Family: LILIACEAE

Vernacular names — Sanskrit, Kumari; Hindi, Ghikanuar; English, Indian aloes; Bengali, Ghrit Kumari; Nepalese, Kunhur; Sinhalese, Komarika; Japanese, Aloe ferox; Chinese, Luhui; German, Aloe; Unani, Ghikuar; Arabian, Subara; Persian, Faigra; Malaysian, Kattavala; Tamil, Rattabolam.

Habitat — Many varieties are cultivated throughout India, some grow wild on the coasts of Bombay, Gujarat, and South India.

Parts used — Whole plant, dried juice of leaves, pulp, and root.

Pharmacognostical characteristics — It is a perennial plant with glaucous-green sessile leaves, scaly, and branched. The margin is spiny and toothed. Flowers are pendulous, imbricate, and yellow. The dried juice of the leaves is dark chocolate-brown to black in color. The odor is characteristic, while the taste is nauseous and bitter.[68]

Transection of an aloe leaf shows: (1) A strongly cuticularized epidermis with numerous stomata on both surfaces; (2) parenchymatous region containing chlorophyll, starch, and needles of calcium oxalate; (3) a central region consisting of large mucilage-containing parenchymatous cells; (4) double row of vascular bundles which lie at the junction of two previous layers and have well-marked pericycle and endodermis. The drug-forming juice is contained in the large pericycle cells and sometimes in the adjacent parenchyma; when the leaves are cut, the aloetic juice flows out. No pressure should be applied or the aloe will be contaminated with mucilage.[69]

Powder, when mounted in glycerine, shows numerous crystalline particles embedded in a brownish matrix; or greenish-brown glossy masses, transparent in thin fragments like Cape aloes; or hard, dark-brown opaque masses with uneven porous fracture as in Socotrine aloes; or dark reddish-brown opaque masses with nearly smooth and slightly porous fracture as in Zanzibar aloes.[68]

Ayurvedic description — *Rasa* — *madhur, tikta; Guna* — *guru, snigdha, pichila; Veerya* — *sheeta; Vipak* — *katu.*

Action and uses — *Tri dosha har, soth har, badana samak, brana ropan, daha smak, bhadan, vrisya, awart jank, grav srawk;* used in *udar roga* like *gulma piliha* and *yakrit bridhi.*

Chemical constituents — Aloes contain C-glycosides and resins. They are anthraquinone-producing plants and the content of anthraquinones is subject to seasonal variation.[69] The barbaloin content of Indian aloe was found to be as low as 4.24%. Purgative properties of aloes are due to presence of three pentosides: barbaloin, isobarbaloin, and β-barbaloin.[70]

Pharmacological action — Stomachic, tonic in small doses, in large doses purgative and indirectly emmenagogue and anthelmintic.[3]

Medicinal properties — Dried juice of *A. vera* is cathartic and given in constipation. Fresh juice is cooling. Pulp is useful in menstrual suppression.[50] Root is given in colic.[1] The expressed juice of *A. vera* in the form of an ointment in vaseline has been found to hasten healing of wounds of thermal burns and radiation injury in albino rats.[71] It has been found that aloe compound is very useful in cases of functional sterility and disturbed menstrual function.[72] It is a favorite remedy for intestinal worms in children. Dissolved in alcohol, it is used as hair dye to stimulate hair growth. The Ayurvedic drug known as "kumari asava" is useful in general debility, cough, asthma, consumption, piles, epilepsy, and colic.

Doses — Infusion — 12 to 20 ml; sara — 0.2 to 0.5 g.

ALPINIA GALANGA Willd.
(Syn. *A. CALCARATA* Rosc.)

Family: SCITAMINIACEAE

Vernacular names — Sanskrit, Malayavach; Hindi, Kulanjan; English, Java glangal; Bengali, Kulinjan; Nepalese, Kulinjan; Sinhalese, A ratta(maha) Japanese, Koozuku; Unani, Khlanjan; Arabian, Khalanjan; Persian, Kharvadrad; Burmese, Podogoji; Malaysian, Tauncher; Tamil, Arathai; Greater glangal or galanga major.

Habitat — Eastern Himalayas, South India, Bengal, and Malabar region.

Parts used — Rhizome and fruits.

Morphological characteristics — A perennial plant, 6 to 7 ft high. Leaves lanceolate, smooth, margins white. Flowers in dense panicles, small, greenish white, lip veined with red, obovate clawed. Fruit orange-red.

Ayurvedic description — *Rasa — katu; Guna — laghu, teekshn; Veerya — ushna; Vipak — katu.*

Action and uses — *Kapha vat samak, sita prasaman, lakhan utejek, bala prad, utejek, mukha sodhan, rochan, dipan, anuloman, rakt var nasak, mutral, vajekaran.*

Chemical constituents — Rhizomes contain a pale yellow volatile essential oil with a pleasant odor obtained on distillation. This oil contains 48% of methyl cinnamate, 20 to 30% of cincole, camphor, and probably *d*-pinene. Alcoholic extract of the rhizome showed the presence of tannin, phlobaphenes, and starch. The ash contained chloride, phosphate, sulfate, potassium, and manganese. Flavones were absent.[73]

Pharmacological action — Aromatic, stimulant, and bitter; stomachic and carminative.[3,50] Steam-volatile portion of extract of rhizome stimulated the bronchial glands directly, but the nonvolatile portion acts reflexly through the gastric mucosa.[74] The alcoholic extract of rhizome produced hypothermia in mice. It also potentiated amphetamine toxicity in grouped mice.[11]

Medicinal properties and uses — The rhizome is very useful in rheumatism and catarrhal affections, especially bronchial catarrh.[26] It is given in doses of 5 to 10 gr in controlling urine and in diabetics. The rhizome is used as a deodorizer of foul smells from the mouth and other parts of the body. Its paste made with a bland oil is applied over acne and other skin diseases. The seeds have the same uses as the rhizome. Used as flavoring agent.

Doses — Root (rhizome) powder — 1 to 2 gr.

ALSTONIA SCHOLARIS (Linn.) R. Br   *Family:* APOCYNACEAE

Vernacular names — Sanskrit, Saptaparna; Hindi, Chhatium; English, Dita bark; Bengali, Chhatim; Nepalese, Chhatiwon; Sinhalese, Rukattana; Japanese, Shimasoksi; German, Dita-Riude; Unani, Sattona; Burmese, Letitok; Malaysian, Pulai; Tamil, Ez.

Habitat — *A. scholaris* is a tall evergreen tree widely cultivated throughout India and found in the sub-Himalayan tract from the Jumuna eastward ascending to 3000 ft. It is also found in abundance in Bengal and South India.

Parts used — Bark, leaves, milky juice, and flowers.

Pharmacognostical characteristics — Leaves occur about seven in a cluster, 4 to 6 in. long and 1 to $1^1/_2$ in. broad, yielding milky juice. Flowers greenish white and odorous. Fruits about 1 ft long, compressed on both sides. Seeds small and white in color. Bark is yellow and exudes a milky juice when injured.

Ayurvedic description — *Rasa* — *tikta, katu; Guna* — *lagu, snigdha, seara; Veerya* — *ushna; Vipaak* — *katu.*

Action and uses — *Hirdya, deepan, varnaghan, raktaamayaghan, krimighna, kushtaghna, jeern jwaraghan, gulmaghna, grahanirogahara.*

Chemical constituents — Pure alkaloids isolated from the bark are ditamine, echitamine, echitenine, and echitamidine. The alcoholic extract of the bark by cold percolation yielded three products, all nonnitrogenous and neutral in character. These have been identified as α-amyrin mp 184°C, lupeol, mp 212 to 213°C, and a mixture of these two products.[75] The leaves also contain an alkaloid,[10,73] indole alkaloid A. The total alkaloid content of Indian bark is stated to be 0.16 to 0.27% of the hydrochloride of the chief alkaloid echitamine.[77] From the mother liquor of echitamine, Goodson has isolated small quantities of another crystalline alkaloid, echitamidine.[78]

Among the nonalkaloidal constituents, Goodson[79] has isolated two isomeric lactones, $C_9H_{14}O_2$, of mp 103 and 107°C. The bark is also rich in sterols.[80]

The latex is found to contain 2.8 to 7.9% caoutchouc. The coagulum contains caoutchouc, 12.9 to 26.5%, and resins, 69.0 to 78.7%.[81]

Pharmacological action — Stimulant, carminative, stomachic, bitter tonic, astringent, aphrodisiac, expectorant, febrifuge, alterative and antiperiodic.[1,3]

Medicinal properties and uses — It is considered a good substitute for cinchona and quinine for the treatment of intermittent and remittent fevers; its powder is given in doses of 1.5 to 4 gr; extract 2 to 10 minims; an infusion of the bark is given in fever, dyspepsia, debility, skin diseases, liver complaints, chronic diarrhea and dysentery in doses of 28 to 56 ml twice or three times a day. The bark is also a useful emmenagogue, astringent, galactagogue, and anthelmintic.

Doses — infusion — 12 to 20 ml; decoction — 56 to 112 ml; powder — 10 to 25 gr; milky juice — 5 to 15 gr.

ALTHAEA OFFICINALIS Linn.   *Family:* MALVACEAE

Vernacular names — Sanskrit, Khatmi; Hindi, Gulkhairo; English, Marsh mallow; Japanese, Usubentachiasi; German, Eibisch; Unani, Khatmi; Arabian, Bazr-ul-Khtmk; Persian, Tukme-katmi.

Habitat — Found in Punjab and Kashmir.

Parts used — Flowers, carpels, leaves, root, and seeds.

Ayurvedic description — *Rasa—madhur; Guna—snigdha, pichil, guru; Veerya—ushna; Vipak — madhur.*

Action and uses — *Vat-pitta samak, kaphanisarak, sothahar, badanasak, anuloman snehan, mutral.*

Chemical constituents — The root contains mucilage (35%) and starch (37%). Its flowers yield a red dye which may be used as an indicator in acidimetry and alkalimetry.[84] Seeds contain 11.9% drying oil.[44] The plant yields fatty oil and phytosterin.[85]

Pharmacological action — Seeds are demulcent, diuretic, and febrifuge. Roots are astringent and demulcent.

Medicinal properties and uses — Taken internally, flowers are expectorant and the root is a demulcent. Carpels are very useful in urinary complaints and cough. Decoction of the root is used as an emollient enema in irritability of the vagina or rectum. The plant is reported to be an antidote against snake bite.

Doses — $1/2$ to 2 drams.

ALTINGIA EXCELSA Noronha *Family:* HAMAMELIDACEAE

Vernacular names — Sanskrit, Sill haka; Hindi, Silaras; English, Storax; Bengali, Silaras; Nepalese, Patharkumkum; German, Rasamala; Unani, Silaras; Arabian, Mia-e-saela; Persian, Mia-e-Sacla.

Habitat — This is a magnificent tree of the Indian Archipelago, common in Assam and Bhutan. Also found in Burma.

Parts used — Resin.

Ayurvedic description — *Rasa — tikta, katu, madhur; Guna — snigdha, laghu ; Veerya — ushna ; Vipak — katu* .

Action and uses — *Kapha-vatta samak, bran ropan, badanasthapan, sesham har, mutral, virisya, artow jann, jwaraghan balya.*

Chemical constituents — Storex is a mixture of cinnamic acid, cinnamic aldehyde, benzaldehyde, vanilion, styrol, and styracin.[50]

Pharmacological action — Stimulant, expectorant, anodyne, carminative, antiphlogistic, stomachic, antiscorbutic, and antiseptic.[1]

Medicinal properties and uses — The resin is useful in affections of the throat and skin diseases like scabies and leukoderma; smeared over the abdomen of children to relieve colic pains; applied in cases of orchiditis over the inflamed testicle; and useful in early stages of hydrocele.[1]

Doses — 0.5 to 1 gr.

AMOMUM SUBULATUM Roxb.   *Family:* ZINGIBERACEAE

Vernacular names — Sanskrit, Elabari; Hindi, Bari elachi; English, Greater cardamon; Bengali, Bara elachi; Nepalese, Eela; Sinhalese, Maha ensal; German, Kardmemeu; Unani, Elayechi kalan; Arabian, Qakilh-ahe-Kibar; Persian, Heel-e-e kalan; Burmese, Ersatz pala; Malaysian, Peri elav.

Habitat — This species is generally cultivated in the eastern Himalayas, Nepal, Bengal, Sikkim, and Assam.

Parts used — Seed and oil.

Pharmacognostical characteristics — The dark red-brown globose capsules 1 in. long contain many seeds in each cell, held together by a viscid sugary pulp. Leaves are 1 to 2 ft long and 3 to 4 ft wide; flower is white.

Ayurvedic description — *Rasa — katu, tikta; Guna — lagu, roosha; Veerya — ushana; Vipak — madhur.*

Action and uses — *Tri dosha samak, dipan, pachan, anuloman, badana samak, brana ropan.*

Pharmacological action — Seeds are stomachic, carminative, cardiac stimulant, expectorant.

Medicinal properties and uses — They are fragrant adjuncts to other stimulants, bitters, and purgatives. An oil extracted from them is applied to the eyelids to allay inflammation.[3] It is an agreeable aromatic stimulant and is used for flavoring. Decoction of cardamom is used as a gargle in affections of the teeth and gums. In combination with the seeds of melon it is used as a diuretic in cases of kidney stones. It is invaluable in certain disorders of the digestive system marked by scanty and vesical secretion from the intestine, promotes elimination of bile, and is useful in liver affection such as congestion of the liver. It is also useful in neuralgia, gonorrhea; it is used as an aphrodisiac, and also is used as an antidote to snake bite and scorpion sting.[50]

Doses — 0.6 to 1 g.

AMORPHOPHALLUS CAMPANULATUS (Roxb.) Blume   Family: ARACEAE

Vernacular names — Sanskrit, Kunda; Hindi, Jangli Suran; English, Telgu potato; Bengali, Ol; Nepalese, Kundapupa; Sinhalese, Kindran; Unani, Zamin Qand; Burmese, Zamin Qand; Tamil, Karunai.

Habitat — Found in India and Nepal. Widely cultivated.

Parts used — Corm or tubers.

Morphological characteristics — Tuberous herb, tubers depressed globose, much warted, 20 to 25 cm in diameter, dark brown; leaves large, 30 to 90 cm broad, appearing after the flower; spathe 15 to 22 cm broad, green, usually with white spots below, greenish purple above, rough and dark, purple within toward the base; female inflorescence cylindric and male subturbinate.

Ayurvedic description — *Rasa — katu, kasaya; Guna — laghu, roosha, teekshna; Veerya — ushna; Vipak — katu.*

Action and uses — *Kapha vatta samak, sotha har, badanasthapan. Dipan, pachan, anuloman, sula prasaman, arsoghan, yakrita-utajek, krimighan, virisya, artaw janan, balya, rasayan.*

Chemical constituents — The corm contains moisture (78.7%), protein (1.2%), fat (0.1%), carbohydrates (18.4%), mineral matter (0.8%), calcium (0.05%), phosphorus (0.02%), iron (0.4 mg), vitamin A (434 IU), vitamin B (20 IU/100 g),[82] and an enzyme. The yield of starch is less than 4.4%.[50]

Pharmacological action — Stomachic, carminative, and tonic.[1]

Medicinal properties and uses — It is used in piles and given as a restorative in dyspepsia and debility. It is a hot carminative in the form of a pickle. Root is used in boils and ophthalmia, also as an emmenagogue. Juice acrid, much used in acute rheumatism.[50]

Doses — Powder — 5 to 10 gr; confection — 2.5 to 10 ml.

ANACYCLUS PYRETHRUM DC.   *Family:* COMPOSITAE

Vernacular names — Sanskrit, Akarava; Hindi, Akarkara; English, Pellitory root; Nepalese, Akarkara; Sinhalese, Akkarapatta; German, Speidetwurzel, Zaharwurzel; Unani, Akarkarha; Arabian, Aaqir-e-garha; Persian, Bekh-e-tarkhoon; Tamil, Akkarakaram.

Habitat — Indigenous to North Africa (Algeria), but reported in Bengal as an escape.

Parts used — Root.

Morphological characteristics — A perennial herb, root 3 to 4 in. long and $1/2$ to $3/4$ in. in diameter. Flowers white or pink having yellow dot in the middle.

Ayurvedic description — *Rasa — katu; Guna — roosha, tikshna; Veerya — ushna; Viapk — katu.*

Action and uses — *Kapha-vat samak, utajak, badana sthapan, sothahar, rakt sodhak sothahar, kaphaghan, kantya, vajkarn.*

Chemical constituents — The roots contain an essential oil and an alkaloid "pyrethrin" or pellitorine,[50] which is physiologically analogous to piperine.

Pharmacological actions — Cardial, stimulant, and sialagogue.

Medicinal properties and uses — A decoction of the root is useful as a gargle in carious teeth, toothache, sore throat, and tonsillitis. Root is a valuable sialagogue and is regarded as a tonic to the nervous system. Powdered root is given in honey or as snuff for epilepsy, useful in rheumatism.[1]

Doses — Root powder — 0.5 to 1 gr.

ANANAS COMOSUS (Stickm.) Merr.   *Family:* BROMELIACEAE
(Syn. *A. SATIVUS* Schult. f.)

Vernacular names — Sanskrit, Ananas; Hindi, Ananas; English, Pineapple; Bengali, Anaras; Nepalese, Bhaikatacher; Sinhalese, Annasi; Japanese, Hakatamenanasu; German, Ananas; Unani, Ananas; Persian, Aainunnas; Tamil, Annasipazham.

Habitat — Cultivated in various parts of India, especially in eastern and South India. Found in tropical and subtropical America and the West Indies.

Parts used — Ripe and unripe fruits and leaves.

Morphological characteristics — Leaves 2 to 3 ft long, with prickly margins and spiny tips. Inflorescence from the center of rosette, terminal, sessile, or peduncled. Perianth biseriate; outer series of three sepals; inner series of three petals, free or somewhat connate. Stamens three to six, free or somewhat connate with the petals. Ovary inferior, three celled, style with three stigmas. The fruit, a baccate syncarp, which has a rough surface and a crown of small leaves, has succulent flesh of yellow to light-orange color. Seeds albuminous.

Ayurvedic description — *Rasa* — *madhura* (ripe fruit), *amla* (unripe); *Guna* — *guru, snigdha; Veerya* — *sheeta; Vipak* — *madhura.*

Action and uses — *Vata-pitta samak, rochan dipan, pachan, anuloman, krimighan, rakta pitta samak, hirdya, asmarivedan jwaraghan, mutral, balaya.*

Chemical constituents — Juice of the pineapple contains 4.3% of total sugar; acid (0.3 to 0.9%), moisture (86.5%), protein (0.6%), fat (0.1%), carbohydrates (12.0%), mineral matter (0.5%), calcium (0.02%), phosphorus (.01%), iron (0.9 mg), vitamin A (60 IU), vitamin C (63 mg/100 g).[82] Fruit contains a digestive ferment, bromelin, closely related to trypsin. The plant is reported to contain arsenic.[1]

Pharmacological action — Febrifuge, alterative, and bitter tonic.

Medicinal properties — Fresh juice of the leaves or leaves themselves are powerfully purgative and anthelmintic and vermicide. Juice of the ripe fruit is antiscorbutic, diuretic, diaphoretic, aperient, refrigerant, and helps in the digestion of albuminous substances. Juice of the unripe fruit is acrid, styptic, powerful diuretic, anthelmintic, emmenagogue, and in large quantities it is abortifacient.[1,3] It is also given to allay gastric irritability in fever and in jaundice. Garg et al.[83] reported that the juice of unripe fruit showed promising antifertility activity in rats.

Doses — Fruit juice — 14 to 28 ml; leaf juice — 1 to 3 drams.

ANDROGRAPHIS PANICULATA (Burm. f.) Nees   Family: ACANTHACEAE

Vernacular names — Sanskrit, Kirta; Hindi, Kiryata, Kalmegh; English, King of bitters; Bengali, Kalmegh; Sinhalese, Hinbinkohomba; German, Andrographis Kraut; Unani, Kalmegh; Arabian, Qasab-uz-zarirah; Persian, Naynahudandi.

Habitat — Found throughout the plains of India.

Parts used — Whole herb.

Morphological characteristics — Leaves lanceolate, 2.5 to 7.5 cm long and about 12 mm wide; flowers small and solitary; corolla is purplish, hairy, dotted, irregular bilabiate. The fruit is a capsule 18 to 22 mm long and 3 mm wide containing 6 to 12 ovoid rugose, glabrous, somewhat flattened seeds. Stem and branches acutely quadrangular, particularly in the upper region with four bulges arising on the four corners, jointed and nearly glabrous, branches cross-armed, four sided, spreading or horizontal; cultivated as a rainy season crop.[84]

Pharmacognostical characteristics — Epidermis of stem has glandular and nonglandular hairs and stomata of caryophyllaceous type. The cortex consists of collenchymatous strands, below the epidermis and in-between the bulges, four to five layers of parenchyma cells, and a distinct layer of endodermis. The phloem is characterized by the presence of acicular fibers in old plants. The vascular system is represented by an ectophloic siphonostele. Cystoliths are present in the epidermal and cortical region of stem and epidermis of the leaves, bracts, bracteoles, and sepals of flower. Small acicular crystals of calcium oxalate occur in the pith and cortical cells of stem and leaf.[85]

Ayurvedic description — *Rasa — tikta; Guna — laghu, rooksha, teekshna; Veerya — ushna; Vipak — katu.*

Action and uses — *Kapha-pitta har, dipan, yakrita utegek, rechen, krimighan, rakt sodhak, kutaghan, jwaraghan.*

Chemical constituents — Leaves contain two bitter substances and traces of an essential oil. The first is yellow crystals with the formula $C_{19}H_{28}O_5$ and mp 206°C; the second one is in an amorphous form, and is named kalmeghin $C_{19}H_{51}O_5$, mp 185°C. A new natural flavone, 5-hydroxyl-7,8, 2′3′-tetramethoxy flavone, $C_{10}H_{18}O_7$ and mp 150 to 151°C, was obtained from roots of *A. paniculata*.[87] Singh et al.[84] reinvestigated *A. paniculata* and found two compounds, A and B, with $C_{23}H_{36}O_3$, mp 101°C, and $C_{23}H_{36}O_8$, mp 170°C, respectively.

Pharmacological action — Roots and leaves are stomachic, tonic, antipyretic, alterative, anthelmintic, febrifuge, and cholagogue.[1,3]

Medicinal properties and uses — The leaf juice is a household remedy for flatulence, loss of appetite, bowel complaints of children, diarrhea, dysentery, dyspepsia, and general debility; it is preferably given with the addition of aromatics. Decoction or infusion of the leaves gives good results in sluggish liver, neuralgia, in general debility, in convalescence after fevers, and in advanced stages of dysentery.

Doses — Infusion — 14 to 28 ml; tincture — 2.5 to 5 ml; juice of leaves and stem — 1 to 4 ml.

ANGELICA GLAUCA Edgew.   *Family:* UMBELLIFERAE

Vernacular names — Sanskrit, Choraka; Hindi, Bhataur-chora; Nepalese, Chameha; Germany, Angelika.

Habitat — Found in western Himalayas from Kashmir to Simla.

Parts used — Root.

Morphological characteristics — A perennial herb, leaves large; leaflets three, oval or obovate, irregular, sharp dentate; flowers white or blue; fruits 13 mm long and 6 mm broad, compressed and smooth.

Pharmacognostical characteristics — Parenchymatous cells in the transection of the root are loaded with starch granules. A few patches of fibers are seen in the wood zone. Collenchymatous patches are present beneath each groove in transverse section of stem. The phloem tissue is capped by a zone of hard bast fiber. Vascular bundles are arranged in two to three irregular rings. A large number of trichomes are seen on the lower surface of the leaf.[91]

Chemical constituents — Large, dry root contains 1.3% essential oil as compared to 0.35 to 1.0% from foreign specimen. It contains lactones, sesquiterpenes, and also *d*-α-cadinene, umbelliprenin, and a terpene alcohol.[178]

Pharmacological action — Cordial and stimulant.[50]

Medicinal properties and uses — It is a good cordial stimulant and is used in flatulence, constipation, and dyspepsia. The aromatic root is a flavoring agent.[90]

ANNONA SQUAMOSA Linn. *Family:* ANNONACEAE

Vernacular names — Sanskrit, Sitaphalam; Hindi, Sharifa; English, Custard apple; Bengali, Ata; Sinhalese, Atta; Japanese, Banreishi; Chinese, Phanlechi; German, Zuckerapfel; Unani, Sharifa; Arabian, Sharifa; Persian, Sharifa; Burmese, Amesa; French, Allier; Tamil, Siththa.

Habitat — An American tree about 20 ft high; now cultivated and naturalized in India.

Parts used — Leaves, bark, root, seeds, and fruit.

Morphological characteristics — A tree about 20 ft high. Leaves $1\frac{1}{2}$ to 3 by $\frac{3}{4}$ to $1\frac{1}{2}$ in. oblong, lanceolate or elliptic, obtuse or subacute, glaucous, and pubescent beneath when young; lateral nerves 8 to 11 pairs. Flowers solitary, leaf opposed or 2 to 4 on short, extraaxillary branchlets. External petals about 1 by $\frac{1}{4}$ in., the internal minute or sometimes wanting. Fruit globose, 2 to 4 in. diameter, tuberculate, yellowish brown; pulp dense. The seeds are many, brownish yellow, black, smooth, and oblong.

Ayurvedic description — *Rasa — madhura; Guna — laghu, snigdha; Veerya — sheeta; Vipaka — madhura.*

Action and uses — *Vata pitta samak, hirdya, rakta pita samak, virisya, mutral, daaha prasaman, balya, brinhan.*

Chemical constituents — Pulp contains moisture (62 to 64%) and reducing sugar 6.55%.[1] Seeds yields an oil and resin; seeds, leaves, and immature fruit contain an acrid principle, which is insecticidal. Fruit is rich in vitamin C. The aporphone alkaloids anonaine roemerine, noreorydine, corydine (reported to have anticancer activity) norisocorydine, isocorydine, and glaucine have been isolated from *A. squamosa*.[92]

Pharmacological action — Astringent, tonic, anthelmintic, purgative, and diuretic. Ether extract of the seed was toxic to *Musca nebulo* and *Tribolium castaneum* adults. An insoluble resin was formed when petroleum ether was added to the ether extract and this resin was found to be six times more toxic as contact poison against *M. nebulo*.[93] The seeds were found to have little oxytocic activity, although some uterotonic activity was observed in isolated tissue experiments.[94] The alcoholic extract of the aerial parts of *A. squamosa* was tested to have anticancer activity against human epidermal carcinoma of the nasopharynx in tissue culture.[11]

Medicinal properties and uses — Ripe fruit bruised and mixed with salt is applied to malignant tumors to hasten suppuration. Leaves made into a paste without adding water are applied to unhealing ulcers; seeds applied to uterus cause abortion. Fresh leaves after crushing, applied to the nostril, cut short fits of hysteria and fainting; powder of seeds is a good hair wash; unripe fruit is given in diarrhea, dysentery and atonic dyspepsia. Bark is used as tonic. Root is given a violent purgative.[1,3,26]

Doses — Fruits — 24 to 48 gr; root powder — 2 to 5 gr.

ANTHOCEPHALUS INDICUS A. Rich.   *Family:* RUBIACEAE
(Syn. *ANTHOCEPHALUS CADAMBA* Miq.)

Vernacular names — Sanskrit, Kadamba; Hindi, Kadamba; English, Wild cinchona; Bengali, Kadami; Nepalese, Kadam; Sinhalese, Ambulbakmi; German, Kadamb; Unani, Kadam; Arabian, Qadam; Burmese, Mau.

Habitat — Sub-Himalayan tract from Nepal eastward to Burma and in south in Northern Circars and Western Ghats.

Parts used — Fruit, leaves, and bark.

Morphological characteristics — A large, deciduous tree with a straight stem about 30 ft high and girth up to 5 to 7 ft. The rounded crown bears drooping branches and yellow flowers in globose heads. The wood is yellowish white or cream colored. The wood is anatomically characterized by inconspicuous, wide growth rings, medium or small vessels, and relatively conspicuous medullary rays which appear as flecks on the radial surface.

Ayurvedic description — *Rasa — katu, tikta, kasasya; Guna — laghu, rooksha; Veerya — sheeta; Vipak — katu.*

Action and uses — *Tridosha nask, badana nasak, sothahar, brnaropan, sodhan, deepan, pachan, grahi, trisanigraha mutragann, asmarinasak, sukra sodhak, rakt stamvan, kasha har.*

Chemical constituents — Bark contains an astringent principle similar to cinchio-tannic acid and the drug contains a ready-formed oxidation product of the nature of cinchona red. The flowers yield essential oil and the bark contains alkaloids, steroids, fats and oils, and reducing sugars.[178]

Pharmacological action — Bark is tonic, febrifuge, and astringent. Fruit is refrigerant.

Medicinal properties and uses — Decoction of the bark is given in fevers, and decoction of the leaves is used as a gargle in aphthae and stomatitis. In inflammation of the eye, the bark juice given with equal quantity of lime juice, opium, and alum is applied around the orbit; juice of the fruit with cumin and sugar is given to children in gastric irritability; used in snake bite.[1,3]

Doses — Bark powder — 6 to 12 g; infusion of fruit — 14 to 24 ml; leaves infusion — 14 to 24 ml.

APIUM GRAVEOLENS Linn.

Family: UMBELLIFERAE

Vernacular names — Sanskrit, Ajmoda; Hindi, Ajmoda; English, Celery; Bengali, Randhuni; Nepalese, Ajumu; Sinhalese, Asamodagam; Japanese, Serori; German, Sellerie; Unani, Ajmoda; Arabian, Bazr-ul-karafs; Persian, Tukhm-e-karafs.

Habitat — Found at the base of the northwestern Himalayas and outlying hills in the Punjab and in western India.

Parts used — Root and seeds.

Pharmacognostical characteristics — It is a biennial erect herb about 2 to 3 ft high. Leaves pinnate, arising from the roots, deeply lobed; segments coarsely toothed at the apex; lower stalks very short. The fruit is small, about 1 mm long and 1 mm in diameter, and contains minute seeds. The epicarp is interspersed with oil ducts.

Ayurvedic description — *Rasa—katu; Guna—teekshna; Veerya—ushna; Vipaka—katu.*

Action and uses — *Vidanhi, hirdya, vrishya, balaya, netrarogohara, krimighan, hikkaashamanana.*

Chemical constituents — Its leaves contain moisture (81.3%), carbohydrates (8.6%), fat (0.6%), protein (6.0%), calcium (0.23%), iron (6.3 mg), vitamin A (5800 to 7500 IU), and vitamin C (62 mg/100 g); the stalks contain moisture (93.5%), carbohydrates (3.5%), protein (0.8%), calcium (0.03%), iron (4.8 mg), and vitamin C (6 mg/100 g).[98] The herb is also reported to contain the glucoside apiin.[99]

The celery fruits yield 2 to 3% of pale-yellow volatile oil with a persistent odor, characteristic of the plant. The fruits also yield 17% of a fatty oil called oil of celery. Chemical constituents of the essential oil are *d*-limonene, *d*-selinene, sesquiterpene alcohols, sedanolide, and sedanomic acid anhydride.[100] Chemical examination of the seeds showed the presence of anthoxanthins which consisted almost completely of glycosides which have been separated into two entities called graveobioside A and graveobioside B. Both of them yielded apiose and glucose on hydrolysis, but the aglycone of graveobioside A was found to be luteolin and that of graveobioside B, chrysoeriol. It was seen that glycosides are apiose glucosides.[102]

Psoralen, xanthotoxin, angelicin, and bergapten are obtained from edible vegetable matter, all with photodynamical activity. Bergapten is responsible for skin disorders among the cultivators. Seeds contain 10% of insoluble carbohydrates and 0.8% of an essential oil.[178]

Pharmacological action — Tonic, carminative, diuretic, emmenagogue, and stimulant. Essential oil was found to possess tranquilizing and anticonvulsant activity on mice.[101] An alkaloid fraction of seeds showed tranquilizing activity like the essential oil.[103]

Medicinal properties and uses — The root is alterative and diuretic; it is given in dropsy and colic. The seeds are carminative, diuretic, stomachic, aphrodisiac, and tonic. Its oil is also used as an antispasmodic and nerve stimulant. It has been successfully employed in rheumatoid arthritis and probably acts as an intestinal antiseptic. As antispasmodics, they are used in bronchitis, asthma, and to some extent for liver and spleen diseases. In India celery leaves and stalk are used as salad and in the preparation of soup.[1,3]

Doses — Seed powder — 1 to 3 gr.

AQUILARIA AGALLOCHA Roxb. *Family:* THYMELAEACEAE

Vernacular names — Sanskrit, Agaru; Hindi, Agar; English, Aloe wood; Bengali, Agaru; Nepalese, Agur; Sinhalese, Agil; Japanese, Chinko; Chinese, Chin heang; German, Adperhopz; French, Agalloche; Unani, Agar; Arabian, Ud-e-gharqee; Persian, Ud; Burmese, Akyan; Malaysian, Kaya gehru.

Habitat — Distributed only on the eastern side of India, and in Bhutan, parts of Bengal, and particularly in Assam on the hill forest of Khasia, Garo, Naga, Cachar, and Sylhet. In Burma it occurs on the Martaban Hills of south Tenasserin.

Parts used — Wood, oil.

Pharmacognostical characteristics — *A. agallocha* is a large evergreen tree over 60 to 70 ft high and 5 to 8 ft in girth with a moderately straight stem. The wood is soft, light, and elastic. It is white to pale yellowish white, and has no characteristic odor. Anatomically it is distinguished by the absence of growth rings, medium-sized to small vessels mostly in radial rows of three to four, and many longitudinal interxylary strands of phloem consisting of soft bast and some fibers. Agar, which consists of irregular patches of dark wood highly charged with oleoresin is found in the interior of comparatively old trees. The agar-bearing tree has a somewhat diseased appearance, and oleoresin is usually found where the branches fork out from the stem.

Ayurvedic description — *Rasa — katu, tikta; Guna — laghu, rooksha, teekshna; Veerya — ushna; Vipaaka — katu.*

Action and uses — *Kafa bat samak, shit prashaman, shiro birachana, mukha sodhak, sotha har, badana samaka, kushtaghna, baji karan, rasayan.*

Chemical constituents — Aloe wood is highly charged with resinous matter and contains 48% alcohol-soluble matter. After saponification of the alcoholic extract, benzyl acetone, another unidentified ketone $C_{14}H_{20}O_2$, a sesquiterpene alcohol, and some acid (including hydrocinnamic acid) have been obtained. The sesquiterpene alcohol possesses the characteristic odor of the wood. It also contains a volatile essential oil.[104]

Pharmacological action — It is a stimulant, cardial, tonic, and carminative. When given internally it acts as a cholagogue and deobstruent.

Medicinal properties and uses — It is an important ingredient of various cosmetics and liniments, and is used in many skin diseases.[1,3] It is an ingredient in various nervine tonics and in carminative and stimulant preparations. It is used in gout and rheumatism; to check vomiting; and also in snake bite. As an anodyne fumigation it is used to relieve pain in surgical wounds and ulcers. A paste of agaru and Isvari with brandy is applied to the chest in bronchitis of children and to the head in headache. Ayurvedic drugs containing *agaru* given in doses of 1.5 to 4 g serve as a nervine tonic in seminal debility, giddiness, and leukorrhea.

Doses — Resin — 1.5 to 4 g; oil — 1 to 5 minims.

ARECA CATECHU Linn. Family: PALMAE

Vernacular names — Sanskrit, Pooga; Hindi, Supari; English, Areca nut; Bengali, Superi; Nepalese, Gowe; Sinhalese, Puvak; Japanese, Binro; German, Betelnusse; French, Arequier-Nox d'arec; Unani, Supari; Arabian, Fufal; Persian, Popal; Burmese, Kunsi; Malaysian, Pinang; Tamil, Kamuku.

Habitat — Cultivated throughout tropical India. It flourishes in the dry plateau of Mysore, Canara, Malabar, southern India, and Assam.

Parts used — Seeds, root, and tender leaves.

Morphological characteristics — *A. catechu* is a handsome palm with a tall, slender graceful stem crowned by a tuft of large, elegant-looking leaves. Leaves pinnate, leaflets numerous, 30 to 60 cm, spathe glabrous, compressed; spadix much branched, male flowers many, minute; female flowers solitary, large; fruit ovoid, about 1.5 to 2 in. across and 2.2 to 2.5 in. long and is bright orange when fully ripe. The pericarp is hard and fibrous, and the kernel (seed), called the areca nut, is about 20 to 25 mm in diameter and grayish brown in color. It is hard and is covered with a network of paler depressed lines.[28]

Ayurvedic description — *Rasa — kasaya, madhur; Guna — guru, rooksha; Veerya — sheeta; Vipak — katu.*

Action and uses — *Kapha pitta samak, bran ropan, rakt pitta samak, lala srab nasak, dipan, krimighan, shukra sthamvan.*

Chemical constituents — Ripe and semiripe areca nuts contain[26] a large number of amino acids both in the free and combined state. The salient features of amino acid makeup are an insignificant quantity of tryptophan and methionin, presence of high percentage of proline both in free and combined forms, and relative increase of free tryosine and phenylalanine and of combined arginine in the semiripe and ripe nuts. Earlier investigations revealed the presence of five alkaloids, viz., arecoline, arecaidine, guvaoline, guvacine, and arecolidine, but a recently detailed analysis of young and mature nuts showed the presence of minor amounts of + catechin and high amounts of procyanidins.[105] Arecoline, a substitute for pilocarpine, has been isolated from the nuts. They also contain β-sitosterol and leukocyanidins and tannins which exhibit antibacterial and antifungal properties.[178]

Pharmacological action — Stimulant, astringent and taeniafuge.[50]

Medicinal properties and uses — Poultice or juice of the leaves mixed with some bland oil is applied to the loins in lumbago. The nut is laxative, carminative, aphrodisiac, and a nervine tonic.[3] The ripe, dry nut is masticatory; it sweetens the breath, removes bad taste from the mouth, strengthens the gum, and checks perspiration. Its powder is a well-known anthelmintic, especially for roundworms and tapeworms. The nut is prescribed in calculous affections, urinary disorders, heartburn of pregnancy, and vaginal discharges. Decoction of the root is a reputed cure for sore lips.

Doses — Powder — 1 to 2 gr; fruit extract — 1 to 2 ml; tincture — 5 to 10 ml.

ARGEMONE MEXICANA Linn. *Family:* PAPAVERACEAE

Vernacular names — Sanskrit, Satyanasi; Hindi, Piladhatura; English, Yellow thistle, Mexican poppy, Prickly poppy; Bengali, Shialkanta; Nepalese, Thakal; Sinhalese, Rankirgokatu; Japanese, A zamigeshi; German, Stachel Mohn; Unani, Satiyanasi; Arabian, Shajzatyssoom; Persian, Badanjane Dashti; Burmese, Khyoa; Tamil, Kudy tupoodu.

Habitat — Grows wild all over India, especially from Bengal to Punjab and Himachal.

Parts used — Milky juice of fresh plant, seeds and mixed oil of fresh seeds and fresh root.

Morphological characteristics — A robust, prickly, herbaceous annual 1 to 4 ft high with spreading branches. Leaves thistle-like, 3 to 7 cm, sessile, $1/2$ amplexicaul, sinuate-pinnatifid, variegate green and white. Flowers 1 to 3 in. across, yellow. Sepals horned at the top, prickly. Capsule $3/4$ to $1 1/2$ in. long, elliptic or oblong, prickly. Seeds blackish-brown, round, netted.[28]

Ayurvedic description — *Rasa — tikta; Guna — laghu, rooksha; Veerya — sheeta; Vipak — katu.*

Action and uses — *Kapha pitta har, bran sodhan, bran ropan, kustaghan, sotha har, badana sthapan, vishaghan, kafa nisark, mutral, visan jwaraghan, daah prasaman.*

Chemical constituents — The seeds yield 22 to 33% of a nauseous, bitter, nonedible oil. The stem and root of the plant contain 0.125% percent of total alkaloids, having 0.041% berberine and 0.084% of protopine. Besides the alkaloids other constituents consist of 1.10% tannins, 1.75% resin, and unidentified toxic principle in the oil.[106] The crushed seeds were exposed for a week and fractioned by hydrogen chloride, resulting in seven fractions which were all toxic.[107] Willaman et al.[10] have reported the following alkaloids: α-allocryptopine from roots and whole plant above ground; berberine from roots, stem, leaf capsule, and seed; chelerythrine from roots; coptisine from stem, leaf, capsule, and seed; protopine from whole plant and roots; sanguinarine from roots, stem, leaf, capsule, seed, and seedlings; and an unnamed alkaloid from seeds.

Pharmacological action — Root is diuretic, alterative, anodyne, and hypnotic. Seeds are laxative, emetic, expectorant, demulcent, and narcotic. Oil is purgative.

Medicinal properties and uses — Its juice is diuretic and alterative; it is given in dropsy, jaundice, skin diseases, and gonorrhea. Being mildly corrosive the juice is applied to blisters, rheumatic pains, ulcers, scabies, herpetic eruptions, and warts. The root is alterative and stimulant; its decoction is given in gonorrhea, gleet, and vascular calculous and skin diseases. It is also given for tapeworm. A decoction of the root is an eye wash, a mouth wash, and a lotion for inflammatory swelling.[1,3] Seeds are given in cough, catarrhal affections, pulmonary diseases, asthma, and whooping cough. The oil extracted from the seeds is purgative, narcotic, and demulcent. In syphilis and leprosy it is used as an alterative. It is applied locally over skin diseases. Alcoholic extract of the whole plant showed antiviral activity against Ranikhat disease virus.[51]

Doses — Root powder — 1 to 3 gr; seed powder — 1 to 3 gr; oil — 1 to 2 ml.

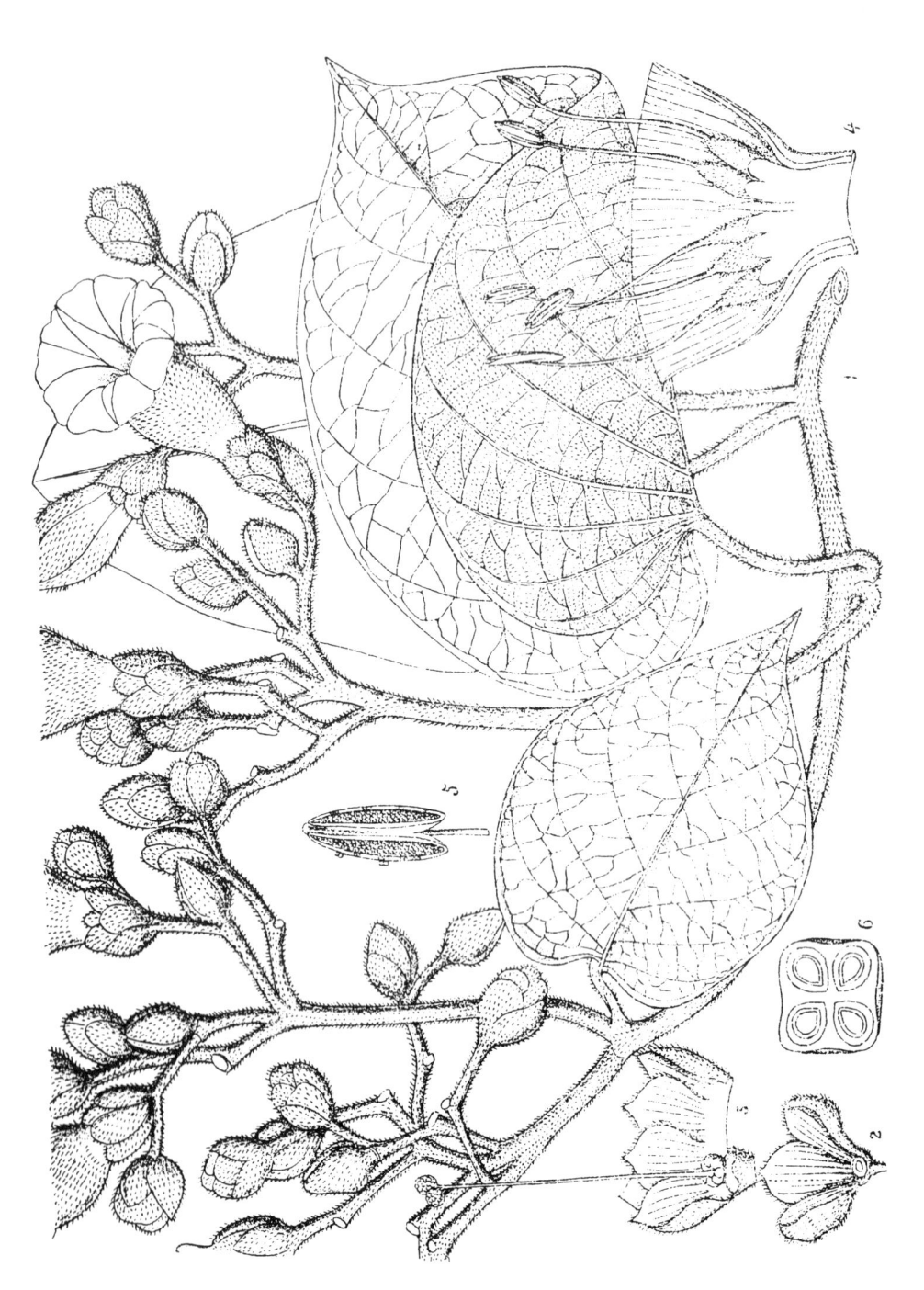

ARGYREIA SPECIOSA Sweet.

ARGYREIA SPECIOSA Sweet. *Family:* CONVOLVULACEAE
(Syn. *LETTSOMIA NERVOSA* Roxb.)

Vernacular names — Sanskrit, Vridha daraka; Hindi, Samudra Shokha; English, Elephant creeper; Bengali, Bijarka; Nepalese, Samudra phool; Sinhalese, Vriddadaru; Unani, Samudar sokh; Tamil, Kadarpalai.

Habitat — Throughout India except in dry, western regions up to 1000 ft elevation, often cultivated.

Parts used — Root and seeds.

Morphological characteristics — Leaves ovate or cordate, up to 12 in. long, glabrous above, white tomentose beneath, long stalked. Flowers in cymes; bracts large, thin, veined, pubescent outside, glabrous inside, calyx white, tomentose outside; corolla funnel shaped, silky pubescent outside, tube somewhat inflated, rose-purple, about 2 in. long. Root cylindrical, 1 to 1.5 cm thick; brown, smooth, round wood is scant, flexible, and smooth, latex oozes at cuts. Seeds are enclosed in a stone, pale yellow-brown globose apiculate, indehiscent berry 1.2 to 2 cm in diameter containing four erect, curved embryos with corrugated cotyledons or two seeds embedded in a meaty pulp.

Ayurvedic description — *Rasa — katu, tikta, kasaya; Guna — laghu, snigdha; Veerya — ushna; Vipak — madhur.*

Action and uses — *Kapha vat samak, bran pachan, daran, sodhan, ropan, nari balya, dipan, pachan, ampachan, anulomon, rachan, hiridya sothahar, surkrjanan, pramehangan, balya, rasayan.*

Chemical constituents — The plant contains tannin and amber-colored resin, soluble in ether, benzole; partly soluble in alkalis; and fatty oil.[103]

The seeds have shown the presence of alkaloids, viz., chanoclavine, ergine, ergonovine, and isoergine by various workers.[10]

Pharmacological action — Alterative, aphrodisiac, antiphlogistic, antiseptic, tonic, and emollient.

Medicinal properties and uses — Powder of the root is given with "ghee" as an alterative; in elephantiasis the powder is given with rice water. In inflammation of the joints it is given with milk and a little castor oil. A paste of the roots made with rice water is applied over rheumatic swelling and rubbed over the body to reduce obesity. The whole plant is reported to have antiseptic properties.[1] The leaves are antiphlogistic; they are applied over skin diseases and wounds;[109] the silky side of the leaf is applied over tumors, boils, sores, and carbuncles;, as an irritant to promote maturation and suppuration.[50] The leaves are also used for extracting guinea worms. A drop of the leaf juice is used in otitis.

Doses — Powder — 1 to 3 gr; seed powder — 0.5 to 1.5 gr; decoction of the root — 14 to 28 ml.

ARISTOLOCHIA BRACTEATA Retz.   *Family:* ARISTOLOCHIACEAE

Vernacular names — Sanskrit, Kitamari; Hindi, Kiramari; English, Wormkiller; French, Aristoloch; Unani, Kirmar; Tamil, Adu.

Habitat — Grows along the banks of the Ganges and in southern India.

Parts used — Whole herb, seeds, and leaves.

Morphological characteristics — Leaves glaucous, heart shaped or kidney shaped, up to 3 in. broad. Flowers axillary, solitary; bract roundish; perianth tube greenish, dilated at base, expanding in an inflated obliquely 2-lipped, dark purple, trumpet-shaped mouth. Capsules 6-valved, 12-ribbed; seeds numerous, thin, heart shaped.

Ayurvedic description[26] — *Rasa* — *tikta; Guna* — *laghu, rooksha, teekshna; Veerya* — *ushna; Vipak* — *katu.*

Action and uses — *Kapha vat samak, bran sodhan, dipan, rachan, krimighan, kafaghana, sotha har, jwarnasak, rajorodha nasak.*

Chemical constituents — Stem and root contain the alkaloid aristolochic acid.[110] From the leaves and fruits, ceryl alcohol and β-sitosterol have been isolated in addition to potassium chloride and aristolochic acid, a yellow microcrystalline compound, mp 275 to 277°C with the molecular formula $C_{17}H_{11}O_7N$. Alcoholic extract of roots yielded another crystalline mass, potassium nitrate.[111] Chopra et al.[178] reported the presence of aristolochic acid and aristo red, the two crystalline principles of the roots.

Pharmacological action — Purgative, emmenagogue, alterative, antiperiodic, and anthelmintic, abortifacient.

Medicinal properties and uses — Dry powdered root increases the contraction of the uterus during labor and is used as a substitute for ergot. With castor oil it is given in colic and tormina, amenorrhea, dysmenorrhea, intermittent fever, and worms. It is also given in syphilis, gonorrhea and skin disease. Its juice is applied to foul and neglected ulcers to destroy insect larvae.[1,3]

Doses — Powder — 1.5 to 3 g; infusion — 15 to 30 ml; decoction — 14 to 28 ml; seeds — 2 to 5 gr.

ARISTOLOCHIA INDICA Linn. Family: ARISTOLOCHIACEAE

Vernacular names — Sanskrit, Ishwari, Sunanda; Hindi, Isharmul; English, Indian birthwort; Bengali, Ishwarmul; Sinhalese, Sapsanda; German, Indische Ostertuzei; Unani, Isharmool; Arabian and Persian, Zaravandhindi; Tamil, Isvaramulisedi.

Habitat — A twining perennial, found all over tropical regions of India, viz., Bengal, Konkan, Kerala, and the Coromandal coast.

Parts used — Root, rhizome, and leaves.

Pharmacognostical characteristics—Leaves simple, alternate, short petioled, blade somewhat wedge-shaped, ovate, very variable in shape and size. Tender leaves are light purplish. Fruit roundish or oblong hexagonal, shallowly grooved; pendulous six-celled, six-valved, septicidal capsule 2.5 to 4 cm long and slightly less broad, contains many triangular seeds. The young roots are light brown and fairly smooth, whereas the older ones are comparatively rough because of the development of cork and lenticels and the presence of scars of rootlets. The cork layer is somewhat friable. In a freshly cut transverse section of the root the entire bark appears as a narrow cream-colored strip surrounding a wide woody core. The wood appears composed of a number of wedge-shaped radiating pieces or plate with their apices toward the center. The pieces of the wood are separated by or alternate with the medullary rays, of which four are broad and start near the center and four or more are narrower and do not reach the center. The wood has a light yellow color and appears highly porous, the pores being sufficiently large to be easily visible with the naked eye; the medullarys rays are soft and cream white in color; there is no pith in the center.

Ayurvedic description — *Rasa* — *katu, tikta, kasaya; Guna* — *laghu, rooksha, teekshna; Veerya* — *ushna; Vipak* — *katu.*

Action and uses — *Tridosha har, vishaghan, brana sodhan, sothahar, badana sthapan, vat samak, dipan, anuloman, grahi, krimighan, soth har, hiridyatajk, rakt sodhak, kapha nisark, mutral, jwaraghan.*

Chemical constituents — An alkaloid aristochine with molecular formula $C_{36}H_{38}O_6N_2$ was reported in the roots.[112] A tetracyclic sesquiterpene ketone called "ishwarone"[113] $C_{15}H_{22}O$, bp 57°C, two new sesquiterpene hydrocarbons, ishwarane $C_{15}H_{24}$, bp 80 to 82°C/1 mm, aristolochene $C_{16}H_{24}$, bp 85°C/0.5 mm,[114] and a new tetracyclic sesquiterpene alcohol ishwarol, $C_{15}H_{24}O$, bp 110°C/1 mm were isolated from the roots.[115]

Pharmacologic action — Root is tonic, stimulant, emmenagogue, alexiteric, and antiarthritic. Leaves are stomachic, tonic, and antiperiodic.[1,3,26]

The juice of *A. indica* was found to have a relaxant effect on the rhythmic contractions of the uterus of rats and human beings. It had a nonspecific antispasmodic effect on spasms induced by acetylocholine, barium chloride, and pitocin at a concentration of 4 mg/ml.[116]

Medicinal properties and uses — Root is a good antidote to snake bite or poisonous insects and scorpion; it is used both externally and internally; it makes the part bitten insensible to the ill effects of poison. Juice of the leaves and bark is chiefly used in bowel complaints of children, cholera, diarrhea, and in intermittent fevers.

Doses — Leaves infusion — 5 to 10 ml; root powder — 1 to 2 gr.

ARTEMISIA MARITIMA Linn. *Family:* COMPOSITAE
(Syn. *A. BREVIFOLIA* Wall.)

Vernacular names — Sanskrit, Chauhar; Hindi, Kirmala; English, Worm seed; Nepalese, Titepati; Chinese, Chinai; German, Meerstrand Beiful; Unani, Zhech; Arabian, Shik; Afsantin, al-baher; Persian, Dirmanah; Pushtu, Tarkh.

Habitat — Many species grow abundantly in the western Himalayas from Kumaon to Kashmir, in Kurran Valley, Baluchistan, and in Afghanistan.

Parts used — Flowers, buds, and leaves.

Morphological characteristics — A shrub with stout, branched, rootstock; stems up to 3.5 ft high, slender, much branched from the bases, striate. Leaves 1 to 1.5 in., often quite white, 2-pinnatisect, segments numerous, small, spreading, linear, obtuse; upper leaves simple, linear. Flowerheads homogamous, numerous, ellipsoid, oblong or ovoid, about $1/_{10}$ in. long, 3 to 10 flowered, in spicate fascicles in the axil of a small linear leaf. Receptacle naked.[28]

Ayurvedic description — *Rasa—tikta, katu; Guna—laghu, rooksha, teekshna; Veerya—ushna; Vipak—Katu.*

Action and uses — *Kapha-vata samak, bran ropan dipan, vataanuloman krimighan, swashav rehan, swashar, kaphanisarak, mutral, vagekaran, artawbjanan, sheeta prasaman jwaroghan, lakhan.*

Chemical constituents — Important constituents of *A. maritima* are an essential oil having an odor like cajuput oil and camphor, containing cineole, thujone, camphene, etc.; a lactone, santonin, which is responsible for anthelmintic properties; and a second crystalline artemisin, closely related to santonin.[69]

Pharmacological action — Antiperiodic, aperient, stomachic, tonic, anthelmintic, and cardiac stimulant.

Medicinal properties and uses — Its infusion is given as an anthelmintic and cardiac stimulant. As a discutient and antiseptic, a fomentation of the herb is applied over inflammations, tumors, and foul ulcers.

Doses — Plants, powder — 3 to 6 gr.

ARTEMISIA VULGARIS Linn.   *Family:* COMPOSITAE

Vernacular names — Sanskrit, Nagadamni; Hindi, Nagadouna; English, Mugwort, Indian worm-wood; Bengali, Nagadona; Nepalese, Titepatti; Japanese, Yomogi; German, Beiful; Unani, Nagdaun; Arabian, Halyoon; Persian, Marchobah; Tamil, Masipathiri.

Habitat — Western Himalayas, Assam, western and southern India.

Parts used — Leaves and flowering tops.

Morphological characteristics — An aromatic perennial, shrub-like, 2 to 8 ft high, pubescent or tomentose. Stems branched, red, ribbed, erect, rough. Upper leaves smaller than lower, 3-lobed or entire; lanceolate; lower leaves 2 to 4 in. long; ovate, pinnately lobed, hairy on both surfaces. Flower heads small, reddish, yellow, or whitish drooping, in branched, wooly spikes, heterogamous, minute.[28] A very variable plant regarding the leaves and flower heads.

Ayurvedic description — *Rasa* — *tikta, kasaya; Guna* — *laghu, rooksha, teekshna; Veerya* — *ushna; Vipak* — *katu.*

Action and uses — *Tridosha samak, sothahar, badana sthapan, dipan, pachan, anuloman, pitta sark, hiridya, raktsodhak, jwaraghan.*

Chemical constituents — The plant yields about 0.2% volatile oil which contains thujone, borneol, etc.[99,100]

Pharmacological action — Anthelmintic, antiseptic, expectorant, larvicide.[1119]

Medicinal properties and uses — The herb is antispasmodic, expectorant, stomachic, tonic, aperient, and anthelmintic; it is prescribed particularly for obstructed menses, hysteria, and preventing abortion in the form of an infusion.[1] As an anthelmintic a strong decoction is given; a weak decoction is given to children with measles; a few drops of juice of the leaves are prescribed for whooping cough. A powder of the leaves is given for checking hemorrhage, metorrhagia, dysentery, and intestinal and urinary troubles. A paste or powder of the leaves is applied over skin diseases.

Doses — Infusion — 14 to 28 ml; powder — 0.5 to 1 gr.

ARTOCARPUS INTEGRIFOLIA Linn.   Family: MORACEAE
(Syn. A. HETEROPHYLLUS Lam.)

Vernacular names — Sanskrit, Panasa; Hindi, Katahar; English, Indian Jack tree; Bengali, Kanthel; Nepalese, Rukhakathar; Sinhalese, Cos; Chinese, Palomi; German, Indischerbrod Baum; Unani, Kathal; Persian, Chakki; Tamil, Pala.

Habitat — Cultivated all over India.

Parts used — Fruits, seeds, leaves, root, and milky juice of the plant.

Morphological characteristics — Medium to large tree, flowers unisexual; fruits are oval or oblong, 1 to 2 ft in length and 6 to 12 in. broad. The skin is studded with short spines, is pale green when immature, becoming greenish yellow to brownish when ripe. Fruits contain a large number of seeds, each enclosed in a yellowish juicy sheath. The wood is yellow when freshly cut, but gradually turns light brown on exposure.

Ayurvedic description — *Rasa* — *madhur, kasaya; Guna* — *guru, snigdha; Veerya* — *sheeta; Vipak* — *madhur.*

Action and uses — *Kapha vatta wardhak, pitta samak, sothahar, bran pachan, sthamvan, rakt sthamvan, suka wardhak, balya, uringhan, visaghan.*

Chemical constituents — The fruit contains 77.2% moisture, 1.9% protein, 0.1% fat, 18.9% carbohydrate, 1.1% fiber, 0.8% mineral water, 0.02% calcium, 0.03% phosphorus, 0.5 mg iron; vitamin A, 540 IU; vitamin C, 10 mg/100 g. The seeds contain starch. The bark and the fruits exude a milky latex (moisture and water soluble, 65.9 to 96.0%; caoutchouc, 2.3 to 2.9%). The coagulum contains caoutchouc (6.0 to 10.0%, resins (82.6 to 86.4%), and insolubles (3.9 to 8.1%).[82] Wood yields coloring matter morin and cyanomaclurin,[44] bark conatins 3.3% tannins,[122] crystalline steroketon, artostenone isolated from latex.[123] Artostenone has been converted to artosterone, a compound with highly androgenic properties.[124]

Pharmacological action — Ripe fruit is demulcent, nutritive, and laxative. Unripe fruit is astringent.[3] Various chemical and biological tests demonstrated the presence of an acetylcholine-like substance in the seeds of *A. integrifolia*. There was evidence of the presence of another active substance which had positive inotropic and chronotropic effects on the frog's heart.[125]

Medicinal properties and uses — Seeds of ripe fruit when roasted in hot ashes are very palatable and nutritious. Milky juice of the plant alone or mixed with vinegar when applied externally to glandular swelling and abscesses promotes absorption or suppuration; it is also used in snake bite. Root is used in diarrhea. Tender leaves and root are useful in skin diseases. An alcoholic drink can be made from jack fruit.

Doses — Decoction — 56 to 112 ml.

ASPARAGUS ADSCENDENS Roxb. *Family:* LILIACEAE

Vernacular names — Sanskrit, Musli; Hindi, Safed musli; Nepalese, Musali; Sinhalese, Hirth wariya; Unani, Satawar Misri; Arabian, Shaqoqule-hindi; Persian, Shaqoqule-hindi.

Habitat — Punjab, western Himalayas, up to 5300 ft, Gujerat, Bombay, Rohilkhand, and the central Mussori Hills.

Parts used — Tuberous root or rhizome decorticated.[26]

Pharmacognostical characteristics — Root 30 to 100 cm long, 0.7 to 1 cm thick, scars and protuberances of the lateral rootlets seen all over external surface, hard, odorless, and sweetish in taste. A transection of the root shows an outer cortex with characteristics similar to *A. racemosus*. Sclerenchyma is present in the inner cortex. Pith is completely or partially lignified. Vascular bundles 34 to 37 at upper end and 45 to 50 at the middle level.[26]

Ayurvedic description — *Rasa — madhur, tikta; Guna — guru, snigdha; Veerya — ushna; Vipak — madhur.*

Action and uses — *Tridosha samak, virisya, sukral, balya, brinhan, rasayan.*

Chemical constituents — The underground parts contain asparagin, albuminous matter, mucilage, and cellulose.

Pharmacological action — Nutritive, tonic, galactagogue, and demulcent.[3]

Medicinal properties and uses — Tubers boiled in milk and sugar are used in spermatorrhea, gleet, and chronic leukorrhea; also in diarrhea, dysentery, and general debility. A compound powder containing many other ingredients is given as a nutritive tonic with milk in cases of seminal weakness and impotence. They have excellent cooling and demulcent properties.[1]

Doses — Powder — 3 to 6 gr.

ASPARAGUS RACEMOSUS Willd.   Family: LILIACEAE

Vernacular names — Sanskrit, Shatavari; Hindi, Satavari; English, Asparagus; Bengali, Satmuli; Nepalese, Kurilo; Sinhalese, Hatavari; German, Spargel; French, Asperge; Unani, Satavar; Arabian, Shaqaqul; Persian, Shaqaqul; Tibetan, Rt sh-babrgya; Burmese, Kan yo-mi.

Habitat — Climber found all over India, especially in North India.

Parts used — Roots and leaves.

Pharmacognostical characteristics — The adventitious roots arising at a point become fleshy, tuberous, and fasciculated. These roots are swollen in the middle for food storage and tapering toward the base. The color of the root is silver white or light ash.

Diagnostic characters of transverse section of *A. racemosus* are[126] presence of piliferous layer, often ruptured, outer cortex of 7 to 10 layers of suberized thick-walled cells; inner cortex 18 to 24 layers in upper level, 42 to 46 layers in middle level. Sclerenchyma is scattered in upper level only, stone cells and raphides present, vascular bundles number 30 to 35 in upper level and 35 to 45 in the middle. Tracheids, metaxylem vessels, and fibrous pith cells are present.

Ayurvedic description — *Rasa — madhura, tikta; Guna — guru, snigdha; Veerya — sheeta; Vipak — madhura.*

Action and uses — *Vat pitta samak, balya, dasha samak, brana ropan, medhya, badana sthapan, nari baldayak, dipan, anuloman, grahi hiridya, shukal, sthanyajan, balya, rasayan, chachusaya.*

Chemical constituents — The alcohol, ethyl acetate, and acetone extract of powdered dry roots of *A. racemosus* yielded pharmacologically active (antioxytocic) saponins.[26]

Pharmacological action — Root is antidiarrhetic, refrigerant, diuretic, nutritive, tonic, galactagogue, aphrodisiac, and antispasmodic. Alcoholic extract of the aerial parts of *A. racemosus* showed anticancer activity in human epidermal carcinoma of the nasopharynx in tissue culture.[51] The aqueous solution of the crude alcoholic extract of the roots also had an inhibitory effect on the growth of *E. histolytica in vitro*.[127]

Medicinal properties and uses — It is useful in dysentery, diarrhea, tumors, inflammations, biliousness, diseases of the blood, kidney, liver complaints; it is also useful in rheumatism, dyspepsia, gleet, and gonorrhea.[1] The roots have been found to have galactogogue action in buffaloes in which the milk yield was found to be significantly increased after the uses of the drug. The aqueous extracts of both fresh and dried roots were found to have amylase and lipase activities.[128] The juice of this drug taken with milk is useful in gonorrhea.

Doses — Infusion — 12 to 20 ml; decoction — 56 to 112 ml; powder — 20 to 30 gr.

ATROPA ACUMINATA Royle ex Lindley   *Family:* SOLANACEAE

Vernacular names — Sanskrit, Suchi; Hindi, Sag-angur; Bengali, Yabruj; German, Tolkkivacha; Unani, Luffah; Arabian, Yabriyssanum; Persian, Luffah; English, Indian belladonna.

Habitat — Grows wild and also cultivated in forests of western Himalayas, extending from Simla to Kashmir.

Parts used — Roots and leaves.

Morphological characteristics — A perennial herb; about 42 in. high; stem has purplish streaks at base. Leaves eliptic-lanceolate tapering at both ends, 5 to 8.5 by 2 to 3.5 in., entire, petiole purplish at the base. Flowers 1 to 2 in the axil of the lower leaves of flowering branch and 4 to 5 more in a cyme on the secondary branches in the axil of upper leaves. Corolla campanulate, yellow with greenish veins. Fruit a berry.[28]

Ayurvedic description — *Rasa — tikta, katu; Guna — laghu, rooksha; Veerya — ushna; Vipak — katu.*

Action and uses — *Kapha vata har, pitta wardhak, madak, utejak, poralapajanya, dipan, lala prasak, shula prasaman, awasadak, kashahar, swas har, stamvan, sukra sosan kandughan.*

Chemical constituents — It contains alkaloids, hyoscyamine, and atropine; the roots contain some volatile bases which are therapeutically inactive.[1]

Pharmacological action — Leaves and roots are narcotic, sedative, diuretic, antispasmodic, and mydriatic and used as an anodyne. Berries are poisonous.[50]

Medicinal properties and uses — This plant has the same medicinal properties as the European *Atropa belladonna*. The leaves and root are used as a sedative, antispasmodic, narcotic, and mydriatic; they are valuable antidotes in opium and muscarine poisoning. The dried leaves are smoked as an antispasmodic. The extract prepared from the leaves causes the pupils of the eyes to dilate and is used in ophthalmic surgery. The root is generally used externally; it is used as an anodyne for rheumatism, neuralgia, lumbago, and local inflammations.

Doses — Powder — 25 to 50 mg; tincture — 5 to 30 minims.

AVERRHOA CARAMBOLA Linn.　　　　　　　　　　　Family: OXALIDACEAE

Vernacular names — Sanskrit, Karmaranga; Hindi, Kamrak; English, Chinese gooseberry; Bengali, Kamranga; Sinhalese, Kamaranga; Japanese, Goren; Chinese, Maping long; Unani, Kamrak; Burmese, Saunggya; Malaysian, Kembola; Tamil, Gyvirali.

Habitat — Cultivated in the hotter parts of India.

Parts used — Leaves, root, and fruits.

Morphological characteristics — A small, handsome, evergreen tree 15 to 30 ft high with flowers springing from the bark, and acutely five angled, ovoid fruits 3 to 4 in. long, watery, and translucent. The wood is white, turning reddish.

Ayurvedic description — *Rasa* — *amla, madhura, kasaya; Guna* — *laghu, rooksha; Veerya* — *sheet* (ripe), *ushna* (unripe); *Vipak* — *amla* (unripe), *madhur* (ripe).

Action and uses — *Vat kapha samak, ruchiprada, deepan, grahi, sonit sthapan, daah prasaman, jwaraghan, rakt pitta samak, grahani roga nasak.*

Chemical constituents — The fruits of the sweet variety have been found to contain 93.9% moisture, 0.5% protein, 0.2% fat, 4.8% carbohydrate, 0.2% mineral matter, 0.6 mg. iron, 240 IU/100 g vitamin A. The acid potassium oxalate content is given as 0.8%.[50,82]

Pharmacological action — Laxative, refrigerant, antiscorbutic, and antidysenteric.

Medicinal properties and uses — There are two varieties, sweet and sour. Juice of the fruits is an excellent cooling drink in relieving thirst during fevers. Ripe fruit is a good remedy for bleeding piles, particularly the internal piles.

Doses — Infusion — 3 to 5 drams.

AZADIRACHTA INDICA A. Juss.   *Family:* MELIACEAE
(Syn. *MELIA AZADIRACHTA* Linn.)

Vernacular names — Sanskrit, Arishta; Hindi, Nim; English, Persian lilac; Bengali, Nimb; Nepalese, Neeha; Sinhalese, Kohomba; Chinese, Chado; German, Indischer Zedrach; French, Margousier; Unani, Neem; Arabian, Harbit; Persian, Neeb; Tamil, Veppilai.

Habitat — A large evergreen glabrous tree found all over India; also cultivated.

Parts used — Root bark, young fruit, nut or seeds, leaves, flowers, bark, gum, and toddy.

Morphological characteristics — Leaves alternate, exstipulate, imparipinnate, leaflets 9 to

15 in. long, subopposite, lanceolate, closely clustered toward the ends of branches. The trunk and older branches are covered with moderately thick, brown, rough, longitudinally and obliquely furrowed bark. The tender parts are glabrous, with a thick waxy bloom or coating. In fresh bark the outer part is fairly thick with its peripheral region rose or purplish red, the middle part lustrous starchy white, and the inner part tangentially lamellated in transverse section. The inner part of the rind consists of a few rows of regular thin-walled rectangular cork cells. The officinal part consists almost exclusively of secondary bast or radial segment of bast alternating with medullary rays. Each radial segment consists of tangential zones of thin-walled phloem elements, alternating with bands of sclerenchyma. The phloem parenchyma cells contain starch; a few scattered secretory cavities are present in the phloem.[133]

Ayurvedic description — *Rasa* — *tikta, kasaya; Guna* — *laghu; Veerya* — *sheeta; Vipak* — *katu.*

Action and uses — *Kapha pitta samak, brana sodhan putihar, dah prasaman, kustaghan, badana sthapan, grahi, krimighan, daahsamak, rakt sodhak, aama pachan, jwaraghana.*

Chemical constituents — The petroleum ether extract of dry flowers gave a waxy substance which yielded β-sitosterol. The flavonoids of neem flowers consist of three related compounds, kemferol, quarcetin, and myricetin. They were present mostly as glycosides extractable with hot alcohol. In addition, some other compounds were also obtained, viz., (1) green amorphous bitter toxic substance; (2) nonacosane $C_{29}H_{60}$, mp 64 to 66°C; (3) highly pungent essential oil; and (4) a sesquiterpene and a fatty acid fraction.[26] The trunk bark contained nimbin (0.04%), nimbinin (0.001%), nimbosterol (0.03%), essential oil (0.02%), tannin (6.0%), and a bitter principle margosine.[133] The principle constituents of the neem oil were sulfur, a very bitter yellow substance which was suspected to be an alkaloid, resins, glycosides, and fatty acids.[134]

The crushed neem seeds yielded a white nonbitter sulfur-containing product $C_{13}H_{27}O_7S_2$ which was levorotatory.[26]

The petroleum ether extract of all the flowers yields a water-insoluble amorphous bitter of acidic character. The petroleum ether-soluble fraction yielded nimbosterol (β-sitosterol) mp 137°C, which was present in other parts of the plant as well.

The starch of gum indicated a nonreducing white powder after purification by precipitation with alcohol. On hydrolysis the gum yielded L-arabinose, L-fucose, D-galactose, and D-glucuronic acid. The aldobiuronic acid component of the gum was shown to be 4-*O*(D-glucopyranosyl uronic acid)-D-galactopyranose. The presence of free amino acids in the stem exudate was of considerable importance. The heartwood showed the presence of tannins, inorganic calcium potassium, and iron salts in the alcohol and acetone extracts. The extracts yielded an ether-soluble crystalline compound, $C_{24}H_{30}O_5$, mp 137 to 138°C, provisionally called "nimatone", on which IR and UV studies have been conducted.[135] The bitter constituent nimbin molecule contains an acetoxy, a lactone, an ester of methoxy, and an aldehyde group.[178] Desacetylnimbin has been isolated from the seed and bark as a natural product.[136] The isolation of quercetin and β-sitosterol from the leaves of *A. indica* has also been reported.[137]

Pharmacological actions — Anthelmintic, antiseptic, bitter, deodorant, diuretic, emmenagogue, and febrifuge.[3] The leaf extract fractions delayed clotting time.[19] Insecticidal and insect-repellant properties of neem have been reported. Neem oil inhibited the growth of all the three strains of *M. tuberculosis* and *M. pyogenes* var. *aureus*.[1] A 10% water extract of leaves of the plant possessed antiviral properties.[138] The leaf extract also showed an antibacterial principle.[178]

Neem oil and two of its bitter principles, nimbidin and nimbidol, exhibited antibacterial, antifungal, spermicidal activity; and was effective in various skin diseases.[139-141]

Medicinal properties and uses — It is very useful in blood disorders, consumption, eye diseases, intermittent fevers, as well as persistent low fever. Oil is very useful in leprosy, scrofulas, skin diseases, ulcers, and wounds. The bark contains a resinous bitter principle and is usually prescribed in the form of a tincture or an infusion. It is also regarded as beneficial in malarial fever.

Doses — Infusion — 3 to 5 drams; powder — 1 to 2 gr; oil — 4 to 10 minims.

BACOPA MONNIERA (L.) Pennell   *Family:* SCROPHULARIACEAE
(Syn. *HERPESTIS MONNIERA* (L.) H.B.K.;
MONNIERA CUNEIFOLIA Mich.)

Vernacular names — Sanskrit, Brahmi; Hindi, Brambhi; Bengali, Brihmi sak; Nepalese, Bramhi; Sinhalese, Lunuvila; Japanese, Otomeazene; Unani, Bramhi; Arabian, Zarazab; Persian, Sard Turkistan; Tamil, Nirbrahmi.

Habitat — This small creeping plant is found in moist and wet places such as on borders of water channels, wells, irrigated fields, etc. in all parts of India.

Parts used — Whole plant root, stalks, and leaves.

Pharmacognostical characteristics — The herb spreads on ground and its stems and small leaves are fleshy. Roots arise on the nodes of the stem, also. Flowers arise in the axil of the leaves and are borne on short pedicels. One of the five petals is larger than others. The corolla is bluish-white in color and 1 cm across.[28]

Ayurvedic description — *Rasa — titka; Guna — laghu, snigdha; Veerya — ushna; Vipak — katu.*

Action and uses — *Kapha vat nasak, sothahar, badana sthapan, visaghana, medhya, dipan, pachan, anuloman, rakt sodhak.*

Chemical constituents — Earlier studies on the chemical consitutents of the plant *Bacopa monniera* indicated the presence of alkaloids, brahmine, and herpestine and a mixture of three alkaloids.[440,441] Later, D-mannitol and a saponin were also reported to be present.[442] Rastogi and Dhar[443] isolated betulic acid, a crystalline saponin and a crystalline product with mp 71°C. This saponin on hydrolysis gave glucose, arabinose, and an aglycone. The saponins designated as bacoside A (mp 251°C decomp.) and bacoside B (mp 203°C decomp.), betulic acid, D-mannitol, stigmastarol, and β-sitosterol were isolated from the whole plant.[444,445]

Pharmacological action — It is a cardiac and nervine tonic; leaves and stalks are diuretic and aperient.[1,3] The crude alcoholic extract of the plant showed a sedative effect on frogs and dogs; it also exerted cardiotonic, vasoconstrictor, and neuromuscular blocking action in frogs. A spasmodic action was seen on rabbits' and guinea pigs' ilea and rats' uteri.[446] The total alkaloidal fraction produced an initial fall followed by a rise in the blood pressure of dogs.[447] The alcoholic extract of the entire plant was found to have anticancer activity against Walker carcinosarcoma 256 (intramuscular) in rats.[11]

Medicinal properties and uses — Leaves and stalks are very useful in the stoppage of urine which is accompanied by obstinate costiveness. A poultice made of the boiled plant is placed on the chest in acute bronchitis and other coughs of children. Leaves also give satisfactory results in cases of asthenia, nervous breakdown, and other low adynamic conditions. It is also given in combination with *ghrita,* a well-known Ayurvedic medicine *(brahmi ghrita)* in cases of hysteria and epilepsy. It is also useful in insanity, neurasthenia, aphonia and hoarseness.[3,50]

Doses — Infusion — 8 to 16 ml; powder — 5 to 10 gr.

BALANITES AEGYPTIACA (Linn.) Delile *Family:* SIMAROUBACEAE
(Syn. *B. ROXBURGHII* Planch.)

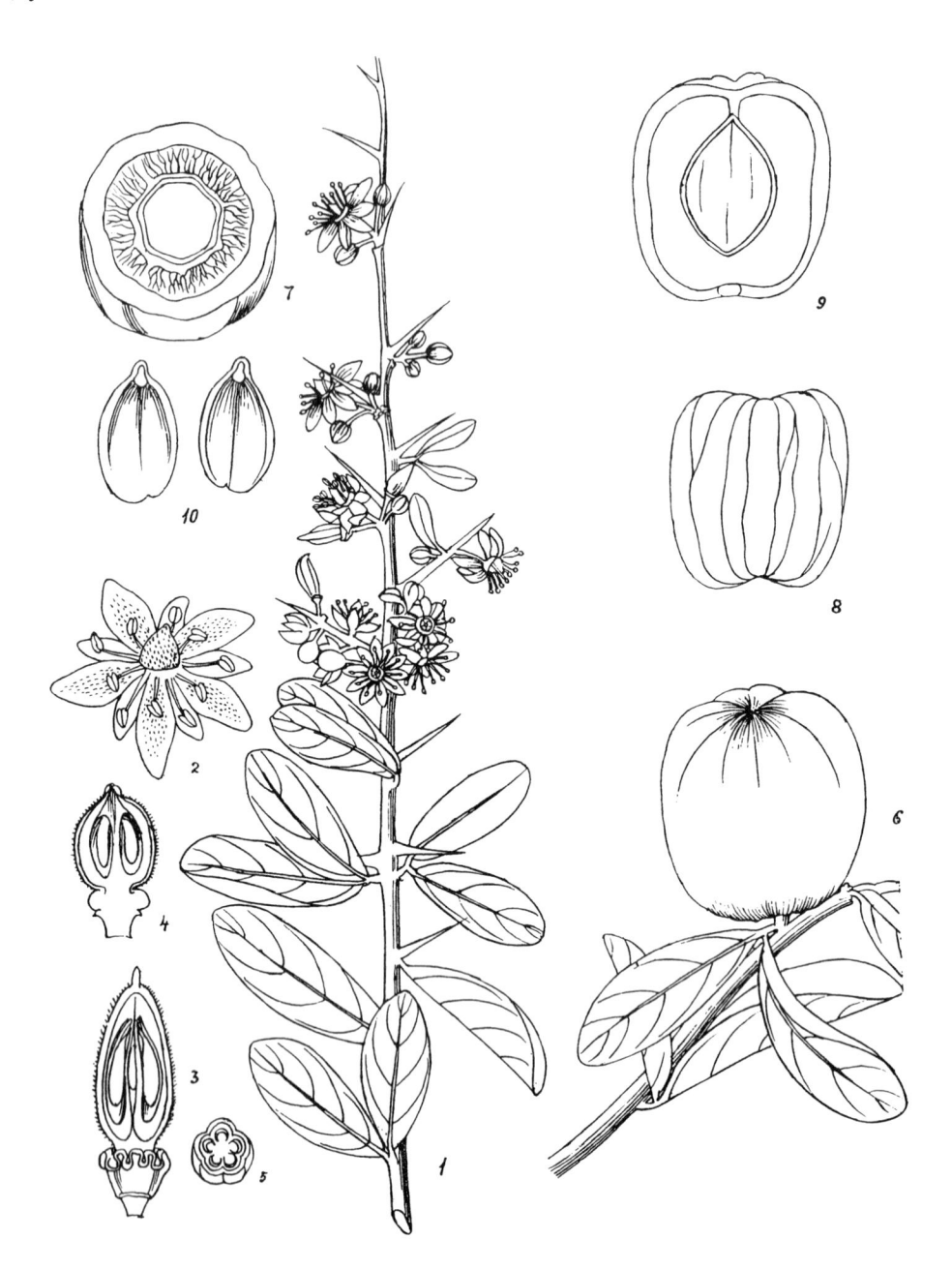

Vernacular names — Sanskrit, Ingudi-vraksha; Hindi, Hingol; Bengali, Hingol; Nepalese, Cheure; Sinhalese, Ingudi; Unani, Hingot.

Habitat — A small evergreen tree or shrub found in Punjab, Delhi to Sikkim, Bengal, Bombay, and South India.

Parts used — Seeds, bark, leaves, and fruit.

Morphological characteristics — A small spiny tree with bifoliate ashy green leaves; leaflets 1 to 2 by $1/2$ to 1 in., elliptic or ovate; petals slightly larger than sepals; fruit an ovoid drupe, faintly five grooved, pale yellow when ripe; pulp with an offensive greasy smell.[28]

Ayurvedic description — *Rasa — tikta, katu; Guna — laghu, snigdh; Veerya — ushna; Vipak — katu.*

Action and uses — *Kapha vat samak, kesya brana ropan, shirobirachan, dipan, sansran, rakt sodhak, kapaha nisarak, kustaghan.*

Chemical constituents — Bark contains some saponin.[142] The seed kernel yields a saponin, a tetraglycoside of a sapogenin;* acid hydrolysis gives nitrogenin, which is an active hemolytic agent.[143] Seeds yield 43% of a bland, yellow, tasteless oil. Pulp of the fruit contains organic acid, saponin (7.2%), mucilage, and sugar.

Pharmacological action — Purgative and anthelmintic.

Medicinal properties and uses — Seeds are given in cough and colic. Bark, unripe fruit, and leaves are pungent, bitter, purgative, and anthelmintic for cattle; its juice is used as a fish poison. Oil expressed from the seeds is applied to burns and freckles. The seed extract is used clinically as a hypotensive active principle.

Doses — Decoction — 25 to 75 ml; pulp of the fruit — 0.5 to 1 gr; fruit powder — 0.5 to 1 gr; oil — 0.5 to 1 ml.

---

* Sapogenin is diosgenin.

BALIOSPERMUM MONTANUM Muell. Arg.   *Family:* EUPHORBIACEAE
(Syn. *B. AXILLARE* Blume)

Vernacular names — Sanskrit, Danti; Hindi, Danti; Bengali, Hakun; Nepalese, Danti; Sinhalese, Detta; Unani, Danti; Tibetan, Da-nti; Burmese, Natcho.

Habitat — Common in outer ranges of Himalayas from Kashmir to Bhutan up to 3000 ft; Assam, Khasia Hills, Bihar, from central and western India to Kerala.

Parts used — Leaves, seeds, and root.

Pharmacognostical characteristics — Leaves of upper parts 2 to 3 in. long and those of lower parts 6 to 12 in. long and divided; flowers unisexual, fruits $1/2$ to 1 in. long, containing three seeds brown in color.

Ayurvedic description — *Rasa* — *katu; Guna* — *guru, rooksha, teekshna; Veerya* — *ushna; Vipak* — *katu.*

Action and uses — *Kapha pitta har, dipan, pittasarak, birachan, krimiginana, rakt sodhak.*

Chemical constituents — Root contains starch, and seeds yield an essential oil.

Pharmacological action — Seeds are used as drastic purgative; externally, seeds act as stimulant and rubifacient.

Medicinal properties and uses — Roots and leaves have similar properties. Root cathartic, used in dropsy, anascara, and jaundice. Decoction of leaves used in asthma. Oil from seeds hydrogogue, cathartic; for external applications in rheumatism. Root and seeds are used as antiseptic and also in snake bite. Leaves are very useful in asthma.[1]

Doses — Root powder — 1 to 3 gr; seed powder — 0.2 gr; leaf decoction — 28 to 56 ml.

BAMBUSA BAMBOS (Linn.) Voss. *Family:* GRAMINEAE
(Syn. *B. ARUNDINACEA* Willd.)

Vernacular names — Sanskrit, Vansa; Hindi, Wans; English, Bamboo; Bengali, Kaoura; Nepalese, Bansalochan; Sinhalese, Una; German, Bambus; Unani, Tabasheer; Arabian, Tabashir; Persian, Tabashir; Burmese, Kyakatwa; French, Bambou commun; Tamil, Mungil.

Habitat — Found throughout the plains and low hills of India, especially in the hill forests of western and southern India, ascending to an altitude of 3000 ft in the Nilgris; often cultivated.

Parts used — The interior stalks or stem of the female plant, containing a concretion of silica called *tabashir* (bamboo manna), young shoot, leaves, seeds, and root.

Morphological characteristics — A bamboo growing 25 to 30 ft or more high in dense tufts. Culms armed with spines, hollow, about 4 in. in diameter. Leaves 4 to 6 in. long and $1/_3$ to $1/_2$ in. broad, subsessile, glabrous, on the upper surface, hirsute below. Inflorescence a large panicle, much branched, its ultimate branches with clusters of spikelets on their nodes. Spikelets lanceolate, mucronate, often absent. Lemmas keeled, ovate, mucronate. Lodicules 3. Caryopsis oblong, $1/_5$ to $1/_4$ in. long.[28] Curious concretions of silica are found in the hollows of the culms; this *tabashir* is widely used in indigenous medicine as a cooling, tonic, stimulant, and as an astringent.

Ayurvedic description — *Rasa* — madhur, kasaya; *Guna* — rooksha, laghu, teekshna; *Veerya* — sheeta; *Vipak* — madhur.

Action and uses — *Kapha pitta samak, pitta wardhak, vat pitta samak, varnya, kustaghan, sothahar, dipan, pachan, krimighan, vidahi, samak, trisanigraha, grahi jwaraghan.*

Chemical constituents — *Tabashir* (*bansalochan* or bamboo manna) contains silica (90%) or silicum as hydrate of silicic acid, peroxide of iron, potash, lime, aluminia, vegetable matter, cholin, betain, nuclease, urease, proteolytic enzyme, diastatic and emulsifying enzyme, and cyanogentic glucoside.[1] An unnamed alkaloid has been reported from *Bambusa* sp.[10]

Pharmacological action — Leaves are emmenagogue and anthelmintic. *Tabashir* (bamboo manna) is stimulant, astringent, febrifuge, tonic, cooling, antispasmodic, and aphrodisiac. Bapat et al. observed that a single oral dose of 25% aqueous extract of leaves caused significant lowering of blood sugar both in normal and alloxan-treated rabbits. The effect persisted for about 96 h.[44] A compound was isolated from the water extract which was soluble in propylene glycol. Its solution also caused lowering of blood sugar in both types of rabbits and the hypoglycemic response was dose dependent.[145] Tewari et al.[146] observed that fresh juice of leaves of the plant had a weak ecbolic action in doses of 1 to 10 mg/ml and maximum activity at 3 mg/ml on isolated human as well as rat uteri. He concluded that uterine stimulation was due to its action on cholinergic receptors.

Medicinal properties and uses — A decoction of the leaves is useful as an emmenagogue, especially to induce lochia after childbirth; as an anthelmintic, the leaf buds are particularly useful for threadworms; the leaf juice is given with aromatics in vomiting of blood. A decoction of the joints or nodes of the bamboo stem is a useful emmenagogue; it is also used as an abortifacient. The decoction gives relief to inflamed joints. In Ayurvedic medicine, the stem and leaves are considered sour, bitter, and useful in diseases of blood, leukoderma, and inflammatory conditions. The sprouts and seeds are acrid, laxative, and said to be beneficial in strangury and urinary discharges. *Tabashir* or *bansalochan*, a silicious crystalline secretion found in the female plants, is a febrifuge, expectorant, tonic, aphrodisiac; it is given in hectic fever, pthisis, asthma, and paralytic complaints. The young shoots contain 0.3% of hydrocyanic acid and are lethal to mosquito larva.[28]

In Unani medicine, the root is said to be tonic. It is burned and applied locally to ringworm infections of the skin, to bleeding gums, and to painful conditions of joints. An infusion made from leaves is used as an eyewash, and given internally for bronchitis, gonorrhea, and fever. The juice of flowers is dropped in the ear for earache and deafness. The leaves, mixed with black pepper and common salt, have been used to check diarrhea in cattle.[1]

Doses — Decoction — 56 to 112 ml; bansalochan — 1 to 3 gr.

BARLERIA PRIONITIS Linn.   Family: ACANTHACEAE

Vernacular names — Sanskrit, Kurantaka; Hindi, Katsareya; Begali, Kantajati; Nepalese, Somolata; Sinhalese, Kattukurandu; Unani, Pialease; Burmese, Leithaywe.

Habitat — Small spiny bush met with in tropical India and found abundant in Bombay, Madras, Assam.

Parts used — Whole plant, especially leaves and root.

Pharmacognostical characteristics — A prickly, variable shrub, spines sharp, white leaves, opposite, entire, elliptic, apex spinous, smooth above, often softly hairy beneath. Flowers sessile, solitary on lower axils, in dense spikes above; bracts leaf-like oblong, spinous pointed; corolla yellow or whitish or buff colored, capsules ovoid oblong. Leaves have a pair of epidermal cells containing calcium carbonate crystals and having caryophyllaceous stomata. The periderm of stem bark is subtended by five to seven layers of tubular collenchyma cells. Granular and needle-shaped calcium oxalate crystals filling up many cells in all tissues of bark are present. In the root bark, yellowish cork and sometimes a new phellogen layer separating the outer and inner cortex and root cells containing blackish-brown contents are present.[147]

Ayurvedic description — *Rasa — tikta, madhur; Guna — laghu, snigdh; Veerya — ushna; Vipak — katu.*

Action and uses — *Kapha vata samak, sothahar bada nasthapan, bran pachan, bran sodhan, kustaghan, kesya, rakta sodhak, sukra sodhan, mutral, kandughan, jawarghana, vishaghan.*

Chemical constituents — The plant is reported to contain alkaloid and is rich in potassium.[50] The leaves of another species, *B. lupulina* Lindl., contain an unnamed alkaloid.[10]

Pharmacological action — Alterative, diuretic, tonic, and febrifuge.[1,3] The alcoholic extract of the entire plant *B. cristata* was found to have hypoglycemic activity in albino rats.[51] The extract obtained from *B. cristata* var. *dictrotoma* showed CNS-depressant activity in mice.[11] It also produced hypothermia in mice.[11] A decoction of the leaves of *B. prionitis* in a dose of 2 to 5 ml/100g was found to have moderate diuretic activity in rats, compared to urea as the standard drug.[148]

Medicinal properties and uses — The leaf juice is given with honey in catarrhal diseases of children; with cumin seeds it is given in spermatorrhea. The juice is applied as a dressing for feet in the rainy season to protect the skin from cracking; with honey it is applied to bleeding gums, and in otitis. A paste of the leaves with salt is applied to strengthen the gums and bring relief to toothache. Reported to be used as an antidote for snake bite and scorpion sting.[1] The juice of the bark is diaphoretic and expectorant; it is given in dropsy. The dried bark is given in whooping cough. The whole plant is diuretic, tonic, febrifuge, and anticatarrhal; in the form of decoctions, it is given in dropsy.

Doses — Infusion — 14 to 28 ml; decoction — 56 to 112 ml.

BARRINGTONIA ACUTANGULA (Linn.) Gaertn.　　　　　　　Family: MYRTACEAE

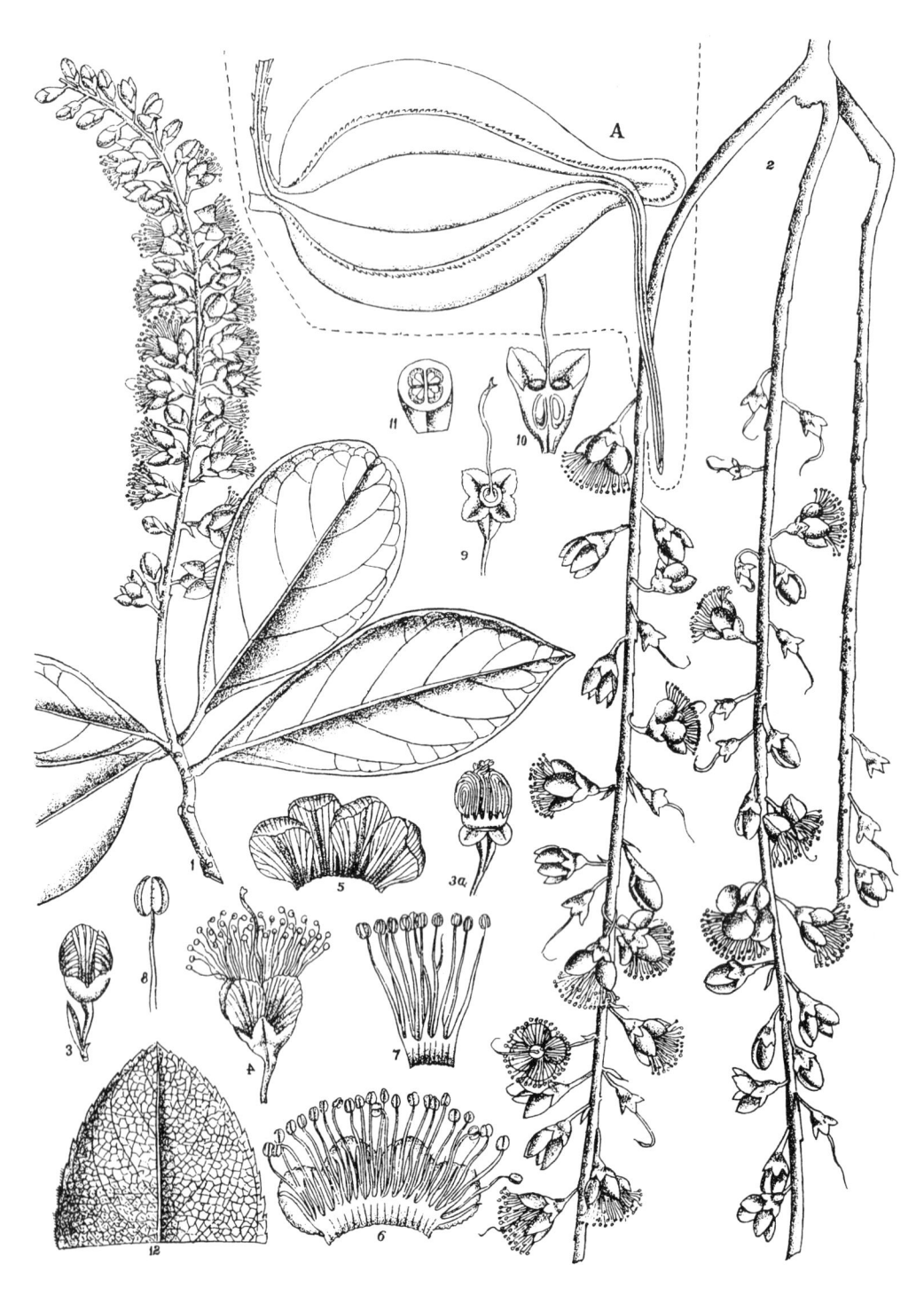

Vernacular names — Sanskrit, Hijjala; Hindi, Hijjal; Bengali, Hijal; Nepalese, Samudraphal; Sinhalese, Elamidola; Unani, Samndraphal; Burmese, Kyailha.
Habitat — Throughout India, especially in sub-Himalayan tracts and in the plains of Bengal.
Parts used — Seed, fruits, root, and leaves.

Morphological characteristics — It is a small- or medium-sized tree up to 40 ft in height. The stem externally is white and soft. Leaves 5 in. long and 2 in. broad, cuneate-elliptic, flowers $1/_3$ to $1/_2$ in. diameter, pink, in long drooping racemes. Fruits 1 to $1 1/_2$ in. by $1/_2$ to 1 in., quadrangular oblong; slightly narrowed toward and subtruncate at each end.[28]

Pharmacognostical characteristics — Leaf — In transverse section the cuticle is thick and epidermal cells are polygonal in shape. Stomata are ranunculaceous. Parenchymatous ground tissue is filled with rosette crystals of calcium oxalate. Root — Cork in transection shows four to six layers of rectangular tangentially elongated cells. Secondary phloem is arranged in wedge-shaped radiating arms. Tyloses in vessels and an eccentric pith of parenchymatous fiber are seen.[149]

Ayurvedic description — *Rasa — tikta, katu, madhur; Guna — laghu, rooksha, teekshna; Veerya — ushna; Vipak — katu.*

Action and uses — *Vat samak, kapha pitta sansodhak, lakhan, siro virachan, krimighan, grahi, raktasodhak, kaphanisark, mutral, kustaghan, jwaraghan.*

Chemical constituents — It contains glucoside, saponin, barrington, starch, protein, cellulose, fat, caoutchouc, and alkaline salts.[1] A new triterpene dicarboxylic acid, i.e., barringtonic acid, has been isolated from the ether extract of the plant.[150] The fruit yielded a new triterpenoid sapogenin — barringtogenol C.[151]

Pharmacological action — Seeds are aromatic, carminative, and emetic. The alcoholic extracts of root and stem bark exhibited CNS-depressant activity and produced hypothermia in mice. The root extract also showed hypoglycemic activity in albino rats, whereas the extract of the stem bark showed antiprotozoal activity against *Entamoeba histolytica*.[51]

Medicinal properties and uses — Fruit is rubbed with the juice of fresh ginger and used in catarrhs of the nose and respiratory passages. Externally it is applied to the chest to relieve pain. Powdered seeds are used as snuff in headache. Juice of the leaves is given in diarrhea and dysentery and used as bitter tonic. The root is cooling, aperient, expectorant, and possesses properties similar to cinchona.[149]

Doses — Fruit powder — 3 to 6 gr (as emetic), 0.5 to 1 gr (in general); root powder — 0.5 to 1 gr; leaf infusion — 7 to 14 ml.

BAUHINIA TOMENTOSA Linn. *Family:* CAESALPINIACEAE

Vernacular names — Sanskrit, Aswamantaka; Hindi, Kachnar; English, Mountain ebony; Bengali, Kanchan; Nepalese, Kavero; Sinhalese, Kha petan; Japanese, Murasakisoshinkvoa; Unani, Kachnal; Tamil, Kattatti.

Habitat — Throughout India; often cultivated.

Parts used — Whole plant, root, bark, young flowers, seeds, and fruits.

Morphological characteristics — A tree, stem bark 3 mm thick, externally rough, yellowish gray, internally smooth; leaves simple, entire, cordate, 5 to 10 cm long, 2 lobed, flowers small, pubescent corymbs, light red-purple, upper petal darker and often lined with cream, and red corolla.

Ayurvedic description — *Rasa* — *kasaaya; Guna* — *rooksha, laghu; Veerya* — *sheeta; Vipak* — *katu.*

Action and uses — *Kapha-pitta samak, brana sodhan, brana-ropan, khustaghan, sothahar, sthamvan, lakhan, rakt sthmavan, gandmala nasak.*

Chemical constituents — Bark contains tannin and gum. Flowers yielded about 4.6% of rutin.[152]

Pharmacological action — Antidysenteric, anthelmintic, fruit is diuretic. Seeds are tonic and aphrodisiac.[3]

Medicinal properties and uses — Decoction of the root bark is useful in inflammation of the liver and as a vermifuge. Buds and young flowers are prescribed in dysenteric affections. Bruised bark ground with rice water made into a paste is applied externally to tumors and wounds such as scrofulous. Fruit is diuretic. Plant used in snake bite and scorpion sting. The seeds are tonic and aphrodisiac.[1,3]

Doses — Powder — 1 to 2 drams; decoction — 56 to 112 ml; infusion of flowers — 3 to 5 drams.

BAUHINIA VARIEGATA Linn.   *Family:* CAESALPINIACEAE

Vernacular names — Sanskrit, Kanchanara; Hindi, Kachnar; Bengali, Rakta Kachan; Nepalese, Kunabhu; Sinhalese, Kobobela; Unani, Kachnal; Malaysian, Chuvanna, mandaram.

Habitat — It is distributed in sub-Himalayan tract, also in dry forests over eastern, central, and South India and in Burma.

Parts used — Bark, root, gum, leaves, seeds, and flowers.

Pharmacognostical characteristics — Its young stem shows four ridges, of which three are more prominent. In transverse section, some epidermal cells elongate to form multicellular nonglandular hairs. The cortex consists of collenchyma followed by parenchyma. The endodermis and pericycle are distinct. The stele is developed below the ridges. The stem becomes circular after secondary growth. The pericycle forms strands of thick-walled fibers which separate, and stone cells are formed in the intervening parenchyma. The bark shows a thick periderm of 15 to 20 layers of cork cells and a wide zone of phelloderm, whose cells become thick walled and sclerenchymatous. The secondary phloem is represented by concentric tangential bands of fibers, sieve tubes, companion cells, and phloem parenchyma separated by funnel-shaped medullary rays. Cortical bundles, fibers of the pericycle, phloem, and stone cells of the phelloderm are the distinguishing features of the bark.[153]

Ayurvedic description — *Rasa—kasaya; Guna—rooksha, lagu; Veerya—sheeta; Vipak—katu.*

Action and uses — *Gundmala nask, kaf pit samak, brana sodhan, khutaghan, stambhan, krimighna anuloman, rakt stambhan, kafghan, mutra sanghraniya.*

Chemical constituents — The seeds yield 16.5% of a pale yellow fatty oil on extraction with petroleum ether, but only about 6.1% when expressed in a hydraulic press.[155] Ramchandran and Joshi,[154] working on allied species, isolated isoquercitrin and astragalin from fresh flowers of *B. purpurea*. The coloring matter was separated which was 3-glucoside. The authors thought the presence of these glucosides may perhaps account for the medicinal value of flowers.

Pharmacological action — Bark is alterative, tonic, and astringent; root is carminative; and flowers are laxative.[3] The alcoholic extract of the stem bark of an allied species, *B. racemosa*, was found to have anticancer activity against human epidermal carcinoma of the nasopharynx in tissue culture. It also possessed CNS-depressant activity. It produced hypothermia in mice.[51]

Medicinal properties and uses — The flower is a good remedy in salivation and sore throat; decoction of the buds is employed in cough, bleeding piles, and menorrhagia. Bark rubbed into an emulsion with rice water with the addition of ginger is administered in scrofulous enlargement of the gland of the neck. Decoction of the bark is a useful wash in ulcers and skin diseases. Dried buds are also useful in diarrhea, worms, piles, and dysentery. Decoction of the root is given in dyspepsia and flatulence; it is also an antifat remedy and therefore valuable for corpulent persons. This plant is used in malaria and is also an antidote to snake poison.

Doses — Powder — 4 to 8 gr; decoction — 56 to 112 ml; infusion of flower — 12 to 20 ml.

BENINCASA HISPIDA (Thunb.) Cogn. Family: CUCURBITACEAE
(Syn. BENINCASA CERIFERA Savi)

Vernacular names — Sanskrit, Kusmanda; Hindi, Golkaddu; English, White pumpkin; Bengali, Kumras; Nepalese, Bhuyphasi; (Newari, Bhuy phasi); Sinhalese, Alupuhul; Japanese, Togan; German, Wachsgurkensamen; French, Courge; Unani, Petha; Arabian, Muhadba; Persian, Kaddu-e-room; Burmese, Kyaukpayan; Malaysian, Kumbalam.

Habitat — It is cultivated for its fruits throughout the plains of India and on the hills up to 4000 ft.

Parts used — Seeds, fruits, and fruit juice.

Pharmacognostical characteristics — It is an extensive, climbing herb. The fruit is a broadly cylindrical or spheroidal gourd, 1 to 1.5 ft long, with white flesh, containing numerous, much compressed, and margined seeds.

Ayurvedic description — *Rasa — madhur; Guna — lagu, snigdh; Veerya — sheeta; Vipak — madhur.*

Action and uses — *Vat pit samak, daha samak, nidrajank, anuloman, medha kar, shukra bardhak, rakt pit samak, krimighna, balya.*

Chemical constituents — The fruit of *B. cerifera* contains moisture (96%), protein (0.4%), fat (0.1%), carbohydrate (3.2%), mineral matter (0.3%), and vitamin B (21 IU/100 g). The seeds yield a pale yellow oil.[98]

Pharmacological action — Fruit is nutritive, tonic, laxative, and diuretic. The seeds are said to possess anthelmintic properties. Its confectionery is alterative, tonic, diuretic, and restorative.

Medicinal properties and uses[1] — Fresh juice of the fruit is administered as a specific in hemoptysis and other hemorrhage from internal organs. It is also very useful in insanity, epilepsy, and other nervous diseases. A confection of squash or white gourd made with several useful ingredients is administered in hemoptysis, phthisis, cough, asthma, ulceration of the lungs, and in hoarseness, in doses of 12 to 24 g. It is a highly nutritious food in wasting diseases such as consumption. The fruit is reported to have a musk odor.

Doses — Fruit juice — 56 to 112 ml.

BERBERIS ARISTATA DC.   *Family:* BERBERIDACEAE
(Syn. *B. ARISTATA* var. *FLORIBUNDA*
   Hook. f. and Thomas;
   B. CORIARIA Royle ex. Lindl.)

Vernacular names — Sanskrit, Daruharidra; Hindi, Darhald; English, Tree turmeric; Bengali, Darhaldi; Nepalese, Chutro; Sinhalese, Daruharida; German, Berberitze; Unani, Daruhaldi; Tibetan, Skyer-rtsa-yun-ha; Persian, Darchoba; Tamil, Maramayal.

Habitat — Found in northwestern Himalayas in open shrubby places at altitudes of 6000 to 10,000 ft from Chota Banghal to Nepal.

Parts used — Fruit, root bark, stem, and wood.

Morphological characteristics — A robust shrub 8 to 10 ft high, basal spines tripartite, upper simple. Leaves 1.5 to 3.0 by 1.3 in., obovate or elliptic, deciduous, entire or spiny serrate, glabrous, green on both sides. Racemes simple, flowers golden yellow, pedicels stout, fruit blue, ovoid.[28]

Ayurvedic description — *Rasa* — *katu, tikta; Guna* — *rooksha; Veerya* — *ushna; Vipak* — *katu.*

Action and uses — *Lekani, shodhni, kapha pitta har, bran sodhan, bran ropan, rakt sodhak, netra roghar, dipan, grahi, rakt sodhak, atisar nasak, varnakari.*[26]

Chemical constituents — Root and wood contain a yellow alkaloid "berberine". From the roots of an allied species *B. asiatica,* Bhakuni et al.[156] isolated berberine, palmitine, jatrorrhizine, columbamine, tetrahydropalmitine, berbamine, oxyberberine, and oxyacanthine. Berberine and palmitine were found as chloride. From *B. lycium* a new compound has been isolated which is identical with the berberine-acetone complex in all respects. This species is considered a good source of berberine.[157] Bark is reported to contain two alkaloids, palmatine chloride and palmatine, and a berberine chloride mixture.[178]

Pharmacological action — Tonic, stomachic, astringent, antiperiodic, diaphoretic, antipyretic, and alterative; root bark is purgative.[3,50] Alcoholic extract of roots of *B. aristata* showed hypoglycemic activity in albino rats. The extracts of *B. aristata, B. asiatica,* and *B. lycium* possessed anticancer activity in human epidermal carcinoma of the nasopharynx in tissue culture.[51] Berberine hydrochloride, an alkaloid isolated from *B. aristata,* had shown anti-inflammatory activity on acute, subacute, and chronic types of inflammations produced by immunological and nonimmunological methods.[158] Berberine is not vibriocidal but prevents the death of cholera-infected rabbits when given early, and is reported to be better than sulfaguanidine.[178]

Medicinal properties and uses — Plant extracts are as valuable as quinine in malarial fevers. They are particularly useful in relieving pyrexia and checking the return of paroxysms of intermittent fevers. Decoction is efficacious as a wash for ulcers and sores. It is used as a purgative for children, blood purifier, alterative, tonic, deobstruent, and febrifuge.

Externally, plant extract "rasant" is chiefly used for eye diseases. Watery solution of rasant is used for washing piles, and glandular swelling.[1]

Doses — Infusion — 12 to 20 ml; decoction — 56 to 112 ml; rasant (extract) — 0.5 to 1gr; fruits — 4 to 12 gr.

BERGENIA LIGULATA (Wall.) Engl.   *Family:* SAXIFRAGACEAE
(Syn. *SAXIFRAGA LIGULATA* Wall.)

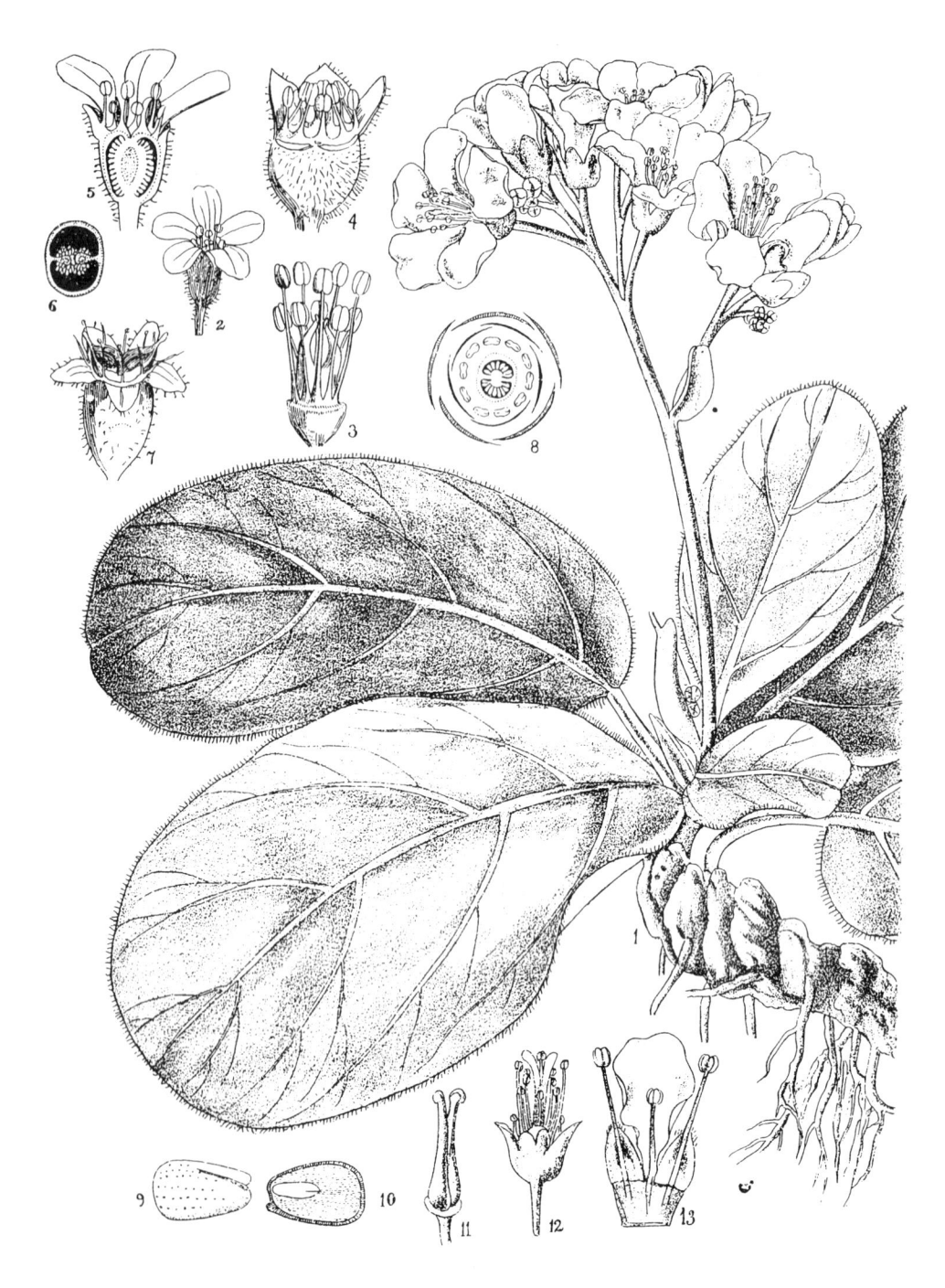

Vernacular names — Sanskrit, Pashanbheda; Hindi, Pakhenbed; Nepalese, Pakanabhada; Sinhalese, Pahanabeya; Japanese, Yukinoshita; German, Steinbrech; Unani, Mukha; Arabian, Junteyenah; Persian, Gashah; Pustu, Kamarghuel.

Habitat — Temperate Himalayas from Kashmir to Bhutan between 7000 and 10,000 ft and Khasia Hills at 4000 ft.

Parts used — Root.

Pharmacognostical characteristics — A perennial herb; rhizome 1 cm in diameter; outer surface brown, rough, inner smooth; leaves ovate or round, 5 to 15 cm long, turning bright red, entire flowers white, pink, or purple, and 3.2 cm in diameter.

The drug *pashanbheda* consists of dried rhizomes of three species of *Bergenia*, viz., *B. ligulata, B. ciliata,* and *B. stracheyi.* The external morphology of all the three species is more or less the same. *B. ligulata* in transection shows a wide cortex of thick parenchymatous cells with starch grains and tanninferous contents and a few rosette crystals of calcium oxalate. The vascular bundles are arranged in a ring. Some of the pith cells in the central portion have pitted walls. The rhizomes of *B. stracheyi* have an irregular outline and the central zone consists of secondary wood in which xylem elements are arranged in rows forming streaks. The central small pith becomes cut off from the vascular zone by the formation of internal cork cambium. No significant anatomical differences are found between the rhizomes of *B. ligulata* and *B. ciliata.*[842]

Ayurvedic description — *Rasa — kasaya, tikta; Guna — laghu, snigdha, teekshna; Veerya — sheeta; Vipak — katu.*

Action and uses — *Tridosha har, sotha har, bran ropan, sthamvan, vadan, rakta pitta samak, hiridya, kafa nisark, asmari bhadan, mutral, jawarghana.*

Chemical constituents — *B. ciliata* roots yield β-sitosterol, bergenin, and galloylated leukoanthocyanidin-4-(2-*O*-galloyl) glucoside. The leaves contained two flavonols, quercetin and kaempferol, along with their 3-rhamnosides, quercitrin and afzelin. The leaves also yield arbutin derivates and β-sitosterol. The roots of *B. ligulata* contain bergenin and β-sitosterol. The roots of *B. stracheyi* yield a new catechin derivative, (+) catechin-3-gallate, in addition to β-sitosterol and bergenin.[843,844]

Pharmacological action — Diuretic, demulcent, and astringent. The root extracts of *B. ciliata* produced a drop in blood pressure of anesthetized dogs and significant diuretic activity in rats.[845] The alcohol extract of the rhizome of *B. ligulata* showed anticancer activity in Walker carcinosarcoma 256 in rats. It also showed antiprotozoal activity against *Entamoeba histolytica*.[51] Alcohol and aqueous extract of *B. ligulata* produced hypotension in rats. The extracts showed insignificant antilithic activity in dissolving stones produced in male rats.[846]

Medicinal properties and uses — Root is a tonic, used in fever, diarrhea, and pulmonary affections and as a lixorbutic. It acts as an antidote to opium. With honey, it is applied to the gum in teething of children to allay irritation. It is also applied in ophthalmia. In the indigenous system of medicine, it is of high repute for dissolving kidney stones.[178]

Doses — 0.5 to 1.5 gr.

BETULA UTILIS D. Don *Family:* CUPULIFERAE

Vernacular names — Sanskrit, Bhurjapatra; Hindi, Bhujpattra; English, Birch; Begali, Bhujipattra; Nepalese, Bhujapattra; Sinhalese, Bhujapat; German, Birke; Unani, Bhojpattar.

Habitat — Occurs in Himalayan regions from Kashmir to Nepal. Reaches the highest limits of tree growth in India.

Parts used — Bark.

Morphological characteristics — Outer bark white, inner layer pink; leaves 5 to 7 cm long, alternate, petiolate, ovate, serrated, and long pointed; male flowers in pendulous spikes and female flowers in erect or pendulous spikes.[28]

Ayurvedic description — *Rasa — kasaya; Guna — lagu, snigdha; Veerya — ushna; Vipak — katu.*

Action and uses — *Tridosha samak, ashepahar, putihar, rakt pitta samk, raktrodhak, kaphaghan, visha nasak, lakhan.*

Chemical constituents — Bark contains betulin and an essential oil.[50]

Pharmacological action — Antiseptic.

Medicinal properties and uses — Infusion or decoction of the bark is antiseptic and carminative. It is useful for washing wounds and ulcers; also applied externally in some skin diseases. It is also given in hysteria.[1]

Doses — Powder — 1 to 3 gr; decoction — 56 to 112 ml.

BLEPHARIS EDULIS Pers.   *Family:* ACANTHACEAE

Vernacular names — Sanskrit, Utangan; Hindi, Utanjan; Nepalese, Seribari; Unani, Utanjan; Arabian, Qariz; Persian, Anjarah.

Habitat — A shrub found in the Punjab and Sind, Iran, and Afghanistan.

Parts used — Seeds, leaves and root.

Ayurvedic description — *Rasa—madhura, tikta; Guna—guru, snigdha; Veerya—ushna; Vipak—madhur.*

Action and uses — *Tridosh har, sukrajann, sukra sthamvan, vajkarn, mutral.*

Chemical constituents — Seeds contain *dl*-allantoin (2.1%) and blepharin, a bitter glycoside (1.2%) which is optically active.[159] It is reported that dihydrofurano-dihydroisocoumarin was first detected as occurring in nature in this plant.[178]

Pharmacological action — Resolvent, diuretic, aphrodisiac, and expectorant.

Medicinal properties and uses — Seed powder of *B. edulis* is recommended for impotence or seminal debility. Seeds are considered useful in diseases of blood, chest, lungs, and liver. The leaves are acrid and are considered tonic, cooling, astringent to the bowels, aphrodisiac, alterative, useful in *tridosh* fever, urinary discharges, leukoderma, and mental derangements.[3] When applied locally they are said to have beneficial effects on wounds and ulcers. In Unani medicine the root is considered diuretic; said to regulate menstruation and beneficial in urinary discharges.[1]

Doses — 3 to 5 gr.

BLUMEA LACERA DC. Family: COMPOSITAE

Vernacular names — Sanskrit, Kukurandru; Hindi, Kakronda; English, Blume; Bengali, Kukursunga; Nepalese, Khichhavwatha; Chinese, Nagi; German, Blumeacampher; Unani, Kukraunda; Arabian, Kamaphilusa; Persian, Karfas roomi; Burmese, Maiyagan; Tamil, Kattumullangi.

Habitat — Found in eastern part of India, Assam, Khasi Mountains; Burma, eastern peninsula, to Singapore and Penang.

Morphological characteristics — A strongly camphoraceous smelling, tomentose, villous, or silky-wooly subshrub. Stems tall, corymbosely branched above. Leaves 4 to 8, elliptic or oblong lanceolate, narrowed into an auricled short petiole. Head sessile or peduncled in clusters on the stout branches. Involucre bracts tomentose, receptacle glabrous, achenes ten-ribbed, silky, pappus red.[28]

Parts used — Whole plant.

Ayurvedic description — *Rasa* — *tikta, kasaya; Guna* — *laghu, rooksha, teekshna; Veerya* — *ushna; Vipak* — *katu.*

Action and uses — *Kapha, pitta samak, sirovirachan, sotha har, rakt rodhak, krimighna, branropan, dipan, anuloman, vamak, jwaraghan, visaghan.*

Chemical constituents — The herb gives 0.085% of essential oil, containing blumea camphor.[1,50]

Pharmacological action — Astringent, stomachic, antispasmodic, emmenagogue, and diuretic.

Medicinal properties and uses — Externally fresh juice of the leaves is dropped into the eyes in chronic purulent discharges. Internally decoction is given for worms. It is very useful in various catarrhal affections.

Doses — Infusion — 14 to 28 ml.

BOERHAAVIA DIFFUSA Linn.   *Family:* NYCTAGINACEAE
(Syn. *B. REPENS* Linn.)

Vernacular names — Sanskrit, Punarnava; Hindi, Godhaparna; English, Spreading hogweed; Bengali, Punarnaba; Nepalese, Punarnwa; Sinhalese, Sarana; Japanese, Benikasumi; German, Punarnava; French, Herlee a cochons; Unani, Bish khapra; Arabian, Hand qoqa; Persian, Denispat; Tamil, Mukkirattai.

Habitat — All over India, abundant during rains. It is a diffuse herb with a stout woody rootstock and erect or procumbent branches.

Parts used — Whole herb and root.

Pharmacognostical characteristics — The tap root is long, tuberous, and tapering. Leaves are opposite, glabrous, margin undulate. Flowers are hermaphrodite and pedicellate. Bracts are deciduous, involucrate. Fruit is detachable, embryo curved, and radicle is long. The epidermis of the stem is single layered with multicellular hairs. The cortex has an outer collenchymatous layer and inner endodermis; the pericycle is many layered. Vascular bundles are collateral, open, endarch, and irregularly distributed. The stem shows anomalous secondary growth, and long needle-like crystals are present. The leaf is dorsiventral and multicellular epidermal hairs are present. Stomata are present on both surfaces; the palisade is one to three layered, with numerous chloroplasts; crystals are present. The epiblema of the root has numerous unicellular hairs. The endodermis and pericycle are single layered. The young root is diarch and tetrarch, and conjunctive tissue is parenchymatous. Polygonal cells containing starch grains are present. Primary xylem occupies the center of the root.[160]

Ayurvedic description — *Rasa — madhura, tikta, kasaya; Guna — lagu, rooksha; Veerya — ushna; Vipak — madhura.*

Action and uses — *Tridosha har, lakhan, shoth har, dipan, anuloman, rachan, bamak, mutral, hirdya, rasayan.*

Chemical constituents — Goshal[161] analyzed the drug and found (1) sulfate of an alkaloid — Basu and co-workers[162] called it punarnavine; (2) an amorphous mass; (3) sulfates, chlorides, and traces of nitrate and chlorates from the ash. Chopra et al.[163] reported the presence of unusually large quantities of potassium nitrate (6.41%). Misra and Tiwari[164] isolated hentriacontane, β-sitosterol and ursolic acid from the roots. Phytochemical tests on *B. repanda*, an allied species, showed the presence of proteins, reducing sugars, and alkaloids.[165] Srivastra et al.[166] confirmed that sterol isolated from the root of *B. diffusa* is β-sitosterol.

Pharmacological action — It is bitter, stomachic, laxative, diuretic, expectorant, diaphoretic, and emetic. Root is purgative, anthelmintic, and febrifuge.[3] Singh and Udupa[167] reported that total alcoholic extract showed significant anti-inflammatory activity against carageenin-induced hind paw edema in rats. The aqueous extract and the alkaloid (punarnavine) inhibited the increased serum aminotransferase activity in arthritic animals, similar to hydrocortisone. They increased the liver ATP phosphohydrolase activity.[168,169] Under experiments the total alcoholic extract showed a cardiotonic effect, an increase in blood pressure, and a relaxant effect. It also showed promising diuretic activity.[179]

Medicinal properties and uses — Decoction of roots is used as diuretic, laxative, and as expectorant. It is also used for asthma, jaundice, anascara, anemia, and for internal inflammation. It is used as an antidote to snake venom.[3,50]

BOMBAX CEIBA Linn.
(Syn. *SALMALIA MALABARICA* Schott and Endl.)

Family: BOMBACACEAE

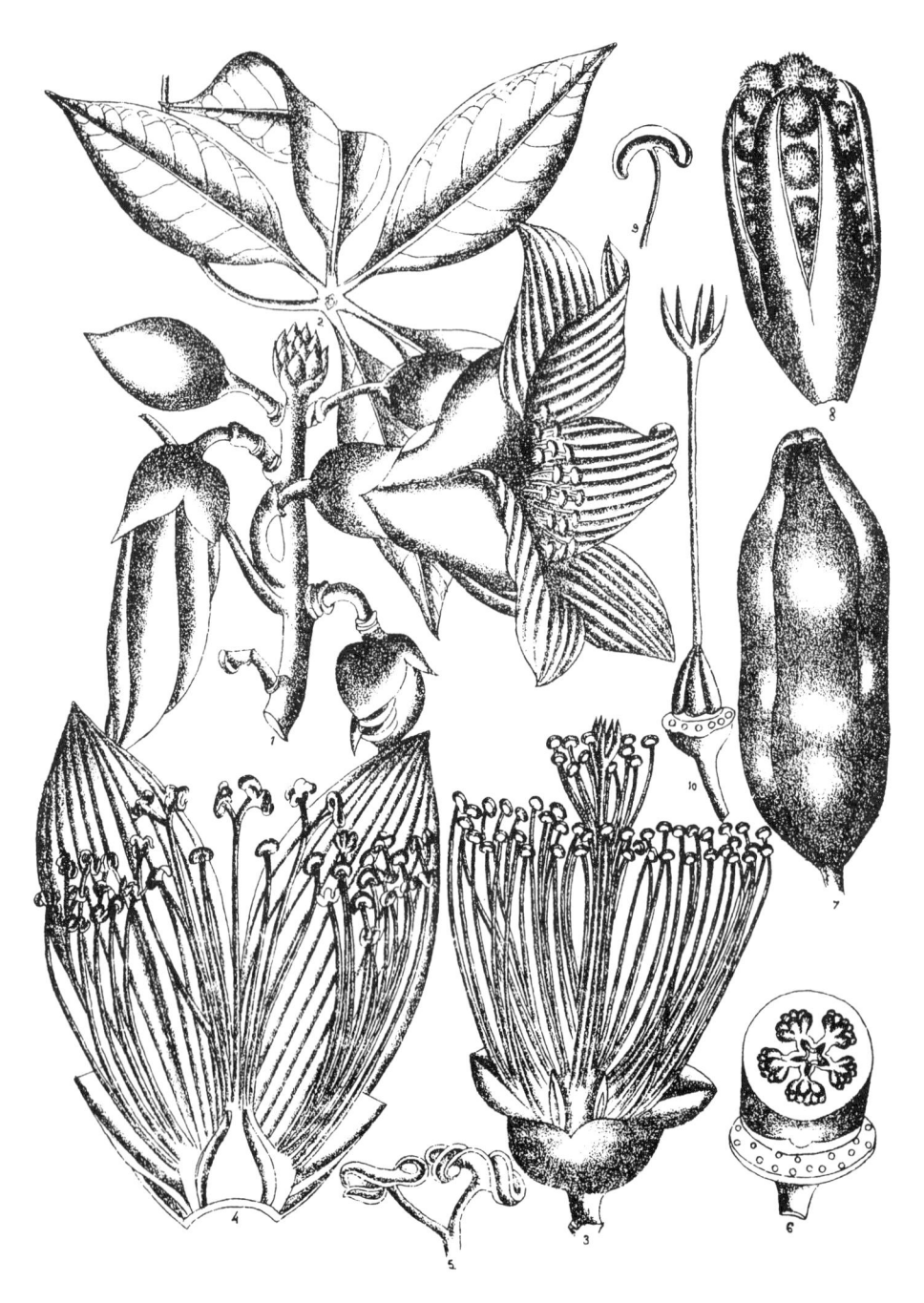

Vernacular names — Sanskrit, Shalmali; Hindi, Simal; English, Silk cotton tree; Bengali, Shimul; Nepalese, Simal; Sinhalese, Imbul; Japanese, Indowatanoki; German, Malabarischer Wollbaum; French, Kapokur; Unani, Shimal; Persian, Sembhal; Tamil, Mulilvau.

Habitat — Throughout the hotter region of India.

Parts used — Seeds, leaves, fruit, flowers and root.

Pharmacognostical characteristics — Leaves 4 to 6 in. long, divided; flowers red, fruits 6 to 7 in. long, containing many black seeds. Transection of the root showed the presence of concentric series of fibrous patches alternating with groups of sieve elements in the secondary phloem. Some other distinguishing features of the plant are: presence of mucilage canals, tannin cells, and cells containing rosette crystals of calcium oxalate in the ground tissue of root and stem.[171]

Ayurvedic description — *Rasa — madhura, kashaya; Guna — snigdha; Veerya — sheeta; Vipak — madhura.*

Action and uses — *Graahi, vrishya, balya, raktsthambhana, vedana asthapana, rasaayani, pramehaghna, pravakika-shamana, daahadhaman.*

Chemical constituents — Seeds yield a good nondrying oil. Gum called *mocharas;* contains tannin and gallic acid.

Glycosides and tannins are reportedly present in root, stem, and leaf of the plant. In addition, the stem showed some alkaloid, while the root exhibited the presence of protein.[171] Mukerje and Roy[172] isolated lupeol and β-sitosterol from the petroleum ether extract of the stem bark. From the root bark, three naphthalene derivatives have been isolated.[173,174]

The flowers contained the β-D-glucoside of β-sitosterol, free β-sitosterol, hentriacontane, hentriacontanol, traces of essential oil, kaempferol, and quercetin.[175]

Pharmacological action — Gum is astringent and styptic. Tap root of young plant is demulcent, tonic, diuretic, and aphrodisiac. Bark is demulcent, diuretic, tonic, and slightly astringent. Root is stimulant. Flowers are laxative and diuretic.[3,6,50]

Hot aqueous extract of seeds showed moderate oxytocic activity on gravid and nongravid isolated rat uteri and on guinea pig and rabbit uterine strips. It was found to have musculotropic action in guinea pig ileum and cardiac stimulant action on frog's heart *in situ*.[176]

Medicinal properties and uses — The bark is mucilaginous; its infusion is demulcent and aphrodisiac. The bark is an efficacious styptic in abnormal uterine bleeding. Its paste is applied over inflammation and skin eruptions.

Mocharas, the gum that exudes from the stem, contains a tannin and gallic acid; it is an efficacious remedy for diarrhea, dysentery, hemoptysis of pulmonary tuberculosis, influenza, and menorrhagia.

A paste of the flowers is applied over boils, sores, and itch. The dried young fruits are given in calculous affections and chronic inflammation and ulceration of the bladder and kidneys. The seeds are given in gonorrhea, gleet, and chronic cystitis.

Doses — Root powder — 6 to 12 gr; infusion of flowers — 12 to 24 gr; fruit powder — 1 to 3 gr; gum — 1 to 3 gr.

BORASSUS FLABELLIFER Linn.
(Syn. *B. FLABELLIFORMIS* Murr.)

Family: PALMAE

Vernacular names — Sanskrit, Tala; Hindi, Tal; English, Palmyra palm; Bengali, Tal; Nepalese, Tad; Sinhalese, Tal; Japanese, Parumirayashi; German, Palmyrapalm; Unani, Tar; Arabian, Tar; Persian, Tal; Burmese, Tan; Tamil, Panai.

Parts used — Root, flowering stalk, juice, bark, fruit juice, and leaves.

Habitat — Grows on dry or sandy localities along river banks, throughout India; common along the coastal areas of the peninsula, Bihar and Bengal.

Morphological characteristics — A tall palm bearing a terminal crown of 3 to 40 large fan-like leaves, 3 to 5 ft in width. The leaf stalks are 2 to 4 in. long, strong fibrous. The stem is black and consists of a hard outer portion mainly composed of stiff longitudinal fibers. In the central portion, the pith is soft and starchy. Male and female flowers are borne on different plants. The fruits are large and fibrous, containing usually three nut-like portions, each of which encloses a seed.

Ayurvedic description — *Rasa* — *madhur; Guna* — *guru; Veerya* — *sheeta; Vipak* — *madhur*.

Action and uses — *Vata pitta samak, rakt sthamvan, daah prasaman, sothahar; bran ropan, tawg dosha har, lakhan, dipan, anuloman, jawa roghan, hirdya, balya, artaw jaan, rakt rodhak, virsya, mastisk bala bardhak.*

Chemical constituents — The fresh sap of these palms, called "nira" or sweet toddy, contains about 12% sucrose; also a good source of vitamin B complex. After fermentation, the juice produces about 3% alcohol and 0.1% of acids during the first 6 to 8 h; alcohol content increases to about 5% thereafter.[177] Constitution of the galactomannan from the kernels of the nut has been reported.[178]

Pharmacological action — Root is cooling and restorative; juice is diuretic, cooling, stimulant, and antiphlogistic when fresh; useful in inflammatory affections and dropsy. Pulp of the unripe fruit is diuretic, nutritive, and demulcent.

Medicinal properties and uses — The leaf juice checks hiccup and relieves gastric catarrh. An extract of the green leaves is given in secondary syphilis. Root decoction is given in gonorrhea and respiratory diseases. The stalk of the dry flowering shoot is very useful for bilious affections and enlarged liver and spleen. Sweetish juice from flowering stalk is cooling and stimulant; if taken regularly, it acts as a laxative. The slightly fermented juice is given in diabetes and chronic gonorrhea and with some aromatics in phthisis. Palm sugar prepared from toddy is used as an antidote in poisoning; it is antibilious and alterative; it is given in disorders of the liver and gleet.

Doses — Infusion — 14 to 28 ml; toddy — 58 to 112 ml; kshar — 1 to 3 gr.

BOSWELLIA SERRATA Roxb.   *Family:* BURSERACEAE

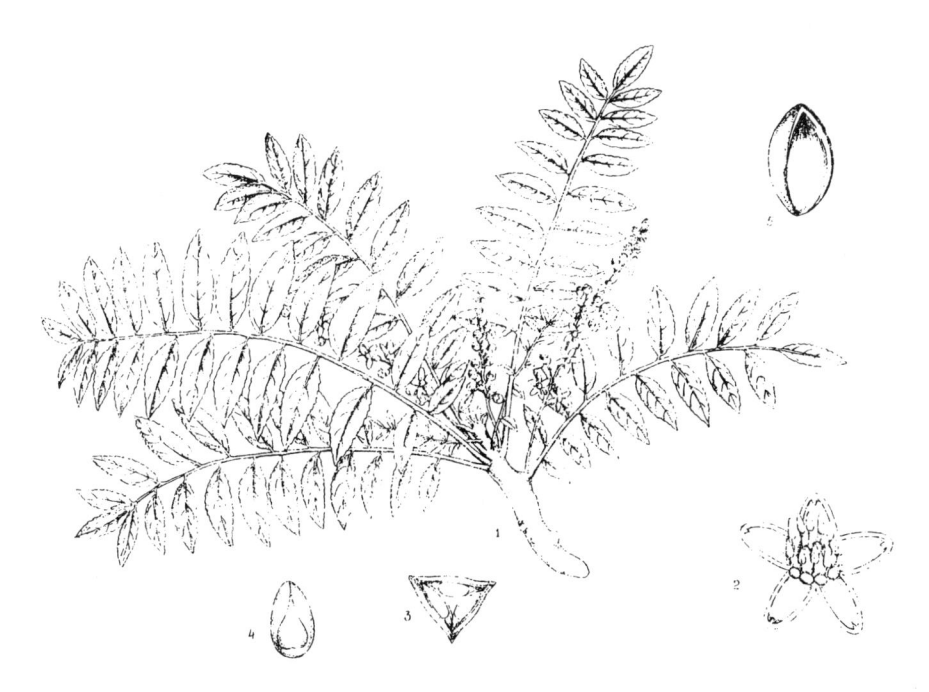

Vernacular names — Sanskrit, Shallaki; Hindi, Luban; English, Indian olibaum; Bengali, Luban; Nepalese, Gobahr shalla; Sinhalese, Kundirikkam; Chinese, Fan Hun Hsiang; German, Salaibaum; Unani, Loban; Arabian, Zarw; Persian, Husn-e-lubban; French, Baswellie-dentelee; Tamil, Kunthreekan.

Habitat — Mountainous tracts of central India, Deccan, Bihar, Orissa, and North Gujrat.

Parts used — Bark and gum.

Morphological characteristics — Bark thick and aromatic. Leaves look like neem leaves. Flowers small, aromatic, and white, ovary divided into three chambers. Fruit contains single compressed seed. When its bark is cut, a secretion exudes which becomes gum-like after exposure to air.

Ayurvedic description — *Rasa* — *kasaya, tikta, madhur; Guna* — *laghu, rooksha; Veerya* — *sheeta; Vipak* — *katu.*

Action and uses — *Kapha-pitta samak, sotha har, bran sodhan, bran ropan, chachusya, dipan, pachan, grahi, vata anuloman, hirdya, mutral, jwaraghan.*

Chemical constituents — It contains gum, resin, and aromatic oil. Phytosterol is reported to contain β-sitosterol.

Pharmacological action — Refrigerant, diuretic, aromatic, demulcent, aperient, alterative, and emmenagogue. Alcoholic extract of the root possessed anticancer activity against human epidermal carcinoma of the nasopharynx in tissue culture. CNS-depressant activity was observed with alcoholic extract of both the root and fruit in mice. The extracts of both fruit and stem showed hypoglycemic activity in albino rats.[51] The nonphenolic fraction of the gum resin exhibited marked sedative and analgesic effects.[179]

Medicinal properties and uses — The gum resin that exudes from the trunk is prescribed in chronic lung diseases, diarrhea, dysentery, pulmonary diseases, menorrhea, dysmenorrhea, gonorrhea, syphilitic affection, piles, and liver disorders. The oil extracted from the gum resin is prescribed with demulcent drink in gonorrhea. A paste made of the gum resin with coconut oil or lemon juice is applied to ulcers, indolent swellings, carbuncles, boils, and ringworm.[3,50]

Doses — Gum resin — 2 to 3 gr; oil — 1 to 1.5 ml; bark decoction — 56 to 112 ml.

BRASSICA JUNCEA (L.) Czern. and Coss *Family:* CRUCIFERAE
(Syn. *B. JUNCEA* Hook. f. and Thoms.
   subsp. *JUNCEA* Prain var. *OLEIERA* Prain)

Vernacular names — Sanskrit, Rajika; Hindi, Sarson, Asal rai; English, Brown mustard; Bengali, Raisarisha; Nepalese, Tu; Sinhalese, Ganaba; German, Grünersenf; Unani, Sarson; Arabian, Sarshaf; Persian, Harful Abayaz.

Habitat — Cultivated in many parts of India.

Parts used — Seeds and oil.

Morphological characteristics — Annual herb. Lower leaves distinctly petiolate, lyrate-pinnatisect, 2- to 3-jugate; terminal lobe largest; upper leaves narrowed in a small petiole. Racemes lax, 20 to 40 flowered. Petals pale yellow. Pods on pedicel 1.6 to 0.6 cm long, erecto-patent or suberect, broadly linear, 3.2 to 5.6 cm long, 2 to 3.5 mm broad, subtetragonous, torulose, gradually attenuate in a seedless beak 6 to 12 mm long. Seeds globose, 1.3 mm diameter; obscurely purple brown or rarely yellow, alveolate.[28]

Three varieties are met with under cultivation in India. They are (1) var. *elata* (Prain) O.E. Schulz, (2) var. *aspera* (Prain) O.E. Schulz, and (3) var. *paevis* (Prain) O.E. Schulz.

Ayurvedic description — *Rasa* — *katu, tikta; Guna* — *laghu, teekshna; Veerya* — *ushna; Vipak* — *katu.*

Action and uses — *Kapha vat samak, sothahar, shul har, krimighan, rakt pitta kopak, sweda jank.*

Chemical constituents — The seeds contain 0.3 to 1.13% essential oil of mustard.[44] The oil has been found to contain 40% of allyl mustard oil, about 50% of crotonyl mustard oil, probably some allyl cyanide, and traces of dimethyl sulfide.[99]

Pharmacological action — Whole plant is bitter, aperient and tonic. The oil is stimulant and counterirritant.

Medicinal properties and uses — Oil extracted from the seeds is used as an external stimulant application in chest affections, especially of children. Oil combined with camphor forms an efficacious embrocation in muscular rheumatism and stiff neck. Mustard poultice prepared with cold water forms an excellent counterirritant; used in many inflammatory and neuralgic affections, in abdominal colic, and obstinate vomiting. In no case is the plaster recommended to be in contact with skin for more than 10 min. The action of this plaster goes deep, for the light volatile oil present in it speedily penetrates into the deeper layers of skin where it sets up an inflammation. The volatile oil acts as a powerful antiseptic; its minimum lethal dose is stated to be $1/2$ minim/kg. It cannot be recommended for internal uses. One or two teaspoons of mustard in water is an efficient emetic to empty the stomach in cases of poisoning. A hot mustard bath is an emmenagogue.[28]

Doses — Seed powder — 1 to 3 gr.

BUCHANANIA LANZAN Spreng.   *Family:* ANACARDIACEAE
(Syn. *B. LATIFOLIA* Roxb.)

Vernacular names — Sanskrit, Piyala; Hindi, Chirongi; English, Cuddapa almond; Bengali, Piyal; Nepalese, Chironji; Sinhalese, Urushi; German, Chirongiol; Unani, Chirongi; Arabian, Habbussimnah; Persian, Naql-e-khajah; Tamil, Charaiparuppu.

Habitat — Throughout greater part of India in dry, deciduous forests, in northwestern India from Sutlej to Nepal ascending to 3000 ft, a dominant forest species in some parts of Deccan plateau.

Parts used — Fruits, seeds, gum, root, leaves, and bark.

Morphological characteristics — A medium-sized tree with conspicuous rough bark and hard wood. Bark gray, 1 in. thick; leaves 6 to 10 in. long and 5 to 6 in. broad, pointed, soft, hairy; flowers small, bluish white; fruits round, compressed, blackish gray. A transverse section of stem shows a single layer of epidermis formed of thick-walled cells, some of which elongate to form nonglandular trichomes. These hairs are uniseriate, 1 to 5 celled, thick walled, with pointed, tapering ends. Primary cortex consists of 5 to 6 layers of parenchymatous cells filled with reddish-brown contents. Collenchyma is 8 to 10 layered with heavily thickened corners. The pericycle consists of an isolated group of fibers. The resin canals are present, scattered in the phloem region and pith. Phloem consists of sieve tubes, companion cells, and parenchyma.

In a mature stem about 8 to 12 mm thick, the periderm including the rhytidoma is about 2 mm thick and is occupied by secondary phloem. The sieve tubes are all crushed in the greater part of the bark forming ceratenchyma. Prismatic crystals of calcium oxalate are abundantly present. Some fibers increase in thickness and form stone cells. Secretory cavities in the phloem region have varying diameters. Cell contents — solitary and prismatic crystals of calcium oxalate are present in the parenchymatous tissue. Starch grains and tanniniferous contents are also present. Secretory cavities contain yellowish contents. Reddish-brown contents are scattered throughout the tissue.[181]

Ayurvedic description — *Rasa — madhur; Guna — snigdha, guru, sar; Veerya — sheeta; Vipak — madhur.*

Action and uses — *Vata pitta samak, kustaghan, sothahar, sarak, trisasamak, snigdha, guru kafanisarak, mutral, vajekarn, daha saman, brindhan, balya.*

Chemical constituents — Albuminoids (28%), mucilage (2.5%), oil (51.5%), fiber and ash (3.5%). Starch (12.1%), protein (21.6%), and sugar (5%). Bark contains 13.4% tannin. Mitra and Mehrotra[181] reported the presence of alkaloid, glycosides, saponins, flavonoids, reducing sugars, sterols, and tannins in different extractives of plant.

Pharmacological action — Demulcent, alterative, laxative, and diuretic.[1,3]

Medicinal properties and uses — Seeds are palatable and nutritious. They yield a gum useful in diarrhea. The kernel, in the form of an ointment, is applied in skin diseases to cure itch and also remove spots and blemishes from the face. Oil applied to glandular swelling of the neck. Oil from kernels used as substitute for almond oil in Ayurvedic medicinal preparations. Gum is used in diarrhea.

Doses — Bark decoction — 56 to 112 ml; seeds — 12 to 24 gr.

BUTEA MONOSPERMA (Lam.) Kuntze
(Syn. B. FRONDOSA Koen. ex Roxb.)
Family: PAPILIONACEAE

Vernacular names — Sanskrit, Palasa; Hindi, Palas; English, Bastard teak; Bengali, Palas; Nepalese, Palas; Sinhalese, Kela; German, Palasbaum; Unani, Palas; Persian, Pullah; Tibetan, Tshos-sin; Burmese, Pauk; Tamil, Palasu.

Habitat — Common throughout the greater part of India, ascending to an altitude of 3000 to 4000 ft.

Parts used — Gum, seeds, flowers, bark, and leaves.

Morphological characteristics — A small- or medium-sized tree with usually crooked trunk. Leaflets 4 to 8 in. long and broad. Flowers scarlet and orange, calyx dark velvety, corolla silvery, tomentose outside, keel much curved, pod 4 to 8 × 1 to 2 in., downy stalked. Seed oval, compressed, brown, 1½ in. long. The tree yields a gum called "Bengal kino".[28]

Ayurvedic description — *Rasa* — katu, tikta, kasaya; *Guna* — laghu, snigdha; *Veerya* — ushana; *Vipak* — katu.

Action and uses — Kapha vat samak, kaoha pitta samak, lakhan, soth har, badana sthapan, dipan, grahi, anuloman, dah prasaman, rat stambhan, rakt sodhan, veersya, rasayan.

Chemical constituents — The root of *B. monosperma* contains glucose, glycine, aglycoside, and an aromatic hydroxy compound.[182] Gupta et al.[183] investigated the flowers of *B. monosperma* which revealed the presence of seven flavonoid glucosides, two of them being butrin and isobutrin isolated earlier from the plant. Three of the glucosides were identified as coreopsin, isocoreopsin, and sulfurein, respectively. Two other new ones were assigned, monospermoside and isomonspermoside structures. All seven are structurally and biogenetically related, three being glucosides of butin, three of butein, and the last sulfurein being aurone glucoside. Seshadri and Trikha[184] isolated tetramers of leukocyanidin having -C-C and -C-O-C- linkages from the gum and bark of the plant. The molecular size of the procyanidins were determined by a study of more stable derivatives, viz., acetate, methyl ether, and methyl ether acetates. The NMR spectra of the methyl ether acetate indicated the ratio of acetoxyl to methoxyl protons to be 1:4 and these supported the structural assignment.

Pharmacological action — The leaves are astringent, diuretic, tonic, and aphrodisiac. Seeds are anthelmintic and laxative.[3,50] Flowers are tonic. Alcoholic extract of seeds exhibited antifertility activity in female mice and rats.[185] Hot petroleum ether extract of seeds also showed a significant antifertility effect in female rats.[186] Palasonin, an active principle isolated from the seeds, exhibited good anthelmintic activity *in vitro* on *Ascaris lumbricoides* and *in vivo* on *Toxicara*, comparing favorably with piperazine and santonin.[187]

Medicinal properties and uses — Leaves are given in diarrhea, heartburn, sweating of phthisis, diabetes, flatulence, colic, piles, and worms; leaves induce a fall in the amount of blood sugar and so are given in glycosuria; its infusion or decoction is given as vaginal douche in leukorrhea and as a mouthwash for septic and congested throat.

The seeds contain 18% of fixed oil called "moodooga" oil; the fresh seeds are a powerful anthelmintic for roundworm — 0.5 to 1.5 gr of the seeds with honey is given three times a day for 3 d.[28] Paste of the seeds is applied on ringworm.

Butea gum is the gum that exudes from the trunk; it is rich in gallic and tannic acids. Being astringent, it is a useful substitute for the true kino; it acts as a mild astringent and therefore it is suitable particularly for children and delicate women suffering from diarrhea and dysentery; given in doses of 2 to 3 gr the gum is given in phthisis and bleeding from the stomach and bladder.[1]

Doses — Seed powder — 0.5 to 1.5; gum — 0.5 to 2 gr.

CAESALPINIA CRISTA Linn. *Family:* CAESALPINIACEAE
(Syn. *C. BONDUCELLA* Flem.;
 C. BONDUC [Linn.] Roxb.)

Vernacular names — Sanskrit, Latakaranja; Hindi, Katkaranj; English, Bonducella nut; Bengali, Nata karanja; Nepalese, Latakas turi; Sinhalese, Kumburuwel; German, Kugelstrauch Samen; Unani, Karanjuba; Arabian, Akitamakat; Persian, Khayah-e-Iblis.

Habitat — Found throughout the hotter parts of India, Burma, and Sri Lanka, particularly along the seacoast.

Parts used — Nuts, root, bark, and leaves.

Pharmacognostical characteristics — Leaves with large, leafy, branched, basal appendages; main leaf axis armed with stout, sharp, recurved spines, divided into four to eight pairs of secondary branches. Leaflets about ten pairs, elliptic-oblong, slightly hairy, tip pointed; 1 to 2 in. long, having pairs of short spines. Flower yellow; perianth clothed with wiry prickles, ovate or oblong, inflated, 2 to 3 in. long, 1 to 2 in. broad. Seeds are hard, globular, shiny, and gray with thick testa; they are exalbuminous. In transverse section, the seeds show a palisade layer composed of verticle, columnar, and laterally closed appressed cells. Thickenings are present on the walls of pallisade cells which in tangential section appear as six to ten denticulate projections into the lumen of cells. Then follows a layer of bearer cells and a thick zone of parenchymatous cells. The majority of bearer cells are T-shaped, nonlignified, and thick walled. Roots or their powder do not show any fluorescence when exposed to ultraviolet light. However, the extract in 1% NaOH solution ethyl alcohol and solvent ether emitted a green fluorescence under ultraviolet light.

Ayurvedic description — *Rasa — katu, tikta; Guna — laghu, rooksha, teekshna; Veerya — ushna; Vipak — katu.*

Action and uses — *Kapha, vat samak, sotha har, badana sthapan, dipan, anuloman, krimighan, rakt sodhak, swashar, mutral, jwaraghan.*

Chemical constituents — The kernel contains fatty oil (20 to 24%); starch, sucrose, two phytosterols, one of them identified as sitosterol, and a hydrocarbon (mp 58 to 59°C) identified as heptocosane.[189] The oil, which is thick and pale yellow with a disagreeable smell, has the following characteristics: sp gr 0.926; saponification value 197.9; iodine value (Hanus), 111.0; acetyl value, 35.6; acid value, 8.5; and unsaponified matter 1.1%. The constituents of fatty acid are palmitic, stearic, lignoceric, oleic, linolenic, and a mixture of unsaturated acid of low molecular weights.[178,189] Ghatak's[190] investigation of seed kernels showed the presence of noncrystalline bitter glycoside bonducin, a neutral saponin, starch, sucrose, an enzyme, and a yellow oil. A white amorphous bitter substance (0.035%) has been reported. An unnamed alkaloid has been isolated from the leaves, stem, twig, and fruit of the plant.[10]

Pharmacological action — Nut and root bark are antiperiodic, antispasmodic, bitter tonic, anthelmintic, and febrifuge. Leaves are deobstrurent and emmenagogue. The powdered seeds exhibited antiestrogenic activity in mice and rabbits. Antifertility action of seeds was noted in mice and rats.[191] The nuts of *C. bonducella* showed significant antidiarrheal activity in mice.[192] The alcoholic extracts of roots and stem showed antiviral activity against vaccinia virus.[51]

Medicinal properties and uses — Seeds and root bark are very useful in simple, continued, and intermittent fevers, asthma, and colic. Seeds are febrifuge and antiperiodic; they are also valuable for dispersing swelling, restraining hemorrhage, and keeping off infectious diseases.[3,50] Tender leaves are efficacious in liver disorders. Oil expressed from them is useful in convulsions and nervous complaints.

Doses — Seed powder — 1 to 2 gr; root powder — 1 to 2 gr; leaf infusion — 12 to 20 ml.

CALLICARPA MACROPHYLLA Vahl.   Family: VERBENACEAE

Vernacular names — Sanskrit, Pringu; Hindi, Pringu; Bengali, Mathara; Nepalese, Pringu; Sinhalese, Ruknal.

Habitat — This is a shrub occurring in the upper Gangetic Plains, Bengal, western Himalayas from Kashmir eastward up to 6000 ft, and in Assam.[26]

Parts used — Root and leaves.

Morphological characteristics — Leaves 6 to 10 in. long, margin dentate, upper part of the leaves smooth and lower is hairy, flowers small, pink. Fruits white and divided into four chambers, each containing seeds.

Ayurvedic description — *Rasa* — *tikta, kasaya, madhur; Guna* — *guru, rooksha; Veerya* — *sheeta; Vipak* — *katu.*

Action and uses — *Tridosha samak, vata pitta samak, daah prasaman, badana sthapan, dipan, anuloman, sthamvan, rakt sodhak, rakt pitta samak, mutra veraje neya, daah prasaman, vishaghan.*

Chemical constituents — The root yields an aromatic oil; it also contains hydrocyanic acid. From the petroleum ether extract of the aerial parts of *C. macrophylla* two new tetracyclic diterpenes were isolated, viz., calliterpenone $C_{20}H_{34}O_3$ mp 153 to 155°C, and calliterpenone monoacetate $C_{22}H_{34}O_4$ mp 124°C.[193]

Pharmacological action — Antipyretic and tonic.

Medicinal properties and uses — Root is used in the treatment of stomach disorders. The leaves are warmed and applied to rheumatic joints to relieve pain. It is also useful in many skin diseases; it is taken internally as a blood purifier.[1,3]

Doses — Powder — 1 to 2 gr.

CALOPHYLLUM INOPHYLLUM Linn.

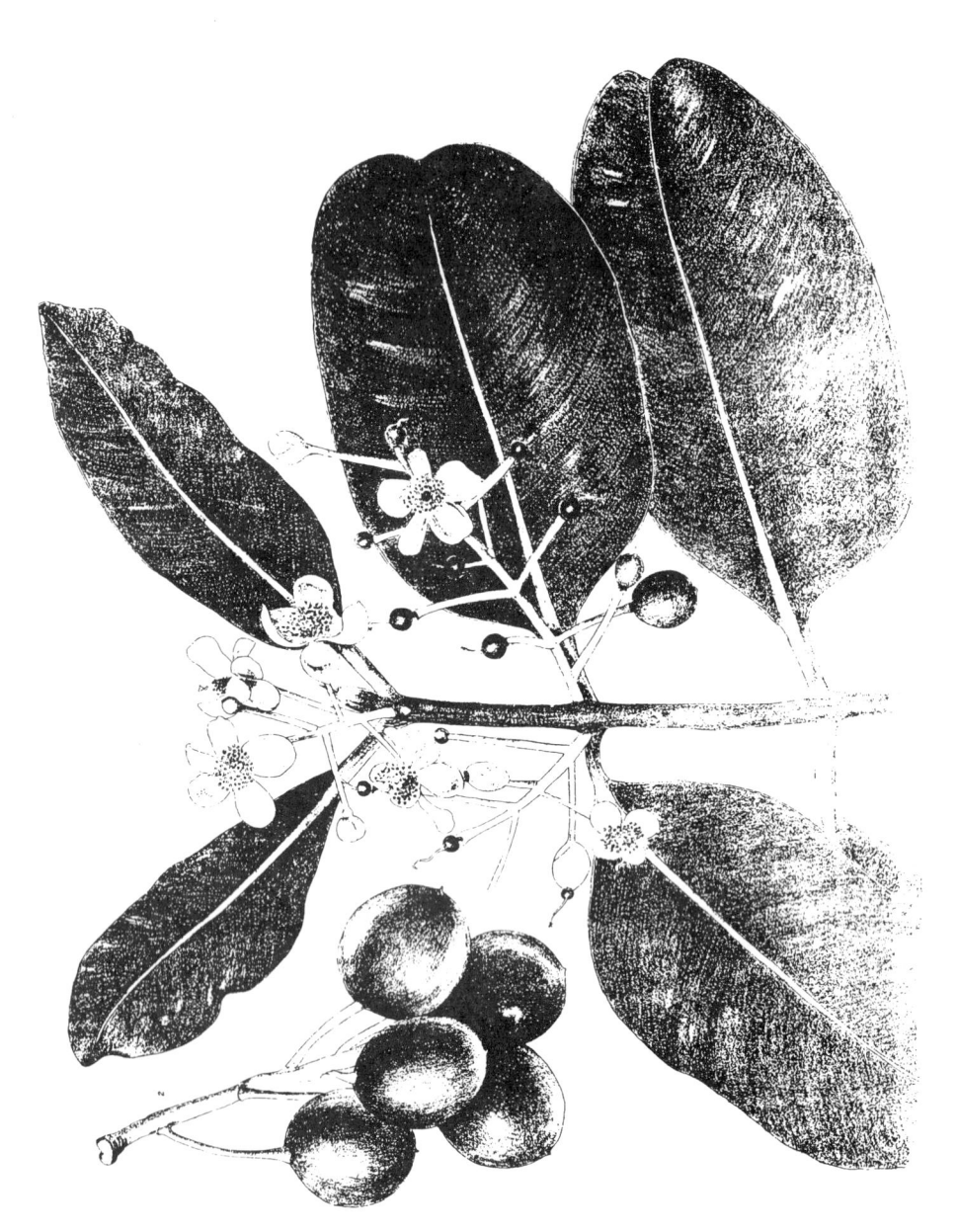

CALOPHYLLUM INOPHYLLUM Linn.   *Family:* GUTTIFERAE

Vernacular names — Sanskrit, Punnaga; Hindi, Surpan champa; English, Pannay tree, Alexandrian laurel; Bengali, Punnag; Nepalese, Luchi; Sinhalese, Domba; Japanese, Terihabata; German, Tacama Hacharz; French, Laurier d'Alexandria; Unani, Sultan champa; Burmese, Pengnyet; Tibetan, Pri-yan-ku; Tamil, Punnai.

Habitat — Cultivated throughout India, especially near the sea, as an ornamental tree.

Parts used — Leaves, bark, gum, and oil.

Morphological characteristics — An exceedingly handsome, moderate-sized tree with bright green juice (of the resin cannals), leaves 4 to 8 in. long, glabrous, shining, oblong. Flowers $3/4$ in. diameter, white, fragrant, in axillary drooping racemes 4 to 6 in. long. Sepals 4, petals 4, stamens in 4 bundles. Drupe globose, 1 in. diameter, yellow and pulpy when quite ripe.[28]

In transection, the stem bark has an outer cork layer composed of 6 to 12 layers of cells followed by the cortex. In the cortex, numerous secretory ducts are scattered which contain gum and resin and these ducts are schizolysigenous in nature. Numerous calcium oxalate crystals and starch grains are also present. Phloem fibers have large lumens, and thickening-pitted parenchymatous cells filled with granular masses are also present. Abundant chambered crystalliferous cells with prismatic crystals of calcium oxalate in the phloem region constitute a highly diagnostic character. No stone cells are seen in the stem bark.

Ayurvedic description — *Rasa — madhur, kasaya; Guna — laghu, rooksha; Veerya — sheeta; Vipak — madhur.*

Action and uses — *Kapha pitta samak, lakhan, bran ropan, sthamvan.*

Chemical constituents — Analysis of the fresh seeds showed moisture (27.23%); ash (1.07%); protein (6.41%); fat (60.72%); and carbohydrate (4.07%).[195] The bark contains (11.9%) tannin.[50] A lactone, calophyllolide, a related acid, calophyllic acid, a new polyene acid, inophyllic acid, and an essential oil fraction have been isolated from the nonglyceridic portion of oil obtained from the nuts of *C. inophyllum* Linn. Inophyllic acid has been reported from the stem bark, also.[196] Mesuaxanthone B and a new xanthone, calophyllin B, were isolated from the heartwood of *C. inophyllum*.[197] Friedelin and three new triterpenes of the friedelin group have been isolated from the leaves of the plant.[198] The androecium of the flowers of *C. inophyllum* had myricetin-7 glucoside along with small quantities of free myricetin and quercetin.[199] The hard testa of the seed coat is reported to contain (±) leukocynidin.[178]

Pharmacological action — Antipyretic, astringent; purgative and alterative.[50] Calophyllolide produced a slight bradycardia in isolated perfused rabbit heart. It was found to be as effective as quinidine in suppressing ventricular ectopic trachycardia resulting from acute myocardial infarction in dogs. The anticoagulant activity of calophyllolide was found to be intermediate in potency between dicoumarol and tromexan.[200]

Medicinal properties and uses — The leaves soaked in water are used for sores; they are used as snuff for relief of vertigo and migraine. The bark is an astringent; it is given for internal hemorrhages; its juice is a strong purgative; its decoction is a useful wash for indolent ulcers; its paste is applied in orchitis.[50] The gum resin from the stem and branches is emetic, purgative, anodyne, vulnerary, and resolvent. The fixed oil extracted from kernels is known as dillo oil; it is used in gonorrhea and gleet, as it has a soothing action on the mucous membrane of the genitourinary organs; it is also used in scabies and other skin diseases. For relief of pain in leprosy, the refined oil is injected intramuscularly.[1,3]

Doses — Decoction — 56 to 112 ml.

CALOTROPIS PROCERA (Ait.) R. Br.

CALOTROPIS GIGANTEA (Linn.) R. Br.

CALOTROPIS GIGANTEA (Linn.) R. Br.   Family: ASCLEPIADACEAE
(Syn. *ASCLEPIAS GIGANTEA* Linn.)

Vernacular names — Sanskrit, Alarka; Hindi, Madar safed; English, Mudar, bowstring hemp; Bengali, Akanda; Nepalese, Baramadha aka; Sinhalese, Elawara; French, Herbe Lirondelle; Unani, Aak safed; Arabian, Ushar-e-Abyaz; Persian, Khark-e-safed; Pustu, Spalwekk; Tamil, Yerukki.

Habitat — Frequently met with throughout India as a weed on fallow land and in waste grounds.

Parts used — Root, root bark, leaves, juice, and flower.

Morphological characteristics — A stout, hoary-tomentose shrub 4 to 10 ft. high with milky juice. Leaves sessile, thick, glaucous green, 4 to 8 in. long, elliptic or obovate-oblong, clothed beneath with fine cottony tomentum. Flowers $1\frac{1}{2}$ to 2 in. diameter. Not scented. Corolla purplish or white, lobes spreading. Coronal scales hairy with two obtuse auricles just below the rounded apex. Follicles 3 to 4 in. long, curved, turgid. Seeds with a tuft of silky hair. The epidermis of the leaf lamina is followed by three layers of closely packed palisade cells filled with chloroplasts. Multicellular thin-walled trichomes are distributed throughout the leaf. A rubiaceous type of stomata is found in the lower epidermis.[201] Epidermal cells contain starch grains, fats, and oils in parenchyma, but tannin is absent.

Ayurvedic description — *Rasa — katu, tikta; Guna — rooksha, laghu, teekshna; Veerya — ushna; Vipak — katu.*

Action and uses — *Badona astapan, sothhar, barna sodhan, kustaghana, dipan, pachan, krimighna.*

Chemical constituents — The leaves of *C. gigantea* contain an active principle, mudarine. Besides this, a yellow bitter acid and resin were also found. In addition, the leaves contain three glucosides, viz., calotropin, uscharin, and calotoxin.[26] Singh and Rastogi[203] observed that asclepin isolated from *Calatropis* spp. was found to be 3-*O*-acetyl-calotropin by various chemical and physical tests.

It will be seen that the root bark from the older plants has a higher percentage of acrid and bitter resinous matter than that from the younger plants. After quantitative experiments on the powdered bark the following results were obtained:

	Percentage of contents	
	From young plants	From old plants
Moisture	12.1	10.2
Spirit extract	15.0	16.2
Soluble in water	7.2	7.5
Resins	7.8	8.7
Total ash	7.0	12.2
Sand	2.8	7.2
Pure ash	4.2	5.0

Pharmacological action — Root bark is alterative, tonic, antispasmodic, expectorant, and emetic. It is also recommended in leprosy, hepatic and splenic enlargements, dropsy, and worms.[3,50] The milky juice of *C. gigantea* showed marked stimulant action on the spontaneous activity of the isolated nongravid rat uterus.[180] The alcoholic extracts of the roots and leaves of *C. gigantea* were found to have anticancer activity against human epidermal carcinoma of the nasopharynx in tissue culture.[11,51]

Medicinal properties and uses — The medicinal properties of this plant are similar to those of *C. procera*. The milky juice is used as a blistering agent. The root bark is very useful in acute and subacute dysentery. Its tincture and powder were used in bronchitis and dysentery and were found efficacious.[1,3] The drug acts like digitalis on the heart. The taste of root bark of both species is mucilaginous and bitter, and the odor is peculiar. Flowers are digestive, stomachic and tonic. The milky juice is a violent purgative and gastrointestinal irritant. All parts of the plant have alterative properties when taken in small doses.[1,3,50]

Doses — Tincture — 14 to 28 ml; powder — 0.5 to 1 gr.

CALOTROPIS PROCERA (Ait.) R. Br.* Family: ASCLEPIADACEAE

Vernacular names — Sanskrit, Arka; Hindi, Madar; English, Mudar; Bengali, Akanda; Nepalese, Aka; Sinhalese, Wara; French, Aeribe hirondeille, Arbre-a-sofa; Unani, Aak; Arabian, Ushar; Persian, Khark; Tibetan, A-rka; Burmese, Mehabin; Malaysian, Erikka.

Habitat — Found in northern, western and central India from Sind and in Punjab, Bengal, Bihar, and Bombay, and in the drier climate of the Deccan. This is the smaller white-flowered variety.

Parts used — Root, root bark, leaves, juice of the plant, and flowers.

Morphological characteristics — A shrub very similar in foliage and general appearance to *C. gigantea*, but not usually over 4 ft high. Flowers about 1 in. across, scented. Corolla lobes erect, whitish with purplish blotch on the upper half. Coronal scales glabrous or pubescent, the apex bifid and without auricles. Follicles as in *C. gigantea*.[202] The leaves are thick, glaucous, clothed beneath with fine cottony tomentum.

Pharmacognostical characteristics — The quantitative values, physical constants, and anatomical characteristics of both species are very much similar but are different in the case of fluorescence characteristics which are useful in the identification of the two species.

Fluorescence Characteristics of Leaf Powder of *C. gigantea* and *C. procera*

	C. gigantea	*C. procera*
Dry powder	Greenish blue	Gray
Powder treated with nitrocellulose	Orange	Dirty orange
Powder treated with IN methanolic NaOH	Brown	Yellowish green
Powder treated with IN methanolic NaOH and mounted in nitrocellulose	Yellowish brown	Pale green

Ayurvedic description — *Rasa* — katu, tikta, madhura; *Guna* — sara, singdha, laghu; *Veerya* — ushna; *Vipaka* — katu.

Action and uses — *Krimighna, kushtaghna, arshoghna, vishahara.*

Chemical constituents — The milky juice contains a protocolytic enzyme and a toxic substance.[204] It also contains a highly active rennin.[205] The root bark contains a bitter yellow resin but no alkaloid.[206] Shukla and Krishnamurti[207] reported the presence of a powerful bacteriolytic agent capable of lysing *Micrococcus lysodeikticus* in the latex of *C. procera*. The enzyme had maximum activity at pH 5 to 5.4 at 45°C. Singh and Rastogi[203] isolated asclepin from *Calotropis* sp., which was found to be 3-*O*-acetyl-calotropin. Leaves and stalks contain calotropin and calotropagenin. The latex contains useharin, calotoxin, and calactin,[50] as well as five cardiac steroid glucosides with the same aglycone calotropagenin.[178]

Pharmacology — The crude latex of *C. procera* and its protein fractions were found to possess high fibrinolytic and anticoagulant activity both in rabbit and human plasma.[209] The aqueous and alcoholic extracts of the roots of *C. procera* produced slight depression followed by stimulation of the rate and force of myocardial contraction on isolated frogs' hearts. Both the extracts showed persistent rise in blood pressure of dog. Both the extracts showed vermicidal action on earthworms.[208]

Medicinal properties and uses — Flowers have detergent properties. Root bark is administered for dysentery, in the form of paste, it is applied to elephantiasis. Tincture of leaves is used in intermittent fevers. Latex is irritant and used as purgative. Powdered flowers are prescribed in cold, cough, asthma, and indigestion.[50]

* See plate on page 92.

CANNABIS INDICA Linn.   *Family:* URTICACEAE
(Syn. *CANNABIS SATIVA* Linn.)

Vernacular names — Sanskrit, Bhanga; Hindi, Ganja; English, Indian hemp; Bengali, Bhang; Nepalese, Bhang; Sinhalese, Kansa; Chinese, Tang Ma; German, Hanf; French, Chanvre indien; Unani, Bhang; Arabian, Quinnab; Persian, Falak sair; Burmese, Bhenbin; Tamil, Kanja.

Habitat — The plant is a native of Persia, western and central Asia; now it is cultivated throughout India. Found wild on the eastern Himalayas from Kashmir to east of Assam. The plant is nitrophile, blooming well in nitrogenous waste and near human habitation. The life cycle of the plant is short (4 to 6 months), seedlings sprout in rains, flowering and fruiting completed by late September.[210]

Parts used — Dried flowering and fruiting tops of pistillate plant, leaves, seeds, and resinous exudation of cannabis for *ganja, bhang,* and *charas.*

Pharmacognostical characteristics — The leaves of the female plants are longer than the leaves of the male plant. The stem is longitudinally furrowed, pubescent, light green when young and becomes light purplish when mature. Leaves are palmately compound, while the leaflets are linear-lanceolate with serrate margin. The odor is agreeable and aromatic, the taste is pungent. Bracts are foliaceous and the ovary is one chambered containing a single campylatropous ovule. The stem has both glandular and nonglandular types of trichomes. The stele has a continuous ring of vascular bundles. Lactiferous vessels with yellowish-brown contents are present in the pith. Female plants have more secretory cells in the phloem region. The leaf has numerous glandular trichomes. In the midrib, the secretory cells in the phloem region are filled with resin. Rosette crystals of calcium oxalate are clearly visible in the parenchyma. Abundant starch grains are present in the stem and leaf. A transection of the flowers shows six lignified vascular bundles in the perianth.[211]

Ayurvedic description — *Rasa* — *tikta; Guna* — *laghu, teeshan, rooksha; Veerya* — *ushana; Vipak* — *katu.*

Action and uses — *Vat kapha har, pitta wardhak, badana nasak, madakari, nidrajanan, dipan, pachan, grahi, shula prasaman.*

Chemical constituents — The following constituents have been found to be present in cannabis; cannabinol-6-9-trimethyl-3-pentyl 6*H*-dibenzene[*b,d*]-pyran-1-ol, which is pharmacologically inactive; tetrahydrocannabinol, 6a,7,8,10a-tetrahydro-6,6,9-trimethyl-3-pentyl-6*H*-dibenzo[*b,d*]-pyran-1-ol; which is euphorically active; cannabidiol, which has no pharmacological activity; cannabidiolic acid, which has sedative and antibacterial activities; and cannabigerol, in which no activity is reported.[212] The resin content in crude "charas", leaves, and stem of *C. indica* was found to be 29.23, 7.25, and 3.25%, respectively.[213]

Pharmacological action — The whole plant is intoxicating, stomachic, antispasmodic, analgesic, stimulant, aphrodisiac, and sedative.[1] The pure resin produced an inhibitory action on the respiratory movements of dogs. It relaxed the plain muscles of rabbit intestine and rat uterus.[213] It produced initial excitement followed by depression. It has been found to increase 5-HT content of rat brain and to antagonize the peripheral actions of 5-HT, to which the psychotomimetic effect of the resin has been attributed.[214] The activity of the alkali-insoluble fraction was found greater than that of homotetrahydracannabinol, which indicated that the fraction contains some optical isomers of tetrahydrocannabinol which are more active.[215]

The slow and prolonged hypotensive action of cannabis and its interaction with catecholamines in the peripheral system suggests the possibility of an interaction with the brain amides being responsible for the behavioral effects observed.[216]

The toxic effects of marijuana reported in human beings are psychotic episodes like nausea, dysphoria, tremors, etc., which usually subside with the passing of time. Toxic effects of prolonged cannabis use in chronic users include lassitude, indifference, lack of productive activity, insomnia, headache, nystagmus, increased susceptibility to infections, gastrointestinal disturbances, sexual impotence, and personality changes.[26]

Medicinal properties and uses — The leaves are administered to induce sleep where opium cannot be used; they are also used in dysmenorrhea. A paste of the fresh leaves is used to resolve tumors. A leaf poultice is also applied to the eyes in ophthalmia. The dried pistillate flowering tops which are coated with resinous exudation are known as *ganja;* it is one of the best anodynes, hypnotic, and antispasmodic. It is locally used to relieve pain in itching skin diseases.

Charas, the resinous exudation that collects on the leaves and flowering tops of plants, is the active principle of hemp; it is of great value in malarial and periodical headaches, acute mania, insanity delirium, whooping cough, cough of phthisis, asthma, nervous vomiting, nervous exhaustion, and dysuria.

Doses — Powder — 2 to 4 gr (for children); 15 to 20 gr (for adults); *ganja* — 1 to 2 gr; *charas* — $1/2$ gr.

CAPSICUM ANNUM Linn.   Family: SOLANACEAE

Vernacular names — Sanskrit, Katuvira; Hindi, Lalmirichi; English, Spanish pepper; Bengali, Lanka maric; Nepalese, Malabhata; Sinhalese, Meris; Japanese, Togarashi; Chinese, Franchiao; German, Paptika; French, Pimment annuel; Unani, Mirch surkh; Arabian, Fibile-abmar; Persian, Mirch surkh; Burmese, Nayusi; Malaysian, Chakeai.

Habitat — Largely cultivated throughout India for its fruits.

Parts used — Fruits.

Morphological characteristics — Leaves entire; flowers pediceled, axillary, white corolla; fruit 5 to 8 cm long, conical in shape, and dark red when ripe.

Ayurvedic description — *Rasa* — *katu; Guna* — *laghu, rooksha, teekshna; Veerya* — *ushna; Vipaka* — *katu.*

Action and uses — *Kapha vat samak, pitta-wardhak, lakhan, nidrajanan, lalanisark, dipan, pachan, anuloman, vidahi, vajikar.*

Chemical constituents — Capsaicin, a volatile alkaloid; capsaicin, a crystalline acrid substance; solanine, a volatile oil; fixed oil, fatty acid; resin; red coloring matter, and ash (4 to 5%). Its pungency and acridity are due to the capsaicin which is concentrated in the inner walls of the fruit.[178] Willaman and Li[10] reported the presence of alkaloids salanidine and solasodine from embryo and endosperm of the seeds.

Pharmacological action — A powerful local irritant; heart and general stimulant; stomachic and tonic.[50]

Medicinal properties and uses — Externally its paste is used as a rubefacient and a local stimulant. In chronic lumbago, a plaster of capsicum with garlic, pepper, and "silaras" is an efficent stimulant and rubefacient. It is also very useful in the form of tincture in dyspepsia, loss of appetite, and flatulence. The whole plant cooked up in milk is successfully applied to reduce swellings and hardened tumors. Fresh fruit made into a paste in combination with mustard seed is used as a counterirritant.[1] A pill made of capsicum, ginger, and rhubarb is carminative and may be advantageously employed in atonic dyspepsia.

Doses — Seed powder — 0.1 to 1 gr.

CARICA PAPAYA Linn.   *Family:* CARICACEAE

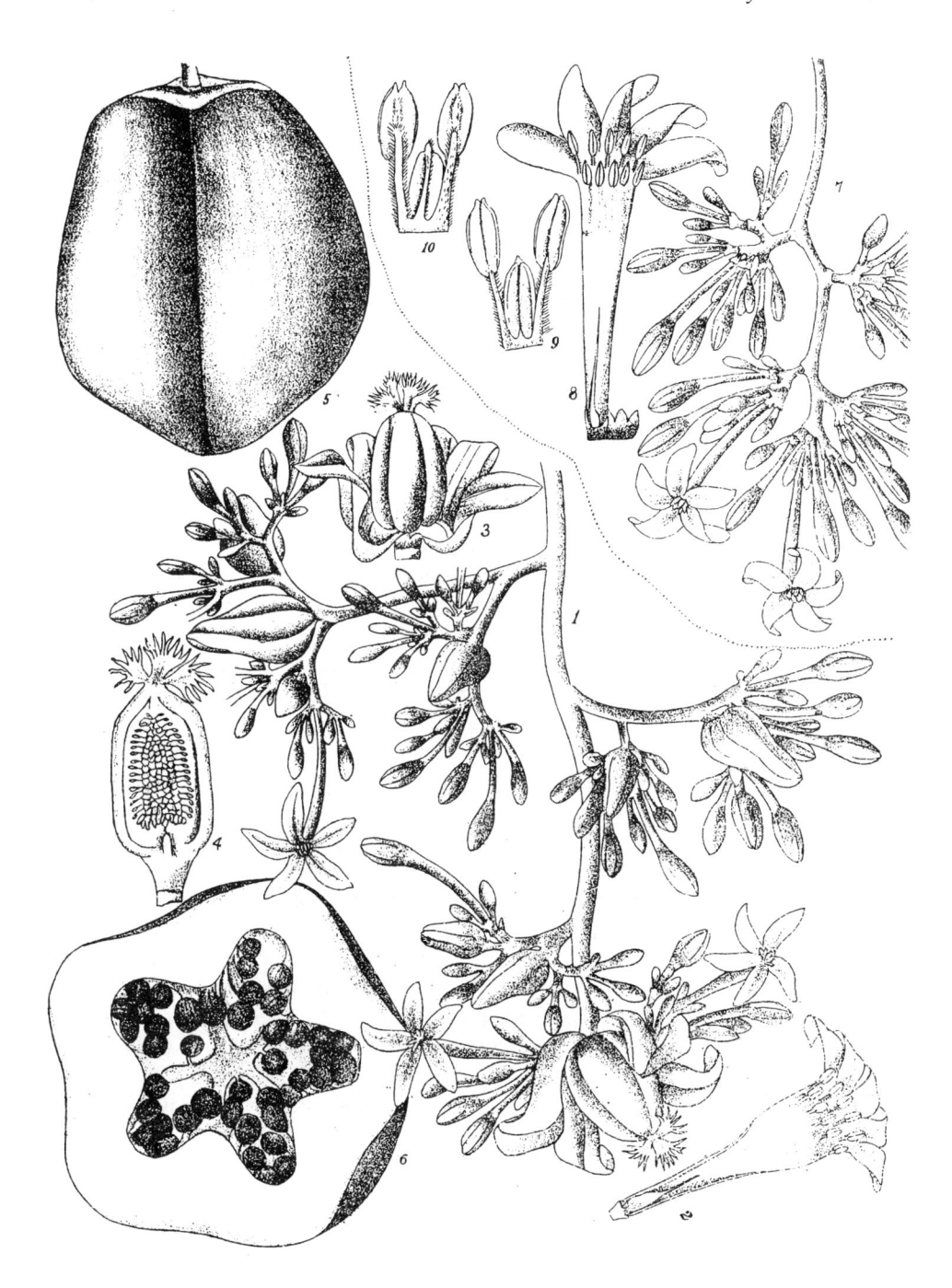

Vernacular names — Sanskrit, Arandkharkati; Hindi, Popaiyah; English, Papaw; Bengali, Papeya; Sinhalese, Pepol; Japanese, Papaya; Chinese, Kaydudu; German, Melonenbaum; Unani, Arandkharbuza; Arabian and Persian, Amba hindi; French, Popayer commun; Tamil, Pappali.

Habitat — Commonly cultivated in gardens throughout India.

Parts used — Milky juice, seeds, and pulp.

Morphological characteristics — A subherbaceous tree, leaves divided into seven parts, deeply lobed. Flowers dioecious but occasionally monoecious; plant bearing hermaphrodite flowers have also been recorded. Fruits green when unripe and become yellow when ripe; contain brown-colored seeds.

Ayurvedic description — *Rasa — katu, tikt; Guna — laghu, rooksha, teekshna; Veerya — ushna; Vipak — katu.*

Action and uses — *Kapha vata samak, pitta nasak, sothahar, badana, sthapan, dipan, pachan, vatanuloman, kafa nisark, mutral, balya.*

Chemical constituents — In the early stages, the fruit of *C. papaya* gives a white, milky juice which contains an albuminoid, a digestive enzyme papain or papayotin. A milky juice comes from the rind, which becomes yellow or orange when ripe. Pulp of the fresh fruit contains a caoutchouc-like substance, a soft yellow resin, fat, albuminoids, sugar, pectin, citric, tartaric, and malic acid, and dextrin. Dried fruit contains a large amount of ash (8.4%) which contains soda, potash and phosphoric acid. Seeds contain an oil. Leaves contain an alkaloid called carpaine and a glucoside named carposide.

From the seeds of *C. papaya* a substance named carpasemine[222] has been isolated, having a molecular formula $C_8H_{10}N_2S$, mp 165°C. Nineteen different carotenoids have been identified in the fruit, the chief one being cryptoxanthene (48%).[223] The alkaloid pseudocarpaine isolated from the plant showed that it was dimeric under mass spectrum.[224]

Pharmacological action — Digestive, anthelmintic, stomachic, and emmenagogue.

Petroleum ether extract of the pulp exerted significant antifertility activity in female albino rats.[225] The latex of green fruit possesses high oxytocic activity. The seeds decreased fertility of albino mice but were highly toxic.[226,227] Aqueous extract of the latex exhibited anticoagulant activity against plasma and whole blood. No clotting was observed even after 24 h.[228]

An anticoagulant principle isolated from the latex was found to increase the prothrombin time and coagulation time in dogs, rabbits, rats, and mice.[229] The alkaloidal solution showed depressant action on heart, blood pressure and intestine.[230] Benzythiourea compound isolated from seeds showed anthelmintic activity when tested against *Ascaris lumbricoides*.[231]

Medicinal properties and uses — Juice of the green fruits acts as an emmenagogue and in large doses as an ecbolic. Fresh milky juice is rubefacient. Ripe fruit is digestive and alterative. The green fruit is a laxative and diuretic. The leaves are anthelmintic and febrifuge; they are given in beriberi; they contain the alkaloid carpine which physiologically has the same effect as digitalis; it is a potent amebicide and is very efficacious for the treatment of amebic dysentery. Lightly bruised or roasted leaves are applied as a galactagogue to the breast of nursing mothers.

The seeds contain the glucoside caricin; they are anthelmintic, emmenagogue, and carminative; they are given with honey for expelling roundworms. Their juice is given in dyspepsia, bleeding piles, and enlargement of the liver and spleen. Paste of the seeds is also applied in skin diseases like ringworm.[1,3,50]

Doses — Powder of dried leaves — 1 to 2 gr; dried latex — 2 to 4 gr; seed powder — 0.5 to 1 g.

CARUM CARVI Linn.   Family: UMBELLIFERAE

Vernacular names — Sanskrit, Krishna jira; Hindi, Shiajire; English, Caraway; Bengali, Jira; Nepalese, Hakugajee; Sinhalese, Kaludaru; Japanese, Himeuikyo; German, Gemeiner Kummel; Unani, Siyah Jira; Arabian, Kammeasvad; Persian, Jirah-e-siyah; Burmese, Satmung; Malaysian, Karinchirakam; French, Cuminnoir; Pustu, Siyah daru; Tamil, Karuncheeragam.

Habitat — The plant is widely distributed in the temperate regions of both hemispheres. It is found wild in northern Himalayan regions. It is cultivated in the plains as a cold season crop and in the hills of Kashmir, Kumaon, Garhwal, and Chamba as a summer crop.

Parts used — Fruits.

Pharmacognostical characteristics — A perennial or biennal herb 1 to 3 ft high, with a thick tuberous root, compound leaves with linear segments, and small white flowers borne in dense umbels.

The brown cremocarps (3 to 7 mm long and 1.5 to 2 mm in diameter), which are aromatic, are oblong, laterally compressed, slightly curved, tapering toward both ends.

Ayurvedic description — *Rasa — katu; Guna — lagu, rooksha; Veerya — ushna; Vipak — katu.*

Action and uses — *Kaf bat samak, dipan, pachan, grahi, shoth har, sthanyajana, hirdya.*

Chemical constituents — The fruits have the following composition: moisture (11.5 to 15.5%); ash (5.5 to 6.7%); water soluble ash (2.0 to 2.2%); volatile oil (2.7 to 8.2%); fixed oil and resin (6.2 to 10.1%); crude fiber (17.5 to 22.3%); and nitrogen (5.9 to 6.4%). The B.P. specified an ash content of not more than 9% with acid insoluble content less than 1.5%. The U.S.P. requires an ash content of not more than 8%.

Caraway oil, distilled from fresh fruits is a colorless or pale yellow oil, sp gr 15, 0.907 to 0.920. The volatile oil contains a mixture of ketone, carvone (sp gr 0.850), a terpene formerly called carvene but now recognized to be di-limonene, and traces of carvacol. Pure carvone ($C_{10}H_{14}O$) is prepared by decomposing the crystalline compound of carvone with hydrogen sulfide.

Atal and Sood[238] obtained oil from Lahul *Carum carvi* which they considered was superior from its physico-chemical contents. Carvone and limonene were isolated which constituted the major bulk of oil. Lahul caraway yielded 5 to 7% of oil and had 63.3% carvone content as compared to 53.0% of the pharmacopoeial standard.

The dried, exhausted and pulverized caraway chaff contains 20 to 23.5% crude protein (of which 75 to 85% is digestible) and 14 to 16% fat.

Pharmacological action — Fruits are carminative, aromatic, stomachic, stimulant, and lactagogue. They are cooling in effect.

Medicinal properties and uses — Caraway oil is used chiefly for flavoring purposes and in medicine as a carminative. It is also used to correct the nauseating and griping effects of medicines. For the treatment of scabies, a solution containing 5 parts each of alcohol and oil of caraway in 75 parts of castor oil is recommended. It is a mild carminative, occasionally used in flatulent colic and as adjuvant or corrective for medicines.

Doses — Powder — 0.5 to 1 g.

CARUM COPTICUM Benth.   Family: UMBELLIFERAE
(Syn. *TRACHYSPERMUM AMMI* [Linn.] Sprag.;
PTYCHOTIS AJOWAN DC.)

Vernacular names — Sanskrit, Yamani; Hindi, Ajowan; English, Omum (seeds); Bengali, Jowan; Nepalese, Emu; Sinhalese, Assamodam; Japanese, Ayowan; German, Ajowan Kummel; Unani, Ajowan; Arabian, Kamun-e-mulooki; Persian, Nan khah.

Habitat — This plant is largely cultivated in India, and is particularly abundant in and around Indore.

Parts used — Fruits and root.

Ayurvedic description — *Rasa — katu, tikta; Guna — laghu, rooksha, teekshna; Veerya — ushana; Vipak — katu.*

Action and uses — *Kapha vat samak, pitta wardhak, badana sthapk, soth har, anuloman, vishaghan, dipan, pachan, shula prasaman, krimighan, swas har, mutral.*

Chemical constituents — It yields aromatic volatile oil and a crystalline substance — steroptin — which collects on the surface of distilled water; also cumene and a terpene, "thymene". The steroptin known as crude thymol is identical with English thymol. The seeds of *Carum copticum*[1] contain the antiseptic thymol and they yield 2 to 3% of an essential oil which is official as oil of *ajowan* and contains not less than 40 to 50% of thymol $C_{10}H_4O$.

Pharmacological action — Seeds are stimulant, carminative, tonic, aromatic, antispasmodic, and antiseptic.

Medicinal properties and uses — Seeds are useful in flatulence, indigestion, colic, atonic dyspepsia, diarrhea, cholera, and spasmodic affections of the bowels. Oil is also used in flatulent colic, atonic dyspepsia or diarrhea, hysteria, and indigestion. The chief importance of *ajowan* seeds is for production of thymol, which is used in medicine in a number of preparations. Leaves of the tender plant are used as vermicide; its juice is given for worms.[3]

Doses — Powder — 3 to 6 g; arka — 28 to 56 ml; extract — 125 mg.

CASSIA ABSUS Linn.   *Family:* LEGUMINOSAE

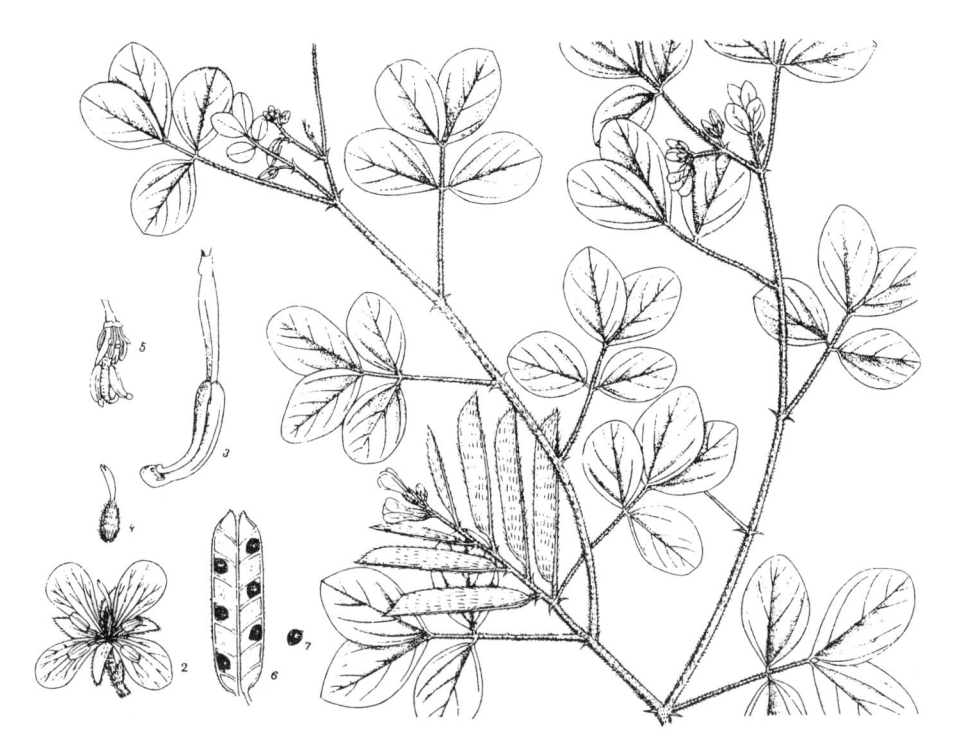

Vernacular names — Sanskrit, Chaksu; Hindi, Chaksu; Bengali, Chakus; Sinhalese, Butora; Unani, Chakau; Arabian, Tashmizaj; Persian, Chashmiza.

Habitat — An erect annual herb growing in lower parts of western Himalayas.

Parts used — Leaves and seeds.

Morphological characteristics — Leaves compound, leaflets 1 to 2 in. long. Flowers reddish yellow, fruit 1 to 1.5 in. long and slightly curved. Seeds five, compressed, shining, and brownish black.

Ayurvedic description — *Rasa* — *kasaya, tikta; Guna* — *rooksha; Veerya* — *sheeta; Vipak* — *katu.*

Action and uses — *Kapha-pitta samak, chachusya, lakhan, grahi, rakt sthamvan, mutral, visaghan.*

Chemical constituents — Two water-soluble isomeric quaternary bases, chaksine and isochaksine ($C_{11}H_{21}O_3N_3$), have been isolated in the form of carbonates from the seed kernel in 1.5% yield. The seed oil contains oleic acid (16.32%), linoleic acid (47.32%), linolenic acid (0.41%), hydroxy acids (0.75%), palmitic acid (6.28%), stearic (8.10%), lignoceric acid (0.82%), unsaponifiable matter (8.40%), glycerol (10.40%), and unidentified matter (1.20%).[239] β-Sitosterol has been also identified in *chaksu* oil.[178]

Pharmacological action — Bitter and astringent. Pradhan et al.[240] reported that the alkaloid chaksine isolated from seeds produced a fall in blood pressure mainly by dilation of capillaries in anesthetized cats and dogs. It stimulated their respiration, contractions of uterus, intestine, bladder, and musculature of blood vessels. It depressed parasympathetic nerve endings of intestine and bladder. It had weak curariform activity. The alkaloid is reported to have antibacterial properties.[178]

Medicinal properties and uses — The leaves are bitter and are used as astringent. It is also considered to be a remedy for cough. The seeds are used in ringworm and other skin diseases, in conjunctivitis, and in ophthalmia.[1,67]

Doses — Powder — 1 to 3 gr.

CASSIA ANGUSTIFOLIA Wahl.   Family: LEGUMINOSAE

Vernacular names — Sanskrit, Markandika; Hindi, Hindi sana; English, Indian or Tinnevelly senna; Bengali, Sonamukhi; Nepalese, Sanai; Sinhalese, Senehakola; Japanese, Senna; German, Sennes Blätter; Unani, Sena; Arabian, Sena; Persian, Sena-e-makki; Tamil, Nilavarai.

Habitat — Cultivated in South India, Punjab, and Gujarat.

Parts used — Leaves and pods.

Morphological and pharmacognostical characteristics — A bushy, herbaceous plant. Leaves are paripinnate and leaflets 1 to 2 in. long, 0.2 to 0.6 m wide, glabrous, and of yellowish-green color. Flowers in racemes. Pods broadly oblong, slightly curved; valves papery smooth, 1.4 to 2.8 in. long, about 0.8 in. wide, greenish brown to dark brown in color, and contain obvate, dark brown, smooth seeds. Epidermal hairs are sparse, about six epidermal cells apart. Stomata have two to three subsidiary cells, respectively, and are in the ratio of 7:3. Vein islet no. 19.5 to 22.5, stomatal index 17.1 to 20 W.

Ayurvedic description — *Rasa — tikta, katu; Guna — laghu, rooksha, teekshna; Veerya — ushna; Vipak — katu.*

Action and uses — *Kapha vat samak, lakhan, vamak, anuloman, sansarn, krimighan, rakt sodhak.*

Chemical constituents — Senna leaves contain about 1.3 to 1.5% of anthraquinone derivates which are present in both the free and the combined state. Stoll et al.[241] obtained two crystalline glucosides which they call senoside A and B, believed to be the laxative principles. Fairbairn and Saleh[242] attribute 30% of the activity of the drug to a third glucoside. Senna also contains the yellow flavonal coloring matter, kaempferol, its glycoside, and isorhamnetin; also, a sterol and its glycoside, mucilage, calcium oxalate, and resin. Constituents of the pods are similar to those of the leaves. Sennoside content varies from about 1.2 to 2.5% in the Tinnevelly and from 2.5 to 4.5% in Alexandrian pods.[69] The pods are reported to be less griping than the leaves as they contain less resin. Two new glycosides of rhein and chrysophanic acid and the presence of aloe-emodin or emodin glucoside are reported.[178]

Pharmacological action — Purgative.

Medicinal properties and uses — Senna leaves are a sure and safe purgative, even for children and weak and elderly persons; they are effective in constipation, billiousness, gout, and rheumatism; they are given in the form of an infusion or decoction. The leaves may cause nausea and griping and therefore they should be taken with some aromatic; senna is also used as an anthelmintic for intestinal worms and as a mild liver stimulant. It should not be given in inflammatory conditions of the alimentary canal, fever, piles, menorrhagia, prolapse of the uterus or rectum, and pregnancy. A paste of the dried leaves is given with vinegar for certain skin diseases; the paste is also useful for removing pimples.

Senna pods are used as a purgative; but they are milder and slower in action than the leaves.

Doses — Powder — 1 to 2 gr; decoction — 28 to 56 ml; compound infusion — 28 to 56 ml; tincture — 5 to 20 ml.

105

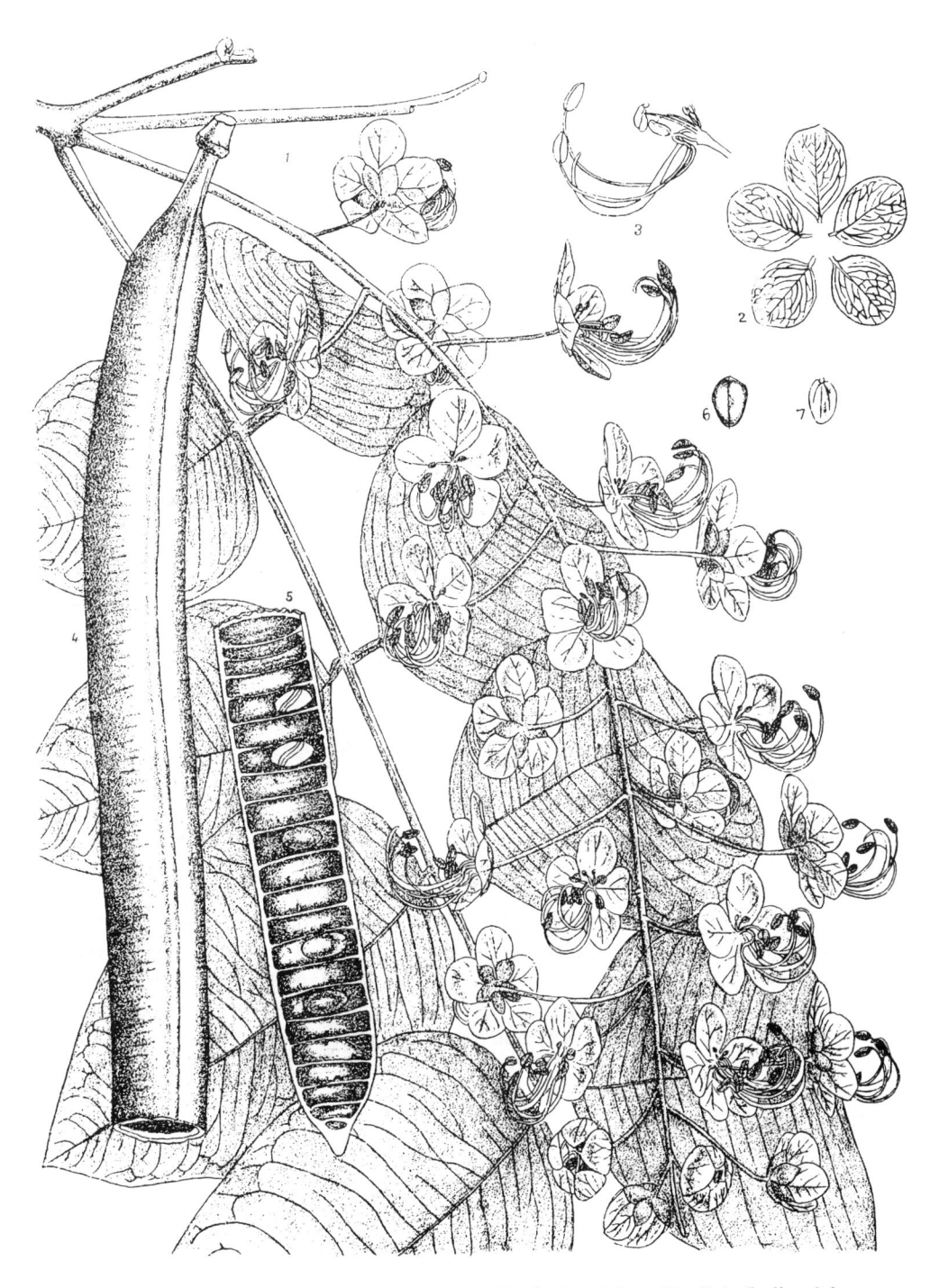

CASSIA FISTULA Linn.   *Family:* CAESALPINIACEAE

Vernacular names — Sanskrit, Argbhada; Hindi, Amulthus; English, Indian laburnum; Bengali, Sonndali; Nepalese, Rajbrichya; Sinhalese, Khela; Japanese, Ebisugusa; Chinese, Chiieh-ming-tzu; German, Rohrkassie; French, Cassie purgative; Unani, Amaltas; Arabian, Kheyarshanbar; Persian, Kheyar Shanbar; Tibetan, Don-ka; Burmese, Gnookye; Tamil, Sarakonrai.

Habitat — The tree is found throughout India from Punjab to Kanya-kumari in all deciduous forests and tracts from sea level to about 4000 ft elevation in the outer Himalayas. It is one of the loveliest flowering trees, and is frequently planted in gardens and along avenues.

Parts used — Fruit pulp, root bark, flowers, pods, leaves, and root.

Pharmacognostical characteristics — *C. fistula* is a medium-sized deciduous tree growing about 10 to 35 ft with a straight trunk and spreading branches; bearing alternate pinnately compound leaves 20 to 45 cm long, having three to eight pairs of large ovate or ovate-oblong acute coriaceous leaflets. It bark is smooth, of light ash color. Fruit is a pendulous, cylindrical, nearly straight, dark brown or brownish black pod. Fruits contain 40 to 100 seeds, broadly ovate, smooth light to dark or reddish brown, slightly flattened, about 8 mm long.

The leaflets are in four to eight pairs and when fresh are coriaceous; on drying they become papery. Unicellular, nonlignified, pointed trichomes are present on upper epidermis. Stomata are absent on the upper epidermis. Transection through the midrib shows a layer of epidermal cells with pointed trichomes. A group of bundles is also seen in the center of the midrib. The rest of the midrib is occupied by thin-walled parenchymatous cells, some of which contain prismatic crystals of calcium oxalate, and others contain starch grains.[243] The leaves of *C. fistula* are dorsiventral in transection, with two to three rows of palisade cells below the epidermis. The upper epidermis possesses unicellular, nonlignified, long, coiled trichomes nearer the midrib. The lower epidermis has unicellular, warty, dagger-shaped trichomes. The lower epidermal cells are papillose. Stout prisms of calcium oxalate are present, but only a few clusters are seen. The midrib possesses an arc of radiate xylem followed by phloem. The pericycle fibers form a complete circle, enclosing two separate, inverted, vascular bundles above the normal xylem. The stomatal index, palisade ration, and vein-islet number are different from the other species of *Cassia*.[244]

Ayurvedic description — *Rasa* — *madhura, tikta; Guna* — *guru, mridu, snigdha; Veerya* — *sheeta; Vipaka* — *katu*.

Action and uses — *Krimighan, shoolaghna, kaphodarahara, pramehaghna, gulm hara, jwarahara, kandoohara, kushta hara, vistambhaghna, hridrogaghna, yakta pitla hara, shotha hara, vedanaastha pana.*

Chemical constituents — Root bark, besides tannin, contains phlobaphenes and an oxyanthraquinone substance. Pulp contains a small amount of volatile oil, three waxy substances, and a resinous substance. Ceryl alcohol, kaempferol, rhein, and a new bi-anthraquinone glycoside, fistulin, were isolated from the ethanolic extract of the flowers of *C. fistula*. The bark and heartwood of the plant contained fistucacidin, 3,4,7,8,3-pentahydroxyflavan, along with harbaloin and rhein. Leaves contain glycosides which have been identified as senoside A and senoside B, free rhein, and rheinglucoside.[246] The pod pulp exhibited the presence of rhein glucoside.[247] The stem bark contained tannins, lupeol, β-sitosterol, and hexacosanol.[248] The flowers of *C. fistula* have been found to contain kaempferol and proanthocyanidin, and the bark yielded a leukopelargonidin trimer with no free glycol unit.[249] Pods on isolation gave a new anthraquinone named fistulic acid from alcoholic extract.[250]

Pharmacological action — Pulp, root bark, flowers, pods, and leaves possess purgative properties. The root acts as a purgative, tonic, and febrifuge. The fruit is cathartic. The aqueous extract of the pulp of *C. fistula* exhibited a slightly lower antibacterial activity than its dealcoholized extracts as observed by the inhibition of the growth of *Microccus pyogenes* var. *aureus, M. pyogenes* var. *albus, M. citreus, Corynebacterium diphtheriae, Bacillus megatherium, Salmonella typhus, S. paratyphi, S. schottmulleri,* and *Escherichia coli*.[245]

Medicinal properties and uses — The leaves are emolient; their juice or their paste is a very useful dressing for ringworm and chilblains, for relief from rheumatism, and facial paralysis. Pulp of pods is an agreeable laxative safe for children and pregnant women. Root is useful in fever, heart diseases, retained excretions, and in biliousness.

Doses — Bark decoction — 56 to 112 ml; seed powder — 1 to 2 gr.

CASSIA OCCIDENTALIS Linn. Family: LEGUMINOSAE

Vernacular names — Sanskrit, Kasmard; Hindi, Kasaundi; English, Negro coffee; Bengali, Kasmard; Nepalese, Chikchhika; Sinhalese, Penitora; German, Rinde-Tedegoso; French, Cassier; Unani, Kasaundi; Burmese, Kalan; Malaysian, Kachang kola; Tamil, Payavarai.

Habitat — A common weed scattered from the Himalayas to West Bengal, South India, Burma, and Sri Lanka.

Parts used — Leaves, seeds, and roots.

Morphological and pharmacognostical characteristics — Leaves abruptly pinnate, 10 to 20 cm long, alternate. Leaflets three to five pairs, opposite, unequal, the lowermost ovate and smallest. Flowers pediceled, yellowish. The pedicels are longer than the peduncles. In flower stage they are about 5 mm long, but elongate to 1.2 cm or more and spreading in fruit. Bracts 12 mm long, thin, ovate, acuminate, whitish with a pink tinge, caducous. Calyx 6 to 8 or up to 10 mm long, the sepals membranous, whitish or cream yellow with a slight pink tinge, oblong, obtuse, and imbricate. Petals five, free, subequal, the lower two being smaller than the other three and closer together, about 1.2 cm long, ovate, oblong, obtuse, yellowish, veined, imbricate in bud. Stamens free. Fruit a slightly falcate, nearly cylindrical to compressed, transversely septate, distinctly straight or recurved glabrous pod, 10 to 12 cm or more long and 5 mm thick. Seeds hard, smooth, ovoid, gray, dark, olive green, or pale brown.[28]

Ayurvedic description — *Rasa* — *tikta, madhur; Guna* — *rooksha, laghu, teekshna; Veerya* — *ushna; Vipak* — *katu.*

Action and uses — *Kapha vat samak, pittasark, khustaghan, vishaghan, badanasak, dipan, vata anuloman, pitta sark, rachan, swahar, jwaraghana.*

Chemical constituents — Seeds contain fatty matter, tannic acid, sugar, gum, starch, cellulose, achrosine, traces of calcium sulfate and phosphate, sodium chloride, magnesium sulfate, iron, silica, malic acid, and chrysophanic acid. Leaves contain carthartin, a coloring matter, and salts. The seeds of *C. occidentalis* were also found to contain rhein, aloe-emodin, and chrysophanol.[251] Extractive of *C. occidentalis* furnished, in addition to emodin and physcion, two unidentified pigments having mp 214 to 216 and 243 to 245°C; chrysophanol; α-3-sitosterol and a new xanthone, cassiollin, identified as 1,7-dihydroxy-5-methoxy carbonyl-3-methyl xanthone.[252] Isolation of physcion (3-methyl-6-methoxy-1,8-dihydroxyanthraquinone) has also been reported from the plant.[178]

Pharmacological action — Purgative, febrifuge, diuretic, and antiperiodic.[3]

Medicinal properties and uses — Infusion of the root is considered as an antidote to various poisons; it is given in fevers and neuralgia. Decoction of the leaves, flowers, and root is highly valuable in hysteria to relieve the spasm; also useful in relieving flatulence of the dyspeptic in nervous women. Seeds are useful in cough and whooping cough.

Doses — Leaf juice — 5 to 15 ml; seed powder — 1 to 2 g; root decoction — 56 to 112 ml.

CASSIA TORA Linn.   *Family:* LEGUMINOSAE
(Syn. *C. TOROIDES* Roxb.;
    C. FOETIDA Pers;
    C. OBTUSIFOLIA Linn.)

Vernacular names — Sanskrit, Chakramanda; Hindi, Chakunda; English, Fetid cassia; Bengali, Chakunda; Nepalese, Tapre; Sinhalese, Pititora; Japanese, Ehisugusa; Chinese, Cheuh Ming; Unani, Panuwar; Arabian, Sanjbasoyah; Persian, Sanjbasoyah; Tibetan, Thal-ka; Burmese, Dan-kilay-iwai; Tamil, Thararai.

Habitat — Annual herb growing on dry soil in Bengal and throughout the tropical parts of ndia.

Parts used — Leaves, root and seeds.

Morphological characteristics — Annual herb with erect stem, compound leaves, but does not have well-developed taproot system. Leaves pinnate with glands, leaflets, 3 pairs, ovate. Flowers small, yellow. Pod slender, long, 4 sided, curved, sharp pointed, 12 to 20 cm long, seeds 20 to 30, grayish or pale brown, about 2 mm long.

Young root in transverse section shows single-layered epidermis, followed by cortex, endodermis, and pericycle enclosing a diarch stele. Mature root shows phelloderm of 8 to 12 layers of parenchymatous cells. Crystals of calcium oxalate of the rosette and prismatic type are present in phelloderm parenchyma. The secondary phloem shows sieve tubes, companion cells and phloem parenchyma. The wood is composed of vessels, tracheids, fibers, medullary rays, and some parenchymatous cells. In the young stem, the cortex contains a rosette of small prismatic crystals of calcium oxalate. Vascular bundles are of the collateral type. In the mature stem, tanninferous contents occur in cork and outer phelloderm. The narrow medullary rays have some prismatic crystals.[253]

Ayurvedic description — *Rasa* — *katu; Guna* — *laghu, rooksha; Veerya* — *ushna; Vipak* — *katu.*

Action and uses — *Kapha-vatta samak, lakhan, vishaghan, kustaghan, anuloman, krimighan, rachan, hridya, kaphanisark vishaghan, oja wardhak, medohar, dadru har.*

Chemical constituents — Chemical examination of the ash of stem and leaf of *C. tora* showed the presence of sulfate, phosphate, calcium, iron, magnesium, sodium, and potassium.[254] The seeds were found to contain rhein, aloe-emodin, and chrysophenol as indicated by thin layer chromatography studies.[251] Leaves also contain a flavanol glucoside, chlorophyll, and gum-suspending agent for calomel, kaolin, and lactone. Seeds contain oxytocic principle and three crystalline components.[178]

Phamacological action — Antiperiodic, aperient, alterative, and anthelmintic.[1,3]

Medicinal properties and uses — The leaves are generally given to children having intestinal disorders; its decoction is a mild laxative in doses of 5 to 15 ml, especially for children having fever while teething. They are given in skin diseases as an alterative. Poultice of the leaves is used locally in gout, sciatica, and pains in the joints. Seeds contain chrysopanic acid and so are very valuable for the treatment of skin diseases like ringworm, scabies, and eczema. The paste of the root made with lime juice is a specific remedy for ringworm. Root is used as antidote against snake bite.[1]

Doses — Seed powder — 1 to 3 gr; infusion — 5 to 15 ml.

CEDRUS DEODARA Loud.   Family: PINACEAE
(Syn. *C. LIBANI* Barrel var. *DEODARA* Hook. f.;
    PINUS DEODARA Roxb. S. & B.)

Vernacular names — Sanskrit, Devadaru; Hindi, Deodar; English, Deodar; Bengali, Toon; Nepalese, Debadaru; Sinhalese, Devadara; Japanese, Himarayosugi; French, Cedre deodar; Unani, Deodar; Arabian, Shaj-ul-jim; Persian, Deodar; Tibetan, Than-sin; Tamil, Devadaru.

Habitat — Found all over northern Himalayas, viz., Kashmir to Nainital, Nepal. Also cultivated in forest blanks.

Parts used — Wood, bark, and leaves.

Pharmacognostical characteristics — Bark of *C. deodara* is cracked and thick walled; leaves green, long, and pointed. Flowers appear in bunch, greenish or yellow in color; fruits, when ripe, blackish in color and containing long seeds inside the cones.

Ayurvedic description — *Rasa* — *tikta, katu; Guna* — *lagu, snigdha; Veerya* — *ushana; Vipak* — *katu.*

Action and uses — *Kapha vata samak, sothhar, badana samak, khutaghan, bran sodhan, bran ropan, lakhan, prameghan.*

Chemical constituents — The essential oil from the wood of Himalayan *deodar* contained *p*-methylacetophenone, *p*-methyl-Δ3 tetrahydroacetophenone, and atlantone. Two new sesquiterpenes, designated as α- and β-himachalene, constitute the major portion of the oil. Himachalol, a new sesquiterpene alcohol related to himachalenes, has also been isolated.[255] Pande et al.[256] reported the isolation and characterization of pure atlantone. The minor constituent of the essential oil was shown to be the *cis*-isomer. Intraconversion of the isomers has also been reported. Nonbitter variety needles are reported to contain 150 to 250 mg% of vitamin C and essential oil. The presence of cedrol has been estimated in the cedarwood oil.[178]

Pharmacological action — Wood is carminative; bark is a powerful astringent and febrifuge; leaves have mild terebinthinate properties.[3] Dhar et al.[51] reported that alcoholic extract of the stem exhibited anticancer activity against human epidermal carcinoma of the nasopharynx in tissue culture.

Medicinal properties and uses — The bark is a good remedy in remittent and intermittent fevers, diarrhea, and dysentery; it is astringent. Its powder is applied with much benefit in the treatment of ulcers. It is considered especially useful in bilious fevers and inveterate diarrhea arising from atony of the muscular fiber. Wood is diaphoretic, diuretic, carminative, useful in fever, flatulence, pulmonary and urinary disorders, rheumatism, piles, and kidney stones, antidote to snake bite; oil is diaphoretic, used in skin diseases, and for ulcers.[1]

Doses — Wood powder — 3 to 6 gr; decoction — 28 to 56 ml; turpentine — 10 to 40 minims.

CELASTRUS PANICULATUS Willd.   *Family:* CELASTRACEAE
(Syn. *C. MONTANA* Roth;
    C. MULTIFLORA Roxb.;
    C. NUTANS Roxb.)

Vernacular names — Sanskrit, Kanguni; Hindi, Malakanguni; English, Staff tree; Bengali, Latafatki; Nepalese, Malkangun; Sinhalese, Duhudu; German, Dudukol Celasterol; Unani, Malakanguni; Arabian, Tilaqoon; Persian, Malkangni; Burmese, Myinkoung nayoung; Tamil, Valuzhuvai.

Habitat — Found all over the hilly district of the Himalayas up to an altitude of 6000 ft. It grows in hilly parts of Bombay, Gujarat, Central India, and Madras.

Parts used — Seeds, leaves, and oil.

Pharmacognostical characteristics — It is a large, deciduous climbing shrub with yellow fruits. The seeds are brown and are covered by a scarlet aril.

Ayurvedic description — *Rasa — katu, teekta; Guna — teekshna, snigdha, sar; Veerya — ushna; Vipaka — katu.*

Action and uses — *Kaf bat samak, badana samak, dipan, anuloman, khutaghna vajikaran.*

Chemical constituents — The seeds yield a brownish-yellow oil (52.2%) with an unpleasant taste. Gunde and Hilditch[257] reported an appreciable proportion of acetic and benzoic acids in the oil of seeds and fruit coat, in addition to the usual higher fatty acids. Shah et al.[258] isolated a crystalline material believed to be a tetracasanol and sterol. Basu and Pabrai[259] reported two alkaloids from oil cake, viz., celastrine $C_{19}H_{25}NO_3$, mp 26°C, and paniculatine.

An orange-red semisolid fat (30%) is obtained on extraction of the fleshy arils with petroleum. This contains 6.46% unsaponifiable matter from which a phytosterol, celastrol — $C_{27}H_{46}O_3$, mp 142°C — has been isolated. In addition, a highly colored resinous substance is present in the unsaponifiable matter, which has not been identified.

Pharmacological action — Oil is rubefacient; seeds are alterative, stimulant, and nervine tonic. Seeds and oil stimulate intellect and sharpen memory.[1,3] The seed oil produced sedation and anticonvulsant activity in rats and produced a gradual fall in blood pressure in cats.[260] Aerial parts of the plant indicated antiviral activity against Ranikhet disease virus.[11] An active fraction isolated from the oil by the countercurrent distribution method had a tranquilizing effect on rats, mice, monkeys, and cats.[261]

Medicinal properties and uses — Decoction of the seeds with or without the addition of some aromatics is recommended in rheumatism, gout, paralysis, and leprosy. Oil with benzoin, cloves, nutmeg, and mace is a sovereign remedy in beriberi and a powerful stimulant in doses of 10 to 15 minim. The leaves are emmenagogue, and the leaf sap is used as an antidote for opium poisoning. The bark is abortifacient, and the seeds are bitter, laxative, emetic, and tonic.

Doses — Oil — 1 to 2 ml; powder — 1 to 3 gr; decoction — 28 to 56 ml.

CENTRATHERUM ANTHELMINTICUM (Willd.) Kuntze   *Family:* COMPOSITAE
(Syn. *VERNONIA ANTHELMINTICUM* Willd.)

Vernacular names — Sanskrit, Aranyajira; Hindi, Kaijiri; Nepalese, Jangalijera; Sinhalese, Sanninayan; Unani, Kaliziri; Arabian, Kamun-e-Barri; Persian, Shahzira.

Habitat — Distributed throughout India.

Parts used — All parts of the plant and seeds.

Morphological characteristics — A robust, branched, glandular-pubescent annual 2 to 3 ft high. Leaves 3 to 8 in., membranous, lanceolate or ovate-lanceolate, coarsely serrate. Heads $1/2$ to $3/4$ in. diameter, subcorymbose. Involucral bracts linear, innermost usually the longest and with purplish tips. Achenes $1/5$ in., narrowed towards the base, ten-ribbed, black hairy, pappus reddish, outer row of short, rigid, persistent scales, inner hairs deciduous.[28]

Ayurvedic description — *Rasa — katu, tikta; Guna — laghu, tikshna; Veerya — ushna; Vipak — katu.*

Action and uses — *Kapha vat samak, sothahar, badana sthapan, krimighan, kustaghan, dipan, vaman karak, rakt sodhak, jwaraghana.*

Chemical constituents — The petroleum ether extract of powdered seeds consisted mainly of a fixed oil and a very small amount of an essential oil (about 0.02%). The chlorophyll extract contained a bitter substance. The alcoholic extract consisted mainly of resin. There was no alkaloid present. Majumdar[262] extracted the seed with petroleum ether and obtained 17.33% of oil which had unsaponifiable matter; 1.68% of this fraction contained brassicasterol, stigmosterol, and a sterol. Other fractions contained stearic, palmitic, myristic, oleic, monohydroxyoleic acids, and two noncrystalline bitter principles or resins. Vidyarthi[263] reported that oil contains resin (2%), myristic (7.4%), palmitic (7%), stearic (5.9%), oleic (5.7%), linoleic (9.6%), and vernolic acid (52.4%).[50]

Seeds are reported to contain an oxygenated acid fraction in the oil, which is identical with 12,13-epoxy-9-octadecenoic acid.[178]

Pharmacological action — Seeds are anthelmintic, tonic, stomachic, and diuretic.[13]

Medicinal properties and uses — Seeds are credited with anthelmintic properties and are effective against threadworms even if their administration is not followed by a purgative. It is also very useful in skin diseases and in scorpion sting.[1]

Doses — Seed powder — 6 to 12 gr; leaf infusion — 14 to 28 ml.

CHENOPODIUM AMBROSIOIDES Linn.   Family: CHENOPODIACEAE

Vernacular names — Sanskrit, Sugandhavastuk; Hindi, Suganhavastuk; English, Mexican tea; Bengali, Bathu Sag; Nepalese, Hyang hamo; Japanese, Amerikaaritasa; German, Scho Kraut; Unani, Bathua; Arabian, Qataf; Persian, Sarmaq.

Habitat — Found in South India, Bengal, Sylhet, Madras, and Bombay.

Parts used — Leaves, fruit, and seeds.

Morphological characteristics — An erect, much-branched herb, 2 to 4 ft high, with aromatic glandular hairs, leaves short, petioled, oblong, or lanceolate; flowers small green, in axillary and terminal panicled leafy spikes; fruits are somewhat glabrous, slightly compressed, with a thin pericarp surrounding the seeds. Seeds are small, brown, smooth, and shining and possess a bitter pungent taste. A volatile oil of medicinal value is found in the glandular hairs, especially of the pericarp of the fruit.

Ayurvedic description — *Rasa — madhur; Guna — laghu, snigdha; Veerya — sheeta; Vipak — katu.*

Action and uses — *Tridoshahar, sotha har, kustaghan, dipan, pachan, anuloman, pitta sarak, krimighan kafanisark, mutral, jwaraghan, sukal, balya.*

Chemical constituents — The fresh vegetable contains moisture (86.59%), and dried material contains ether extract (5.14%), albuminoids (18.18%), nitrogen (2.91%), soluble carbohydrate (59.23%), woody fiber (7.31%), and ash (10.14%). The oil obtained from Indian *C. ambrosioides* yields 40 to 50% ascaridole content. The hydrocarbon fraction of Indian chenopodium oil is *p*-cymene with a small amount of dextrarotatory terpene.[50,178]

Medicinal properties and uses — The plant is anthelmintic and the volatile oil obtained from it is generally employed in medicine. The oil is effective against many forms of intestinal parasites. It has also been found useful in the treatment of amebic dysentery; also useful against roundworm and hookworm.[1]

Doses — Oil — 3 to 15 minims.

CICHORIUM INTYBUS Linn.   *Family:* COMPOSITAE

Vernacular names — Sanskrit, Kasni; Hindi, Kasni; English, Chicory; Bengali, Kasni; Chinese, Ku-T'sail; German, Zichorie; Unani, Kasni; Arabian, Hindaba; Persian, Kasani.

Habitat — Found in northwestern parts of India, also cultivated in Bombay area.

Parts used — Seed, root, and flowers.

Morphological characteristics — A glabrous, erect perennial herb with stout stem, peniculately branched, leaves coriaceous, radicle, and lower sinuate-toothed. Flowers blue, fruit achenes, root tuberous.

Ayurvedic description — *Rasa — tikta; Guna — laghu, rooksha; Veerya — sheeta; Vipak — katu.*

Action and uses — *Kapha pitta har, daah samak, sothhar, nidrajank, dipan, hirdya, rakt sodhak, artwajank, jawarghan.*

Chemical constituents — Seeds yield a bland oil. Chicory is devoid of caffeine and tannins.[264] It gives a characteristic odor on roasting. The volatile matter contains acetaldehyde, acetone, diacetyl, diketopentane, furfuraldehyde, 5-hydroxy-methyl furfuraldehyde maltol, furan, methyl and furfuryl alcohols, and acetic, pyruvic, lactic, pyromucic, and palmitic acids, together with traces of phenol and neutral oil.[265] Nietzki[266] isolated a glycoside cichorin, $C_{32}H_{32}O_{19}$, mp 215 to 220°C, from the flowers. Dutt and Misra[267] obtained 13.8% grayish-white ash on burning the seeds. The ash consisted of 17.5% water-soluble and 82.5% water-insoluble inorganic material containing mainly potassium, sodium (traces), calcium, aluminum sulfates, phosphates, chlorides, carbonates, and silica. It also contained semidrying oil, a mixture of unsaturated oleic and linolic acids, and saturated stearic and palmitic acids. The unsaponifiable matter contained phytosterol, mp 131 to 133°C. On analysis, the fresh root showed water (77.00%), gummy matter (7.5%), glucose (1.1%), bitter extractive (40%), fat (0.6%), cellulose, inulin, and fiber (9.0%), and ash (0.8%). The ash of the root and leaves is rich in potash. The bitter principle is a glycoside of fructose and pyrocatechuic acid. The juice of the roots is reported to contain a stearin, mannite, and tartaric acid. Betaine and choline are also present in small quantities. During storage of chicory roots, inulin is partially converted to inulide and fructose, indicating the presence of an enzyme. Another enzyme, inulocoagulase, which coagulates inulin in the expressed juice of the roots, is also reported to be present.[265]

Pharmacological action — Digestive, stomachic, emmenagogue,[50] and alexiteric.

Medicinal properties and uses — Seeds in the form of decoction or powder are used in obstructed or disordered menstruation. Root is used as an adulterant or substitute for coffee. Infusion is useful in obstructions of the liver and in checking bilious enlargement of the spleen with general dropsy.

Doses — Seed decoction — 28 to 56 ml; fluid extract of root — 5 to 10 ml; root powder — 3 to 6 gr.

CINNAMOMUM CAMPHORA Nees and Eberm. Family: LAURACEAE

Vernacular names — Sanskrit, Karpoor; Hindi, Kapur; English, Camphor; Bengali, Kapur; Nepalese, Kapoor; Sinhalese, Vandu Kapur; Japanese, Kusunoki; Chinese, Chang; German, Kampher; Unani, Kafoor; Arabian, Kaefoor; Persian, Kapoor; French, Camphre; Burmese, Payo; Malaysian, Karpooram.

Habitat — A large, handsome, evergreen tree, native of China, Japan, and Formosa, and introduced into and cultivated in many other countries including India at Dehradun, Saharanpur, Calcutta, Nilgiris, and Mysore.

Parts used — The concrete volatile oil, i.e., camphor obtained by distillation of the wood of the plant and purified by sublimation.

Pharmacognostical characteristics — The tree attains a height of about 100 ft with a girth of 6 to 20 ft. The leaves are leathery, shining, 2 to 4 in. long, and aromatic. The fruits are dark green, ovoid, rather dry, globose, and about 0.3 in. in diameter When ripe they turn black. The odor of bruised leaves of the camphor-yielding plant resembles that of camphor. Camphor is formed in the oil cells distributed in all parts of plant. These cells begin to form early in the growth organs, and are filled with yellow oil from which camphor is slowly deposited. The formation of camphor is brought about through the agency of an enzyme present in the growing parts of the tree particularly in the tissue within the cambium region. Each layer of wood, as it is formed, is enriched by camphor.[1]

Ayurvedic description — *Rasa — tikta, katu, madhur; Guna — lagu, teekshna; Veerya — sheeta; Vipak — katu.*

Action and uses — *Tridosha har, badana, samak, achapa har, kash har, kantya swadajank, baji karan, chashya, dant sodhak.*

Chemical constituents — Camphor treated with chloride of zinc and distilled is converted into cymene, a substance contained in many essential oils. When it is treated with nitric acid it oxidizes and forms camphoric acid, a crystalline body, odorless and soluble in alcohol, ether, and fatty oil. All parts of *C. camphora* on distillation yield a semisolid oil from which camphor can be separated by a mechanical process. Seed fat contains glycerides of lauric, capric, and oleic acid.[178]

Pharmacological action — Camphor is a local irritant with a benumbing influence upon the peripheral sensory nerve. Camphor is diaphoretic, stimulant, antiseptic, antispasmodic, internally expectorant, sedative, temporary aphrodisiac, narcotic, internally carminative, and externally anodyne, carminative.[50]

Medicinal properties and uses — Camphor is esteemed as an analeptic in various cardiac depressions and has been used in the treatment of myocarditis. In doses of 0.2 gr, it is very useful in hysteria and nervousness and is used in the treament of serious diarrhea. It is also employed in external application as an counterirritant in the treatment of muscular strains, rheumatic conditions, and inflammations. In combination with menthol or phenol it relieves itching of the skin. It is good in typhus, confluent smallpox, and all fevers and eruptions of the typhoid class; also in febrile delirium, whooping cough, hiccup, spasmodic asthma, dysmenorrhea, puerperal mania, chorea, epilepsy, atonic gout, melancholia, toothache, and chronic bronchitis. In uterine pains, the liniment of camphor is rubbed on the abdomen.[1]

Doses — 0.2 — 0.3 gr.

CINNAMOMUM TAMALA Nees and Eberm.   Family: LAURACEAE

Vernacular names — Sanskrit, Tejpatra; Hindi, Tejpat; English, Cassia cinnamon; Bengali, Tejpat; Nepalese, Tejpat; Sinhalese, Tejpatra; Japanese, Tamara nikkei; German, Zimtbaum; French, Cannelle; Unani, Tezpat; Arabian, Sazaj hindi; Persian, Sazaj hind; Burmese, Thitchubo; Tamil, Perialavangapallai.

Habitat — In tropical and subtropical Himalayas, Khasi and Jaintia Hills, and Bangladesh.

Parts used — Leaf, bark, and oil.

Morphological characteristics — Moderately sized evergreen tree. Leaves glabrous, usually 10 to 13 cm long, very variable in breadth, rarely alternate, shiny above, leathery, rarely elliptical and obtuse, 3-nerved from the base; flowers unisexual, numerous, 0.5 to 0.6 cm long; fruit 1.25 cm long, penducle and calyx small; drupe ovoid.

Ayurvedic description — *Rasa — katu, tikta, madhur; Guna — laghu, rooksha, teekshna; Veerya — ushna; Vipak — katu.*

Action and uses — *Kapha vatta samak, pitta wardhak, pitta samak, utegek, lakhan, dipan, pachan, vattanulomon, grahi, rakt sodhak, vajekarn.*

Chemical constituents — The leaves yield an essential oil which is soluble in 1.2 vol of 70% alcohol. The oil resembles cinnamon leaf oil and contains d-α-phellandrene and 78% eugenol.[50] The essential oil from the bark is pale yellow and contains 70 to 85% aldehyde.[1]

Pharmacological action — Carminative, stimulant, diuretic, diaphoretic, deobstruent, and lactagogue. Oil is a powerful stimulant.

Medicinal properties and uses — It is used in anorexia, bladder disorders, dryness of mouth, coryza, diarrhea, nausea, and spermatorrhea. The bark in the form of infusion, decoction, or powder is prescribed in bowel complaints such as dyspepsia, flatulence, and vomiting.

Doses — Leaf powder — 1.5 to 3 g.

CINNAMOMUM ZEYLANICUM Breyn.
(Syn. *CINNAMOMUM CASSIA* Blume)

Family: LAURACEAE

Vernacular names — Sanskrit, Twak; Hindi, Dalchini; English, Cinnamon; Bengali, Daruchini; Nepalese, Newari-Dalchini; Sinhalese, Kurandu; Japanese, Nikkei; German, Zimt; Unani, Darchini; Arabian, Darasini; Persian, Darchini; Burmese, Timboti kyoboi; Malaysian, Kulit manis; French, Cannelle; Tamil, Pattai.

Habitat — Indigenous to Ceylon, southern India, and growing wild in Western Ghats from Konkan southward and in the forests of Burma.

Parts used — Dried inner bark of the shoots from truncated stalks and essential oil.

Pharmacognostical characteristics — The leaves are 4 to 7 in. long, leathery and shining, green on the upper surface when mature. They have a spicy odor when bruised and a "hot" taste. The flowers have a fetid, disagreeable smell. The fruit is dark purple and one seeded. It has a terebinthine odor when opened, and is somewhat similar to that of the juniper berry. The bark of the tender shoots and stems is smooth and pale, while that of old aged branches is rough and brown. Thickness of the bark is 1 mm or less, taste strongly pungent, aromatic, and sweet. Cork cells present in thin-walled, tangentially elongated phellogen are not clear. A layer of stone cells is continuous, and pericyclic fibers are present in groups in the outside cell layer.

Ayurvedic description — *Rasa* — katu, tikta, madhur; *Guna* — laghu, rooksha, teekshna; *Veerya* — ushana; *Vipak* — katu.

Action and uses — *Kaf bat samak, pit bardhak, badana nasak, dipan, pachan, hridya, yachama nasak, grahi, garvasay san kochak, bajkaran.*

Chemical constituents — Cinnamon bark oil contains cinnamaldehyde (60 to 75%); euginol, benzaldehyde, methyl amyl ketone, phellandrene; pinene, cymene, nonyl aldehyde, linalool cumic aldehyde, carophyllene, and ester of isobutyric acid.[265] The British Pharmacopoeia prescribed 50 to 65% cinnamic aldehyde content. Green leaves yield dark-colored oil on distillation, which differs from cinnamon bark oil. It contains 70 to 80% of eugenol with traces of cinnamic aldehyde, pinene, and linalol.

Pharmacological action — Bark is carminative, antispasmodic, aromatic, stimulant, hemostatic, astringent, antiseptic, stomachic, and germicide. Oil is a vascular and nervine stimulant; in large doses it is an irritant and narcotic poison.

Medicinal properties and uses — Infusion, decoction, or powder of the bark is effective in bowel complaints such as dyspepsia, flatulency, diarrhea, and vomiting. It is frequently employed as an adjunct to bitter tonic and purgatives. As a stimulant of the uterine muscular fiber it is employed in menorrhagia and protracted labor due to defective uterine contractions. The crystalline cinnamic acid is antitubercular and is used as injection in phthisis.[1,3]

Doses — Powder cinnamon — 0.5 to 1.5 g; oil — 5 to 15 minims.

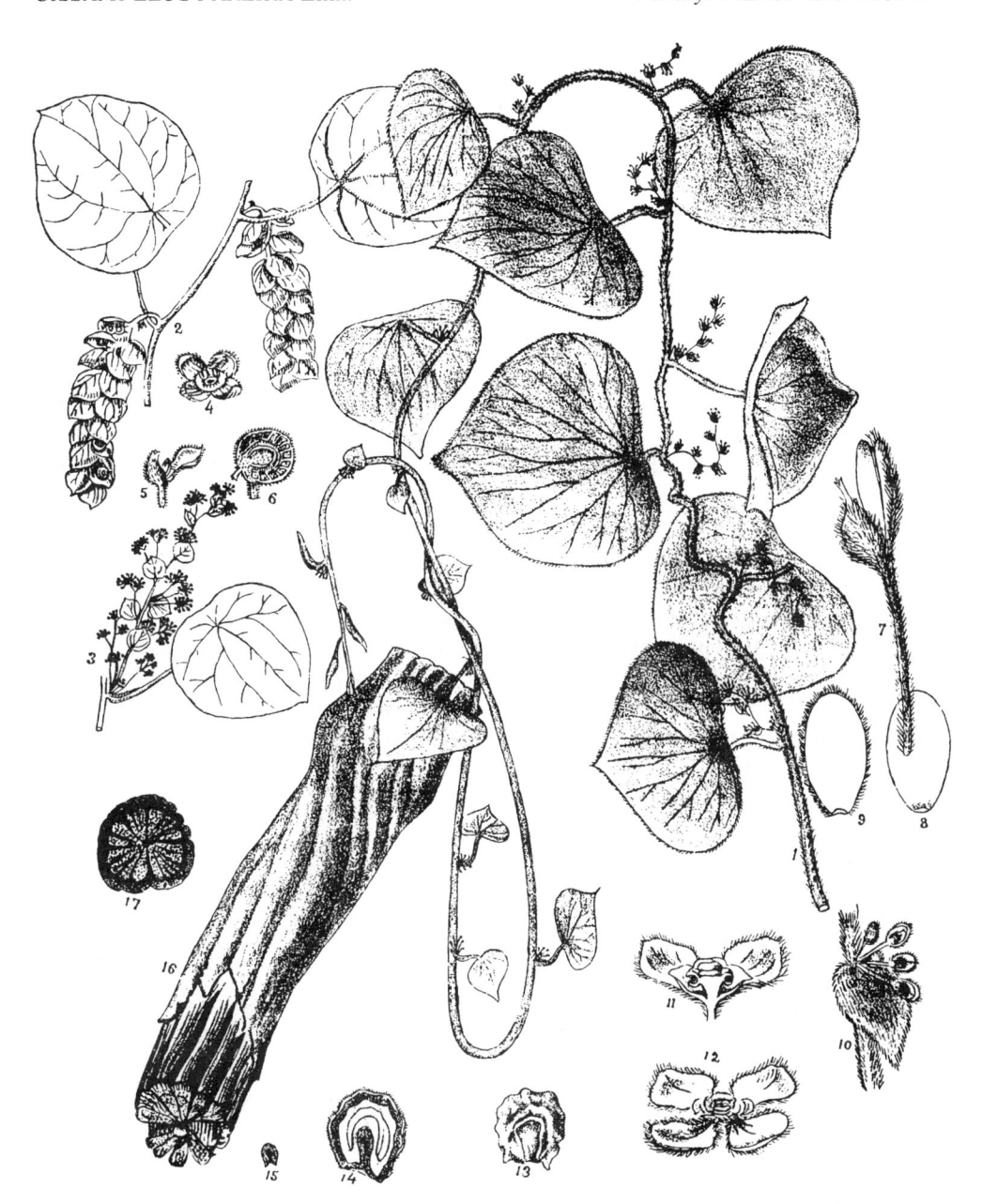

CISSAMPELOS PAREIRA Linn. *Family:* MENISPERMACEAE

Vernacular names — Sanskrit, Laghu Patha; Hindi, Harjori; English, Velvet leaf; Bengali, Nirbisi; Nepalese, Paha; Sinhalese, Diyamitta; Japanese, Pareira; German, Talsche Pareivawurzel; Unani, Harjor; Tibetan, Pa-tha; Malaysian, Akarmumpanang.

Habitat — Grows in tropical and subtropical parts of India, from Sind and the Punjab to South India and Sri Lanka.

Parts used — Root, bark, and leaves.

Pharmacognostical characteristics — It is a climbing shrub, leaves alternate, globular, cordate, or kidney shaped. Flowers greenish, dioecious; male inflorescence hairy, many flowered, cyme 1 to 2 in. long; female inflorescence a raceme with large kidney-shaped or glabular bracts. Drupes globose, hairy, scarlet, very small.

Ayurvedic description — *Rasa* — *tikta; Guna* — *laghu, teeshna; Veerya* — *ushna; Vipak* — *katu*.

Action and uses — *Tridos samak, brana ropan, bisoaghana, dipan, pachan, anuloman, grahi, krimighan, rakt sodhak, shoth har, dah prasman, stanya sodhan, visaghna, balya*.

Chemical constituents — The following alkaloids have been separated and crystallized from *C. pareira*.[268]

1. Cycleanine, $C_{38}H_{42}N_2O_6$, mp 272 to 273°C
2. Hayatidin, $C_{39}H_{44}N_2O_7$, mp 179 to 180°C
3. Hayatinin, $C_{37}H_{40}N_2O_6$, mp 231 to 232°C
4. Hayatin, mp 265 to 268°C
5. Orange base
6. A number of water-soluble bases, viz., (1) menismin iodide, (2) cissamin chloride, (3) parerin chloride

Pharmacological action — Mild stomachic, bitter tonic, diuretic, and antilithic. Hayatin and its various derivatives were found to possess neuromuscular blocking action. Low doses of hayatin methiodide had little effect on blood pressure, but higher doses caused a fall in intact spinal dogs, rabbits, and cats. The drug had little action on heart, isolated or *in situ*.[269] Hayatin methiodide and methochloride possessed almost an equal degree of curariform activity as compared to *d*-tubocurarine chloride.[270] Hayatin methochloride possessed similar muscle-relaxant properties to *d*-tubocurarine chloride in cats, dogs, mice, and rabbits.[271]

Medicinal properties and uses — The dried roots are stomachic, alterative, and astringent; they are given with some aromatics in bowel complaints, dyspepsia, diarrhea, prolapse of the uterus, internal inflammations, and urinary diseases. Decoction of the root is used as a blood purifier in skin diseases and syphilis.

Doses — Decoction — 28 to 56 ml; root powder — 0.5 to 1.5 g; liquid extract — 2 to 8 ml.

CISSUS QUADRANGULARIS Wall.   Family: VITACEAE
(Syn. *VITIS QUADRANGULARIS* Wall.;
HELIOTROPIUM INDICUM Linn.)

Vernacular names — Sanskrit, Asthisanhari; Hindi, Harjora; English, Bone setter; Bengali, Hasijora; Nepalese, Hadachud; Sinhalese, Hirassa; Chinese, Dixanh young; Unani, Hadjor.

Habitat — Found throughout the hotter parts of India and Sri Lanka.

Parts used — Leaves and young shoots, roots.

Morphological characteristics — A long, slender climber, with membranous leaves. Flowers small, greenish white; fruits round, about 6 mm in size, red when ripe containing one seed. Fresh stem is thick, fleshy, succulent, quadrangular with four-angled internodes and contracted nodes. Fracture is short and fibrous and fractured surface is greenish yellow in color; odor is distinct with acrid taste. A transection of an internode shows a four-angled outline. Epidermis of single row of cells is covered with thick cuticle and tubular cells. Numerous stomata are found on the epidermis. Cortex is composed of thin-walled parenchymatous cells. Some cells show chloroplasts, starch grains, or raphides of calcium oxalate. Secretory cells are seen embedded in cortex. Three to four layers of sclerenchymatous cells are seen at the corners internal to epidermis. Three to four layers of cork cells are seen next. Vascular bundles are called collateral, open, and arranged in a ring around a central pith.[272]

Ayurvedic description — *Rasa — madhur; Guna — laghu, rooksha; Veerya — ushna; Vipak — amala.*

Action and uses — *Kapha vat samak, pitta-wardhak, sthamavan, sandhaniya, dipan, pachan, anuloman, krimighan rakt sodhak, rakta sthmavak, veerisy.*

Chemical constituents — Dry plants contain moisture (13.1%), proteins (12.8%), fat and wax (1.0%), carbohydrate (36.6%), ash (18.2%), mucilages and pectin (1.2%). A yellow wax and tartaric acid as the acid potassium salt are present. The plant contains a large amount of vitamin C (398 mg/100 g). The plant has a steroidal principle, mp 134 to 136°C, which is further separated into two principles I and II with molecular formula $C_{27}H_{45}O$ and $C_{23}H_{41}O$, respectively.[273] The plant is reported to contain 3-ketosteroid, a steroidal principle.[178]

Pharmacological action — Alterative and stomachic. Subbu[274-276] found the drug acts on cell membranes by inhibiting the permeability of calcium ions into the cell substance. The total extract was found to hasten fracture healing by reducing the total convalescence period by 33% in experimental rats and dogs.[277] Prasad and Udupa[278] reported that the plant extract has been found to influence fracture healing indirectly only. The plant has exhibited acetylcholine-like action, analogous to muscarine and nicotine. In atropinized hearts, the drug showed cardiotonic activity.[178]

Medicinal properties and uses — Juice of the stem is used in irregular menstruation and scurvy. The stem is given internally and applied topically for fracture of bones. Stem beaten in to a paste is given in asthma. Powdered root is used for the fracture of bones with the same effect as plaster.[1,3] The total ash obtained from young shoots and leaves is given in dyspepsia, indigestion, and bowel complaints. The juice of the stem is said to be useful in otorrhea and epistaxis. The stem prepared in boiling lime water is a useful stomachic.

Doses — Infusion — 1 to 2 ml.

CITRULLUS COLOCYNTHIS Schrad. Family: CUCURBITACEAE

Vernacular names — Sanskrit, Indravaruni; Hindi, Indrayan; English, Bitter apple; Bengali, Rakhalsa; Nepalese, Endrayani; Sinhalese, A Gonkekiri; Japanese, Koroshinto; Chinese, Hsikua; German, Koloquinte; French, Conchomlere amer; Unani, Indrayan; Arabian, Hanzal; Persian, Kharpazah; Tibetan, Bi-sa-ia; Burmese, Kaya-si; Malaysian, Paikummetu; Tamil, Attuthmalti.

Habitat — Found wild in wastelands almost throughout India, particularly central and southern India, and on the Coromandal coast. Colocynth is not systematically cultivated anywhere in India.

Parts used — Fruit deprived of its rind, root, dried pulp of the fruit freed from seeds.

Pharmacognostical characteristics — A prostrate scabrous herb with perennial root and simple or 2-fid tendrils. Leaves $1^1/_2$ to $2^1/_2 \times 1$ to 2 in., pale green above, ashy beneath. Three to four lobed. Corolla $1/_4$ in. long, pale yellow. Fruit globular, slightly depressed, 2 to 3 in. diameter, variegated green and white, glabrous when ripe, filled with dry, spongy, very bitter pulp; epicarp thin. Seed $1/_6$ to $1/_4$ in. long, pale brown, not margined.[28]

Ayurvedic description — *Rasa* — *tikta; Guna* — *laghu, rooksha, teekshana; Veerya* — *ushna; Vipak* — *katu.*

Action and uses — *Kaf pita har, rakt sodhak, brana-sodhan, sothahar, keshya, prameha-nashak, garbhashyaya-sankochak, jawar-ghana, vishaghana.*

Chemical constituents — The pulp of the fruit contains a bitter substance colocynthin, and colocythetin.[50] The fruit juice shows the presence of α-elatrine, citrulluin, citrullene, and citrulluric acid. Fresh mesocarp of fruit and seeds contain glucose and α-pinosterol and crystalline bitters. *p*-Hydroxybenzyl methyl ester from unripe fruit has been isolated. Bitter oil "citbittol" has been isolated from the peel-free flesh of ripe fruit.[178]

Roots are reported to contain α-elatrin, hentriacontane, and saponins. Seeds contain fixed oil, a phytosterolin, two phytosterols, two hydrocarbons, a saponin, alkaloid, glycoside, tannin.[50]

Pharmacological action — It is a drastic hydrogogue, cathartic, and diuretic; in large doses it is emetic and a gastrointestinal irritant; in small doses it is expectorant and alterative. Coloside A, a glycoside isolated from the pulp of the fruit, showed antihistaminic and anticetylcholine-like activity on isolated rabbit intestine and guinea pig ileum. It failed to stimulate the isolated rat uterus but revealed purgative properties.[279] Trease and Evans[69] reported that activity of colocynth is contained in an ether-chloroform soluble resin. Among the several components of the pulp is cucurbitacin E or α-elaterin in the form of glycoside, which is reported to have tumor-necrosing activity. Choline and two alkaloids have also been reported. Being a powerful cathartic, it is not now used in standard medicine. There is, however, considerable interest in its constituents due to their necrotic activity.

Medicinal properties and uses — The pulp of colocynth in large doses causes violent griping, prostration, and sometimes bloody discharges. Even in moderate doses it is seldom prescribed except as an adjuvant to other cathartics. In the form of solid extract, it enters into many of the purgative pills of modern pharmacy. The roots of the plant also possess purgative properties and are used in ascites, jaundice, urinary diseases, and rheumatism. Internal pulp only of the dried, peeled fruit is official in the British Pharmacopoeia; it is useful in biliousness, fever, intestinal parasites, constipation, hepatic and abdominal, visceral and also cerebral congestion, and dropsy. Oil from the seed is used for snakebites, scorpion-stings, any bowel complaints, epilepsy, and also for the growth and blackening of the hair.[1]

In small doses, it is useful in colic, neuralgia, and sciatica and also to relieve pain of the glaucoma. In rheumatism equal parts of the root and long pepper are given in pill form. A paste of the root is applied to the enlarged abdomen of children. The powder is often used as an insecticide. It should be avoided in pregnancy and in irritable conditions of the intestinal canal.

Doses — Root powder — 0.2 to 0.4 gr; fruit powder — 0.1 to 0.4 gr; infusion of root — 3 to 10 ml.

CLERODENDRUM INFORTUNATUM Linn.

Family: VERBENACEAE

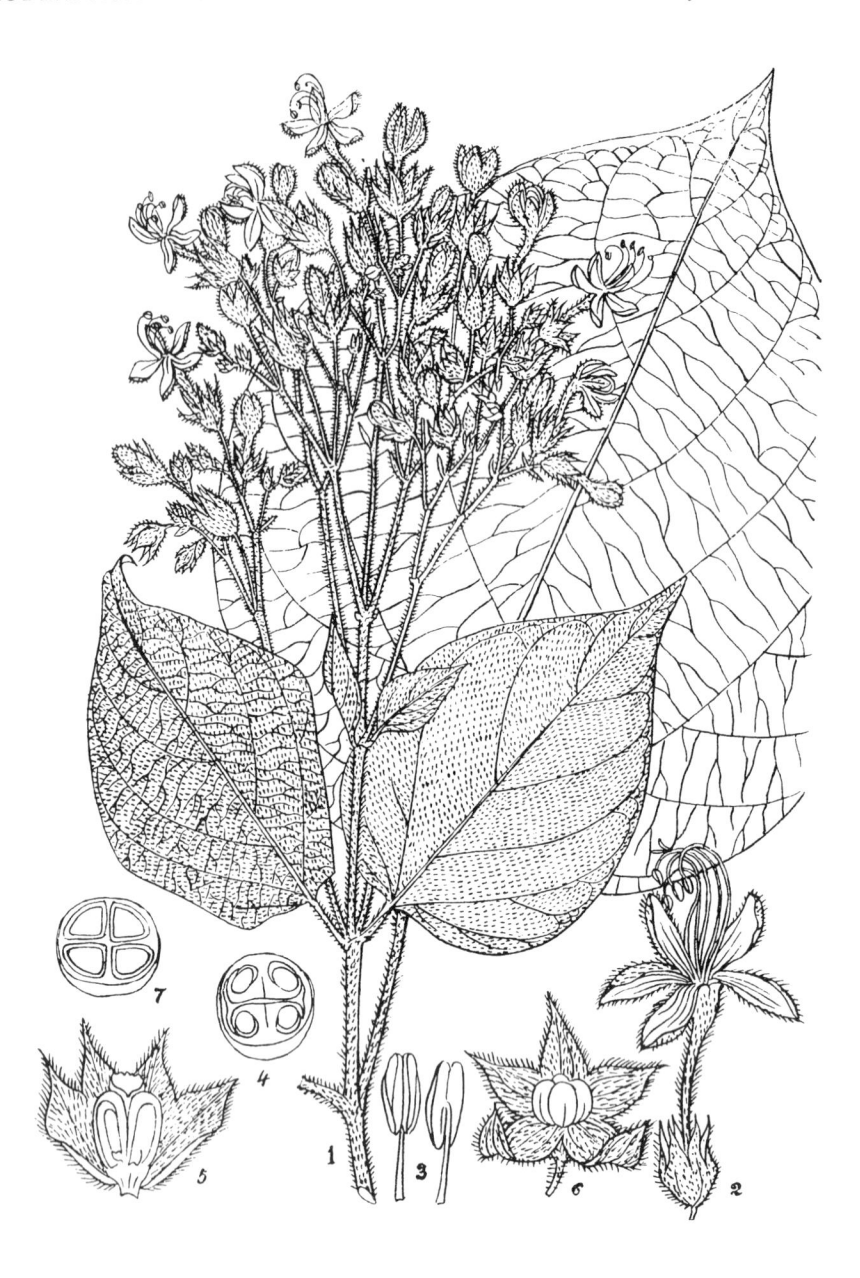

Vernacular names — Sanskrit, Bhandira; Hindi, Bhant; Bengali, Bhat; Nepalese, Chitu; Sinhalese, Caspinna; Unani, Bhant; Burmese, Kivalamon.

Habitat — Throughout India.

Parts used — Leaves and root.

Morphological characteristics — A shrub with pinkish-white flowers, grows commonly in wastelands. Leaves round, ovate to oblong; hairy, up to 10 in. long, 8 in. broad. Flowers in large, terminal, erect panicles, calyx persistent; corolla white or tinged red; drupe black, enclosed wholly by the enlarged, red calyx.

Ayurvedic description — *Rasa — tikta, katu, kasaya, madhur; Guna — rooksha, laghu; Veerya — ushna; Vipak — katu.*

Action and uses — *Anuloman, pitta sark, krimighan, jwaraghana.*

Chemical constituents — Analysis of the leaves showed ash (8.0%), protein (21.2%), crude fiber (14.8%), reducing sugar (3.0%), and total sugar (17.0%). From the petroleum ether extract (3.85%) of the air-dried leaf powder a bitter principle, clerodin, $C_{13}H_{18}O_3$, was isolated.[50,178] Leaves also contain a fixed oil which consists of glycerides of lenolenic, oleic, stearic, and lignoceric acid.[280] The leaves also contain proteinase and peptidase.[281] With hot petroleum and benzene, the roots yield lupeol and β-sitosterol, successively.[178]

Pharmacological action — Bitter tonic, antiperiodic, vermifuge, laxative, and cholagogue.

Medicinal properties and uses — The leaves and root are used externally for tumors and certain skin diseases. Fresh leaf juice is used as an injection into the rectum for ascarids. The root relieves congestion and torpidity of the bowels.[1,3]

Doses — Root powder — 0.5 to 1 gr; leaf juice — 28 to 56 ml.

CLITORIA TERNATEA Linn.   *Family:* PAPILIONACEAE

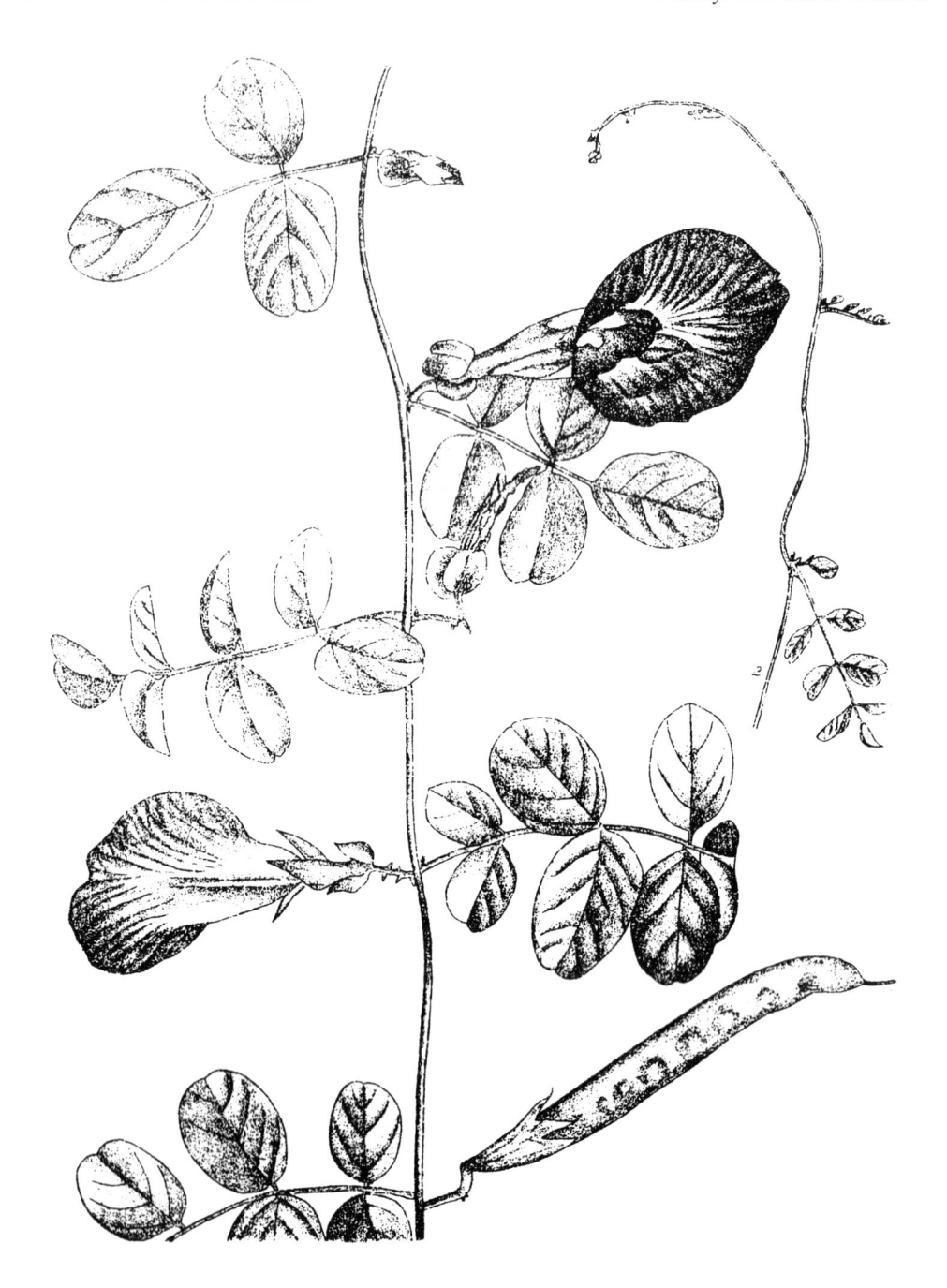

Vernacular names — Sanskrit, Aparajita; Hindi, Aparajita; English, Butterfly pea; Bengali, Nila aparajita; Nepalese, Aparajita; Sinhalese, Katarodu; Japanese, Chomama; Tibetan, A-paradzi-ta; Unani, Heyat; Arabian, Buzrula; Persian, Tukhm-i-bikhe-hyata; French, Clitore-deternate; Malaysian, Aral; Tamil, Kakkanam.

Habitat — *C. ternatea* is a common garden plant found all over India, especially in southern India, and bears blue and white flowers.

Parts used — Root, bark, seeds, and leaves.

Morphological characteristics — It is a beautiful climber. Leaves pinnate, leaflets 5 to 7, ovate. Flowers solitary, blue or white. Pods linear, oblong, straight, thin; many seeds; globose or compressed.

Ayurvedic description — *Rasa — kashaaya, katu, tikta; Guna — laghu; Veerya — sheeta; Vipaka — katu.*

Action and uses — *Kanthya, budhiprada, medihakara, chakshushya, mootradoshara, jawarghana, vishaghana, unmadahara, udar rog har.*

Chemical constituents — Root bark contains starch, tannin, and resins. The seed contains a fixed oil and a bitter resinous principle. The analytical values for dry matter, digestible proteins, total digestible nutrients, and starch are, respectively, 89.41, 10.98, 52.97, 38.81, and 13.8 %. The seeds also yield a blue dye. The blue dye from the corollas of the flowers can be used as a substitute for litmus, besides taraxerol from the ether extract of the roots.[280] Another product, identified as taraxerone a crystalline substance with the molecular formula $C_{30}H_{48}O$, mp 239 to 241°C, has also been isolated by Banerjee and Chakravarti.[281] Both seeds and root bark contain tannin.[282] Seeds are reported to have given positive tests for nucleoprotein which show the presence of leucine, isoeleucine, valine, alanine, glycine, arginine, glutamic acid, aspartic acid, and tyrosine on hydrolysis.

The leaves also yield an 8-lactone, aparajitin. The blue and white varieties contain an ester and resin glycoside.[178]

Pharmacological action — Fresh root has a bitter taste; it is aperient and diuretic. Seeds have a powerful cathartic action like Jalap. Root bark is demulcent, diuretic, laxative, and vermifuge.

Medicinal properties and uses — Nadkarni[3] reported that fresh root has an acrid, bitter taste and is aperient, laxative, and diuretic. Seeds have powerful cathartic action. Root bark is demulcent and also laxative. Powder of the seeds mixed with ginger is laxative and is given to check excessive perspiration of hectic fever. The infusion of the leaves is a very useful wash for ulcers. An alcoholic extract of the root in doses of 0.6 gr is useful in ascites and enlargement of the abdominal viscera. For relief of hemicrania, the juice of the white flowers is used. Infusion of the root bark is also used in gonorrhea and irritation of the bladder and urethra. In the Philippine Islands, roots are used in poultices for swollen joint. It is also a valuable tonic given in general and seminal debility and urinary affections.

Doses — Infusion — 28 to 56 ml; powder root — 0.6 to 2 gr; alcoholic extract — 0.3 to 0.6 gr.

COCCINIA INDICA Wight and Arun   *Family:* CUCURBITACEAE
(Syn. *C. GRANDIS* [Linn.] Voight;
     C. CORDIFOLIA Cogn.)

Vernacular names — Sanskrit, Bimba; Hindi, Kandurikibel; Bengali, Tela kucha; Nepalese, Kotusi; Sinhalese, Kowakka; Unani, Kebar; Arabian and Persian, Kebar; Burmese, Kenbung; Malaysian, Pepasan; Tamil, Kovai.

Habitat — Grows wild abundantly in Bengal and in most parts of India.

Parts used — Leaves, root, fruit, and bark.

Morphological characteristics — Root long tuberous; fruits 1 to 2 in. long and $1/_2$ to 1 in. in diameter. The fruit is smooth and bright green with white stripes when immature, becoming bright scarlet when ripe.

Pharmacognostical characteristics — Root — The fresh taproot is thick, tuberous, long, tapering, tortuous, with few fibrous rootlets attached. The soft flexible roots break with fibrous fracture. In transection the root shows parenchyma full of starch grains. Phelloderm consists of three to four rows of cells. A few pericylic fibers and stone cells are present. Occasionally, tyloses is found in some vessels. Powder of root is yellowish white in color. Leaf — The leaves are studded with numerous small round disks or papillae above. They are palmately five-nerved from a cordate base. In transection the lamina shows upper epidermis covered with cuticle and having a layer of polygonal cells with wavy walls. Numerous stalkless glands with bicellular heads are found in the depressions of the upper epidermis. On the lower surface there are numerous stomata of the ranunculaceous type. Uniseriate multicellular trichomes occur around the margin of the leaf.[283]

Ayurvedic description — *Rasa* — *tikta, kasaya; Guna* — *laghu, rooksha, teekshna; Veerya* — *ushna; Vipak* — *katu.*

Action and uses — *Kapha pitta har, bran ropan, sotha har, dipan, bawan, rachak, rakt sodhak, kapha nisark, jawarghana, swedagan.*

Chemical constituents — Fruits contain moisture (93.1%), protein (1.2%), fat (0.1%), fiber (1.6%), carbohydrates (3.5%), mineral matter (0.5%), calcium (0.04%), phosphorous (0.03%), iron (1.4 mg/100 g), carotene evaluated as vitamin A (260 IU/100 g), vitamin C (28 mg/100 g).[82] The plant contains enzymes, hormones, and traces of inert alkaloid. The juice shows the presence of an amylase.[50] Roots contain lupeol acetate, β-amyrin acetate, and β-sitosterol. Young fruits of the bitter variety contain lupeol and β-amyrin, as well as cucurbitacin β-glycoside.[178]

Pharmacological action — Alterative. Dried bark is cathartic; leaves and stem are antispasmodic and expectorant. The plant reduces blood sugar in patients suffering from diabetes mellitus. Gupta[284] observed that water-soluble fraction of the alcoholic extract of the root had hypoglycemic properties on alloxan diabetic rabbits. Ethanolic and aqueous extracts showed hypoglycemic principles which are reported to be orally active and comparable with tolbutamide.[178] The plant extract caused appreciable inhibition of the hypoglycemic response of the anterior pituitary extract in glucose-fed albino rats.[178] The plants possess antiprotozoal properties against *Entamoeba histolytica.*[11]

Medicinal properties and uses — Fresh juice from tuberous roots is given either alone or in combination with some metallic preparations in early cases of diabetes, intermittent glycosuria, enlarged glands, and in skin diseases. Leaves are also applied to skin eruptions such as those of small pox; and the plant is generally used as a tincture internally in gonorrhea. Decoction of the leaves and stem is useful in bronchial catarrh and bronchitis.[1,3]

Doses — Infusion — 10 to 20 ml; root powder — 3 to 6 gr.

COCOS NUCIFERA Linn.   *Family:* PALMAE

Vernacular names — Sanskrit, Narikela; Hindi, Nariyal; English, Coconut plant; Bengali, Narikal; Nepalese, Neikyaa; Sinhalese, Pol; Japanese, Yashi; German, Echte Kokospalme; French, Coctier; Unani, Narial; Arabian, Narjeel; Persian, Nargeel; Burmese, Ondi; Tamil, Tengumaram.

Habitat — Cultivated extensively in South India, Bengal, Malabar, and the Coromandal coast.

Parts used — Flowers, root, fruit, oil, and ash.

Morphological characteristics — Wood hard, red outside, reddish brown but softer inside; leaves pinnate, 3.5 to 4.67 m; leaflet 61 to 91 × 5 cm linear, lanceolate, acuminate, bright green, smooth, shining; petiole 91 to 152 cm stout. Spathe 45 to 61 cm, narrowly oblong, tapering at both ends, springing from within the leaves; spadix about 46 cm when in flower, extending to 91 to 122 cm when in fruit, monoecious, divided into numerous drooping spikes, bearing at base; female with few male flowers, upper parts being densely covered with male flowers; flowers straw colored, male 1.9 cm, female 2.5 cm. Fruit yellowish green, ovoid, drupe 23 to 38 cm, fibrous, trigonous; endocarp hard ovoid, 10 to 12 cm diameter, endosperm oily, mucilagenous with high glucose content.[28]

Ayurvedic description — *Rasa — madhur; Guna — guru, snigdha; Veerya — sheeta; Vipak — madur.*

Action and uses — *Tridosha samak, daah samak, khustaghan, bran ropan, pitta samak, anuloman, shul prasaman, agni depan, rakt pitta samak, mutral, vajikarn, pitta samak, balya.*

Chemical constituents — Coconut milk of the green fruit contains histidine, arginine, lysine, tyrosine, tryptophan, proleine, leucine, and alanine. Oil content ranges from 57 to 75%, consisting of lauric, myristic, and fatty acids. It has mixed glycerides, phytosterol, and squalene. Coconut water contains the vitamin B group. A mannan from the kernel and water-soluble galactamannan have been reported.[178]

Pharmacological action — Refrigerant, diuretic, antiseptic, and laxative.

Medicinal properties and uses — The ash of the bark of the stem is an antiseptic and dentifrice; it is used for scabies and toothache. The root is a valuable diuretic; it is given in uterine diseases, gleet, bronchitis, liver complaints, dysentery, and blenorrhagia in the form of decoction; the young roots are astringent; their infusion or decoction is used as a gargle in sore throat. The freshly tapped juice from the flower stalk is refrigerant and diuretic; the fermented juice is laxative.[1,3]

The young nuts are astringent; they are useful in sore throats of children; the liquid inside these tender fruits is refrigerant; it is given in thirst, fever, and urinary disorders; it is a blood purifier and also checks vomiting. The milky juice expressed from the pulp of the immature nut is nutritive, anthelmintic, and diuretic; it is given in malnutrition, general debility, phthisis, thirst, and fevers; the meat of the nut taken with castor oil is vermifuge, especially for tapeworm. The oil extracted from the flesh of the ripe nut is used as a substitute for codliver oil. The oil is an effective dressing for burns and scalds and is an efficacious hair oil. The tarlike fluid obtained from the red-hot shell of a ripe nut is a rubefacient; it is a household remedy for ringworms, itch, and other skin diseases.[1]

Doses — Fruit — 25 to 35 gr; oil — 1 to 1.5 ml.

COLCHICUM LUTEUM Baker
Family: LILIACEAE

Vernacular names — Sanskrit, Hiranyatutha; Hindi, Harantutiya; English, Golden Collyrium; Nepalese, Nilotutho; German, Gelbe Herbstzeitlose; Unani, Suranjanitalkh; Arabian, Suranjan; Persian, Suranjan.

Habitat — Found in the temperate western Himalayas from Kashmir to Chamba at altitudes of 3000 to 8000 ft, usually in open grassy places.

Parts used — Corms and seeds.

Morphological characteristics — The corms are somewhat conical or broadly ovoid or elongated and planoconvex in section, brownish to brownish gray in color, and are either translucent or opaque. The flat side is longitudinally grooved. The fresh corm is 15 to 35 mm in length and 10 to 20 mm in diameter. Leaves few, lorate, linear oblong or obtuse, appearing with flowers, golden yellow, 1 to 2. The plant flowers soon after the snow melts at higher altitudes.[28]

Ayurvedic description — *Rasa — tikta; Guna — laghu, rooksha; Veerya — ushna; Vipak — katu.*

Action and use — *Kapha vatta samak, badanasthapan, bran sodhan, ropan, vatta samak, madak, awasadak, dipan, pitta sark, vamak, rakt sodhak, vajikarana, mutral, kustaghan, balya.*

Chemical constituents — Indian colchicum corms contain abundant starch and alkaloid, colchicine 0.21 to 0.25% of dried corm. The seeds contain 0.41 to 0.43% alkaloid. Colchicine, $C_{22}H_{25}O_6$, occurs in the form of yellow flakes, crystals, or as a whitish-yellow amorphous powder, which darkens on exposure to light.[69,285]

Pharmacological action — Alterative.[3]

Medicinal properties and uses — It is used as a carminative, laxative, aphrodisiac, alterative, and aperient, and is given for gout, rheumatism, and diseases of the liver and spleen. Externally it is applied in paste to lessen inflammation and pain.[1]

Doses — Powder — 125 to 300 mg; extract — 25 to 50 mg.

COMMIPHORA MUKUL (Hooker, Stedor) Engl.   *Family:* BURSERACEAE
(Syn. *BALSAMODENDRON MUKUL* Hook.)

Vernacular names — Sanskrit, Goggulu; Hindi, Gugal; English, Gum gugal; Bengali, Muku; Nepalese, Gungu; Sinhalese, Gugal; German, Myrrhe; Unani, Muquil; Arabian, Muquila-raque; Persian, Bajahundana; Tibetan, Gu-Gu-La.

Habitat — The tree grows in Sind, Rajastan, Bangladesh, Hyderabad, Assam, Madhya Pradesh, and Karnataka.

Parts used — Gum resin.

Pharmacognostical characteristics — It is a small thorny tree 4 to 6 ft in height, trunk knotty, outer bark greenish yellow; each branch ends in a sharp spine. Leaves alternate, simple or 3-foliate; margin smooth, shiny. Flowers unisexual or bisexual, few, solitary or in clusters; calyx tubular, glandular, petals 4. Drupe fleshy, roundish. The oleo gum resin exudes from the stem of *C. mukul*.

Ayurvedic description — *Rasa* — *tikta, katu, madu, kasaya; Guna* — *laghu, teeshna, snigdha, pichila, sar; Veerya* — *ushana; Vipak* — *katu*.

Action and uses — *Tridosh har, soth har, badana samak, barn sodhan, hirdya, mutral, khutaghan, dipan, arsoghona, rasayan.* Used in all types of *vat bayadh*.

Chemical constituents — The gum resin of *C. mukul* was found to contain 4.6% foreign impurities, 3.2% gum, 19.5% mineral matter consisting chiefly of silicon dioxide, calcium, magnesium, iron and aluminum. It also contained 1.45% essential oil.[286] The essential oil was found to contain 6.4% myrcene, 11% dimyrcene, and some polymyrcene.[287] On systematic chromatographic separation, the petroleum ether extract of the gum resin yielded sesamin, cholesterol, and a few other steroids.[26]

Guggulu has been found to be a complex mixture of a variety of organic compounds and inorganic ions. The material was neutral in character and free organic acids were found only to the extent of less than 0.5%. On steam distillation it yielded an essential oil (ll.5%) which was conveniently segregated into hexane-soluble (9 to 11%), ethyl acetate-soluble (32 to 35%), and ethyl acetate-insoluble (54 to 59%) portions. Hexane and ethyl acetate-soluble fractions were found to be a complex blend of steroidal ketones, alcohols, and aliphatic triols. Many of these alcohols occurred as esters of ferulic acid. A number of steroids have been isolated and characterized.[288] The structure elucidation of steroidal constituents, viz., Z-guggulsterone and *E*-guggulsterol, I, II, and III have been established. Partial synthesis of guggulsterol II from diosgenin has also been reported. In addition, diterpenoid constituents cembrene A and mubulol were isolated and their structures elucidated. Some fatty tetrols were also isolated.[289,290]

Pharmacological action — Demulcent, aperient, alterative, carminative, antisposmodic, and emmenagogue. It also acts as an astringent.[1,3]

The oleoresin portion of the plant was found to be a highly potent anti-inflammatory agent, as compared to hydrocortisone and butazolodin against Brownlee's formaldehyde-induced arthritis in albino rats.[291]

The alcoholic extract of *C. mukul*, when administered orally to Indian domestic pigs kept on a standard atherogenic diet for 6 months, was found to reduce significantly the total serum cholesterol. The drug also reduced effectively the serum β-lipoprotein fraction and significantly altered the lipoprotein ratio.[292] Preliminary studies carried out on monkeys fed on cholesterol diet indicated that fraction A of the petroleum ether extract of *C. mukul* could effectively lower serum total lipids, cholesterol, phospholipids, and triglycerides. The hypolipemic activity of fraction A of *C. mukul* in the monkey was noted to be comparable to that of Atromid-S.[293] A fall in the total serum cholesterol and serum lipid-phosphorus was noted in all patients of hypercholesterolemia associated with obesity, ischemic heart disease, hypertension, diabetes, etc.[294-297] When crude *guggulu* was given orally, a fall in the total serum cholesterol and serum lipid-phosphorus was noted in all cases in clinical trials. In chronic endometritis, amenorrhea, and menorrhagia it is particularly valued. It is said to improve the general condition of the patients with leprosy, relieves lassitude, gives a sense of well-being, and relieves nervous pains.

Tripathi et al.[297] reported extensive clinical studies indicating that oral administration of crude *guggulu* could effectively lower serum turbidity and also prolong coagulation time.

Medicinal properties and uses — It has no action on the unbroken skin, but on abraded skin and mucous membranes it acts as an astringent and antiseptic. When taken internally, it is reported to possess appetizing, carminative, antisuppurative, aphrodisiac, and emmenagogue properties.[1]

COPTIS TEETA Wall.   *Family:* RANUNCULACEAE

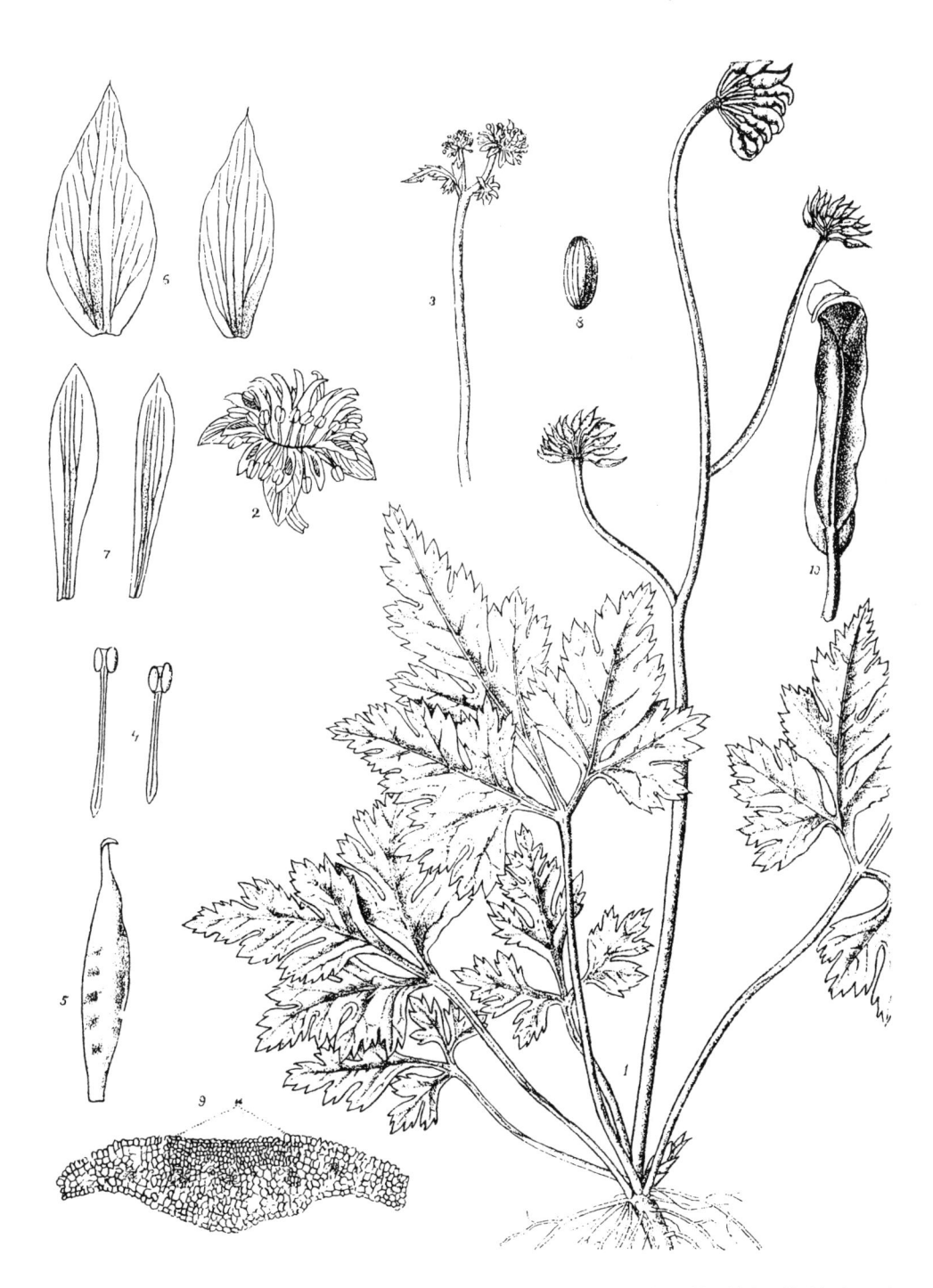

Vernacular names — Sanskrit, Mishamitika; Hindi, Mamira; English, Gold thread; Bengali, Tita; Sinhalese, Pitakarosan; Unani, Momiran; Arabian, Mamiran-e-seeni; Persian, Mamiran-e-chini; Tamil, Pitharoghan.
    Habitat — Found growing wild in the Mishmi Mountains, east of upper Assam.
    Parts used — Rhizome.

Ayurvedic description — *Rasa* — *tikta; Guna* — *rooksha; Veerya* — *ushana; Vipak* — *katu.*

Action and uses — *Tridoshahar, lakhan, pitta samak, sotha har, dipan, pachan, anuloman, kaphaghan, netra saktiwardhak, ama pachan, jwaraghan.*

Chemical constituents — Rhizome contains a yellow, bitter principle berberine to the extent of 8.5% soluble in water and alcohol. Chatterjee et al.[298] isolated berberine (9.0 %), coptine (0.08%), palmatine (traces), coptisine (0.02%), and jatrorrhizine (0.01%), reported to have been isolated for the first time from *Coptis.*

Pharmacological action — A bitter tonic and febrifuge.

Medicinal properties and uses — As a bitter tonic, it increases appetite, restores digestive powers, and removes flatulence and visceral obstructions.[1] It is very useful in jaundice, debility, convalescence after fevers, debilitating diseases, and atonic dyspepsia. In catarrhal and rheumatic conjunctivitis, this rhizome made into paste is used as a collyrium or as a salve for the eyes. In China it is used as an antidiabetic.[3]

Doses — Powder — 0.7 to 10 gr; tincture — 2.5 to 5 ml; infusion — 28 to 56 ml.

CORDIA MYXA Roxb. *Family:* BORAGINACEAE
(Syn. *C. DICHOTOMA* Forst. f.[*C. OBLIQUA* Willd.;
C. MYXA] C. B. Clark. non Linn.)[28]

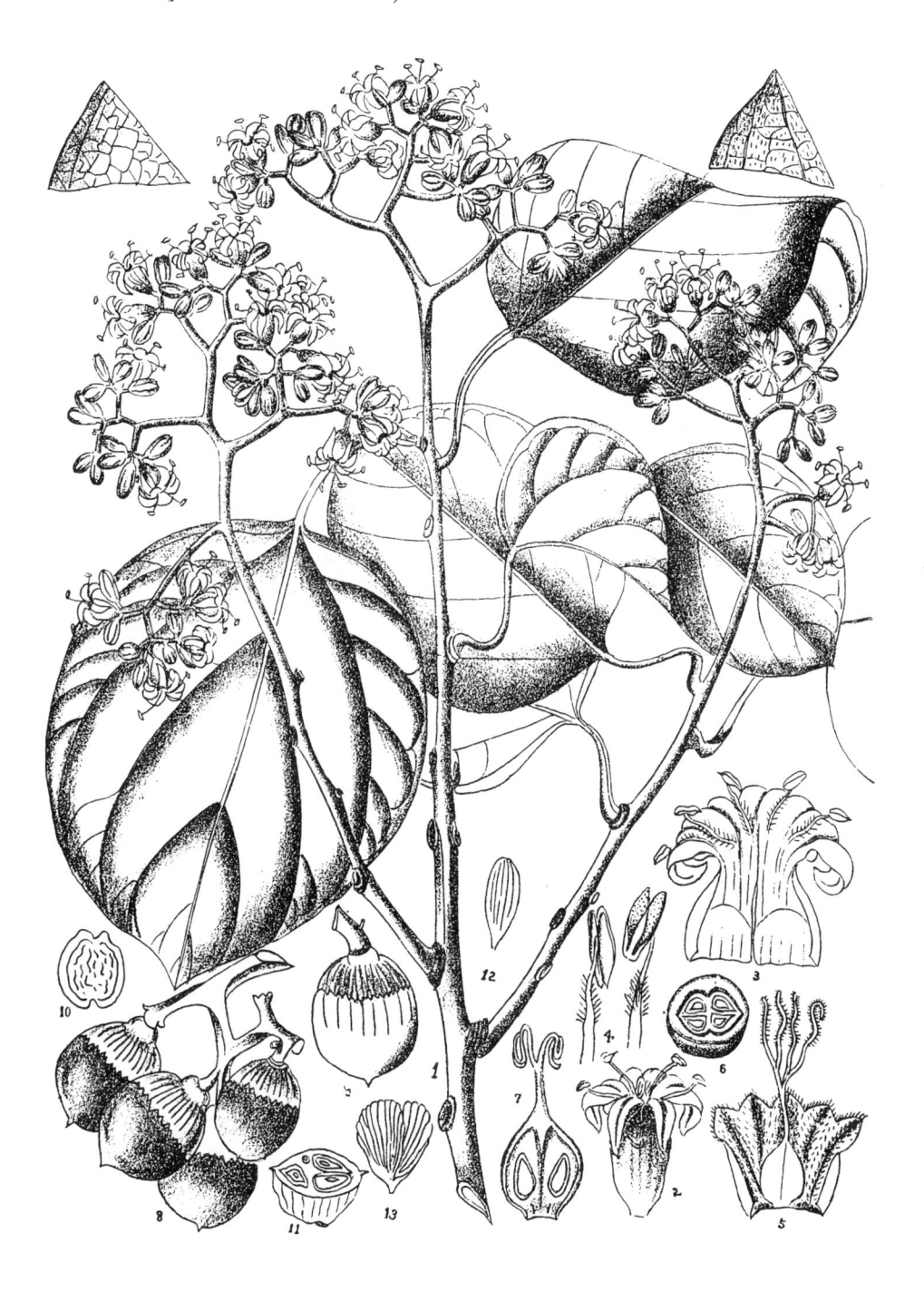

Vernacular names — Sanskrit, Sleshmataka; Hindi, Bara lasora; English, Sebesten plum; Bengali, Bahubar; Nepalese, Amly; Sinhalese, Lolu; German, Cordia; French, Seleastan; Unani, Lasorah; Arabian, Daleque; Persian, Siupastan; Burmese, Tana.

Habitat — Grows all over India and cultivated in Bengal.

Parts used — Fruit, its mucillage, kernel, and bark.

Morphological characteristics — The fruit of this arboreous plant is 0.5 to 1.0 in. long; yellow-brown, pink, or nearly black when ripe. The wood is clear yellow when freshly cut, changing rapidly on exposure to olive or bluish gray and finally to brown or grayish brown.[28]

Ayurvedic description — *Rasa* — *madhur; Guna* — *snigdha, guru, pichil; Veeyra* — *sheeta; Vipak* — *makhur.*

Action and uses — *Vatta pitta samak, kaphawardhak, vishaghan, bran ropan, grahi.*

Chemical constituents — Pulp of the fruit contains sugar, gum, extractive matter, and ash; bark contains a principle allied to "cathartin"; it also contains 2% tannin. The fatty acid composition of the fixed oil from the seed contains β-sitosterol, mp 126 to 127°C; benzopate, mp 144 to 145°C; and digitomide, mp 221°C.[299]

Pharmacological action — Fruit is demulcent; bark is mild astringent and tonic.[50]

Medicinal properties and uses — Fruit is very mucilaginous and esteemed in cough and diseases of the chest and is given in bilious affection as a laxative. Bark infusion is used as gargle. Kernels mixed with some bland oil is a good remedy for ringworms. Juice obtained from the bark and administered in coconut milk relieves severe colic pains. Chopra et al.[1] described the fruit as astringent, demulcent, diuretic, expectorant, and also used in affections of urinary passages and diseases of lungs and spleen. Decoction of the bark is used in dyspepsia and fevers, and leaves are applied to ulcers and headache. The whole plant is used in snakebite.

Doses — Bark decoction — 56 to 112 ml; syrup — 14 to 28 ml.

CORIANDRUM SATIVUM Linn. *Family:* UMBELLIFERAE

Vernacular names — Sanskrit, Dhanyaka; Hindi, Dhania; English, Coriander; Bengali, Dhane; Nepalese, Dhaniya; Sinhalese, Kottamalli; German, Gemeiner coriender; French, Biles cereales; Unani, Dhaniya; Arabian, Kazbarah; Persian, Kashniz; Burmese, Naunau; Malaysian, Ketumbah; Tamil, Koththamalli.

Habitat — It is a herbaceous plant extensively cultivated in all parts of India for its fruit and seeds.

Parts used — Fruit and leaves.

Pharmacognostical characteristics — An annual herb, 1 to 3 ft high, with small, white, or pinkish-purple flowers borne on compound terminal umbels. The lower leaves are broad, with crenately lobed margins, while the upper ones are narrow, finely cut with linear lobes. The fruits are globular and ribbed, yellow-brown in color, and range in size from 2.0 to 3.5 mm diameter. When pressed, they separate into two mericarps, each containing a seed.

The coriander fruit of commerce consists of the whole cremocarp having two hemispherical mericarps united by their margins. The Indian variety is oval. The apex has two divergent styles. The fruit bears ten primary, wavy ridges alternating with eight more prominent secondary ridges. The fruits have an aromatic odor and spicy taste. Under the microscope, a transection of the fruit shows only two mature vittae in each mericarp. Within the vittae-bearing region of the mesocarp a thick layer of sclerenchyma is formed, which consists of pitted fusiform cells. Traversing the band of sclerenchyma are small vascular strands composed of a small group of spiral vessels. The testa is composed of brown, flattened cells. The endosperm is curved and consists of parenchymatous cells containing fixed oil and aleurone grains, having rosettes of calcium oxalate.[69]

Ayurvedic description — *Rasa* — *kasaya, tikta, madhura, katu; Guna* — *laghu, snigdha; Veerya* — *ushna; Vipak* — *madhura.*

Action and uses — *Tridoshahar, snigdh ushna, pitta samak, dipan, pachan, grahi, daha nasek, shulohar, shothhar, rakt pit samak.*

Chemical constituents — Fruits contain moisture (11.2%), protein (14.1%), fat (ether extract) (16.1%), carbohydrate (21.6%), fiber (32.6%), mineral matter (4.4%), calcium (0.6%), phosphorus (0.37%), iron (17.9 mg%/100 g.[1] The leaves are a rich source of vitamin C (250 mg/100 g) and carotene (250 mg/100g). The aromatic odor and taste of coriander fruits is due to volatile oil (1.0%) which is pale yellow liquid. The seeds contain 19 to 21% of fatty oil having dark, brownish-green color and an odor similar to that of coriander oil. The distilled oil contains 65 to 70% of (+)-linalol (coriandrol) and pinene.[69]

Pharmacological action — Fruit is aromatic, stimulant, carminative, stomachic, antibilious, refrigerant, tonic, diuretic, and aphrodisiac. Fresh leaves are pungent and aromatic.[1]

Medicinal properties and uses — Coriander oil is very useful in flatulence, colic, rheumatism, and neuralgia. The dried fruit has similar effects; it is generally used as infusion or decoction in sore throats, flatulence, indigestion, vomiting and other intestinal disorders, common catarrh, and bilious complaints. In combination with cardamom and caraway, it forms a good carminative. The plant is alleged to have antiseptic, antitubercular properties and is an antidote for snakebite and scorpion stings.[1]

Doses — Infusion — 12 to 20 ml; oil — 1 to 4 minims; powder — 4 to 8 gr.

COSTUS SPECIOSUS Sm.

Family: SCITAMINACEAE

Vernacular names — Sanskrit, Kemuka; Hindi, Keu; English, Costus; Bengali, Keu; Sinhalese, Tebu; German, Pritge Kostwurz; Unani, Keu; French, Costus elegant.

Habitat — An elegant climbing plant found plentifully in Bengal and Kashmir, Himachal, and Uttar Pradesh.

Parts Used — Root and tuber.

Morphological characteristics — Leaves large, lanceolate, about 1 ft long. The bracts are red, the flowers large, with white limbs and yellow centers. Rhizomes are tuberous and insipid, fibrous. Vascular bundles are amphicribal and starch grains measure 6 to 16 μ. In transection of the root, cork cambium is seen arising in the third layer of cortex and forms three to five layers of cork cells.[301]

Ayurvedic description — *Rasa — katu; Guna — laghu; Veerya — ushna; Vipak — katu.*

Action and uses — *Kantikara, shukrala, vata raktaghan, visha hara, kan dooghna kantharogaghna, kustahara, vishaghna, kaasaghna, hikkaa shmani, jwaraghna.*

Chemical constituents — Tubers contain a large amount of starch. Diosgenin and tigogenin in the free state were isolated from the petroleum ether extract of dried rhizomes.[302] Two saponins were isolated from the chloroform extract of the dried, powdered rhizomes: saponin A, isolated as granules with mp 289 to 290°C, on acid hydrolysis gave glucose; and β-sitosterol and saponin B, also as granules, mp 305 to 307°C, on acid hydrolysis yielded diosgenin, glucose, and rhamnose.[303,304]

Pharmacological action — Root is bitter, astringent, stimulant, digestive, anthelmentic, depurative, and aphrodisiac. Saponins seem to have significant anti-inflammatory and anti-arthritic activity against acute pedal inflammation induced by carrageenin (1%) and formalin (1%) in albino rats. The saponins also increased the weight of uterus in spayed albino rats. They also produced proliferative changes in uterus and vagina similar to those produced by stilbesterol.[304,305]

The alkaloids isolated from the plant possessed anticholinesterase activity both *in vitro* and *in vivo*. Like physostigmine they were equally effective against true and pseudocholinesterase. The use of the plant in eye diseases and as a depurative might be due to its anticholinesterase activity.[306]

Medicinal properties and uses — Root of *C. speciosus* is useful in catarrhal fevers, coughs, dyspepsia, worms, skin diseases, and snakebites. Tuber is cooked and made into a syrup or preserve which is very wholesome.[3]

Doses — 2 to 5 gr.

CRINUM DEFIXUM Ker-Gawl.   *Family:* AMARYLLIDEAE
(Syn. *CRINUM LATIFOLIUM* Linn.;
   C. ASIATICUM Linn.;
   C. BRACTEATUM Wild.;
   C. TOXICARIUM Roxb.,
   AMARYLLIS ZEYLANICUM Linn.)

Vernacular names — Sanskrit, sudarshan; Hindi, Chinder; English, Poison bulb; Bengali, Banakanur; Nepalese, Sudarshan, Sinhalese, Godamanel; Japanese, Indohamayu; German, Gift-Zwiebel; Unani, sukhdarshan.

Habitat — Grows on swampy river banks all over India.

Parts used — Leaves and bulbs.

Morphological characteristics — A stout herb; rootstock bulbous. Leaves fleshy, 3 to 4 ft long, 5 to 7 in. wide. Inflorescence an umbel; provided with two spathaceous bracts; stalk compressed, solid, 1.5 to 2 ft long; flowers numerous, large, white perianth salver shaped, lobes six, linear; stamens six, often reddish; bractoles linear.

Ayurvedic description — *Rasa* — *madhur; Guna* — *rooksha, tikshana; Veerya* — *ushna; Vipak* — *madhur.*

Action and uses — *Kapha vata samak, sothadar badanasthapan, branropan vamak, rechan, sothahar, kustaghan, jwaraghana.*

Chemical constituents — An alkaloid lycorine, mp 272 to 276°C, has been isolated from the bulbs and seeds of *C. defixum*.[307,308]

Pharmacological action — Emetic, diaphoretic, and purgative. Pharmacologically, lycorine is a weak alkaloid.[178]

Medicinal properties and uses — The bruised hot leaves, coated with castor oil or mustard oil, are applied to whitlow, inflamed ends of toes and fingers, inflamed joints, and sprains. The warm leaf juice with the addition of a little common salt is used for ear complaints. The juice can be used as an emetic in place of ipecacuanha. The bulb is also a good substitute for ipecacuanha; it is a bitter tonic, laxative, emetic, nauseant, and diaphoretic; it is used in biliousness and urinary complaints. The roasted bulb is used as a rubefacient and in rheumatism.[3,50]

Doses — Juice of the bulb — 56 to 112 ml (as emetic); infusion — 28 to 56 ml.

CROCUS SATIVUS Linn. *Family:* IRIDACEAE

Vernacular names — Sanskrit, Kumkuma; Hindi, Kesar; English, Saffron; Bengali, Jafran; Nepalese, Kesar; Sinhalese, Kumkumapu: Japanese, Safuran; Chinese, Fan Hunghau; German, Safran; French, Safran; Unani, Zafran; Arabian, Kurkum; Persian, Zaffran; Tibetan, Gur-gum; Burmese, Thanwai; Tamil, kungampoo.

Habitat — Cultivated in Kashmir and Kishtwar (Jammu) and in Nepal. On commercial scale, it is grown in Spain, France, Italy, Greece, Turkey, and China.

Parts used — Dried stigmas and tops of the styles of the flowers of *C. sativus*.

Morphological characteristics — A small, bulbous perennial, 6 to 10 in. high; flowers scented blue or lavender appear before leaves. Tripartite, subclavate, orange-red stigmas along with the style tops of the flower yield the saffron. Under the microscope, the stigma appears either united in three or isolated. Each stigma is about 20 to 25 mm long and has the shape of a slender funnel, the rim of which is dentate or fimbricate.[28]

Ayurvedic description — *Rasa* — *katu, tikta; Guna* — *snigdha, laghu; Veerya* — *ushna; Vipak* — *katu.*

Action and uses — *Tridoshahar, varyna sothahar, dipan, pachan, grahi, hridya, vajikarn, jwaraghan, rasayan.*

Chemical constituents — Saffron contains three crystalline coloring matters: (1) α-crocetin ($C_{24}H_{28}O_5$, mp 272 to 73°C) constituting 0.1%, (2) β-crocetin ($C_{25}H_{30}O_5$, mp 205 to 206°C) constituting 0.7%, and (3) γ-crocetin ($C_{26}H_{32}O_5$, mp 202 to 3°C), constituting 0.3%.[44] A fatty oil (8 to 13.4%), an essential oil 1.31%, and a bitter substance have been reported. The bulbs contain saponins which are toxic to young animals but not to old ones.[1]

According to Trease and Evans,[69] saffron contains a number of carotenoid pigments. A hypothetic protocrocin of the fresh plant is decomposed on drying into one molecule of crocin (a colored glycoside) and two molecules of picrocrocin (a colorless bitter glycoside). Crocin on hydrolysis yields gentiobiose and crocetin, while picrocrocin yields glucose and safronal. The latter substance is largely responsible for the characteristic odor of the drug. The essential oil from stigmas and petals contains 34 or more components, mainly terpenes, terpene alcohols, and esters. The flowers are reported to contain riboflavin and thiamine.[178]

Pharmacological action — Stimulant, aphrodisiac, stomachic, anodyne, and antispasmodic. Trease and Evans,[69] however, reported little or no such action by saffron.

Medicinal properties and uses — It is a nerve sedative, mild narcotic, diuretic, and emmenagogue; in small doses it is a mild stimulant; in large doses it is an aphrodisiac and narcotic. It is a popular remedy for promoting menstruation; generally used in fever, hysteria, melancholia, leukorrhea, and in piles. A paste of saffron is used as a dressing for bruises, superficial sores, rheumatic and neuralgic pains, and congestion of the chest.[1,3]

Doses — Tincture — 5 to 20 minims; infusion — 28 to 112 ml.

CROTON OBLONGIFOLIUS Roxb.   *Family:* EUPHORBIACEAE

Vernacular names — Sanskrit, Nagdanti; Hindi, Hakum; Chucka Bengali, Baragaccha; Unani, Armal; Burmese, Theyin; Persian, Kote.

Habitat — Found in Bengal, Bihar, South India, Deccan, Burma, Sri Lanka, and Bangladesh.[19]

Parts used — Bark, root, fruits, and leaves.

Morphological characteristics — A small deciduous tree; young parts, inflorescence, and ovary clothed with minute, orbicular, silvery scales. Leaves crowded toward the ends of branchlets, coriaceous 4 to 12 in. long, oblong-lanceolate, subacute, crenate or serrate, penninerved, glabrous when mature; base usually acute without apparent glands above the petioles. Flowers $1/3$ in. diameter; yellowish green, in erect fascicled racemes 5 to 12 in. long; stamens about 12; filaments hairy below. Ovary lepidote; styles 3, each 2-partite. Capsule $1/3$ to $1/2$ in. diameter, subglobuse, slightly 3-lobed, covered with minute orbicular scales, splitting into 2-valved cocci. Seeds $1/3 \times 1/4$ in., ellipsoid, rounded, and quite smooth on the back.[28]

Ayurvedic description — *Rasa—katu; Guna—guru, rooksha, teekshna; Veerya—ushna; Vipak—katu.*

Action and uses — *Kapha pitta har, sothahar, badanasthapan, dipan, pitta sark, virachan, krimighan, rakt sodhak, sodhak, swashar, kustaghan.*

Chemical constituents — Root contains resin and starch. Seeds yield fatty oil. Oblongifoliol ($C_{25}H_{22}O_2$) and deoxyoblongifoliol ($C_{20}H_{32}O$) were isolated from the bark of *C. oblongifolius*. Besides diterpenes, a triterpene acid, acetyl aleuritolic acid, has been isolated from the bark.[309,310]

Pharmacological action — Bark and root are alterative and cholagogue. Seeds are purgative.[3]

Medicinal properties and uses — Bark used in external applications for sprains and internally useful in liver diseases. Bark, root, fruits, and seeds are purgative and also given in snakebite. Externally bark and roots are applied in the form of poultice or paste on swellings and act as an antipyretic.[50]

Doses — Root powder — 1 to 3 gr; seeds — 50 to 125 mg; leaf decoction — 56 to 112 ml.

CROTON TIGLIUM Linn.   Family: EUPHORBIACEAE

Vernacular names — Sanskrit, Jayapala; Hindi, Jamalgota; English, Purgative croton; Bengali, Jaipal; Nepalese, Ajayapal; Sinhalese, Jayapal; Japanese, Hazu; German, Krotonol; Unani, Jamalgota; Arabian, Habul-salatina; Persian, Tukham-e-bed-e-khatai; Burmese, Kanako; Malaysian, Bori; French, Huile dectiglium; Tamil, Nervalam.

Habitat — Found throughout India, plentiful in Assam and in Bangladesh.

Parts used — Seeds and fixed oil.

Morphological characteristics — A small evergreen tree, with young shoots stellately hairy. Leaves membranous up to 7 in. long, elliptic or ovate, acuminate, serrate with 3 to 5 basal nerves, glabrous when mature. Flowers small, green, in clusters in terminal subglabrous racemes about 3 in. long. Stamens 15 to 20, ovary stellately hispid, styles long, 2-partite capsule $3/4$ to 1 in. long, white, turbinately ovoid, 3-lobed. Seeds $1/2$ in., slightly compressed, ellipsoid, with 8 raised lines.[28]

Ayurvedic description — *Rasa — katu; Guna — guru, snigdha, teekshna; Veerya — ushna; Vipak — katu.*

Action and uses — *Kapha vata har, lakhan, vidahi, dipan, rachan, krimighan, virachan, shothahar, jwaraghan, visaghan.*

Chemical constituents — Seeds contain a fatty fixed oil. Analysis of the oil gave toxic resin (3.4%), oleic acid (37.0%), linolic acid (19.0%), arachidic acid (1.5%), stearic acid (0.3%), palmitic acid (0.9%), myristic acid (7.5%), louric, tiglic, valeric, and butyric acids in traces, acetic acid (0.6%), and formic acid (0.8%). Seed kernels contain 43 to 63% croton oil. The oil contains a toxic resin; the seed kernels contain two toxic proteins, croton globulin and croton albumin, sucrose, and a glycoside, crotonoside.[50] Trease confirmed that the oil contains croton resin; also "crotin", a mixture of croton globulin and croton albumin comparable with ricin. The oil also contains diesters of the tetracylic diterpene phorbol; acids involved are acetic, a short chain acid, and capric, lauric, and palmitic as long chain acids. These compounds are cocarcinogens and also possess inflammatory and vesicant properties. The plant also contains alkaloids. The seeds contain two tumor-promoting principles.[178]

Pharmacological action — Seeds and oil — drastic purgative, irritant, rubefacient, cathartic, fish poison, and indicated in snakebites. Wood — diaphoretic in small doses, but purgative and emetic in large doses. Purgative effect may also follow the application of oil to the skin.[1]

Medicinal properties and uses — All parts of the plant possess drastic purgative properties. The wood, medicinally known as "lignum pavanae" is, in small quantities, diuretic, mildly emetic, and powerfully diaphoretic. The seeds and oil extracted from them are a drastic purgative; they contain dangerous toxic and purgative properties and therefore can be used internally only after extracting the poison. This is done by first removing the testa and the embryo from the seeds and then boiling the kernel twice or three times in milk. These seeds are given in apoplexy, insanity, convulsions, obstinate constipation, intestinal worms, ascites, dropsy, enlargement of the abdominal viscera, tympanitis, colic, calculous affections, and gout. Externally the oil is stimulant and a powerful irritant and rubefacient. It is used as a liniment and a good stimulant embrocation in infantile bronchitis, asthma, paralysis, gout, chronic rheumatism, arthritis, indolent tumors, laryngitis, and neuralgia.[1]

Doses — Oil — $1/2$ to 1 drop; liniment — 2 to 5 gr; seed powder — 2 to 5 gr.

CUCUMIS SATIVUS Linn.   Family: CUCURBITACEAE

Vernacular names — Sanskrit, Trapusha; Hindi, Kankri; English, Common cucumber; Bengali, Khira; Napalese, Tushi; Sinhalese, Pepingcha; Japanese, Kyuri; German, Gurke; Unani, Khira; Arabian, Qasd; Persian, Kheyar; Burmese, Thagwa; Tamil, Kar-Karikay.

Habitat — Found wild in the Himalayas from Kumaon to Sikkim; but it is cultivated throughout India.

Parts used — Seeds and leaves.

Morphological characteristics — A hispidly hairy climber. Leaves about $4^1/_2$ in. diameter, membranous, deeply cordate, angled or shallowly 3 to 5 lobed, both sides hairy with softish hairs, but the upper with thickened bases, ribs beneath scabrid or hispid, margin denticulate. Flowers $^3/_4$ to 1 in. across, males clustered. Fruit cylinderic covered with harsh hairs or soft spines, and finally more or less muricate.[28]

Ayurvedic description — *Rasa — madhur; Guna — laghu, rooksha; Veerya — sheeta; Vipak — madhur.*

Action and uses — *Pitta samak, vatta-kapha wardhak, daah prasaman, rakt pitta samak, mutral, balya.*

Chemical constituents — Seeds contain crude protein (42%) and fat (42.5%). The ash is rich in phosphate (P_2O_5 [0.62%]). The oil extracted from the seed is clear and light yellow. It has the following characteristics: sp gr 0.9130; acid val. 0.22; saponification value 193.0; iodine value (Wijs), 114.9; acetyl value 3.1; soluble fatty acids (as butyric acid), 0.4 and unsaponified matter 0.91%. The fatty acid components are palmitic (0.63%), stearic (16.2%), linoleic (40.11%), and oleic (38.70%). Analysis of the seed cake gave the following values: moisture (8.13%), proteins (72.53%), ash (9.7%), crude fiber (1.0%), carbohydrates (8.64%).[311] Analysis of the fruit gave the following values: moisture (96.4%), protein (0.4%), fat (0.1%), carbohydrates (2.8%), mineral matter (0.3%), calcium (0.01%), phosphorus (0.03%), iron (1.5 mg/100 g), vitamin B_1 (30 IU/100 g), vitamin C (7 mg/100 g). Enzyme crepsin,[312] proteolytic enzymes, ascorbic acid, oxidase, and succinic and malic dehydrogenases are present in fruit. Odorous principle is extractable with alcohol.[313]

Pharmacological action — Fruit is nutritive and demulcent. Seeds are cooling, diuretic, and tonic.[50]

Medicinal properties and uses — Fruits are used as a vegetable. Leaves boiled and mixed with cumin seeds are administered in throat affection. Powdered and mixed with sugar they are powerfully diuretic; as a refrigerant, it is given in remitted and inflammatory fevers.[1]

Doses — Seed powder — 1 to 2 gr; leaf powder — 1 to 2 gr; decoction or infusion — 14 to 28 ml.

CUMINUM CYMINUM Linn. *Family:* UMBELLIFERAE

Vernacular names — Sanskrit, Jeeraka; Hindi, Safed Jeera; English, Cumin seed; Bengali, Safed Jeera; Nepalese, Gee; Sinhalese, Sudu duru; Japanese, Kumin; German, Kreuz Kummel; Unani, Zira-safed; Arabian, Kamun-e-Abyaz; Persian, Zeera-e-safed; Tibetan, Go-snod; Burmese, Ziya; French, Anisacre; Tamil, Chirakam.

Habitat — The plant is grown extensively in southeastern Europe and North Africa, India, and China. It is cultivated in almost all the states in India except Bengal and Assam.

Parts used — Fruit or seed.

Pharmacognostical characteristics — A small, annual herb about 1 ft high, with a much-branched angular or striated stem, bearing 2- or 3-partite linear leaves, bluish-green in color, and having sheathing bases. The flowers are white or rose colored, borne in compound umbels. The fruits are grayish, about $1/4$ in. long, tapering towards both base and apex, and compressed laterally with ridges covered by papillose hairs. Trease and Evans[69] described the fruit as about 6 mm long and resembling caraway at first glance. The mericarps are, however, straighter than those of caraway and are densely covered with short, bristly hairs. Whole cremocarps attached to short pedicels occur as well as mesocarps. Each mericarp has four dorsal vittae and two commissural ones. The odor and taste are coarser than those of caraway.

Ayurvedic description — *Rasa — katu; Guna — laghu, rooksha; Veerya, ushna; Vipak — katu.*

Action and uses — *Bat kaf samak, pit bardhak, dipan, pachan, bataanuloman, grahi, mutral, verisya, jawar nasak.*

Chemical constituents — Analysis of seeds gave the following results: moisture (11.9%), protein (18.7%), ether extract (15.0%), carbohydrates (36.6%), fiber (12.0%), mineral matter (5.8%), calcium (1.08%), phosphorus (0.49%), iron (31.0 mg/100 g), carotene calculated as vitamin A (870 IU/100 g), and vitamin C (3 mg/100 g)[312]

The seeds yield on distillation a volatile oil (oil content, 2.0 to 4.0%) with an unpleasant taste. The oil is colorless or yellow when fresh.

The chief constituent of the volatile oil is cumaldehyde, $C_{10}H_{12}O$ (*p*-isopropyl) benzaldehyde, (bp 235°C), which forms nearly 20 to 40% of the oil. Besides the aldehyde, the oil contains *p*-cymene, pinene, dipentene, cumene, cuminic alcohol, β-phellandrene, and α-terpenol. Cumene (C_9H_{12}, bp 152°C) is produced by distillating cuminic acid with lime or baryta. Cumaldehyde is used in perfumery, and cumene for sterilizing catgut. According to Trease and Evans,[69] cumin yields 2.5 to 4.0% volatile oil which contains 25 to 35% of aldehydes (cuminic aldehyde), pinene, and α-terpinol.

Pharmacological action — Seeds are carminative, aromatic, stomachic, stimulant, and astringent.

Medicinal properties and uses — Seeds have cooling effect and therefore form an ingredient of most prescriptions for gonorrhea, chronic diarrhea, and dyspepsia. It is also useful in hoarseness of voice, dyspepsia, and chronic diarrhea in a dose of 1 to 2 gr. Cumin oil can be readily converted artificially into thymol; thymol is used as an anthelmintic against hookworm infections and is also an antiseptic, forming part of many proprietary preparations.

Doses — 3 to 6 gr.

CURCULIGO ORCHIODES Gaertn. Family: AMARYLLIDACEAE

Vernacular names — Sanskrit, Talamulika; Hindi, Kali musli; English, Black musalie, Bengali, Talamusli; Nepalese, Musali; Sinhalese, Hin-bin-tal; Japanese, Kinbai zassa; Unani, Musli siah; Persian, Musli siah.

Habitat — Occurring wild in sandy regions of hotter parts of Indian and Sri Lanka.

Parts used — Tuberous root; rhizome.

Morphological characteristics — Rhizome is stout, 6 to 14 cm long, with fleshy lateral root. Leaves linear, pleated, channeled, membranous, glabrous, or sparsely softly hairy, up to 12 in. long, base sheathing. Flowers starlike, bright yellow, or short, clavate shape, distichous, the lowest bisexual, the upper male, capsules 1 to 4 seeded with a slender beak.

In a transverse section of rhizome, cork is seen to be composed of five layers of rectangular, tangentially elongated, yellowish-brown, suberized cells containing numerous globules of an oily substance. The phellogen consists of one or two layers of rectangular cells. Cortical cells contain starch grains and acicular crystals of calcium oxalate. Secretory cells are of schizogenous origin. Crystals, secretory cells, and starch grains also occur in the stellar parenchyma.[314]

Ayurvedic description — *Rasa — madhur, tikta; Guna — guru, snigdha; Veerya — ushna; Vipak — madhur.*

Action and uses — *Vrishya, shukral, brihman, rasaayan, mootrala, mahaakustaghna.*

Chemical constituents — Resin, tannin (1.45%); mucilage, fat starch (48.48%), and ash (8.60%) containing oxalate of calcium.[1]

Pharmacological action — Bitter aromatic tonic, demulcent, diuretic, and aphrodisiac. Roots are alterative.[50]

Medicinal properties and uses — It is given with some aromatics in piles, diarrhea, jaundice, and asthma; and used as a poultice for itch and in some skin diseases. As an aphrodisiac, the powder is given with sugar and "ghee"; also to remove stones from the urinary system. Root stock is alterative and carminative; it is useful in biliousness, diseases of the blood, bronchitis, ophthalmia, lumbago, and in gonorrhea.

One to two ounces of powdered rhizome in warm milk and sugar is prescribed during convalescence after acute illness.[1,3]

Doses — Powder — 3 to 6 gr.

CURCUMA AMADA Roxb. *Family:* ZINGIBERACEAE

Vernacular names — Sanskrit, Karpura haridra; Hindi, Amahahaldi; English, Mango ginger; Bengali, Amada; Nepalese, Kapuhalu; Sinhalese, Ambakaha; German, Mangeingwer; Unani, Ambahaldi; Arabian, Dar-e-huld; Persian, Dar-e-ghoba.

Habitat — Grown in Bengal and the hills of the west coast of India.

Parts used — Rhizomes.

Morphological characteristics — Leaves 2 to 3 ft long, flower light yellow. The tuber looks like rhizome having odor of mango. The rhizome pieces are 3 to 10 cm long and 1 to 3.5 cm broad, laterally compressed, and pale brown in color. The roots are 1.0 to 15.0 cm long and 0.5 to 2 mm in diameter and of brownish color. Under a microscope there is an outer zone of cork, followed by cortex, the outer zone containing numerous oil cells and starch grains. The inner zone consists of closed vascular bundles. Endodermis is present. The vascular bundles are arranged in a ring immediately within the pericycle and some are also scattered. The center of the stele is parenchymatous, containing abundant starch grains and oil cells. The root shows epidermis, followed by a wide zone of cortex containing air canals, and thick-walled endodermis. The pericycle encloses the stele consisting of a number of xylem and phloem strands alternating radially.[315]

Ayurvedic description — *Rasa* — *madhur; Guna* — *rooksha, laghu; Veerya* — *sheeta; Vipaka* — *katu.*

Action and uses — *Sarvakandoovinaashini.*

Chemical constituents — Rhizomes contain essential oil (1.1%); resin, sugar, gum, starch, albuminoids, and organic acid; essential oil containing d-α-pinene (18%), ocimene (47.2%), linalool (11.2%), linalyl acetate (9.1%), safrole (9.3%).[316]

Pharmacological action — Carminative, cooling, aromatic, bitter, stomachic, and astringent.[3] Pachauri and Mukerjee[317] observed that ethereal extract of *C. amada* in a dose of 1 g given orally in the form of an aqueous suspension could lower the blood cholesterol level in rabbits. Dhar et al.[51] reported that extracts of rhizomes of *C. amada* had shown CNS-depressant activity in mice.

Medicinal properties and uses — Rhizomes are used externally in the form of paste over contusions and sprains and skin diseases. Combined with other medicines, it is used for improving the quality of blood.

Doses — Powder — 2 to 4 gr.

CURCUMA ANGUSTIFOLIA Roxb.   Family: ZINGIBERACEAE

Vernacular names — Sanskrit, Tavakshiri; Hindi, Tikora; English, Curcuma starch; Bengali, Tikhur; Sinhalese, Hulankeeriya; German, Schmal-blattrige Kurkume; Unani, Tikhur; Arabian, Nishasta; Persian, Nishasta.

Habitat — Found in the hilly tracts of central provinces, Bengal, Bombay, Madras, and some of the lower Himalayan ranges.

Parts used — Tuberous root.

Morphological characteristics — A small herb; leaves lanceolate; flower yellow, crowned in stalked spike. Medium-sized rhizome.

Ayurvedic description — *Rasa — madhur; Guna — laghu, snigdha; Veerya — sheeta; Vipak — madhur.*

Action and uses — *Vata-pitta samak, sanehan, anuloman, grahi, hirdya, rakta pitta samak, virisya mutral, jawarghan, daah prasaman, balya.*

Chemical constituents — Rhizomes contain starch, sugar, gum, and fat.

Pharmacological action — Cooling, demulcent, and nutritious. Reported to have antitubercular properties.[1]

Medicinal properties and uses — Starch is highly valued as an article of diet. It is very useful in cases of dysentery, dysuria, and gonorrhea; also recommended in typhoid fevers, ulceration of the bowels and bladder. In cases of difficult and painful micturition it is best administered in the form of thin extract prepared like barley water. Starch is used as substitute for the true arrowroot powder.

Dose — 12 to 24 gr.

CURCUMA LONGA Linn. *Family:* ZINGIBERACEAE
(Syn. *C. DOMESTICA* Valeton)

Vernacular names — Sanskrit, Haridra; Hindi, Haldi; English, Turmeric; Bengali, Haldu; Nepalese, Haku halu; Sinhalese, Kaha; Japanese, Ukon; Chinese, Yii-chin; German, Kurkuma Gelbwurzel; Unani, Haldi; Arabian, Arqussofar; Persian, Zardchob; Tibetan, Skyer-rtsa; Burmese, Hsanwen; Malaysian, Kooneit; Tamil, Manjal.

Habitat — The plant is cultivated throughout India but predominantly in Madras, Bengal, and Bombay.

Parts used — Rhizomes.

Pharmacognostical characteristics — It is a perennial herb 2 to 3 ft high with a short stem and tufted leaves. Leaves very large, in tufts up to 1.2 m or more long, including the petiole. Blade is as long as the petiole, oblong-lanceolate tapering to the base. Inflorescence spike 10 to 15 cm long; length of peduncle 15 cm or more, concealed by the sheathing petiole; flowering bracts pale green or tinged with pink.

The rhizomes are ovate, oblong, pyriform, or cylindrical and often short branched. Externally yellowish to yellowish brown. The internal color varies from yellow to yellow orange, waxy. Taste warmly aromatic and bitter, odor aromatic. The transverse section of the rhizomes is characterized by the presence of mostly thin-walled, rounded, parenchyma cells, scattered vascular bundles, definite endodermis, and a few layers of cork developed under the epidermis and scattered oleo-resin cells with brownish contents. The cells of the ground tissue are also filled with many starch grains. Epidermis is thin walled, consisting of cubical cells of various dimensions. The cork cambium is developed from the subepidermal layers and even after the development of the cork the epidermis is retained. Cork is generally composed of four to six layers of thin-walled brick-shaped parenchymatous cells. The parenchyma of the pith and cortex contains curcumin and is filled with starch grains.

Oil cells have suberized walls and contain either orange-yellow globules of a volatile oil or amorphous resinous masses. Vascular bundles are scattered and are collateral, with phloem toward the periphery and few xylem elements. Endodermis is composed of thin-walled barrel-shaped cells. The vessels have mainly spiral thickenings and only a few have reticulate and annular structure.[318]

Ayurvedic description — *Rasa* — *tikta, katu; Guna* — *rooksha, laghu; Veerya* — *ushna; Vipak* — *katu.*

Action and uses — *Varnya, twagrogahana, mehaghna, rakta do shahara, kandooghns, paandughna, vishaghana, visodhini, aruchighan, shoshaghna, sheeta-pittahar (Utricaria).*

Chemical constituents — Analysis of Indian turmeric gave the following values: moisture (13.1%), protein (6.3%), fat (5.1%), mineral matter (3.5%), fiber (2.6%), carbohydrates (69.4%), and carotene calculated as vitamin A (50 IU/100 g).[319] Essential oil (5.8%) obtained by steam distillations of dry rhizome had the following: α-phellandrene (1%), sabinene (0.6%), cineol (1%), borneol (0.5%), zingiberene (25%), and sesquiterpenes (turmerones) (53.0%). A ketone, $C_{13}H_{20}O$, and an alcohol identified as tolylmethyl carbinol have been obtained from the volatile distillate. The crystalline coloring matter curcumin (yield 0.6%, mp 180 to 183°C) is a diferuloyl methane of the formula $C_{21}H_{20}O_6$. It dissolves in concentrated sulfuric acid, giving a yellow-red coloration.[50] Subramanian and Rao[320] reported that the coloring matter is curcumin belonging to the dicinnomoyl methane group. The rhizome also yields an aromatic oil called turmeric oil. Chromatographic study showed the presence of curcuminoides in rhizomes.[178]

Pharmacological action — Aromatic, stimulant, tonic, and carminative. Internally juice is anthelmintic.

Arora et al.[321,322] reported that petroleum ether extract and two of its fractions, i.e., A (viscous oil) and B (crystalline solid) of the rhizomes of *C. longa*, were found to have significant anti-inflammatory activity in rats which compared favorably with hydrocortisone acetate and phenylbutazone. The volatile oil also exhibited a marked anti-inflammatory effect in experimentally induced edema and arthritis.[26] Petroleum ether, and alcoholic and aqueous extracts of the rhizome showed 80, 60, and 100% antifertility activity in albino rats. Curcumin was found to cause a sharp but transient fall of blood pressure. Mukerji et al.[323] found that curcuma powder increased appreciably the mucin content of gastric juice in rabbits. Dhar et al.[51] reported the antiprotozoal activity of alcoholic extract of the rhizome of *C. longa* against *Entamoeba histolytica*. The essential oil and the pigment curcumin are cholagogue and choleretic acid is antibacterial.[178]

Medicinal properties and uses — The rhizome is a household remedy: it is aromatic and carminative for relief of catarrhal cough; a decoction of fresh rhizome is given with coriander and cinnamon. In dose of 2 to 3 ml, rhizome is given in intermittent fever, dropsy, jaundice, liver disorders, and urinary diseases. In the form of paste it is applied externally in wounds and inflammatory troubles of the joint. Turmeric alone or combined with the pulp of neem leaves is used in ringworms, obstinate itching, eczema, and other parasitic skin diseases. In Ayurvedic and Unani practice it is used as a stomachic, tonic, blood purifier, antiperiodic, and alterative. Turmeric mixed in hot milk is a household remedy and beneficial in the common cold.[1,3]

Doses — Infusion — 4 to 8 ml; powder — 1 to 4 gr.

CURCUMA ZEDOARIA Rosc. *Family:* ZINGIBERACEAE

Vernacular names — Sanskrit, Shati; Hindi, Kachur; English, Round zedoary; Bengali, Sati; Nepalese, Kachur; Sinhalese, Harankaha; Japanese, Gajutsu; German, Zittwer; French, Zeodaire long; Unani, Jadwarkhatai; Arabian and Persian, Jadwar khatai; Burmese, Thanuwan; Tamil, Kitchilipazham.

Habitat — Found in the forest of the eastern Himalayas, Chittagong, Bengal, and Kerala. It is cultivated in Kerala, Karnatka, Madras, and other places providing congenial climatic conditions.

Parts used — Rootstock and leaves.

Morphological characteristics — Rootstock tuberous, ovoid, having many palmately attached sessile, cylindrical, accessory tubers up to 2.5 cm in diameter as well as many smaller oblong to fusiform succulent root tubers at the tips of oblong, stout, fleshy, fibrous roots. There is no distinct aerial stem, but the shoot, which may reach a height of 3 ft, has a pseudo-stem formed of the long and closely overlapping sheathing bases of four to six broad, lanceolate, finely acuminate smooth leaves tapering to the base and often colored purplish along the center. The inflorescence is a spathe from about 15 to 30 cm in length. It arises from the rhizome, separate from the leafy shoot. It is closely enveloped by a few green leaf-sheaths in its basal half. Its distal half terminates as a tufted spike covered with several imbricating, oblong bracts of which each of the lower ones subtends 3 or 4 sessile yellow flowers, while the upper bracts are sterile and colored.

Ayurvedic description — *Rasa* — *katu, tikta; Guna* — *laghu, teekshna; Veerya* — *ushna; Vipak* — *katu.*

Action and uses — *Kapha vat samak, sotha har, kustaghan, rochan, anuloman, krimighan, rakta sodhak, swashar, artawjanan, vajikaran kustaghan, jwaraghan.*

Chemical constituents — The rootstock contains an essential oil, a bitter soft resin, organic acids, gum, starch, sugar, curcumin arabins, albuminoids, crude fiber, and ash. Analysis of a commercial sample of starch gave the values as moisture (13.1%), ash (1.51%), and starch (82.6%). Nearly a third of starch is amylose.[331] The starch grains resemble those of arrowroot starch. The dry rhizomes on steam distillation yield 1 to 1.5% essential oil. Its odor reminds one of ginger oil, but differs from it by a camphor-like odor due to the presence of cineol. It is soluble in 1.5 to 2 vol of 80% alcohol. The lower-boiling fraction of the oil contains cineol; the higher boiling fraction contains sesquiterpene alcohol which imparts a characteristic aroma to the oil. The roots also contain the yellow coloring matter curcumin, starch, and sugars. Rao et al.[332] studied a commercial sample of oil that contained pinene (1.5%), camphene (3.5%), cineol (9.6%), camphor (4.2%), and borneol (1.5%).

Pharmacological action — Stimulant, carminative, expectorant, demulcent, diuretic, and rubefacient.[1]

Medicinal properties and uses — The juice of leaves is given in dropsy. Their paste is locally used in lymphangitis. The rhizomes are a blood purifier; it is given in flatulence, dyspepsia, and as a corrective of purgatives; its decoction with the addition of long pepper, cinnamon, and honey is a household remedy for fever, cold, cough, and bronchitis; the rhizome is also used to check leukorroheal and gonorrheal discharges. It is chewed for improving the sticky taste of the mouth, for clearing the throat, and for smoothing the irritation in the upper parts of the windpipe. A paste of the rhizome made with alum is applied to sprains and bruises.[1,3]

Doses — Infusion — 14 to 28 ml; powder — 1 to 2 gr.

CUSCUTA REFLEXA Roxb.   *Family:* CONVOLVULACEAE

Vernacular names — Sanskrit, Amaravela; Hindi, Akasbel; English, Dodder; Bengali, Algusi; Nepalese, Akashbel; Unani, Akasbel; Arabian, Aftimun-e-hindi; Persian, Darkht-e-pechan; Tamil, Avvaiyar-Kundal.

Habitat — Common throughout India; abundant in Bengal. It has no underground root, but grows only as a parasitic twiner on other plants, trees, or shrubs, and hence is called *akasbel* (sky twiner).

Parts used — Seeds, stem, and fruits.

Morphological characteristics — An extensive parasitic climber, yellow or reddish stem, flowers small, in clusters or racemes, fragrant, waxy, white, and shortly stalked.

Ayurvedic description — *Rasa — tikta, kaasaya; Guna — laghu, rooksha, picila; Veerya — sheeta; Vipak — katu.*

Action and uses — *Kapha pitta har, badana nasak, sotha har, kesya, dipan, pachan, grahi, mutral, jawaraghan, rakta sodhak, krimighan.*

Chemical constituents — Aggarwal and Dutt[324] isolated 0.2% of cuscutin, a coloring matter, and a lactone which they named cuscutalin. Later β-sitosterol and an unidentified flavonoid pigment were isolated.[325] Isolation of dulcitol (6%) from *C. reflexa* growing on *Melia azedirach, Eugenia jambolana,* and *Anacardium occidentale* and mannitol (5%), an isomer of dulcitol from *C. reflexa* parasitic on *Santalum album*, were reported.[326,327] Some workers found that the flavonoid kaempferol and its 3-glucoside constitute the coloring matter, but *C. reflexa* parasitic on *Glycosmis triphylla* contained luteolin and no kaempferol.[328] Mangiferin was found to occur in *C. reflexa* growing on *Mangifera indica* (mango tree).[329] This shows that the host plant has a direct influence on the chemical components of the parasite.

Pharmacological action — Carminative, anthelmintic, and alterative. Freshly prepared aqueous and alcoholic extracts of the plant exhibited relaxant and spasmolytic action on the small intestine of guinea pig and rabbit. In atropinized rabbit's auricles and frog's heart it produced marked cardiotonic activity.[330]

Medicinal properties and uses — Cold infusion of the seeds is given as a depurative and carminative in pains and stomachaches. As poultice it is also applied locally. Decoction of stem is useful in constipation, flatulence, liver complaints, and bilious affections. Externally they are used against itch and other skin diseases.

Doses — Infusion — 14 to 24 ml; seed powder — 1 to 3 gr.

CYNODON DACTYLON (Linn.) Pers.   *Family:* GRAMINEAE

Vernacular names — Sanskrit, Doorwa; Hindi, Doorval, Dhub; English, Dog grass; Bengali, Durba; Nepalese, Situ; Sinhalese, Eethana; Japanese, Kyogishiba; German, Wucherndeu Hundszahn; French, Chiendent; Unani, Doob; Arabian, Ushle; Persian, Murgh; Tibetan, Du-rba; Tamil, Arugu.

Habitat — This elegant perennial grass grows throughout India, up to 7500 ft.

Parts used — Roots, herb.

Morphological characteristics — A perennial, glabrous grass, stems slender, creeping, rooting at the nodes, developing matted tufts with short, erect branches. Leaves 1 to 4 in. long, very narow, flat, glaucous, arranged in two rows; sheaths with or without hairs, tight. Inflorescence consisting of 2 to 6 spikes radiating from a slender green or purplish stalk; stalk of the spikes slender compressed or angular.

Ayurvedic description — *Rasa — madhur, kasaya, tikta; Guna — laghu, snigdha; Veerya — sheeta; Vipak — madhur.*

Action and uses — *Tridosha har, sthamvan bran ropan, daha prasaman, rakt sthamvan, raktsodhak, prajasthmvak, mutral, kustaghan, jeveniya, visaghan.*

Chemical constituents — Analysis of green grass gave the following average composition (on dry matter basis): crude protein (10.46%), fiber (28.17%), N-free extract (47.81%), ether extract (1.80%), and total ash (11.75%).

Pharmacological action — Fresh juice is demulcent, astringent, and diuretic. The plant is hemostatic and laxative. An alkaloid isolated from plant caused a slowing of blood flow in the mesenteric capillaries of rats and mice. It reduced the blood sugar level in rabbits by 15% and also reduced clotting and bleeding time in rabbits.[1] The alcoholic extract of the entire plant was found to have antiviral activity against vaccinia virus.[51]

Medicinal properties and uses — The grass is diuretic, astringent, and styptic; it is used in dropsy, hematuria, and vomiting. As an astringent, its juice is given in chronic diarrhea and dysentery; the juice is also given in hysteria and insanity; an infusion of the grass with milk is prescribed for bleeding piles, irritation of the urinary organs, and vomiting. The plant or its paste is also used locally in gout and rheumatism. Decoction of the roots is diuretic; it s given in visical calculus, secondary syphilis, irritation of the bladder, bleeding piles, and dropsy. Reported to be useful in catarrh and ophthalmia.[50]

Doses — Infusion — 14 to 28 ml; powder — 1 to 3 gr; decoction — 58 to 112 ml.

CYPERUS ROTUNDUS Linn.　　　　　　　　　　*Family:* CYPERACEAE

Vernacular names — Sanskrit, Mustaka; Hindi, Korehijar; English, Nut grass; Bengali, Moothoo; Nepalese, Kashur; Sinhalese, Kalanduru; Japanese, Hamasuge; Chinese, Hiang Fou; German, Grasmandel; French, Souchet; Unani, Nagarmotha; Arabian, Saadkufi; Persian, Mushk-e-zamin; Tibetan, Gla-gan; Burmese, Rumput halia hitam.

Habitat — It is found throughout the plains of India, especially South India.

Parts used — Tuber or bulbous root.

Pharmacognostical characteristics — Perennial herb, leaves narrowly linear, finely acuminate, flat, one-nerved, 4.8 mm broad, mostly crowded near the base of stem. Flowers in simple or compound umbel which consists of short spikes of three to ten slender, spreading, red-brown spikelets. Sometimes contracted into a head of only one spikelet. Root fibers clothed with flexuous hairs. Inner surface white.[28]

The transverse section of the tuber is characterized by a thick-walled endodermis dividing the cortical portion and the central ground tissue. The cortical cells are with intercellular spaces. A hypodermis consisting of 2 to 3 layers of walled cells is present. The epidermis consists of typical parenchymatous cells with brownish pigments. The cortical cells are rounded, thin walled, parenchymatous, and are present in somewhat compact masses a few layers thick under the hypodermis, and toward the endodermis they form a typical parenchymatous tissue with large intercellular spaces. There is no cortical vascular bundle. Vascular bundles are closely scattered in the pith, internal to the endodermis; vascular bundles are more or less roundish in shape or a bit tapering on one side. In transverse section parenchyma of the pith also contains oleoresin cells and starch grains.

Ayurvedic description — *Rasa — katu, tikta, kasaya; Guna — laghu, rooksha; Veerya — sheeta; Vipak — katu.*

Action and uses — *Kaph pitta samak, lakhan, astanyajanan, dipan, pachan, grahi, krimighan, mutral, visaghan.*

Chemical constituents — The tubers contain fat, sugar, gum, carbohydrates, essential oil, albuminous matter, starch, fiber, and ash. There are traces of an alkaloid. The essential oil from *C. rotundus* contained at least 27 components comprising sesquiterpene hydrocarbons, sesquiterpene epoxides, sesquiterpene ketones, monoterpene and aliphatic alcohols, and some unidentified constituents.[333]

Pharmacological action — Stimulant, tonic, demulcent, diuretic, anthelmintic, stomachic, carminative, diaphoretic, astringent, emmenagogue, and vermifuge.[1,3] Petroleum ether extract of the tubers showed anti-inflammatory activity against carageenin-induced edema in albino rats.[334,335] The essential oil obtained from tubers exhibited estrogenic activity and its extract showed tranquilizing activity in rats.[337,338] The essential oil from tubers showed antibiotic activity.[178]

Medicinal properties and uses — Infusions of the tubers are useful in fever, diarrhea, dysentery, dyspepsia, vomiting, and cholera. Fresh tubers in the form of paste are applied to the breast as a galactagogue. Paste is used in scorpion stings, also.[1]

Doses — Powder — 1 to 3 gr; decoction — 56 to 112 ml.

DATURA METAL Linn. *Family:* SOLANACEAE
(Syn. *DATURA ALBA* Nees)

Vernacular names — Sanskrit, Dattura; Hindi, Safed dhatura; English, Thornapple; Bengali, Dhuttra; Nepalese, Dhatur; Sinhalese, Attana; Japanese, Yoshuchosen asaga; Chinese, Chan kiue Tse; German, Weichhaariger Stechapfel; French, Pomme epineuse; Arabian, Jauz-ul-maasil; Persian, Taaturah; Burmese, Padyain; Malaysian, Bori; Tamil, Umaththtai.

Habitat — This species has several varieties distinguished by the color of the flowers as white, purple, etc. These grow commonly in waste places throughout India, from Kashmir to Malabar.

Parts used — Whole plant, leaves, seeds, roots, and fruits.

Pharmacognostical characteristics — Leaves triangular-ovate in outline, unequal at the base; flowers 7 in. long, often double or triple, white, violaceous, reddish purple or purple on the outside and white within; fruit globose, tuberculate or muricate, borne on a short, thick peduncle. Capsule covered with fewer, shorter, stouter, often bluntish spines; capsule dehisces irregularly exposing a mass of closely packed, light brown, flat seeds which nearly fill the interior.[28]

Ayurvedic description — *Rasa* — *madhur (kasaya); Guna* — *laghu; Veerya* — *sheeta; Vipak* — *katu.*

Action and uses — *Madakara, agnivardhaka, nidranasnee, mootrala, jawaraghana, kushthaghana, varnahara, vishahara, vednaa shamanee.*

Chemical constituents — Leaves of *D. alba* chiefly contain an alkaloid-hyoscine (average 0.55%); the same alkaloid is found in all other parts of the plant. The seeds contain hyoscine (0.5% average) and small amounts of hyoscyamine and atropine, besides small quantities of norhyoscyamine.[50,178]

Pharmacological action — The plant as a whole is narcotic, anodyne, and has antispasmodic properties, analogous to those of belladona. Alcoholic extract of the entire plant of *D. metal* showed anthelmintic activity against *Ascardi galli*. It also possessed anticancer activity against human epidermal carcinoma of the nasopharynx in tissue culture.[51]

Medicinal properties and uses — Datura leaves, when applied in the form of poltice, relieve pain by acting as antispasmodic in rheumatic swellings of the joints, lumbago, sciatica, neuralgia, painful tumors, nodes, and glandular inflammations. Dried leaves and seeds are used in asthma, and whooping cough. The linament (prepared by macerating for 7 d 28 g of bruised seeds in sesamum or any bland oil) is useful in relieving the pain of painful or difficult menstruation, and in some painful affections of the uterus.[1,3]

Doses — Leaf powder — 1 to 3 gr; seeds — 1 to 2 gr.

DAUCUS CAROTA Linn. *Family:* UMBELLIFERAE

Vernacular names — Sanskrit, Garijara; Hindi, Gajar; English, Carrot; Bengali, Gajar; Nepalese, Gajar; Sinhalese, Kerat; Japanese, Ninjin; German, Karotte; Unani, Gajar; Arabian, Gazar; Persian, Gazar; French, Garotte cultive.

Habitat — Indigenous to Kashmir and western Himalayas. Now largely cultivated in India for culinary purposes.

Parts used — Root and fruits.

Ayurvedic description — *Rasa* — *madhur, tikta; Guna* — *laghu, teekshna, snigdha; Veerya* — *ushna; Vipak* — *madhur, tikta.*

Action and uses — *Tridosha samak, sothahar, bran ropan, dipan, shanhan, grahi, anuloman, hirdya, rakt sodhak, rakt pitta samak, sothahar, kapha nisark, vajikaran, asmari vedan, mutral, artwajanan.*

Chemical constituents — The seeds of *D. carota* contain fixed oil as well as volatile oil. The water-soluble fraction of the alcoholic extract contains some glycosidal principle.[339] Two quaternary bases have been isolated from the water-soluble fraction of the alcoholic extract of seeds of *D. carota.*[340]

Green carrots yield a glycosidal bitter principle (0.23%) and an ethereal oil (0.03%), but no tannins or saponins. The presence of proto-alkaloids, pyrolidine, and daucine has been reported. They also yield luteolin 7-glucoside. Occurrence of mevalonic acid has been reported. The essential oil of seeds contains daucol and carotal (11.3 and 70%, respectively).[178] Root contains carotin, hydrocarotin, sugar, starch, pectin, malic acid, lignin, albumen, salts, and volatile oil.

Pharmacological action — Carrot has a beneficial influence on the kidneys and dropsy, and prevents the brick-dust sediment sometimes found in the urine. Seeds are aromatic, stimulant, and carminative.[3] Alcoholic and aqueous extracts of seeds showed encouraging antifertility by preventing implantation in rats. The extracts also exerted some abortifacient activity.[341] Aqueous extract showed antiestrogenic, antiprogestation, antihistaminic, and antiacetylcholine activity.[342] Water-soluble fraction of alcoholic extract of seeds was found to have marked transient hypotensive effect in anesthetized dogs.[339] A quaternary base identified as choline chlorine showed cholinergic activity, and the tertiary base possessed relaxant action on isolated smooth muscles of ileum and uterus of different animals.[343,344] The essential oil produced transient fall of arterial blood pressure in the anesthetized dog without affecting the respiration in lower doses.[345]

Medicinal properties and uses — Carrots clean the blood and are recommended in chronic diarrhea. A decoction of carrot is useful in jaundice. Seeds are used as aphrodisiac and nervine tonic. Seeds are also useful for producing abortion.

Doses — Infusion — 14 to 28 ml; floral extract — 5 to 30 minims; seed powder — 2 to 3 gr.

DELPHINIUM DENUDATUM   *Family:* RANUNCULACEAE
Wall. ex Hook. f.& Thoms.

Vernacular names — Sanskrit, Nirvishi; Hindi, Nirbishi; Nepalese, Nerbesa; Sinhalese, Nerivisha; German, Ritterspoon; Unani, Jadwar; Arabian, Jadwar; Persian, Zarwar.

Habitat — West temperate Himalayas ascending up to 10,000 ft.

Parts used — Root (tubers) and seeds.

Morphological characteristics — A perennial herb, leaves small like coriander leaves, divided, flowers small, blue; fruits contain 1 to 7 seeds; root conical, 1 to 1.5 in. long, and gray.

Ayurvedic description — *Rasa — tikta; Guna — laghu, rooksha; Veerya — ushana; Vipak — katu.*

Action and uses — *Tridosha samak, sothahar, lakhan, vishaghan, dipan, amapachan, pitta sark, anuloman, hirdya, rakt sodhak, vajkarn, asmari nasak, mutral, jwaraghan.*

Chemical constituents — Some species contain the alkaloids delphinine and staphisagrine, both soluble in alcohol. Willaman and Li[10] have reported the presence of alkaloids, viz., condelphine, denudatidine, denudatine, and isotalatizidine, in the roots isolated by various workers.

Pharmacological action — Stimulant, alterative, stomachic, tonic, and anodyne.

Medicinal properties and uses — Root is alterative, it is given in syphilis and rheumatism. Decoction of the root is used as a tonic in doses of 10 to 20 ml during convalescence from fevers. Root is chewed to cure toothache. It is considered to be a good antidote to poison, particularly snakebite and to that of *Aconitum ferox.*

Doses — Decoction — 10 to 20 ml; powder — 2 to 5 gr.

DENDROBIUM MACRAEI Lindl.   *Family:* ORCHIDACEAE
(Syn. ***EPHEMERANTHA MACRAEI*** (Lindl.)
    Hunt et Sunmeh)

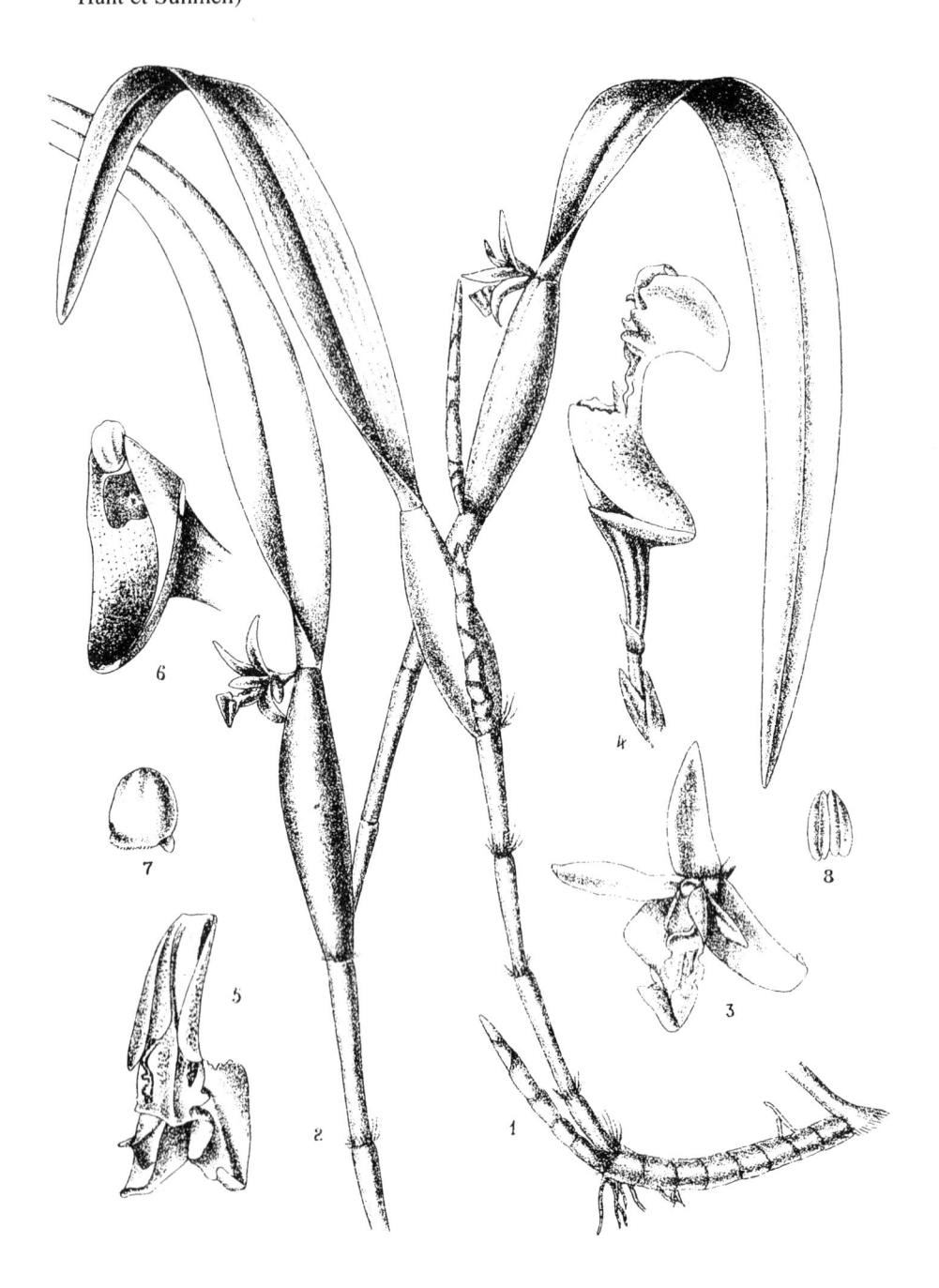

    Vernacular names — Sanskrit, Jivanti; Hindi, Jivanti; Bengali, Jebai; Nepalese, Jivanti; Unani, Jiventi.
    Habitat — A much-branched orchid often found on Jambul trees in Sikkim; Khassia Hills, Konkan, Nilgiris, and on hills of South India.
    Parts used — Whole plant, root, and stem.

Morphological characteristics — Rootstock creeping; leaves 10 to 20 cm long, sessile; flowers racemose, 18 to 25 cm long, white, and pediceled.

Ayurvedic description — *Rasa — madhur; Guna — laghu, snigdha; Veerya — sheeta; Vipak — madhur.*

Action and uses — *Tridosha har, vata pitta samak, daaha prasaman, sanehan, anuloman, grahi, rakt pitta samak, kaphanisarak, virisya, mutral, jwaraghan, rasayan, chachusya.*

Chemical constituents — It contains two resinous principles known as α- and β-jivantic acids and an alkaloid called jibantine. The β acid is bitter and soluble and the α acid is insoluble in ether and slightly bitter. Willaman and Li[10] have also reported the presence of an unnamed alkaloid in the whole plant.

Pharmacological action — Stimulant, demulcent, and tonic. A chemical fraction of the whole plant was found to lower the blood pressure in acute and chronic hypertension induced by noradrenaline infusion in anesthetized cats and dogs and by deoxycorticosterone acetate i.m. in rabbits.[346]

Medicinal properties and uses — As a tonic, it is given in debility due to seminal discharges. The whole plant is cooling, mucilaginous, light, strengthening, and tonic. Chopra et al.[1] reported the use of this drug against snakebite and scorpion sting.

Doses — Powder — 1 to 3 g; decoction — 56 to 112 ml.

DIGITALIS PURPUREA Linn.

Family: SCROPHULARIACEAE

Vernacular names — Sanskrit, Hatapatri; Hindi, Digitalis; English, Foxglove; Sinhalese, Degitalis; Chinese, Mao-ti-huang; German, Roter Fingerhut; Unani, Degitelis; Arabian, Degital; Persian, Degital.

Habitat — Cultivated in Kashmir, Darjeeling district, and the Nilgiris.

Parts used — Leaves.

Morphological characteristics — A biennial herb, it bears during the first year a rosette of radical, rugose, somewhat downy leaves 6 to 12 in. long, ovate to ovate-lanceolate with long, winged petioles. From the center of the leaf rosette arises, in the second year, a single, erect, flowering axis with sessile and subsessile leaves terminating in one-sided raceme of flowers, 2 to 3 in. long, declining; corolla, tubular campanulate, purple, yellow, or white; seeds small, light, and numerous.

Ayurvedic description — *Rasa — tikta; Guna — laghu, rooksha; Veerya — ushna; Vipaka — katu.*

Action and uses — *Kapha vata samak, pitta wardhak, hirdy a sakti wardhak, apeshapa samak, mutral, vajikaran, jwaraghan.*

Chemical constituents — The active constituents of digitalis are several glycosides which are present up to 1.0% in the leaves. These well-defined crystalline glycosides are digitoxin, gitoxin, and gitalin.[69] Digitoxin ($C_{41}H_{64}O_{13}$; mp 255 to 257°C), present in the leaves to the extent of 0.2 to 0.3%, is a colorless, odorless, crystalline substance insoluble in water but soluble in alcohol and chloroform. Gitoxin ($C_{41}H_{64}O_{14}$; mp 226 to 229°C) occurs as white needles and is sparingly soluble in water, alcohol, and chloroform. Gitalin ($C_{35}H_{56}O_{12}$; mp 245°C) is present in digitalis to the extent of 0.3 to 0.9%. It is a white crystalline substance soluble in water, alcohol, and chloroform. The leaves besides the glycosides contain tannins, inositol, luteolin, and gallic, formic, acetic, lactic, succinic, citric, and benzoic acids. The fatty matter of the leaves, consisting of 1.22% contains myristic, palmitic, cerotic, oleic, linoleic, and linolenic acids; melissyl alcohol, sitosterol, and triacontane are present.[44]

The seeds contain 31.4% of an amber-colored fatty oil with a bland taste.

Pharmacological action — Cardiac stimulant, tonic, and diuretic. Bhargava and Gupta[347] observed that digoxin slowed the heart by an action on the medulla. It seemed to block all the incoming influences to the vagal nucleus, including the supramedullary inhibitory influences. The trachycardia appeared to be due to stimulation of the hypothalamic cardioaccelator center.

Medicinal properties and uses — Digitalis is used mainly for its effect on the cardiovascular system, increasing the force of systolic contraction and the efficiency of the decompensated heart. It slows the heart rate and reduces cardiac edema with diuresis. It is used as myocardial stimulant in congestive heart failure, auricular flutter, and rapid auricular fibrillation. It has been recently shown to increase the coagulability of blood and to antagonize the anticoagulant action of heparin in the body. It is a diuretic, useful in dropsy and renal obstructions. In the form of ointment it is useful for cleaning wounds.

Doses — Powder — 0.1 gr; tincture — 5 to 15 minims.

DILLENIA INDICA Linn.   *Family:* DILLENIACEAE
(Syn. *D. SPECIOSA* Griff.)

Vernacular names — Sanskrit, Avartaki; Hindi, Chalta; English, Chalta; Bengali, Chalta; Nepalese, Ramphal; Sinhalese, Hondapara; Japanese, Biwamodoki; Chinese, Dok shan; Unani, Oot; Burmese, Thibuta; Malaysian, Simpoh.

Habitat — A tree growing in the tropical forests of the western peninsula, Bihar, and in the Himalayas from Nepal to Assam and from Sylhat to Sri Lanka.

Parts used — Fruit, bark, and leaves.

Morphological characteristics — Leaf oblong-lanceolate, 8 to 14 in. long and 2 to 4 in. broad, with pointed apex and toothed margin. The upper part of the leaf as well as the veins beneath are covered with hairs. Flowers white and fragrant. Fruit large, 3 to 5 in. diameter, hard, consisting of 5 closely fitted imbricate sepals enclosing numerous seeds embedded in a glutinous pulp. Seeds small, compressed reniform with hairy margins. Diagnostic anatomical features of the stem bark are lignified stone cells in the cortex, mucilaginous and tanniferous cells in the phloem, and medullary ray cells; nonlignified phloem fibers, and acicular cortical and phloem parenchyma cells. Presence of sandy crystals of calcium oxalate in some cortical cells.[348]

Ayurvedic description — *Rasa — madhur, amla, kasaya; Guna — guru; Veerya — sheeta; Vipak — amla.*

Action and uses — *Vat samak, kapha pitta wardhak, rochan, bistavi, grahi, hirdya, kaphanisark, jawaraghan.*

Chemical constituents — Inner kernels consist mostly of pectus matter, of a jelly-like consistency. Fresh ripe fruits contain tannin, glucose, and malic acid.

Pharmacological action — Tonic and laxative. The alcoholic extract of the leaves of *D. indica* possessed CNS-depressant activity in mice.[11]

Medicinal properties and uses — Juice of the fruit mixed with sugar and water is used as a cooling beverage in fevers and as cough mixture. Bark and the leaves are astringent. Fruit is laxative.[3]

DIOSCOREA BULBIFERA Linn.   *Family:* DIOSCOREACEAE
(Syn. *D. CRISPATA* Roxb.;
    D. PULCHELLA Rev.;
    D. SATIVA Thunb.;
    D. VERSICOLOR Buch. Ham ex Wall)

Vernacular names — Sanskrit, Barahi; Hindi, Zamin kand; English, Yam; Bengali, Ratalu; Nepalese, Bhayakur; Sinhalese, Kondol; German, Brotwurzel; Unani, Ratalu; Persian, Zaminqand; Burmese, Kadu-u; Malaysian, Abobo; Tamil, Kai vallikkodi.

Habitat — Common throughout India, ascending up to 6000 ft in the Himalayas.

Parts used — Tuber.

Morphological characteristics — Glabrous climber; tubers knotted, woody, elongated, 0.6 to 1.2 cm thick, externally brown to weak yellowish orange; fractured surface yellowish white to pale yellowish orange; leaves entire, alternate, ovate-lanceolate, long pointed, widely cordate. Flowers small or minute, usually one-sexual, in spikes, racemes, or panicles; Perianth tubular, urceolate, or rotate, 2-seriately 6 cleft, often shortly connate below. Male with stamens inserted at the base of the perianth or its lobes, 3 or 6, or 3 perfect, with 3 alternating staminodes; pistillode sometimes present. Female with 3 to 6 or 0 staminodes, ovary inferior, 3-quetrous, 3-celled, styles 3, short; stigma entire or 2-fid, recurved; ovules 2, superposed in each cell. Fruit a 3-valved capsule or baccate. Seeds with a hard albumen.[26,28]

Ayurvedic description — *Rasa — katu, tikta, madhur; Guna — laghu, snigdha; Veerya — ushna; Vipak — katu.*

Action and uses — *Tridosha har, bran ropan, dipan, anuloman, krimighan, rakt sodhak, veersya, prameghan, kustaghan, balya, rasayan.*

Chemical constituents — Analysis of the tubers gave the following values (on dry basis): albuminoids (7.36 to 13.31%), ash (3.31 to 7.08%), fat (0.75 to 1.28%), carbohydrates (75.11 to 81.39%), fibers (3.28 to 9.64%), and P_2O_5 (0.45 to 0.77%).[50] A novel norditerpene lactone, diosbulbine, has been isolated from the tubers of *D. bulbifera*. Diosbulbine, mp 295°C, analyzed for $C_{19}H_{20}O_6$, showed dimorphic crystalline forms depending on the solvent of crystallization. The structure for the diterpene lactone has been assigned on the basis of spectroscopic and degradative evidence.[349]

Pharmacological action — Alterative, acrid. The acetone extract of rhizomes (10 to 45 mg/kg) inhibited food intake in 90- and 300-min tests in rats. Small doses of methylphenidate and cocaine augmented intake of food.[350]

Medicinal properties and uses — Powder of the tubers applied externally to sores and internally given with cumin and sugar in milk as a remedy for syphilis, piles, and dysentery. They are applied to ulcers.[1]

Doses — Powder — 6 to 12 gr.

DIPTEROCARPUS INDICUS Bedd.   Family: DIPTEROCARPACEAE
(Syn. *DIPTEROCARPUS TURBINATUS* Dyer)

Vernacular names — Sanskrit, Gurjun; Hindi, Garjan; English, Wood oil tree; Bengali, Gurjan; Sinhalese, Hora; German, Gurjunbalsam; Unani, Gurjan.

Habitat — Forests of eastern India from Bengal, Burma, to Singapore and evergreen forests of western Ghats from North Kanara southward up to 3000 ft.

Parts used — Oleoresin.

Morphological characteristics — A lofty tree attaining a height of 120 to 150 ft and a girth of 10 to 15 ft, with a clean, cylindrical bole.

Ayurvedic description — *Rasa — kata, tikta; Guna — laghu, rooksha; Veerya — ushna; Vipak — katu.*

Action and uses — *Kapha-vat samak, khustaghan, mutral, mutra dha samak.*

Chemical constituents — Balsam contains an essential oil; also a dry transparent resin containing a crystallizable acid, garjanic acid, and volatile matter. A pale yellow oil with a balsamic odor is obtained (yield 46%) through steam distillation of the oleoresin. The twig bark contains 7.3% tannin.

Rao et al.[351] found humulene, β-carophyllene, a bicyclic sesquiterpene hydrocarbon, and a sesquiterpene alcohol, the latter two giving cadalene on dehydrogenation, in the essential oil from *D. indicus*.

Pharmacological action — Stimulant, diuretic, and alterative.

Medicinal properties and uses — It is useful in chronic bronchitis. Its essential oil has been successfully administered in the treatment of gleet, gonorrhea in the advanced stage, leukorrhea, other vaginal discharges, leprosy, and other skin diseases. Several species of *Dipterocarpus*, viz., *D. laevis* Ham, *D. alatus* Roxb., and *D. indicus*, grow in Bengal, Bangladesh, Burma, and Siam. They yield an oleoresin extract "*gaurjan* balsam" or "wood oil". The oil has a pale gray or light brown color, may be as thick as honey. It resembles copaiba balsam and has been used as a substitute for oil of copaiba in the treatment of gonorrhea.

Doses — 2 to 6 gr.

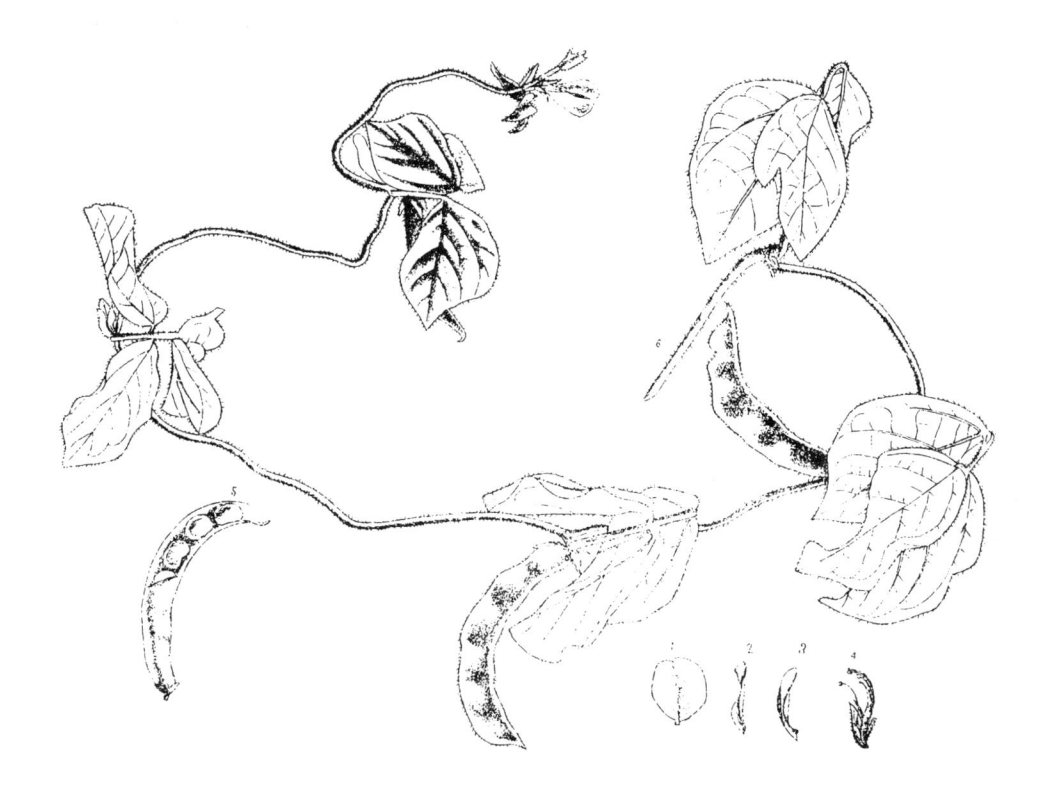

DOLICHOS BIFLORUS Linn. *Family:* PAPILIONACEAE
(Syn. *D. UNIFLORUS* Lam.)

Vernacular names — Sanskrit, Kulitha; Hindi, Kulthee; English, Horse gram; Bengali, Kulti; Nepalese, Kola; Sinhalese, Kollu; German, Pferde Bohne; French, Dolique; Unani, Kulthi; Arabian, Halubulgilt; Persian, Mash-e-hindi; Tamil, Koilu.

Habitat — A common twinning plant grown all over India, especially in Bombay and Madras for its seed or fruit.

Parts used — Seeds.

Morphological characteristics — Leaves trifoliate; stipules oblong; leaflets lanceolate or oblong, entire, downy. Flowers 1 to 3; calyx downy; corolla yellow; standard large. Pod large, compressed, recurved; seeds 5 to 6, compressed, reniform, gray or reddish brown.

Ayurvedic description — *Rasa* — *kasaya; Guna* — *laghu, rooksha, teekshna; Verya* — *ushna; Vipaka* — *amla.*

Action and uses — *Kapha vat samak, rakt pitta kopak, sothabar, vidahi, anuloman, vadan, krimighan, kaphaghan, asmari vadan, mutral, jwaraghan, lakhan.*

Chemical constituents — An analysis of the sample (seed with husk) showed moisture (4.30 to 10.25%), ether extract (0.05 to 1.84%), albuminoids (20.75 to 22.5%) (containing nitrogen [3.32 to 3.56%]), soluble carbohydrates (56.04 to 63.20%), woody fiber (4.85 to 5.50%), and ash (4.20 to 7.45%).[1] Seeds are rich source of urease.[50] It is reported to contain more strepogenin than in casein. Seeds also contain β-sitosterol.[178] A few patients with ascites and edema of legs were treated with aqueous extract of seeds of *D. biflorus,* 8 oz daily in divided doses. The results obtained indicated the diuretic activity of the extract. The preparation was well tolerated.[352]

Pharmacological action — Astringent, diuretic, and tonic.[3]

Medicinal properties and uses — The seeds are astringent to the bowels, antipyretic, diuretic, tonic, anthelmintic, emmenagogue, appetizing, and lithontriptic; it is useful in piles, tumors, bronchitis, heart trouble, kidney stone, enlargement of the spleen, hiccough, asthma, leukoderma, and in abdominal complaints. Decoction is also used in leukorrhea and menstrual cramps. Pulse is eaten extensively in the form of soup and porridge; considered to remove kidney stones.

Doses — Seed powder — 2 to 4 gr; decoction — 14 to 28 ml.

ECLIPTA ALBA Hassk. *Family:* COMPOSITAE

Vernacular names — Sanskrit, Bhringaraj; Hindi, Bungrah; English, Bhringaraj; Bengali, Kesuti; Nepalese, Bhila; Sinhalese, Kikirindi; Japanese, Takasaburo; Chinese, Lichang; Unani, Bungrah; Persian, Jadah; Burmese, Runput beu; Tamil, Kari Salai.

Habitat — This plant is a common weed in moist situations throughout India, ascending up to 6000 ft on the hills.

Parts used — Root and leaves.

Pharmacognostical characteristics — An erect or prostate annual herb, often rooting at the nodes; leaves opposite, sessile, oblong, lanceolate, 1 to 4 in. long; flower-head white, 0.25 to 0.35 in. in diameter. Receptacle is flat. Achenes are narrowly oblong and have ribbed pappus teeth. The leaf epidermis is composed of a single layer of parenchyma cells with characteristic nonglandular trichomes on both the surfaces. In transection the stem is circular in outline with a ring of collateral endarch vascular bundles of varying sizes and central parenchymatous pith. The root has a diarch structure with normal and secondary growth. Prominent multicellular secondary xylem rays are seen. The endodermis is indistinct. Few layers of cork cells are present.[353]

Ayurvedic description — *Rasa — katu, tikta; Guna — rooksha, laghu; Veerya — ushna; Vipak — katu.*

Action and uses — *Kapha vat samak, shoth har, bran sodhan, bran ropan, netra bal bardhak, kesya bardhak, dipan, pachan, yakrit vikar nasak, krimihar, rakta vardak, rasayan.*

Chemical constituents — It contains a large amount of resin and an alkaloidal principal ecliptine. The presence of reducing sugar in the whole plant and sterols in seeds has been observed. Wedelolactone (mp 325 to 330°C) was obtained from leaves and stem.[353] The herb also showed the presence of desmethyl wedelolactone and sulfur-containing peptides. The presence of two thiophene derivatives and a polyacetylenic compound has also been reported. Later a new polythienyle compound (α-terthienylmethanol), besides β-amyrin and stigmasterol, was isolated.[354] Leaves and twigs of the plant are reported to contain an unnamed alkaloid.[10]

Pharmacological action — Juice of the leaves is a hepatic tonic and deobstruent. Root is a tonic and alterative. The alcoholic extract has shown antiviral activity against Ranikhet disease virus.[51]

Medicinal properties and uses — The herb is used as a tonic and deobstruent in hepatic and spleen enlargements and in skin diseases. A preparation obtained from the leaf juice boiled with sesame or coconut oil is used for anointing the head to render the hair black and luxuriant. Fresh juice of the leaves is given in doses of 2 to 8 ml in fevers, dropsy, liver disorders, and rheumatism. A paste of the herb mixed with sesame oil is used over glandular swellings, elephantiasis, and skin diseases. It is also used for headache, and toothache.[1]

Doses — Infusion — 4 to 12 ml; powder — 3 to 6 gr.

ELEPHANTOPUS SCABER Linn.

Family: COMPOSITAE

Vernacular names — Sanskrit, Gojihiva; Hindi, Gobhi; English, Prickly leaves, Elephants foot; Bengali, Gajialata; Nepalese, Gojiho; Sinhalese, Ethadi; Unani, Gobhi; Arabian, Qanhit; Persian, Kalim-e Rumi; Burmese, Katoopine; Malaysian, Tutup leumbi; French, Pied d'elephant.

Habitat — Throughout India in shady places, especially in Bengal.[26]

Parts used — Root and leaves.

Morphological characteristics — A rigid herb with large, obovate-oblong, radical leaves forming a rosette and numerous clusters of flower heads.

Ayurvedic description — *Rasa — tikta, kasaya, madhur; Guna — laghu, rooksha; Veerya — sheeta; Vipak — katu.*

Action and uses — *Kapha pitta samak, sthmavan, krimighan, rakt pittahar, pramehahar, kustaghan, jwaraghan.*

Chemical constituents — Govindachari et al.[355,356] isolated two new sesquiterpene dilactones, viz., deoxyelephantopin and isodeoxyelephantopin, from the plant.

Pharmacological action — Cardiac tonic, astringent, alterative, and febrifuge. Plant is mucilaginous.

Medicinal properties and uses — Decoction of the root and leaves of *E. scabar* along with cumin and buttermilk is given in dysuria and other urethral discharges; also in diarrhea and dysentery. The drug is used in snakebite, also. Roots are given to arrest vomiting; powdered with pepper applied to toothache. Bruised leaves boiled in coconut oil applied to ulcers and eczema. Alcoholic extract of whole shoot shows antibiotic activity.[1]

Doses — Decoction of leaves — 14 to 56 ml.

ELETTARIA CARDAMOMUM Maton   Family: ZINGIBERACEAE

Vernacular names — Sanskrit, Ela Chhoti; Hindi, Chhoti elachi; English, Lesser cardamom; Bengali, Choti elachi; Nepalese, Sukumal; Sinhalese, Heen Ensal; Japanese, Karudemon; German, Kardamome; French, Cardamome; Unani, Elayechi Khurd; Arabian, Qaqelah sighar; Persian, Heel Khurd; Burmese, Palah; Malaysian, Reputage epinvair; Tamil, Elam.

Habitat — Cultivated for its fruit in many parts of western and southern India (forest of Kanara, Mysore, Coorg, Tranvancore, Cochin), Burma, and Sri Lanka.

Parts used — Dried ripe seeds; oil from fruits.

Morphological characteristics — Cardamom is a perennial with thick, fleshy rhizomes and leafy stem 4 to 8 ft in height, with a long, branched inflorescence which arises near the ground. The stem of *E. cardamom* is 7 to 8 ft high, the lower part covered with spongy sheaths. Leaves almost sessile, 15 to 30 in. long, 3 in. wide, oblong-lanceolate. Flowers in 15- to 30-in.-long panicles; bracts persistent, up to 2 in. long. Calyx about 1 in. long; corolla lip white, streaked with violet; capsules oblong, about an inch long, having fine vertical ribs. Rootstock branching, woody or fleshy. Testa with outer epidermis of thick-walled, narrow, elongated cells; below the epidermis a layer of collapsed parenchyma, followed by a single layer, becoming two or three layers in the region of the raphe, composed of large, thin-walled cells and an inner epidermis of thin-walled flattened cells. Inner integument of two layers of cells, and outer layer of palisade sclerenchyma with yellowish to reddish-brown beaker-shaped cells. Strongly thickened on the inner and anticlinal walls, each cell with a small bowl-shaped lumen containing a warty nodule of silica and an inner epidermis of flattened cells. Perisperm of thin-walled cells packed with minute, rounded, polyhedral starch grains containing in a small cavity. Endosperm of thin-walled parenchyma containing protein as a granular hyaline mass in each cell. Embryo of small thin-walled cells containing aleurone grains. Starch absent from endosperm and embryo. Sclerenchyma and large vessels present in the pericarp.

Ayurvedic description — *Rasa—katu, madhur; Guna—lagu, Rooksha; Veerya—sheeta; Vipak — madhur.*

Action and uses — *Tri dosha har, mukha sodhak, depan, pachan, anulomon,* cardiac stimulant, expectorant, *mutra ganan, daha samak, balya.*

Chemical constituents — The seeds contain essential oil to the extent of 4 to 8%. Principal constituents of the oil are cineol, terpineol, terpinene, limonene, sabinene, and terpineol in the form of formic and acetic acids. The aqueous portion of the steam distillate of cardamom contained 0.5% of essential oil with the following constants: sp gr 0.0920, n_D^{25} 1.4606; and $[a]_D^{25}$, 0; cineol content, 80%. Borneol was identified in oil from the aqueous distillate of Malabar cardamom oil. Analysis of cardamom capsules gave the following results: moisture (20.0%), protein (10.2%), ether extract (2.2%), mineral matter (5.4%), crude fiber (20.1%), carbohydrate (42.1%), calcium (0.13%), phosphorus (0.16%), and iron (5.0 mg/100 g).[357]

The following figures may be taken as covering most pure samples: specific gravity 0.923 to 0.945; optical rotation 24° to 48°; refractive index 1.462° to 1.4675°; acid value 1 to 4; ester value 90 to 150; potassium salts, starch, nitrogenous mucilage, yellow coloring matter, ligneous fiber, and ash containing manganese.[1]

Pharmacological action — The seeds are aromatic, pungent, cardiac, tonic, stomachic, laxative, diuretic, and carminative.

Medicinal properties and uses — The seeds are very useful in asthma, bronchitis, piles, strangury, and diseases of the bladder. As diuretic the seeds are given with honey as a corrective; the seeds are administered in flatulence and in griping of purgatives. A decoction of cardamom together with its pericarp and jaggary added is a popular home remedy to relieve giddiness caused by biliousness. A compound powder containing equal parts of cardamom seeds, ginger, cloves, and caraway is a good stomachic in 1.5-gr doses in atonic dyspepsia. it is also very useful to check vomiting.

Doses — Powder seeds — 0.625 to 1.750 gr.

EMBELIA RIBES Burm. f.   *Family:* MYRSINACEAE
(Syn. *E. GLANDULIFERA* Wight)

Vernacular names — Sanskrit, Vidanga; Hindi, Viranga; Bengali, Biranga; Nepalese, Bayubidang; Sinhalese, Walangasal; German, Embelia Fruchte; Unani, Baubring; Arabian, Baran Kabli; Persian, Barang kabli; Tibetan, Bi-dan; Pustu, Babrang; Tamil, Vivlangam.

Habitat — This climber is found in the hilly parts of India from the central and lower Himalayas down to Sri Lanka and Singapore.

Parts used — Berries (fruit), leaves, and root bark.

Pharmacognostical characteristics — A climbing shrub; stems rough, with conical, hard protuberances. Leaves alternate, papery, elliptic, 2 to 3 in. long, 1 in. broad, glandular pits on the under surface; stalk slender, flowers numerous in panicles, white or greenish yellow or greenish white. Fruits small, globose, berries. The fruit of *E. ribes* are grayish black in color, with warty surface; epicarp has rounded cells with wrinkled cuticle; fibrovascular bundles occur irregularly, and endocarp has a single row of palisade-like stone cells.[358,359]

Ayurvedic description — *Rasa — katu; Guna — laghu, rooksha, teekshna; Veerya — ushna; Vipak — katu.*

Action and uses — *Kapha vat samak, sirobirochan, krimighan, dipan, pachan, anuloman rakt sodhak, khustaghana, rasayan.*

Chemical constituents — It contains 2.5 to 3.1% embelin which is its active chemical constituent; an alkaloid (christembine), a resinoid, tannins, and minute quantities of a volatile oil are also present. Embelin (2.5:dihydroxy-3-lauryl-*p*-benzoquinone, mp 142°C) occurs in golden yellow needles insoluble in water and soluble in alcohol, chloroform, and benzene. It is reported to be effective against tapeworm.[50] Rao and Venkateswarlu[360] found that the berries also contained vilangin which contained two units of embelin with a CH_2 bridge. More recently embelin itself has been found by mass spectral analysis to be an inseparable mixture of three components (1) with C_9 side chains (homo embelin), (2) with C_{11} chain (embelin), and (3) with C_{13} chain (rapanone). The first occurs only in traces and the other two are major components, 100 and 69 parts, respectively.[361]

Pharmacological action — Dried berries (seeds) are carminative, anthelmintic, stimulant, and alterative. The pulp is purgative and fresh juice is cooling, diuretic, and laxative. The aqueous extract of the berries of *E. ribes* showed antifertility activity.[362] Clinical studies showed that alcoholic and aqueous extracts were very effective against ascarides.[363]

Medicinal properties and uses — Powered berries are given in the morning on an empty stomach and no food should be taken the rest of the day; to expel the dead worm, a dose of castor oil should be taken the next morning. As alterative, a paste made of the berries is applied to skin diseases such as ringworm; a paste with butter applied to the head relieves headache. Leaves of the plant combined with ginger are used as gargle in sore throat and indolent ulcers of the mouth. A paste of the bark is a valuable application to the chest in lung diseases like pneumonia.

Doses — Powder — 6 to 12 gr (adult), 2 to 3 gr (children); decoction — 14 to 28 ml.

EMBLICA OFFICINALIS Gaertn.   *Family:* EUPHORBIACEAE
(Syn. *PHYLLANTHUS EMBLICA* Linn.)

Vernacular names — Sanskrit, Amalik; Hindi, Amla; English, Emblic myrobalan; Bengali, Amlaki; Nepalese, Amba; Sinhalese, Nelli; Japanese, Amara; Chinese, An Mole; German, Amla; Unani, Aamlah; Arabian, Aamlaj; Persian, Aamlah; Tibetan, Skyu-ru-ra; Burmese, Ziphiyu si; Malaysian, Nellikai, French, Phyllanthe emblic; Tamil, Nelli.

Habitat — *E. officinalis* is found both in the wild and cultivated state, common in the mixed deciduous forests in India, ascending to 4500 ft on the hills.

Parts used — Dried fruit, seed, leaves, root bark, and flowers. Ripe fruit used generally fresh.

Pharmacognostical characteristics — A medium to large tree. Leaves simple, a hundred or more on each branchlet, small, 8 to 10 mm or more long and 2 to 3 mm broad, stipulate, arranged densely on the branchlets in a symmetrically distichous manner, linear oblong or linear elliptic, slightly recurved, entire, obtuse or rounded at base, apex subacute or apiculate, glabrous, light green above, paler and at times puberulous beneath. Flowers are very small, unisexual, 0.5 to 1.5 cm across, greenish yellow. Stem bark is 6 to 12 mm thick, light or ash gray to brownish gray. The fruit is nearly spherical or globular, slightly broader than long, and with small, shallow, conical depressions at either end of its longitudinal axis, especially at the place of attachment of the stalk. Its size varies according to the variety. Normally the fruit is 18 to 25 mm wide at the middle and 15 to 20 mm along the longitudinal axis. The surface of the fruit is smooth and marked with six spaced divisions. Mature fruits have a yellow mesocarp and a yellowish-brown endocarp. The mesocarp is acidulous in fresh fruits and acidulous and astringent in dried fruits. The seed has a cream color when fresh.[108]

Ayurvedic description — *Rasa* — all *rasas* except *lavan; Guna* — *lagu, rooksha; Veerya* — *sheeta; Vipak* — *madhura*.

Action and uses — *Daaha prashamanee, chakshushya, keshya, medhya, rochanee, deepanee, hridya, rasaayana, vrishya, shukrala swedahara, medohara, bhagnasandhaanakra, pramehaghna.*

Chemical constituents — The fruit pulp contains moisture (81.2%), protein (0.5%), fat (0.1%), mineral matter (0.7%), fiber (3.4%), carbohydrates (14.1%), calcium (0.05%), and potassium (0.02%); iron (1.2 mg/100 g), nicotinic acid (0.2 mg/100 g), and vitamin C (600 mg/100 g). Vitamin C content, up to 720 mg/100 g of fresh pulp and 921 mg/100 cc of pressed juice, has been recorded. The fruit is a rich source of pectin.[364] Phyllemblin has been isolated from the fruit pulp. Amla fruit is probably the richest natural source of vitamin C. The fruit juice contains nearly 20 times as much vitamin C as orange juice and a single fruit is equal in antiscorbutic value to one or two oranges. Air-dried fruit contains 13 separable tannins in addition to three or four colloidal complexes.[367]

Fruit contains phyllemblic acid (6.3%), lipids (6.0%), gallic acid (5.0%), and emblicol. Crystalline vitamin C has been isolated from the fruit pericarp in a yield of 70 to 72% of the total. Mucic acid has also been isolated. Phyllembin from the fruit pulp is identified as ethyl gallate. The bark contains a chemical consituent called leukodelphinidin.[178] A tannin containing gallic acid, ellagic acid, and glucose in its molecule and naturally present in the fruit prevents or retards the oxidation of vitamin C and renders the fruit a valuable antiscorbutic in the fresh as well as in the dry condition.

The seeds contain a fixed oil, phosphatides, and a small quantity of essential oil with a characteristic odor. The fixed oil (yield 16%), which is brownish yellow in color, has the following physical and chemical characteristics: sp. gr. 31°, 0.9220; acid value 12.7; saponification value 185, iodine value (Wijs), 139.5; R.M. val. 1.03, acetyl value 2.03; unsaponifiable matter 3.81%, sterol content 2.70%, and saturated fatty acids 7.0%. The components of the oil are linolenic (8.78%), linoleic (44.0%), oleic (28.40%), stearic (2.15%), palmitic (2.99%), and myristic acid (0.95%). Proteolytic and lipolytic enzymes are present in the seeds.

The fruits, bark, and leaves are rich in tannin. Fruit contains 28% of tannin, twig bark (21%), stem bark (8 to 9%), and leaves contain 22% of tannin.[368,369]

Pharmacological action — Fresh fruit is cooling, diuretic, laxative, and stomachic. Bark is astringent. Flowers are cooling and aperient. Phyllembin isolated from fruit pulp has been found to potentiate the action of adrenaline *in vitro* and *in vivo*. It showed a mild depressant action on the CNS and also had spasmolytic activity.[370] The extract of fruit exhibited antibacterial and antiviral properties.[51,371]

Medicinal properties and uses — Fresh fruit is used in inflammation of the lungs and of the eyes as a collyrium. A fixed oil extracted from the fruit is reported to have the property of promoting hair growth. The seeds are used in the treatment of asthma, bronchitis, and biliousness. The dried fruit is useful in hemorrhage, diarrhea, and dysentery. In combination with iron it is used as a remedy for anemia, jaundice, and dyspepsia. Emblic myrobalan is used in many compound preparations. Acute bacillary dysentery may be cured by drinking a syrup (sharbat) of amla with lemon juice. "Triphala" consisting of equal parts of emblic, chebulic, and belleric myrobalan is used as a laxative and in headache, biliousness, dyspepsia, constipation, piles, enlarged liver, and ascites. The exudation from incisions on the fruit is used as an external application for inflammation of the eye. The flowers are cooling, refrigerant, and aperient. The root and bark are astringent. Juice of the bark combined with honey and turmeric is a good remedy for gonorrhea. *Chyvanaprasa,* a popular Ayurvedic preparation, is very useful in anemia, asthma, inflammations of the lungs, and in eye diseases.[1,3]

Doses — Infusion — 10 to 15 ml; powder — 3 to 6 gr.

ERYTHRINA VARIEGATA Linn. *Family:* PAPILIONACEAE
(Syn. *ERYTHRINA INDICA* Lam.;
   E. STRICTA Roxb.;
   E. CORALLODENDRON Linn.)

Vernacular names — Sanskrit, Paribhadra; Hindi, Ferrud; English, Indian coral tree; Bengali, Palidhar palitu-mudar; Nepalese, Phaludo; Sinhalese, Erabaum; Japanese, Deigo; German, Indischer Korallen Baum; Unani, Pangrah; Burmese, Kathit; French, Arbre immorte; Tibetan, Pa-ri; Tamil, Kalyan-morangai.

Habitat — Foot of Himalayas to Sri Lanka, Burma, and Malacca. Common in Bengal and many parts of India, especially in southern India; often grown in gardens as a support for black pepper vine and for providing shade to young cinchona plants.

Parts used — Bark, juice, and leaves.

Morphological characteristics — A tall tree with thin gray bark covered with minute conical, usually black prickles. Petiole 4 to 6 in., mostly unarmed; leaflets membranous, glabrous, 4 to 6 in. long and broad, truncate or broad, rhomboidal at base, leaves trifoliate; leaflets 4 to 6 in. long and nearly as broad; flowers large, coral red, in dense racemes; pods torulose, 6 to 12 in. long, containing 6 to 8 seeds; seeds oblong, smooth, red, dark, to purple or brown.[28] Bark of *E. variegata* var. *orientalis* has no taste. The phloem has sieve tubes, companion cells, phloem fibers, medullary rays, phloem parenchyma, and secretory cells. Calcium oxalate crystals present.[372]

Ayurvedic description — *Rasa — tikta, katu; Guna — laghu; Veerya — ushna; Vipak — katu.*

Action and uses — *Kapha vata samak, sothahar, bran sodhan, korn roghan, nidra janan, dipan, pachan, anuloman, sulher, krimighan, rakt prasadan, kafani sark, vajiekaran, mutrajanan.*

Chemical constituents — Bark contains two resins and bitter poisonous alkaloid erytherine, which exists in the leaves, also. Air-dried powered bark of *E. indica* yielded three identified crystals as β-sitosterol, γ-sitosterol, and δ-sitosterol.[373] From the seeds of *E. variegata* var. *orientalis*, two crystalline fractions, erythraline and free and liberated erysovine, have been isolated.[374] From the trunk bark of *E. variegata* four alkaloids were separated, viz., erysodine, $C_{10}H_{21}NO_3$; erysovine, $C_{18}H_{21}NO_3$; erysonine, $C_{17}H_{19}NO_3$; and hypaphorine, $C_{14}H_{18}N_2O_3$. A new alkaloid erystorine, $C_{19}H_{23}NO_3$, was also isolated.[375]

Pharmacological action — Antibilious, expectorant, febrifuge, and anthelmintic. The alkaloids showed a weak spasmolytic and relaxant action on smooth muscles of guineapig ileum and rat uterus. The alkaloids also showed curarimimetic action on skeletal muscles by both *in vitro* and *in vivo* tests. They also produced depression of the CNS in rats.[376]

Medicinal properties and uses — Decoction of the bark is used in dysentery, in worms, and useful as a collyrium in ophthalmia. Fresh juice of the leaves with a few drops of honey is a good vermifuge; it acts as a cathartic. Crushed leaves are applied hot to rheumatic joints to relieve pain and as poultice they are applied hot and bandaged upon venereal buboes. The drug is used in liver troubles, also. Decoction of the root bark is given in diabetes and is said to reduce the quantity of sugar in urine within a short time. Leaf juice is said to have cured long-standing dysmenorrhea, and also removed sterility in fatty women by gradually reducing fat and producing natural menstrual flow. The juice increases the secretion of milk if taken during the period of lactation. In its action erythrine, the alkaloid, is antagonistic to strychnine and may be used as an antidote to strychnine poisoning.[3]

Doses — Leaf infusion — 10 to 40 ml; decoction of bark — 56 to 112 ml; leaf juice — 15 to 20 ml; bark powder — 2 to 4 g.

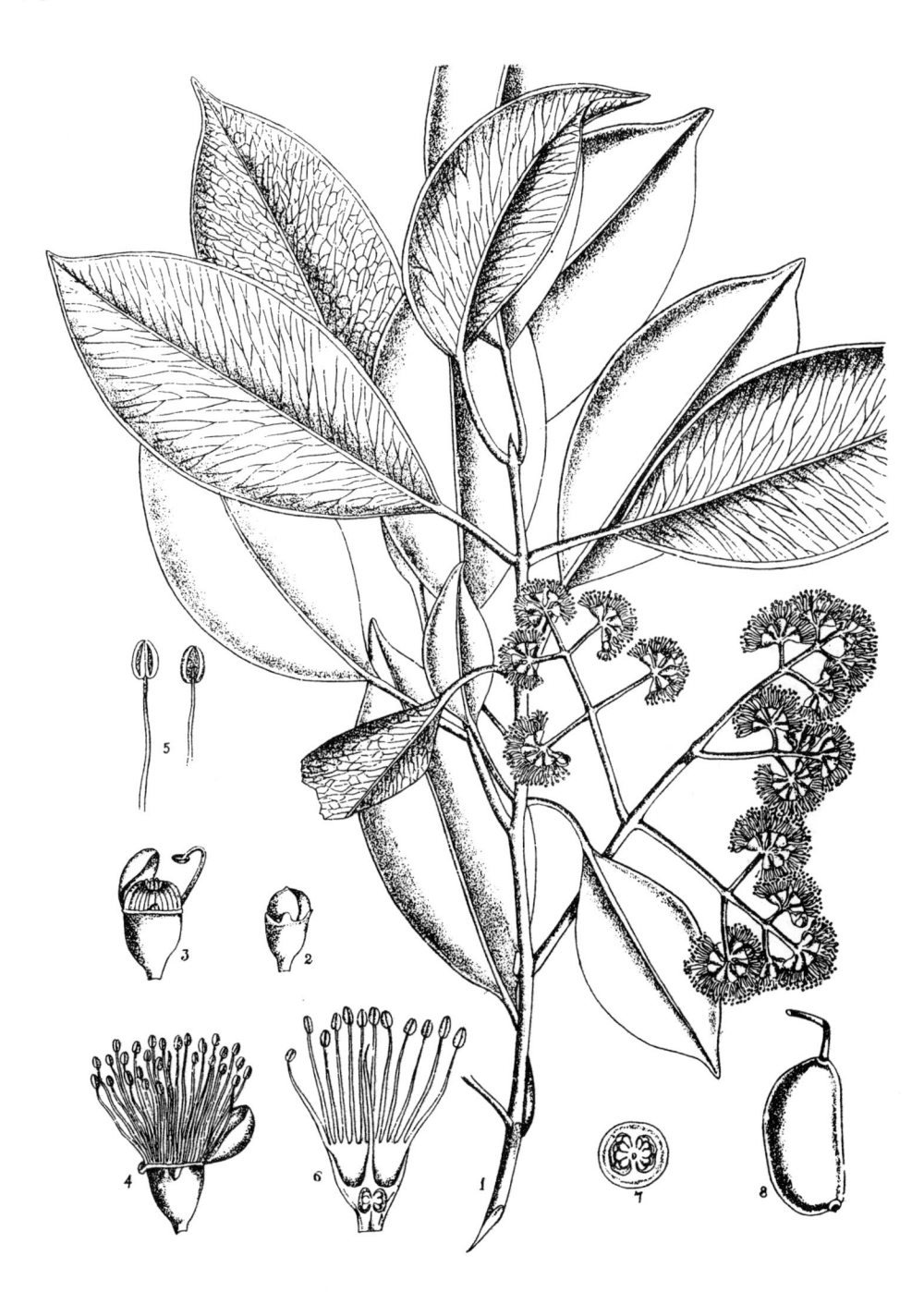

EUGENIA JAMBOLANA Lam.
(Syn. *SYZYGIUM JAMBOLANUM* DC.)

Family: MYRTACEAE

Vernacular names — Sanskrit, Jambu; Hindi, Jamoon; English, Black berry, jambolan; Bengali, Kalajam; Nepalese, Jamun; Sinhalese, Modan; Japanese, Natsume; Chinese, Tsao; German, Gewarz Nelke; French, Pomme rose; Unani, Jamun; Tamil, Naval.

Habitat — Grown throughout the plains from sub-Himalayas to South India.

Parts used — Fruits, leaves, seeds, and bark.

Pharmacognostical characteristics — Fairly large tree, leaves are opposite, oval, and smooth, about 3 to 6 in. long. Flowers white and scented, and fruits 1/3 to 1 1/2 in. long, round, and violet black in color.

Ayurvedic description — *Rasa — kasaya, madur, amal; Guna — laghu, rooksha; Veerya — sheeta; Vipak — madur.*

Action and uses — *Kaf pit samak, stamvan, vat bardhak, rakt stomvan, dipan, pachan, madumeha nasak.*

Chemical constituents — Its flowers contain two triterpene acids, oleanolic acid and crategelic acid, as their respective methyl esters.[377] Seeds contain a glucoside (jamboline), a new phenolic substance, a trace of pale-yellow essential oil, chlorophyll, fat, resin, gallic acid, and albumen. Bark contains tannin (12%) and kino-like gum. The phenolic substance is isolated from *jambu* seeds. Its fruits contain sugar (8.09%), nonreducing sugar (9.26%), and sulfuric acid (1.21%).

Pharmacological action — Bark, leaves, and seeds are astringent. Berry as a whole is astringent. Juice of the fruit is stomachic, astringent, diuretic, and antidiabetic. Vaish[378] reported that fresh powdered seed of *E. jambolana* lowered the level of blood sugar in diabetic rabbits. Aiman[379] confirmed that fruits and seeds were promising hypoglycemic agents. Shrotri reported that aqueous extract of seeds produced 15 to 25% fall in fasting blood sugar in 4 to 5 h after giving a single dose orally.[380]

Medicinal properties and uses — The bark is astringent; its juice is given in 56- to 112-ml doses in chronic diarrhea, dysentery, and menorrhagia. Decoction of the bark is an efficacious mouth-wash and gargle for treating spongy gums, stomatitis, relaxed throat, and other diseases of the mouth. The seeds are very efficacious for diabetes mellitus and for glycosuria; it quickly reduces sugar in urine. Juice of ripe fruit made into vinegar is used as stomachic, carminative, and diuretic.[50] The fresh juice of bark given with goats' milk is prescribed for diarrhea of children.[3]

Doses — Juice — 56 to 112 ml; bark powder — 0.5 to 1 gr; seed powder — 1 to 3 gr.

EUPATORIUM TRIPLINERVE Vahl. *Family:* COMPOSITAE
(Syn. *EUPATORIUM AYAPANA* Vent.)

Vernacular names — Sanskrit, Ayapana; Hindi, Ayapan; Bengali, Ayapan; Unani, Ayapan.

Habitat — Cultivated in various parts of India in damp places. An exotic from Latin America.

Parts used — Whole plant.

Morphological characteristics — A slightly aromtic herb, 3 to 4 ft high, with trailing stems at the nodes, subsessile, lanceolate leaves, and corymbs of bluish-brown head. In transection the stem shows barrel-shaped epidermal cells, few stomata, and few trichomes. Cortex has big intercellular spaces and few starch grains. Few crystals of calcium oxalate are present.

The midrib of the leaf lamina shows an abaxial hump of hypodermal collenchymatous cells. The mesophyll has a broad crescent-shaped meristele embedded in parenchyma and consisting of many narrow collateral bundles. Besides the main vascular tissue there are two subsidiary strands. Phloem has a few starch grains.[381]

Ayurvedic description — *Rasa—tikta, kasaya; Guna—laghu, rooksha; Veerya—ushna; Vipak—katu.*

Action and uses — *Kapha pitta samak, raktrodhak, bran sodhan, bran ropan, visaghan, dipan, pachan, anuloman, grahi, vamak, sheeta jwaraghan, rakt pradarnasak.*

Chemical constituents — It contains an essential oil and a neutral crystalline principle, ayapanin (7-methoxy coumarin; mp 114 to 115°C), and ayapin (6,7-methylendedioxy cumarin; mp 220 to 221°C). Both ayapanin and ayapin are nontoxic and are effective when applied locally. The leaves contain carotene and free vitamin C.[50]

Pharmacological action — Stimulant, tonic, cardiac stimulant, and diaphoretic.

Medicinal properties and uses — Decoction of plant and juice of leaves considered detergent in Philippines and applied to foul ulcers. Leaf decoction is hemostatic. Herb, including its dried leaves, flowering tops, and twigs, is given in the form of an infusion as a bitter tonic, expectorant, diaphoretic for the stomach and bowels, in dyspepsia, cough, and ague. It has also antiscorbutic and alterative properties. An aqueous extract of the leaves and shoots is a cardiac stimulant, increasing the force of the heartbeat but diminishing its frequency.

Doses — Infusion — 14 to 56 ml; powder — $1/2$ to 1 dram.

EUPHORBIA HIRTA Linn.
(Syn. *E. PILULIFERA* Linn.)

Family: EUPHORBIACEAE

Vernacular names — Sanskrit, Dugadhika; Hindi, Dudhi; Bengali, Kharen; Nepalese, Dudajar; Sinhalese, Dadakeeriya; Japanese, Taiwannishikiso; German, Pillenwolfsmilch; Unani, Doodhi; Persian, Sheer-e-geyah; Malaysian, Ambin jantan; Tamil, Ammanpacho haris.

Habitat — Throughout the hotter parts of India.

Parts used — Whole plant.

Morphological characteristics — A small perennial herb with milky latex in all parts of the plant. Stems hairy; leaves small, opposite, elliptic-oblong or oblong lanceolate; fruits small, seeds smooth and blue colored.

Ayurvedic description — *Rasa — madhur; Guna — snigdha; Veerya — sheeta; Vipak — madhur.*

Action and uses — *Vata pitta samak, medhya, trisanigraha, anuloman, rakt pitta samak, sandhan kark, kaphanisark, mutral, virisya, balya, birhan, hirdya.*

Chemical constituents — The drug contains alkaloid and essential oil as well as quercetin, triacontane, jambulon, a phenolic substance, euphosterol, a phytosterol, and phytosterolin; gallic, melissic, palmitic, oleic, and linoleic acids, L-inositol, and the alkaloid xanthorhamnin.[382,383]

Pharmacological action — Anthelmintic and expectorant. Alcoholic extract of entire plant *E. hirta* showed anticancer activity against Friend virus leukemia (solid) in mice. Hypoglycemic activity was observed in albino rats with the extract. The extract also possessed antiprotozoal activity against *Entamoeba histolytica*.[51] One of the active principles causes a spike phase in guinea pig ileum and the other a relaxing action on smooth muscle.[50]

Medicinal properties and uses — The plant as a whole is used in diseases of children in worms, bowel complaints, coughs, etc. Decoction of the plant is given in bronchial affections and asthma. Juice of the plant is also very useful in dysentery and colds. Latex of the plant is used as application for warts.[3,50]

Doses — Infusion — 14 to 28 ml; decoction — 28 to 56 ml.

EUPHORBIA NERIIFOLIA Linn. *Family:* EUPHORBIACEAE

Vernacular names — Sanskrit, Snoohi; Hindi, Sehund; English, Common milk hedge; Bengali, Mansasij; Nepalese, Sihundi; Sinhalese, Daluk; Japanese, Kirinkaku; French, Enpurge; Arabian, Dihu minguta; Burmese, Thasaung; Tamil, Illaikalli.

Habitat — This plant is common in rocky ground throughout the Deccan peninsula and often cultivated for hedges in villages throughout India.

Parts used — Juice and root.

Morphological characteristics — A small shrubbery plant with jointed cylindrical or obscurely five-angled branches bearing short, stipular thorns, more or less confluent in vertical or slightly spiral lines, leaves fleshy, deciduous, obovate-oblong, 6 to 12 in. long, terminal on the branches. The trunk is covered with reticulate bark.

Ayurvedic description — *Rasa — katu; Guna — laghu, snigdha, teekshna; Veerya — ushna; Vipak — katu.*

Action and uses — *Kapha vata har, badanasthapan, lakhan, dipan, rachan, rakt sodhak, kaphanisark, vishaghan.*

Chemical constituents — The latex contains 69 to 93% water solubles and 0.2 to 2.6% caoutchouc.

Pharmacological action — Juice is purgative and expectorant, locally rubefacient. Root is antispasmodic. Iqbal et al.[384] reported that the latex of *E. neriifolia* in distilled water produced a persistent rise in blood pressure of the dog. Intravenous or intracerebroventricular administration produced manifold increase in the rate of respiration in dog. On isolated guinea pig ileum, repeated doses of the latex produced tachyphylaxis of the stimulation action. Rizvi et al.[385] reported that the latex was a powerful contact poison but weaker as a stomach poison. It produced pathological changes in the liver, heart, and kidney.

Medicinal properties and uses — Milky juice exuded from injured fleshy cylindrical stems is used as a drastic purgative in the enlargement of liver and spleen, syphilis, dropsy, general anasarca, and leprosy. Externally the juice is applied to remove warts and similar excrescences; and is dropped into the ear to afford relief in earache. It is also a useful application with butter to unhealthy ulcers and scabies. When it is applied to glandular swelling it prevents suppuration. Root mixed with black pepper is employed in scorpion stings and snakebites both externally and internally. Pulp of the stem mixed with fresh ginger is used to prevent hydrophobia. It is a fish poison.[50]

Doses — Root powder — 3 to 7 gr; stem juice — 15 to 25 ml; leaf juice — 2 to 5 minims; milky juice — 1 to 2 gr.

EVOLVULUS ALSINOIDES Linn.　　　　　　　　　　　　　　　　　Family: CONVOLVULACEAE

Vernacular names — Sanskrit, Vishnukraanti; Hindi, Sankhapuspi; Bengali, Sankhapuspi; Nepalese, Sankhapuspi; Sinhalese, Sakmal; Unani, Sankhahuli; Tamil, Vishnukiranthi.

Habitat — A plant growing as a common weed in open and grassy places throughout India, ascending up to 6000 ft in the Himalayas.

Parts used — The whole herb.

Ayurvedic description — *Rasa* — *kasaya, katu, tikta; Guna* — *guru, snigdha, pichil, sar; Veerya* — *sheeta; Vipak* — *madhur.*

Action and uses — *Vata pitta samak, mastisk samak, nidrajanak, dipan, pachan, anuloman, kaf nisarak, virisya, mana dosha nasak, dhatu vardhak, malsansodhan, rasayan, balya.*

Chemical constituents — The whole plant yields betaine and another water-soluble base, mp 60 to 61°C. Extracts gave positive reactions for alkaloid, sterols, proteins, amino acids, carbohydrates, phenolic compounds, and tannins. The alkaloid was identified as evolvine, which is optically active. The fatty residue was optically inactive.[178]

Pharmacological action — Bitter tonic, alterative, and febrifuge; also anthelmintic and antiphlogistic.[3,50] Baveja and Singla[386] found that a fraction (D) in low doses (0.02 mg) decreased heart rate and force of cardiac contraction in pithed frog and isolated rabbit intestinal loop. In larger doses (0.1 mg) it caused cardiac arrest in diastole. It decreased the amplitude of contraction and increased the tone of intestinal muscle in a concentration of 1:50,000 to 1:250,000. The crude mixture of alkaloid was found toxic to earthworms.

Medicinal properties and uses — The whole plant in the form of decoction or infusion is used in fever, nervous debility, and loss of memory in doses of 56 to 112 ml. It is a good remedy in bowel complaints, especially dysentery. In diarrhea or indigestion, a decoction of the drug with *Ocimum sanctum* is very useful. Leaves made into cigarettes are smoked in chronic bronchitis and asthma. It is also used as anthelmintic and antiphlogistic.[1,3]

Doses — Infusion — 24 to 48 ml; powder — 3 to 6 gr.

FERULA FOETIDA Regel.   *Family:* UMBELLIFERAE

Vernacular names — Sanskrit, Hingu; Hindi, Hingra; English, Asafetida; Bengali, Hingra; Nepalese, Hing; Sinhalese, Perunkayam; German, Stinkasant Teufels Kraut; Unani, Anjadan; Arabian, Anjadan; Persian, Angudan; Tibetan, Sin-kun; Burmese, Singu; Malaysian, Hingu; Tamil, Perunkayam.

Habitat — The plant grows wild in Kashmir, Persia, and Afghanistan.

Parts used — Oleo gum resin, obtained by incision from the root.

Pharmacognostical characteristics — An unpleasant smelling, herbaceous, perennial, 5 to 8 ft high. Leaves downy, at least when young, tripartite, entire or irregular toothed; lower leaves 1 to 2 ft long, stalk sheathing the stem. Flowering umbels arising from the sheath of the lower leaves are simple; terminal umbels branched, leafless, large; flowers yellow. The root in transverse section shows a normal diarch structure. A large number of secretory canals (oleo-gum resin canals) are scattered in the secondary phloem in a concentric way. The parenchymatous cells of the secondary phloem are filled with starch grains. A variable number of vascular strands arise inside the periderm, each of which has its own cambial ring encircling it.

The stem in transection shows an almost circular outline. Collenchymatous patches and some layers of chlorenchyma cells are seen. Secondary canals are present beneath each collenchymatous patch of cells. Vascular bundles are arranged in a ring. A number of collateral medullary bundles are observed in the pith region.[387]

Ayurvedic description — *Rasa — katu; Guna — lagu, snigdha, teekshna, sar; Veerya — ushna; Vipak — katu.*

Action and uses — *Badana sthapan, kapha-vata samak, pittawardhak, shulprasaman, vathar, dipan, pachan, snigdh, anuloman, krimighan, hiridya kandughan, jwaraghan swashar, vajekarn, artawjanan, mutral.*

Chemical constituents — Chopra et al.[50] reported the presence of essential oil, ferulic acid, organic disulfide, and umbelliferone. The constituents of gum are reported to be glucuronic acid, galactose, arabinose, rhamnose, and a protein.[178] Trease reported that asafetida consists of volatile oil, resin, gum, and impurities.[69] The oil has a particularly evil smell and contains sulfur compounds of the formulas $C_7H_{14}S_2$, $C_{16}H_{20}S_2$, $C_8H_{16}S_2$, $C_7H_{14}S_3$, and $C_8H_{16}S_3$; some of these show pesticidal activity. Among the resinous constituents are asaresinol ferulate and free ferulic acid. The drug contains no free umbelliferone. Ferulic acid is closely related to umbellic acid and umbelliferone.[142]

Pharmacological action — Stimulant, carminative, antispasmodic, expectorant, and slightly laxative; also anthelmintic, diuretic, aphrodisiac, and emmenagogue. It is a nervine and pulmonary stimulant and also increases the sexual appetite. Asafetida produced slight inhibitions of growth of *Staphylococcus aureus* and *Shigella sonnei*.[388]

Medicinal properties and uses — It is a valuable remedy for hysteria and nervous disorders of women and children, flatulence, flatulent colic, and spasmodic affections of the bowels especially when connected with hysteria. In nervous palpitations, hypochondriasis, and other affections due to hysteria, whooping cough, pneumonia, and bronchitis in children, asafetida is also given successfully. The volatile oil is rapidly excreted and may be found in the urine, milk, or sweat.[3,50]

Doses — Gum resin — 5 to 10 gr.

FICUS BENGHALENSIS Linn.
(Syn. *F. INDICA* Linn.)

Family: MORACEAE

Vernacular names — Sanskrit, Vata; Hindi, Vada; English, Banyan tree; Bengali, Bargat; Nepalese, Bar; Sinhalese, Nuga; Unani, Bargad; Arabian, Zatuzzawanelia; Persian, Darakht-e-reshah; Burmese, Pyinyuang; French, Figuier due bengal; Tibetan, Nya-gro-dha; Pustu, Bar; Tamil, Alamaram.

Habitat — Found all over India.

Parts used — Milky juice, bark and leaves.

Morphological characteristics — A large tree, juice milky. Bark grayish and smooth, 1.3 cm thick, exfoliating in irregular flakes; wood grayish white, moderately hard with alternate rings of light and dark tissue, heartwood absent; leaves alternate, ovate, or elliptic, entire, obtuse, subcordate or rounded at base, thickly coriaceous, harsh, glabrescent above; nerve basal 3 to 7, lateral 4 to 6 pairs; petioles 2.5 to 5-cm stout; stipule, 1.9 to 2.5 cm, sheathing, decidous, protecting leaf bud; male flowers crowded near mouth of receptacle perianth; 2 to 6-fid, imbricate, stamens 1 to 2, erect in bud; in female flowers, perianth is shorter, style excentric, and the ovule pendulous. Fruit an enlarged hollow cup-shaped closed receptacle, the inner wall studded with crustaceous or fleshy achenes. Albumen scanty.[28]

Ayurvedic description — *Rasa—kasaya; Guna—guru-rooksha; Veerya—sheeta; Vipak—katu.*

Action and uses — *Kapha pitta samak, badana sthapan, bran ropan, rakt rodhak, rakt pitta har, sukra sthamvan, daah prasaman.*

Chemical constituents — Bark and young buds contain about 10% tannin, wax, and caoutchouc; fruit contains oil, albuminoids; carbohydrates, fiber, and ash 5 to 6%. The bark contains a hypoglycemic principle (glycoside). A triterpene, friedelin and β-sitosterol were isolated from the leaves of *F. benghalensis*.[389] Subramanian et al.[390] isolated tigilic acid ester of γ-taraxosterol from the heartwood of the tree. The flavonols of leaves were identified as quercetin-3 galactoside and rutin. Dried bark, on extraction with 95% alcohol, yields an effective glycemic principle.[178]

Pharmacological action — Antiseptic, aphrodisiac, astringent, cooling, and hemostatic. The glycoside isolated from bark produced a hypoglycemic effect in normal rabbits but not in diabetic animals.[391] The flavonoid compounds isolated from the bark were found effective as hypoglycemic agents on oral administration to normal fasting rabbits.[392] The milky latex caused initial lowering of fasting blood sugar in rats.

Medicinal properties and uses — The leaves and buds are astringent; their infusion is given in diarrhea and dysentery. Milky juice and seeds or fruits are useful as external application to pains and bruises, sores and ulcers, in rheumatism, and in lumbago. An infusion of the bark has specific properties in reducing blood sugar in diabetes, dysentery, gonorrhea, and in seminal weakness and is a powerful tonic. Leaves are heated and applied as a poultice to abscesses and wounds to promote suppuration and discharge of pus. Infusion of the small branches is useful in hemoplysis.[1]

Doses — Decoction — 56 to 112 ml; powder — 1 to 2 gr; milk — 0.5 to 1 ml.

FICUS RACEMOSA Linn.   *Family:* MORACEAE
(Syn. *FICUS GLOMERATA* Roxb.)

Vernacular names — Sanskrit, Udumbara; Hindi, Gular; English, Country fig tree, Cluster fig; Bengali, Jayadumar; Nepalese, Durmy; Sinhalese, Attikka; Japanese, Udonge; French, Figuier du dialile; Unani, Goolar; Arabian, Jummaiz; Persian, Samarpasah; Burmese, Thapan; Malaysian, Atti; Pustu, O-rmul; Tamil, Aitt.

Habitat — It is found all over India, but is more common in the outer and sub-Himalayan tracts, Rajputana, Assam, Bengal, central, western, and southern India.

Parts used — Root, root bark, leaves, fruit, and milky juice.

Pharmacognostical characteristics — A moderate-sized to large spreading tree with ovate, ovate-lanceolate, or ellipitic, dark green leaves, fruits red when ripe, 1 to 2 in. in diameter, sub-globose or pyriform, borne in large clusters on short leafless branches emerging from the trunk and the main branches.

The surface of the bark is fairly smooth and soft, and not deeply cracked, fissured, prominently lenticellate, or with exfoliating woody outer rind as in *F. benghalensis*. The thickness of the whole bark varies from less than a quarter of an inch to about three quarters of an inch, depending upon the age of the bark. The surface is extremely thin or thinly papyraceous and somewhat translucent. It is always found cracking in very close, irregular, vertical series and exfoliating in very small, oblong, or circular flakes about 10 in. wide, except in very young bark, which is very thin. The bark of *F. glomerata* is gray, and on drying the pieces become curved. It has an indistinct odor and astringent taste. The cork has three to eight layers of rectangular cells. Individual parenchymatous cells as well as sclereids are arranged in radial files with corresponding phellogen and cork cells. The cortex is wide with numerous sclereids. Phloem has sieve tubes, companion cells, etc. Powder is light pink to light brown in color.[393]

Ayurvedic description — *Rasa* — *kasaya, madhura; Guna* — *gurua, roksha; Veerya* — *sheeta; Vipak* — *katu.*

Action and uses — *Kaf pita samak, soth har, badana-samak, bran-ropana, rakt pita samak, shukrya sthamvak, daha-prasamk.*

Chemical constituents — The leaves contain crude protein (12.36%), ether extract (2.75%), crude fiber (13.03%), total carbohydrates (71.91%), and total ash (12.98%).[394]

Fruit contains moisture (13.6%), albuminoids (7.4%;), fat (5.6%), carbohydrates (49.0%), coloring matter (8.5%), fiber (17.9 and 6.5%); silica (SiO_2) (0.25%), and phosphorus (P_2O_5) (0.91%).

The latex of the tree contains 4.04 to 7.4% caoutchouc and 72.3 to 77.6% resin.

The bark contains 14% tannin. β-Sitosterol and lupeol have been isolated from the petroleum ether extract of the root bark of *F. racamosa*.[395]

Pharmacological action — Bark, leaves, and unripe fruit are astringent, carminative, stomachic, and vermicide. The aqueous extract was found to reduce the blood sugar in normal as well as in alloxan diabetic rabbits.[396]

Medicinal properties and uses — It is astringent and a decoction is used as a wash for wounds. The root is useful in dysentery. Leaves ground to powder and mixed with honey are given in bilious affections. The fruits are astringent, stomachic, and carminative. Milky juice of *F. glomerata* is administered in piles and diarrhea. Fruit is also useful in aphthous complaints, menorrhagia, and hemoptysis. Figs, after boiling in water and straining, are an excellent gargle for sore throat. Bark is used in diabetes in the form of fine powder.

Doses — Powder — 2 to 4 gr; decoction — 56 to 112 ml; infusion of leaves — 6 to l5 ml; milk — 1 to 1.5 ml.

FICUS RELIGIOSA Linn.   Family: MORACEAE

Vernacular names — Sanskrit, Aswatha; Hindi, Pipal; English, Sacred fig; Bengali, Ashwath; Nepalese, Ogalshimu; Sinhalese, Bo; Japanese, Tenjikubodaiju; Chinese, Pou tichou; German, Bobaum Peepal; Unani, Pipal; Arabian, Dar-e-filfil; Persian, Filfil-e-draz; Burmese, Nyaung baudi; Pustu, Pipal; French, Figuier-ou-arbe despagodes; Tamil, Arasu.

Habitat — A large tree found wild and cultivated all over India.

Parts used — Root bark, fruit, seeds, leaves.

Morphological characteristics — Leaves alternate, 8 to 17 cm, ovate, circular, acuminate, glabrous, shining, margins wavy. Bark gray, smooth, wood whitish, moderately hard. The fruit is produced by the union of the cymose inflorescence to form a hollow fleshy axis bearing the flowers on its inner surface. The young fruit is rich in latex but, when mature, no latex is found.[69]

Ayurvedic description — *Rasa—kasaya; Guna—guru, rooksha; Veerya—sheeta; Vipak — katu.*

Action and uses — *Kapha pitta samak, branya branropan, badana sthapan, sothahar, rakt rodhak, sanahan, anuloman, raktsodhak, rakat pit samak, swas har, vaje karn, mutra sangrahaniya.*

Chemical constituents — Bark contains tannin, caoutchouc, and wax. Ambike and Rao[397] isolated β-sitosterol-D-glucoside from the bark of *F. religiosa* as a white crystalline compound, mp 282 to 283°C, molecular formula $C_{36}H_{60}O_6$. Bark is also reported to contain 4% tannin.

Pharmacological actions — Seeds and fruits are cooling, laxative, refrigerant, and alterative. Leaves and young shoots are purgative. Aqueous extract of the bark shows antibacterial activity against *Staphylococcus aureus* and *Escherichia coli*. Pharmacologically, it shows a relaxant and spasmolytic effect on smooth muscles.[178]

Medicinal properties — Bark is cooling and useful in gonorrhea and ulcers, various skin diseases, and scabies. Infusion or decoction is given in cases of hiccup and to stop vomiting sensation. Latex boiled with milk is a good aphrodisiac. A medicated oil made from root bark is applied externally in skin diseases caused by vitiated blood such as eczema, leprosy, and rheumatism. β-Sitosterol-D-glucoside found in the bark has shown hypoglycemic activity, comparing favorably with tolbutamide. Infusion of bark is astringent. Fruit is laxative and digestive.[1,3]

Doses — Infusion — 3 to 5 drams; decoction — 56 to 112 ml; powder — 1 to 2 drams.

FOENICULUM VULGARE Mill.   *Family:* UMBELLIFERAE
(Syn. *F. CAPILLACEUM* Gilib.,
     ANETHUM FOENICULUM Linn.)

Vernacular names — Sanskrit, Satupuspa; Hindi, Saunf; English, Fennel; Bengali, Mauri; Nepalese, Marchyashupp; Sinhalese, Satakuppa; Japanese, Uikyo; Chinese, Hui-hsiang; German, Garten Feuchel; Unani, Saunf; Arabian, Razyanaj; Persian, Badyan; Tamil, Perunchirakam.

Habitat — A native of Europe, but cultivated throughout India.

Parts used — Leaves, seeds, and oil.

Pharmacognostical characteristics — Stem erect, fistular, about 3 ft high, leaves three or four pinnates; segments narrow, linear, glaucous. Flowers small, hermaphrodite in compound umbels. Fruit about $1/_4$ in. long; vittae conspicuous. Shah et al.[398] studied the morphological and anatomical characters of fruit and essential oil obtained from Ootacamund and compared them with the authentic sample. The vascular strands and vittae are much more developed than those of the common fennel. The walls of epicarp cells are beaded and much more thickened than those of the common fennel. Chemical races have been found to occur in the fruit of *F. vulgare*.

Ayurvedic description — *Rasa* — *madhur, tikt, katu; Guna* — *lagu, snigdha, tedeshna; Veerya* — *ushna; Vipak* — *madur.*

Action and uses — *Vata kapha samak, medhya, dristi sakti vardhak, dipan, pachan, anuloman, jawaraghan, daha presaman, aawart janan, sthanjanya, balja.*

Chemical constituents — Fennel fruits yield about 3 to 5% of volatile oil of pleasant aromatic odor, which consists of anethole or anise camphor and variable proportions of a liquid isomeric with oil of turpentine. Small quantities of other substances like fenchone are also present. The fruits of fennel collected from Ootacamund were found to be free of anethol and represented a methyl chavicol race of *F. vulgare*. The volatile content was up to 8.0%.[399]

Pharmacological action — Dried ripe fruit and its essential oil is stimulant, aromatic, carminative, diuretic, emmenagogue, and purgative.

Medicinal properties and uses — The leaves of *F. vulgare* are diuretic, digestive, and appetizing; they are useful in cough, flatulence, colic and thirst. The seeds are sweet, laxative, stomachic, stimulant, and an appetizer. It is given in dysentery, biliousness, diseases of the chest, spleen, and kidneys, headache and griping of the bowels. It also promotes female monthly regularity. The oil is an anthelmintic against hookworms in doses of 4 ml.

Doses — Powder — 10 to 40 gr; oil — 5 to 10 ml; arka — 28 to 56 ml (water distillate).

FUMARIA INDICA (Haussk.) Pungsley   Family: FUMARIACEAE
(Syn. *F. PARVIFLORA* Lamk.)

Vernacular names — Sanskrit, Parpata; Hindi, Pitapapara; English, Common fumitory; Bengali, Pitpapra; Nepalese, Kairuwa; Sinhalese, Patha padagam; German, Erdrauch; Chinese, Tuysha tu chian; Unani, Shahotarah; Arabian, Shahatraj; Persian, Shahatrah; Pustu, Shahtara; Tamil, Thara.

Habitat — An exotic weed, now naturalized and growing throughout India.

Parts used — Whole plant.

Morphological and pharmacognostical characteristics — Leaf pale green, segments flat; racemes scented; flowers pale pink, 0.5 to 1.25 cm; sepals lanceolate, much smaller than the corolla tube; pedicels exceeding the bracts; fruit globose, rugose when dry; rounded at the top with two pits, pale green, much branched; racemes 2.5 cm; fruit one seeded, small globose nutlets.

In mature root with secondary growth, broad medullary rays are observed radiating outward opposite to xylem poles; the secondary wood is laid alternating the xylem pole and is fan-shaped. The secondary wood elements are arranged in radiating rows embedded in parenchyma. Phloem lies outside the xylem and is composed of sieve tubes, companion cells, and phloem parenchyma, followed toward the periphery by a few layers of cork cells.

Transverse section of the stem is pentagonal in outline with fairly prominent angles under which collenchyma lies. The epidermis is a single layer of oblong rectangular cells with outer well-developed tangential wall. The cortex is narrow, composed of three to four layers of chlorenchymatous cells. Endodermis is not discernible. Vascular bundles are present underneath the ridge; the bundles are collateral, endarch, and open. The xylem is composed of vessels, tracheid, fibers, and xylem parenchyma. The protoxylem elements are tracheids with annular or spiral thickening.

Transverse section of leaf shows indistinctly bifacial organization. The upper and lower epidermis are single layered, formed of large hyaline cells, of oblong to rectangular shape; the stomata are of the anomocytic type. Epidermal cell shows the presence of aggregate crystals of calcium oxalate.[400]

Ayurvedic description — *Rasa — tikta; Guna — laghu; Veerya — sheeta; Vipak — katu.*

Action and uses — *Kapha pitta samak, mastisk samak, trisa samak, dipan, grahi, krimighan, rakt sodhak, daah samak, jweraghan, kustaghan.*

Chemical constituents — Pandey et al.[401] isolated seven alkaloids from alcoholic extract of whole plant. These are (1) protopine, $C_{20}H_{19}O_5N$, mp 207 to 209°C; (2) tetrahydrocoptisine; (3) tautomeric form of fumariline, a homogenous gum, $C_{20}H_{17}O_5N$; (4) a racemic mixture of bicuculline and its optical antipode, $C_{20}H_{17}O_6N$ mp 234 to 235°C; (5) bicuculline; (6) fumarilicine, $C_{20}H_{19}O_5N$; and (7) narceimine, $C_{22}H_{21}O_9N$, mp 264 to 265°C.

Pharmacological action — Anthelmintic, aperient, cooling, diaphoretic, diuretic, and febrifuge. Bhattacharya et al.[402] found that the extract of plant showed a relaxant effect and moderate fall of blood pressure on experimental animals. Protopine, the major alkaloid, was found to be approximately equipotent to papaverine.

Medicinal properties and uses — Decoction or infusion of the stem and leaves is given as tonic, anthelmintic, aperient, and alterative; useful in syphilis, scrofula, leprosy, constipation, and dyspepsia due to torpor of the liver or intestine. It is given in ague and jaundice; also in skin diseases to purify the blood.[1,3]

Dose — Decoction — 28 to 56 ml; powder — 4 to 6 gr.

GARCINIA INDICA Choisy *Family:* GUTTIFERAE
(Syn. *G. PURPUREA* Roxb.)

Vernacular names — Sanskrit, Brikshamla; Hindi, Kokam; English, Red mango; Nepalese, Bhakeamilo; Sinhalese, Goraka; German, Kokumol; Unani, Kakum.

Habitat — Konkan, North Kanara, Western Ghats of Bombay, South Kanara, Coorg, Wynaud; often cultivated.

Parts used — Fruit, concrete oil, seeds, and bark.

Morphological characteristics — Tree, usually with yellow juice. Leaves dark green, when young, red when mature, membranous mucronate, rarely obtuse; flowers polygamous; male flowers 4 to 8, in axillary and terminal fascicles; bud as large as pea; sepals orbicular, 4 to 5, outer smaller; petals rather large, 4 to 5 imbricate. Stamens membranous, 12 to 20, forming a short capitate column, anthers sessile on the staminal column or on short, thick filaments. Female flowers solitary, short, terminal, with free or connate staminoids in 4 masses; ovary 4 to 8 celled; stigma sessile, peltate of many lobes; ovule one in each cell, axile fruit spherical, large, purple berry with rough rind; seeds 5 to 8, compressed, enclosed in an acidic pulp.

Ayurvedic description — *Rasa — amla; Guna — laghu, rooksha; Veerya — ushna; Vipak — amla.*

Action and uses — *Vatta samak, kapha pitta, wardhak, pitta samak, bran ropan, rochan, dipan, pachan, grahi vataanuloman, hirdya, jwaroghan, daah prasaman, sandhaniy, ropan.*

Chemical constituents — The seeds contain 30% fat, yielding a pale yellow concrete oil known as kokum oil. Fruits contain cellulose, an extractive, and an insoluble residue.

Pharmacological action — Fruit is cooling, cholagogue, demulcent, emollient, and antiscorbutic. Bark is astringent. Oil is soothing, used in skin diseases.

Medicinal properties and uses — *Kokum* oil is a specific remedy in dysentery, mucus, diarrhea. It is also useful in phthisis and some skin diseases. Externally, oil has a healing property and usefully employed as an application to ulcerations and fissures of the lips. It is considered an excellent substitute for animal fat as a basis for ointment.[1,3] The fruit is refrigerant, dumulcent, antiscorbutic, and astringent. A drink of infusion and its local application all over the body are prescribed in urticaria.[1]

Doses — Oil — 2.5 to 5 ml; decoction of bark — 20 to 30 ml.

GARCINIA MORELLA Desr. Family: GUTTIFERAE

Vernacular names — Sanskrit, Tamal, Kankushtha; Hindi, Gotaghanba; English, Indian gamboge; Bengali, Tamal; Nepalese, Imal; Sinhalese, Gokatu; German, Gummigutt; Unani, Kokum; Burmese, Thamengut.

Habitat — Found in Bangladesh, North Kanara, Western Ghats from Mysore to Travancore, up to 3000 ft.

Parts used — Gum resin.

Morphological characteristics — A medium-sized tree with spreading branches. Leaves 3 to 5 × 1 to $2\frac{1}{2}$ in., elliptic-obovate to ovate lanceolate, narrowed at the base. Flowers tetramerous; petals a little larger than sepals. Male flowers axillary, in fascicles of 2 to 5, subsessile. Stamens 25 to 40, monadelphous, the filaments combined into a subquandrangular central column, but free at the apex; anthers orbicular, flattened, dehiscing transversely. Female flowers larger than the male, solitary, axillary, usually sessile. Staminodes 18 to 30. Ovary smooth, 4-celled; stigma peltate, irregularly lobed and tubercled. Fruit $\frac{3}{4}$ in. diameter, subglobose, slightly 4-lobed, surrounded at the base by the persistent sepals. Seeds 4, ovoid-reniform.[28]

Ayurvedic description — *Rasa — takta, katu; Guna — laghu; Veerya — ushna; Vipak — katu.*

Action and Uses — *Rechan, ruchya, varnavishodhan, shophaghna, gulmaghna aadhmaanhar, varnaghna, shoolshar.*

Chemical constituents — Gamboge is a yellow gum or gum resin that exudes from the tree. Pericarp of the seeds yields morellin. Karanjgoakar et al.[403] isolated a new type of biflavonoid, i.e., morelloflavone, a 3-(8) flavonoyl-flavonone, from the heartwood of *G. morella*. Chopra et al.[178] reported that the pericarp of the seed contained 20% morellin, a highly active compound. Coloring matter consists of isomorellin, desoxymorellin, and dihydroisomorellin.

Pharmacological action — Purgative and vermifuge. Morellin obtained from *G. morella* has been widely used for promoting wound healing and promising results have been reported by Deshpande et al.[404] in clinical trials. The gum resin and seed coat extract showed cathartic activity due to the presence of β-guttiferin and α-l-guttiferin in them.[405]

Medicinal properties and uses — Gum resin is purgative and anthelmintic. It is used in dropsical affections, ammenorrhea, obstinate constipation, and as vermifuge.[3]

GLORIOSA SUPERBA Linn.   Family: LILIACEAE

Vernacular names — Sanskrit, Langalika; Hindi, Kalihari; English, Superb lily, climbing lily; Bengali, Bishalanguli; Sinhalese, Niyangalla; Japanese, Yurigurama; German, Gloriosa Knollen; French, Glorieus du Malaliear; Unani, Muleen; Persian, Buch nak-hindi; Burmese, Simadon.

Habitat — Commonly found in the tropical forests of Bengal, Karnatha, (India), Burma, and Sri Lanka.

Parts used — Leaves and tubers.

Morphological characteristics — It is a tall herbaceous climber. Leaves terminating in tendril-like, long, curling tips. Flowers axillary, long-stalked, large, solitary, nodding; stalks rather stout; floral segments long, narrow, margins crisply waved, at first greenish then a mixture of scarlet and bright yellow. Capsules linear oblong, long. Tubers solid, fleshy, cylindric, white, very long, root fibrous. The rhizomes are cylindrical or slightly flattened with bitter taste. Fibers are absent. Vascular bundles are collateral with slight extensions on both sides of xylem. Starch grains show concentric striations and measure 11 to 38 µm in diameter.

In roots the cork cambium arises in the first layers of cortex and forms three layers of cork cells. During development, cortical parenchyma disintegrates and forms air spaces. Central portion of the wood after lignification becomes delignified to a large extent.[406]

Ayurvedic description — *Rasa — katu, tikta; Guna — laghu, tikshna; Veerya — ushna; Vipak — katu.*

Action and uses — *Kapha vat samak, gravapatak, dipan, pitta sarak, krimighan, rakta sodhak, jwaraghana ballya, rasayan.*

Chemical constituents — Young leaves contain chelidonic acid. Leaves of the Indian plant yield colchicine, dimethylcolchicine, *N*-formyldeacetyl-colchicine, lumicolchicine, and two others. Colchicine and related alkaloids have been identified in the bulbs and seeds.[178]

Pharmacological action — Tonic, antiperiodic, cholagogue, anthelmintic, alterative, and purgative. The fresh juice of plant exhibited stimulant action on isolated uteri of guinea-pig, rabbit, and human being. On isolated rat uterus, it showed papaverine-like action.[407] In high doses, the plant extract initiated labor in pregnant rabbits. The tuber extract shows antibiotic activity against *Staphylococcus aureus*.[408]

Medicinal properties and uses — The leaf juice is used to kill lice in the head hair. The tubers contain the bitter principles superbine and gloriosine, which in large doses are poisonous; however, in small doses, they are used as tonic, antiperiodic, alterative, and purgative. The white flour prepared from the tubers is bitter, alterative, and tonic; it is given with honey in gonorrhea, leprosy, colic, and intestinal worms. For promoting labor pains, a paste of the tuber is applied over the suprapubic region and the vagina; its warm poultice is locally applied in rheumatism and neuralgic pains.[1,3]

Doses — 0.1 to 0.25 gr (as bitter tonic); 0.3 to 0.6 gr (as abortifacient).

GLYCYRRHIZA GLABRA Linn. Family: PAPILIONACEAE

Vernacular names — Sanskrit, Yashtimadhu; Hindi, Mulathee; English, Sweet wood, Licorice; Bengali, Yastomadhu; Nepalese, Estamee; Sinhalese, Velmi; Japanese, Kanzo; Chinese, Kan-ts'ao; German, Sussholz; Unani, Mulathee; Arabian, Aslussos; Persian, Bekh-e-mahak; Burmese, Neekhiyu; Tamil, Athimathuram.

Habitat — Grown in Arabia, Persian Gulf, Afghanistan, Turkistan, Asia Minor, and Siberia. The plant is cultivated for its root in Punjab, sub-Himalayan tracts from Chenab eastward; Sind and Peshawar Valley, Burma, and Andaman Island.

Parts used — Roots.

Pharmacognostical characteristics — The leaves are compound, imparipinnate, alternate, leaflets four to seven pairs, oblong to elliptical lanceolate; flowers violet colored, papilionaceous. Fruit is a compressed legume, l.3 cm long.

The outer surface of licorice root is yellowish brown or dark brown in color, longitudinally wrinkled with patches of cork adhering which are prominent in thicker rhizome. Fracture coarsely fibrous; internal color is yellow; taste sweetish and slightly acrid.

The roots are characterized by the presence of several layers of cork cells with reddish-brown contents. The inner three or four layers have thicker colorless walls. In commercial samples of root, phellogen cells are found to be collapsed. The phellogen consists usually of one to three layers of radially arranged parenchymatous cells, and contain isolated prisms of calcium oxalate. The secondary phloem is broad with wide parenchymatous medullary rays; the phloem fibers have thickened walls, cellulosic in the inner part and lignified in the outer part, and radially arranged in groups of about 10 to 50 fibers, adhering to which are cells of crystal fibers containing monoclinic prisms of calcium oxalate. The xylem structure in young roots is nearly tetrarch and it shows absence of pith. In young roots, four medullary rays are present.[410]

Ayurvedic description — *Rasa*—*mahura; Guna*—*guru, snigdha; Veerya*—*sheeta; Vipak* — *madur*.

Action and uses — *Vat pitta samak, daha, samak, shura bardhak, kapha nisarak, vatanuloman, kantaya, raket sthambk, jawar nasak, jiviniya, sandhaniya, rasayan, balya*.

Chemical constituents — The root extracts contain a sweet crystalline glycoside named glycyrrhizin which gives glycyrrhetinic acid on enzymatic hydrolysis. Root also contains sucrose, starch, and acid resin, a bitter principle, aspargin, malic acid, and some proteinous fatty and inorganic matters.[409]

Roots of *G. glabra* have been reported to contain many phenolic compounds, also, such as flavonoids and their glycosides, coumarin, and cinnamic acid derivatives. In addition, isolation of the glucosides, liquiritin and isoliquiritin, along with their aglucones, liquiritigenin and isoliquiriligenin, was made from Indian samples.[411] Indian samples have been found to contain unusual derivatives of 2-methyl isoflavones.

Licorice of commerce contains 2.2% of isoliquiritin, an anthoxanthin glycoside which imparts to it the yellow color. Roots contain a flavonoside, liquiritoside, of low toxicity; it depresses smooth muscle action. Liquiritigenine chalcone of spasmolytic properties is found in the plant.[178]

Pharmacological action — Tonic, cooling, demulcent, diuretic, emmenagogue, and gentle laxative. Gujral et al.[412] found glycyrrhizin to potentiate the antiarthritic action of hydrocortisone in rats. It also showed anti-inflammatory activity. The recognition of the deoxycorticosterone effects of licorice extracts and glycyrrhetinic acid has led to its use for the treatment of rheumatoid arthritis, Addison's disease, and various inflammatory conditions. Unlike cortisone, licorice may give symptomatic relief from peptic ulcer pain.[69]

Action of glycyrrhizic acid and glycyrrhetinic acid of licorice root resembles that of the adrenal hormones, desoxycorticosterone and hydrocortisone. Chalcone glycosides, isoliquiritin and neoisoliquiritin, have been identified in the plant.[178]

Medicinal properties and uses — Root in infusion, decoction, or extract is useful as a demulcent in inflammatory affection or irritable conditions of the bronchial tubes, bowels, and catarrh of the genitourinary passage, such as cough, hoarseness, sore throat, asthma, and dysuria.

Doses — Root powder — 2 to 4 gr; extract — 1 to 1.5 gr.

GMELINA ARBOREA, Linn.

Family: VERBENACEAE

Vernacular names — Sanskrit, Gambhari; Hindi, Gambhara; Bengali, Gamari; Nepalese, Khamari; Sinhalese, Ethdemata; Unani, Kambhari; Burmese, Yamanai; Malaysian, Kumbula; Tamil, Kumizhnaram.

Habitat — The lower Himalayas, the Nilgiris, and the east and west coasts of India.[26]

Parts used — Root, bark, fruit, and leaves.

Pharmacognostical characteristics — It is an unarmed tree, 60 ft high, with a clear bole of 20 to 30 ft and a girth of 5 to 7 ft. Bark is smooth, whitish gray; leaves opposite, broadly ovate, cordate, glandular; flowers in terminal panicles, brownish yellow; drupe fleshy, ovoid, 1 to 2 seeds.

Prasad[413] described the mature root bark as yellowish in color when fresh. Dried pieces are curved and channeled. Thinner ones form single quills. The external surface is rugged, due to the presence of vertical cracks, ridges, fissures, and numerous lenticles. Taste is mucilaginous, sweetish with slight bitterness. The mature stem bark also occurs as flat and slightly curved pieces. The external surface is slightly rough due to the presence of a few cracks, ridges, etc. Fracture is short and granular.

In transection of mature root, bark has 10 to 18 layers of rectangular cells of cork. Phelloderm is composed of parenchyma and groups of stone cells. The secondary phloem consists of parenchyma, groups of stone cells, sieve tube elements, and medullary rays. In mature stem bark, cork is made up of 12 to 20 layers of slightly thick-walled lignified cells. Phelloderm is parenchymatous and contains stone cells.

Parenchyma cells contain acicular crystals of calcium oxalate. Starch grains, oil globules, and resin are present.

Ayurvedic description — *Rasa* — *tikta, kasaya, madhur; Guna* — *guru; Veerya* — *ushna; Vipaka* — *katu.*

Action and uses — *Tridosh samak, soth har, daha samak, badana samak, dipan, anuloman, rasayan.*

Chemical constituents — The drupes of *G. arborea* are reported to contain butyric acid, resinous and saccharine matter, and tartaric acid in traces. The root and bark contain traces of alkaloid. The root also contains traces of benzoic acid and resinous and saccharine matter. When it is subjected to destructive distillation, the following carbonization products are obtained: charcoal (31.8%), total distillate (47.1%), pyroligneous acid (37.1%), tar (10.0%), acid (4.47%), ester (3.42%), acetone (2.38%), and methanol (1.23%), (dry weight basis). The noncondensable gases contain carbon dioxide (59%), carbon monoxide (31.75%), methane (4.5%), hydrogen (4.15%), and unsaturated hydrocarbons (as ethylene) (0.6%).[414,415]

Rao et al.[416] isolated luteolin from the leaves. Later studies revealed the presence of apigenin, quercetin, hentriacontanol, and β-sitosterol as crystalline components.[417] Govindachari et al.[418] isolated from the heartwood a new long-chain ester, clutyl ferulate, mp 80 to 82°C.

Pharmacological action — Demulcent, stomachic, bitter, tonic, refrigerant, and laxative. Tender leaves are demulcent. Fruit is sweetish-bitter and cooling. Extract of the root is bitter and tonic.

After treatment with the drug, Gaur and Gupta[419] observed increases in the percentage of an α_2 and γ-globulin fractions, gain in body weight, and alertness in physical behavior. The alcoholic extracts of stem bark and wood of stem exhibited hypoglycemic activity in albino rats. The stem bark extract also showed antiviral activity against Ranikhet disease virus.[51]

Medicinal properties and uses — The leaves, root, root bark, and flowers are used in medicine. The drupes, which are sweetish and bitter, are used as an ingredient of refrigerant decoctions for fevers and bilious affections. The leaves are demulcent. A paste of the leaves is rubbed on the head for the relief of pain. The root is bitter tonic, stomachic, laxative, and galactagogue. The root is an ingredient of the Ayurvedic preparation *dasamula*. The bark is a bitter tonic and stomachic, and is considered useful in fever and indigestion. To prevent abortions in the early stage of pregnancy, a powder of the bark and black gingally seeds, *manjista* and *satavari*, is given in milk.

Doses — Root powder — 3 to 6 g; fruit powder — 1 to 3 g.

GOSSYPIUM HERBACEUM Linn.
(Syn. *GOSSYPIUM INDICUM* Lam.)

Family: MALVACEAE

Vernacular names — Sanskrit, Karpas; Hindi, Kapas; English, Indian cotton; Bengali, Karpas; Nepalese, Kapaya; Sinhalese, Kapu; Japanese, Wata; Chinese, Bong; German, Indische Baumwollenstaude; Unani, Kapas; Arabian, Qutn; Persian, Punbah; Burmese, Wah; French, Cotoiner de-I'Inde; Tamil, Karpesi.

Habitat — This is annually cultivated in India and Pakistan.

Parts used — Bark, seeds, leaves, flowers, and root bark.

Morphological characteristics — The plants are shrubs 2 to 4 ft; leaves are heart shaped, 3-lobed, apex pointed, hairy. Flowers solitary; involucre of bracts ovate; corolla yellow with a purple base; 3- to 5-celled capsules, ovoid, apex sharply pointed; numerous seeds, ovoid; completely surrounded with white cotton.

Ayurvedic description — *Rasa — madhura, kasaya; Guna — laghu, snigdha; Veerya — ushna; Vipak — madur.*

Action and uses — *Vat samak, kapha pitta vardhak, bran ropan, sanehan, sansyan, artawajanan.*

Chemical constituents — Bark contains starch and a chromogen gradually changing to bright brownish red. It contains glucose, a yellow resin, a fixed oil, a little tannin, and 6% of ash. Seeds contain an oil (10 to 29%); albuminoids and other nitrogenous substances from 18 to 25%; and lignin, 15 to 25%. The root bark contains yellow or colorless dihydroxybenzoic acid and *gossypetin*.[420] Cotton seed hulls contain an appreciable quantity of leuko-anthocyanidin. Gossypol, a coloring matter, is located in specialized pigment glands in the seeds of *G. arboreum*.[421]

Pharmacological action — Demulcent, laxative, expectorant, aphrodisiac, and galactagogue.

Medicinal properties and uses — The root bark is a substitute for ergot as an abortifacient; it is an effective emmenagogue and is used in uterine disorders. The seeds are laxative, expectorant, antidysenteric, aphrodisiac, demulcent, nervine tonic, galactagogue, and abortifacient; its decoction is given in intermittent fevers and dysentery. Its oil is an embrocation for rheumatic diseases and dressing for freckles, herpes, scabies, and wounds; as a sedative it is locally applied in neuralgic and chronic headache. A poultice of the leaves and seeds is applied to bruises, sores, swelling, burns, and scalds; a hot infusion of the leaves is a useful hip bath in uterine colic.[1,3]

Doses — Root powder — 1 to 4 gr; decoction — 28 to 56 ml; leaf infusion — 5 to 10 ml.

GREWIA HIRSUTA Vahl *Family:* TILIACEAE
(Syn. *G. POLYGAMA* Mast.;
 G. PILOSA Roxb.)

Vernacular names — Sanskrit, Nagbala; Hindi, Gulsakari: Kukurbicha; English, Gulsakri; Bengali, Nagbalu; Nepalese, Nagabala; Sinhalese, Sirivedibevela; Unani, Falsa; Arabian, Falsah; Persian, Palsah.

Habitat — It is a shrub bearing edible fruit, occurring throughout the greater part of India, ascending to 4500 ft in the Himalayas. Especially found in Bihar and Orissa.

Parts used — Root and fruit.

Pharmacognostical characteristics — Leaves of *G. hirsuta* are 2 to 3 in. long and pointed. Flowers internally yellow and externally pink in color. Fruits small and yellow; when ripe split into four parts. Seeds 4 to 6.

The roots are cylindrical; its external surface dark brown with violet tinge and faint longitudinal anastomosing ridges and numerous lenticels. Fracture is hard and fibrous. Odor is indistinct and taste is mucilaginous.

The cork is composed of three to five layers of suberized cells. The phelloderm is a narrow zone of 12 to 16 layers. Groups of fibers occur in this region. The crystal fibers are elongated, thick-walled, and divided by transverse partitions. The xylem shows a diffused porous arrangement and a distinct growth ring near the center. It consists of tracheids, fiber-tracheid fibers, and parenchyma. The ray cells are thick-walled, lignified, having simple pits in their walls. Starch grains, prismatic crystals of calcium oxalate, mucilage and resin are present.[422]

Ayurvedic description — *Rasa* — *madhura, kasaya; Guna* — *guru, snigdha, pichil; Veerya* — *sheeta; Vipak* — *madhura.*

Action and uses — *Vata pitta samak, rakt stamvan, badana samak, daha samak, rakt stamvak, amltanask, anuloman, kaphanisark, mutral, rasayan.*

Chemical constituents — Sarkar and Khanna[423] observed in the air-dried powder of stem the presence of a pale yellow oil, β-sitosterol, α-amyrin, β-amyrin, and faradiol. Paper chromatography of the residue of aqueous extract showed the presence of proline, serine, glutamic acid, phenylalanine, isoleucine, and lysine.

Pharmacological action — Antipyretic, diuretic, and expectorant. Alcoholic extract of whole plant showed diuretic activity in albino rats. It also possessed CNS-depressant activity. Antiviral activity was also observed by the extract against Ranikhet disease virus.[11]

Medicinal properties and uses — The fruit is given in diarrhea and dysentery. A paste of the root is useful application to wounds to hasten suppuration and as a dressing. It is also very useful in heart diseases, spermatorrhea, and leukorrhea.

Doses — Decoction — 56 to 112 ml; powder of root bark — 1 to 3 gr.

GYMNEMA SYLVESTRE R. Br. *Family:* ASCLEPIADACEAE
(Syn. *ASCLEPIAS GEMINATA* Roxb.)

Vernacular names — Sanskrit, Meshasringi; Hindi, Gurmara; Bengali, Meshasringi; Nepalese, Mashasringi; Sinhalese, Masabedda; Unani, Gokhru; Arabian, Kharak; Persian, Khar-e-khasak; Tamil, Sirukurinjan.

Habitat — A climbing plant common in central and southern India; Western Ghats and in the Goa territory.

Parts used — Root and leaves.

Morphological characteristics — Leaves opposite, usually elliptic or ovate (1.25 to 2.0 × 0.5 to 1.25 in.); flowers small, yellow, in umbellate cymes; follicles terete, lanceolate, up to 3 in. in length. The lamina is ovate, elliptic, or ovate-lanceolate with both surfaces pubescent. Hairs, which are all over the surface of the leaf, are nonglandular. There are five vascular bundles, a fan-shaped bundle in the center flanked on either side by two small bundles. The midrib has a ventral bulge. In the lamina, rosette crystals of calcium oxalate are present in idioblasts in the spongy parenchyma.[424]

Ayurvedic description — *Rasa—kasaya, katu; Guna—laghu, rooksha; Veerya—ushna; Vipak—katu.*

Action and uses — *Kaph vat samak, sothahar, badana sthapan, madhumehaghan, mutral, vishaghan.*

Chemical constituents — Dried leaves contain two resins: one is insoluble in alcohol forming the large portion; another is soluble in alcohol; there are also a bitter neutral principle, albuminous and coloring matter, calcium oxalate, pararobin, glucose, carbohydrates, some tartaric acid; an organic acid said to be a glucoside and possessing antisaccharine property; gymnemic acid 6%; cellulose, ash, quercitol.[425] Air-dried leaves yielded after ignition 11.45% of inorganic matter consisting of alkali, phosphoric acid, ferric oxide, and manganese; two hydrocarbons. Chakervarti[426] reported the presence of lupeol, β-amyrin, stigmasterol, and also the gymnemic acid (+)-quercitol in *G. sylvestre*. Saponins are present in the alcoholic extract of the plant. Willaman and Li[10] reported the presence of the alkaloids betain, choline, and trimethylamine in the leaves.

Bark contains a large amount of calcium salts and other crystalline concretions. Gymnemic acid resembles chrysophanic acid, and forms insoluble salts with alkaloids.

Pharmacological action — Astringent, stomachic, tonic, refrigerant, and diuretic. *G. sylvestre* caused insignificant reduction in blood sugar in normal rats but produced marked and significant reduction in anterior pituitary-treated hyperglycemic animals.[427] The body weight of rats treated with *G. sylvestre* increased as compared with controls. The urine output was also greater in rats treated with the drug.[428] The leaf material stimulates insulin secretion, and later studies revealed that it has active principles which have blood sugar-reducing properties. Oral administration caused inhibition of hyperglycemic response of anterior pituitary extract in rats.[178]

Medicinal properties and uses — It is used as a destroyer of glycosuria and in other urinary disorders. Because of its property of abolishing the taste of sugar, it has been given the name of *gur-mar* (sugar-destroying), and it is believed to neutralize the excess of sugar present in the body in diabetes mellitus. Root is very useful remedy for snakebite: its powder is dusted upon the wound or made into a paste with water and applied; its decoction is given internally. A decoction of the leaves is given in fever and cough. Leaves triturated and mixed with castor oil are applied to swollen glands.

Doses — Leaf powder — 2 to 4 gr; leaf decoction — 14 to 28 ml.

HELICTERES ISORA Linn.   Family: STERCULIACEAE

Vernacular names — Sanskrit, Avartani; Hindi, Marorphali; English, East Indian screw tree; Bengali, Atmora; Sinhalese, Liniyagass; Chinese, Caydotron; Unani, Marorphali; Arabian, Iltwa-iltwa; Persian, Pechak; Burmese, Khungiche; Malaysian, Mori; Tamil, Valampurikkai.

Habitat — A tall shrub or a small tree common in central and western India, as far west as Jammu, Sri Lanka.

Parts used — Fruits, seed, root bark, juice.

Morphological characteristics — Leaves obovate or obliquely cordate with serrated margin, scabrous above and pubescent beneath; flowers solitary, 1 to 2 in. long, with red reflexed petals turning pale blue when old; fruits 1 to 2 in. long, greenish brown, beaked, cylindrical with spirally twisted carpels. Bark of the stem gray, young, parts covered with stellate hairs.

Deshmukh and Pandit[429] described the fruit of *H. isora* as short stalked with rough and twisted brown follicles. Each follicle contains 15 to 28 brown cubical seeds. The pericarp bears stellate lignified trichomes and has a number of large lysogenous mucilage cavities. The mesocarp and endocarp consist of fibers. The testa has an outer layer with rectangular thin-walled cells followed by another layer with lignified palisade cells and rows of brown pigment cells. Endosperm and embryo are parenchymatous and contain aleurone grains and oil. Catechol type tannins are present in the aqueous extract.

Ayurvedic description — *Rasa — kasaya; Guna — lagu, snigdha; Veerya — sheeta; Vipak — katu.*

Action and uses — *Tridoshaghan, stamvan, shul prasaman, krimighan, rakt rodhak, mutra sangrahiniya.*

Chemical constituents — Pods contain tannin. Bark showed the presence of chloroplast pigments, phytosterol, a hydroxy-carboxylic acid, an orange-yellow crystalline coloring matter, sugar, saponins, phlobotannins, and lignin.[178]

Kapoor et al.[430] reported the presence of saponins in root and stem bark but their absence in twigs, leaf, and flower. Absence of alkaloids and flavonoids in all these parts of the plant was also reported by them. Atal et al.[431] found no tannins in the whole plant above ground. *H. isora* seeds have been reported to be a new source of diosgenin.[432] Singh et al.[433] reported that the leaves yielded a new ester characterized as tetratriacontanyl-tetratriacontanoate along with tetratriacontanoic acid, tetratriacontanol, and sitosterol.

Pharmacological action — Expectorant, demulcent, mild astringent, and stomachic. The aqueous extract of pods showed a mild stimulant effect on the spontaneous activity of the isolated non-gravid rat uterus.[434] The 50% extract of the whole plant above ground showed spasmogenic action on isolated guinea pig ileum.[11]

Medicinal properties and uses — Fruits are employed in intestinal disturbances such as colic, flatulence, and diarrhea. Pods are used especially in the treatment of chronic dysentery. Decoction or juice of root bark given in diabetes to lessen the quantity of sugar. With castor oil, powder of the seeds forms an excellent application in ottorrhea, ulcers in the ear. The drug is also used in snakebite.[1,3]

Doses — Fruit powder — 1 to 3 gr; decoction — 117 to 234 ml.

HEMIDESMUS INDICUS (L.) Schult. *Family:* PERIPLOCACEAE

Vernacular names — Sanskrit, Sariva; Hindi, Salsa; English, Indian sarsaparilla; Bengali, Anantamul; Nepalese, Sariba; Sinhalese, Irimusu; Japanese, Indosarusa; German, Ostindische Sarsaparilla; French, Salsepareille indienne; Unani, Anantmool; Arabian, Ushbatunnar; Persian, Ushbah-e-hindi; Tamil, Nannari.

Habitat — This plant occurs throughout India, is common in Bengal, Bombay, and extending to Travancore and Sri Lanka.

Parts used — Roots and root bark.

Pharmacognostical characteristics — A perennial creeper or twiner, rootstock woody, fragrant; stems slender, hairless; leaves vary in shape and size, 5 to 10 cm long × 0.5 to 4 cm, dark green often with whitish blotches, pale or whitish, hairy on lower surface. Flowers very small, greenish, in small compact clusters. Fruit 10 to 15 cm long, green, narrow, cylindrical, pointed at tip, in pairs. Seed small, black, with a tuft of white hairs at top. All parts of the plant have white milky juice.

The roots and rootstock of *H. indicus* resemble each other. The root is dull red to dark brown with fine longitudinal wrinkles and transverse cracks, whereas the rootstock has a number of vegetative buds on the surface. Under a microscope, a transverse section of rootstock shows central pith, encircled by an intraxylary phloem, lignified pericyclic fibers, and stone cells. The transection of root does not show these elements.[436,437]

In the leaves, faint, cuticular striations in the form of long, continuous streaks over a few foliar epidermal cells have been observed. The costal cells are more elongated than the intercostals and are arranged in a row. There are no stomata and trichomes in the costal region. The lower epidermis, however, shows abundant stomata and each stoma is flanked by two subsidiary cells.[438]

Ayurvedic description — *Rasa — madhura, tikta; Guna — guru, snigdha; Veerya — sheeta; Vipak — madhura.*

Action and uses — *Tridosha samak, daah saman, soth har, dipan, pachan, anuloman, sthamvan, rakt sodhak, sothahar, jwaraghana, sthanya sodhak, veerysya, rasayan.*

Chemical constituents — The roots yield by steam distillation a steroptene which is supposed to be a volatile acid. The root also contains an essential oil of which about 80% consists of a crystalline material, 2-hydroxy-4-methoxybenzaldehyde. The odor of the drug is due to coumarin. In addition, two sterols were isolated: hemidosterol and hemidesmol. The roots contain resins, tannin, and a very slight amount of a glycoside.

Chopra et al.[178] reported the presence of β-sitosterol, α- and β-amyrins, lupeol, tetracyclic, triterpene alcohols, resin acid, fatty acids, tannins, saponins, a glycoside, and ketone in the roots. Subramanian and Nair[439] reported the presence of hyperoside, isoquercitin, and rutin in the flavonoid glycosides of flowers, but only hyperoside and rutin in those of leaves.

Pharmacological action — It is a valuable alterative, tonic, demulcent, diaphoretic, and diuretic. The aqueous, alcoholic, and stem-distilled fractions of crushed roots of *H. indicus* had no significant diuretic activity.[679] The aqueous ethanolic extract of the whole plant showed antiviral activity against Ranikhet disease virus but was inactive against vaccinia virus.[41]

Medicinal properties and uses — The roots are an excellent blood purifier, diuretic, tonic, and diaphoretic. They are given in loss of appetite, dypepsia, fever, skin diseases, syphilis, leukorrhea, genitourinary diseases, and in chronic cough.[3,50] A paste of the root is applied on swellings and rheumatic joints. Used as a good substitute for true sarsaparilla which is obtained from the genus *Smilax*.[46]

Doses — Root powder — 1 to 4 gr; decoction — 28 to 56 ml.

HIBISCUS ABELMOSCHUS Linn.
(Syn. ABELMOSCHUS MOSCHATUS Medic)

Family: MALVACEAE

Vernacular names — Sanskrit, Lataksturikam; Hindi, Musk dana; English, Musk mallow; Bengali, Kasturi dana; Nepalese, Latakasturi; Sinhalese, Kapukiissa; Unani, Mushkdana; Arabian, Hab-ul-mushk; Burmese, Baluwa; Malaysian, Kapas hantu.

Habitat — Cultivated in many hotter parts of India.

Parts used — Seeds, root, and leaves.

Morphological characteristics — An erect hirsute or hispid herb, with leaves of varying shape, usually palmately 5 to 7 lobed; flowers large, yellow, with crimson center; fruit a capsule or pod, oblong-lanceolate, 1 to 3 in. long, containing large number of small scented seeds; grayish brown, reniform, compressed, 4 mm long, with striations concentric about the hilum.[28]

Ayurvedic description — Rasa — tikta, madhur, katu; Guna — laghu, rooksha, teekshna; Veerya — sheeta; Vipak — katu.

Action and uses — Kapha pitta samak, rochan, dipan, grahi, hiridya utejek, mutral, virisya, chachusya.

Chemical constituents — Seeds contain protein (2.3%); starch (13.35%) and fixed oil, a solid crystalline matter, odoriferous principle, and resin. Fixed oil is a greenish-yellow fluid which solidifies on exposure to the air. Seeds yield 0.2 to 0.5% of an essential oil having an odor resembling a mixture of musk and amber. Seed oil contains 18.9% of linoleic acid. Flowers yield flavonoids, myricetin, and cannabistrin.[178]

Pharmacological action — Seeds are stimulant, carminative, cooling, diuretic, demulcent, stomachic, antispasmodic, and tonic.

Medicinal properties and uses — Seeds in the form of infusion, decoction, or tincture are useful in nervous debility, hysteria and other nervous disorders, atonic dyspepsia, and a few other conditions in which musk is indicated.[1] They enter into the composition of some compound prescriptions recommended for gonorrhea, venereal diseases, and catarrh of the bladder and of the air passage. Seeds rubbed to a paste with milk are used to cure itch. Mucilage prepared from the root and leaves of the plant is recommended in gonorrhea and venereal diseases. Extract of the upper parts of the plants and fruits show insecticidal toxicity.[178]

Doses — Decoction of the seeds — 28 to 56 ml; infusion of the seeds — 38 to 56 ml; tincture — 5 to 10 ml.

HOLARRHENA ANTIDYSENTERICA (Linn.) Wall. Family: APOCYNANCEAE

Vernacular names — Sanskrit, Kutaja; Hindi, Kurchi; English, Kurchi tree, Conessi tree; Bengali, Kureya; Nepalese, Indrojau; Sinhalese, Kelindhal; Japanese, Konetsushi; German, Kurchirinde; Unani, Kura; Arabian, Lisan-el-asofir-elmurr; Persian, Zaban, i-gungishk-i-talk; Tibetan, Dug-mo-nun; French, Ecore-d'Codagapala; Tamil, Kudarapalli.

Habitat — It is found more or less throughout India, ascending to an altitude of 4000 ft in the western Himalayas. There are two forms of *H. antidysenterica*: (1) white and (2) black.

Parts used — Bark, seeds, and leaves.

Pharmacognostical characteristics — A large shrub or a small tree. Leaves somewhat distichously spreading, subsessile, glabrous, pubescent, or tomentose, 6 to $12 \times 1\frac{1}{2}$ to 5 in., ovate to elliptic-oblong, apex rounded or obtusely acuminate, base usually obtuse or rounded. Flowers white, sweetly scented, $\frac{3}{4}$ to $1\frac{1}{2}$ in. in diameter in terminal corymbose cymes 3 to 6 in. wide; corolla tube $\frac{2}{5}$ in. long, pubescent. Follicles slender, divaricate 6 to 16 in. long $\times \frac{1}{6}$ to $\frac{1}{2}$ in. in diameter. Seeds with a long brown coma.[28]

In a transection of the bark, the cork consists of 4 to 12 rows of tangentially elongated cells, radial 15 to 45, tangentical 30 to 60. Phellogen consists of a row of thin-walled tangentially elongated cells. Phelloderm usually wide, parenchymatous, interspersed with strands of stone cells. Stone cells are rectangular to oval, bearing numerous pits and often containing prismatic crystalline calcium oxalate. Nonlignified pericyclic fibers are present in bark up to 5 mm thick. Secondary phloem is wide consisting of sieve tubes, companion cells, phloem parenchyma, and stone cells. Stone cells are in longitudinal rows in a concentric manner associated with crystal sheath-containing prisms.

Medullary rays are mostly bi- or triseriate, rarely uniseriate, becoming wide toward the outer part and consisting of thin-walled radially elongated parenchymatous cells. Medullary ray cells near the stone cells become sclerosed.[448]

Ayurvedic description — *Rasa* — *tikta, kasaya; Guna* — *lagu, rooksh; Veerya* — *sheeta; Vipak* — *katu.*

Action and uses — *Kaf pit smak, branaropan, dipan, stambhan, arsoghan, krimighan, rakt sodhak, rakt stambak.*

Chemical constituents — In a preliminary study, Kapoor et al.[449] confirmed the presence of terpenes, besides alkaloids, and the absence of sterols, saponins, tannins, and flavonoids.

Bark and seeds of *H. antidysenterica* have been subjected to chemical analyses by various workers and the following alkaloids were isolated, as reported by Willaman and Li.[10]

Plant part	Alkaloid	Ref.[a]
Bark	3α-Aminoconan-5-ene	CA.61:5960b
	Concuressine	CA.59:14053b
	Conessimine	CA.61.5960b
Bark, seed	Conessine	BA.32:17562
Bark	Dihydroconcuressine	CA.59:14053b
	Dihydroisoconessine	CA.66:5910b
	Dihydroisoconessimine	CA.61:5960b
Leaf	Holadysamine	SCFB66:1212
	Holadysine	SCFB66:1212
Bark	Holarrhidine	SCFB65:3035
	Holarrimine	CA.53:422g
	Holonamine	CA.61:12057a
	α-Hydroxyconessine	CA.61:12057a
Leaf	Isoconessimine	CA.61:5960b
	Kurchaline	SCFB.66.1212
Bark	Kurchamine	SCFB.65.3035
	Kurchessine	CA.59:14053b

Plant part	Alkaloid	Ref.[a]
	α—Kurchessine	SCFB:65.3035
Leaf	Kurchimine	SCFB:65:3035
	Kurchiphyllamine	SCFB.66:1212
	Kurchiphylline	SCFB.66:1212
Bark	Kurcholessine	Tetra L64:1659
	N_2-methylholarrhimine	Ber.91:1504
	N_3-methylholarrhimine	SCFB.65:3035
	N_{20}-methylholarrhimine	SCFB.65:3035
	N,N^1-Tetramethylholarrhimine	SCFB.65:3035

[a] BA — *Biological Abstracts;* Ber. — *Chemische Berichte Berlin;* CA — *Chemical Abstracts;* SCFB — *Soc Chimique de France* — Paris; Tetra L. — *Tetrahedron Letters* — London.

Pharmacological action — Bark is bitter, stomachic, astringent, powerful antidysenteric, febrifuge, and anthelmintic. Seeds are very bitter, astringent, febrifuge, antidysenteric, anthalmintic, carminative, and also antiperiodic in combination with other antiperiodics. The therapeutic utility of *kurchi* in acute and chronic amebic dysentery has been well known. Various fractions of *H. antidysenterica* showed promising activity against experimental amebiasis in rats and hamsters. Conessine was more potent as an amebicidal agent *in vitro* than the other alkaloids of the plant, viz., conessidine deydrate, conkurchine, holarrhine, and kurchicine.[450]

The fruit extract showed an antiprotozoal effect against *Entamoeba histolytica* strain STA, *Trypanosoma evansi;* an anticancer effect against human epidermoid carcinoma of the nasopharynx in tissue culture; and hypoglycemic activity in rats. The extracts revealed an antispasmodic effect on isolated guineapig ileum. The stem bark showed hypotensive action.[51]

Medicinal properties and uses — Liquid extract of the bark has a good effect in dysentery, acute and chronic, of both children and adults, and also an antipyretic effect. An infusion of the root bark which is very bitter and most unpalatable is useful in the treatment of amebic dysentery. Bark of the stem and root, preferably of the young plants and the seeds, is generally used as a remedy in acute and chronic diarrhea and dysentery. Seeds enter into the composition of many prescriptions for bilious affections, fever, bowel complaints, piles, and intestinal worms. For round- and threadworms, a compound anthelmintic powder is given in doses of 1 to 1.5 gr twice or three times daily for 3 d, followed by castor oil.[1,3]

Doses — Seed powder — 2 to 4 gr; decoction and infusion — 28 to 74 ml; tincture — 2 to 4 ml; powder — 2 to 4 gr.

HYDNOCARPUS LAURIFOLIA (Dennst.) Sleumer   *Family:* FLACOURTIACEAE
(Syn. *HYDNOCARPUS WIGHTIANA* Blume)

Vernacular names — Sanskrit, Tuvaraka; Hindi, Chaulmoogra; English, Jangli almond; Bengali, Chalmoogra; Nepalese, Newari-Chalmugra; Japanese, Daifushi; Chinese, Ta-feng-tzu; German, Chaulmoogra; Unani, Chaulmoogra; Arabian, Uruzmonjra; Persian, Burunjmonjra; Tamil, Neeredimultu.

Habitat — Grows all over the western peninsula, Konkan, along the coast range, Malabar, Travancore, Sri Lanka.

Parts used — Seeds and oil.

Pharmacognostical characteristics — A tree 30 to 50 ft high with brown bark. Leaves 4 to 10 in. long, oblong, more or less serrate, apex, shortly subcurvate-acuminate. Flowers solitary or subcymose. Male flowers: sepals 5, glabrous, greenish white, margin fimbriate. Scales as long as petals, densely ciliate. Stamens 5. Fruit globose 2 to 4 in. in diameter, tomentose, mammillate. Seeds 15 to 20, about $4/_5$ in. long, subavoid, obtusely angular, striate.[28]

Ayurvedic description — *Rasa* — *tikta, kata, kasaya;* *Guna* — *laghu, teeksna, snigdha;* *Veerya* — *ushna;* *Vipak* — *katu.*

Action and uses — *Kaf bat samak, khustaghna, rakt sodhak.*

Chemical constituents — Hydnocarpus oil is a yellow or brownish-yellow oil or soft cream-colored fat, obtained by cold expression from fresh, ripe seeds. It has a slight characteristic odor and a somewhat acrid taste. It is soluble in most organic solvents. The oil consists chiefly of glyceryl esters of two or more new fatty acids, named chaulmoogric and hydnocarpic acid.[1] The chaulmoogric oil also contains small amounts of palmitic acid and another highly unsaturated acid with iodine number 168-3. Ethyl hydnocarpate possesses the therapeutic properties of the oil and is generally preferred in the treatment of leprosy.[50]

The specific gravity of the oil is 0.940 to 0.960. The fatty acid composition of a specimen of the oil was as follows: chaulmoogric (27.0%), hydnocarpic (48.7%), garlic (12.2%), lower homologs of chaulmoogric (alepric, aleprytic, aleprestic, aleprolic, and unidentified acids) (3.4%), oleic (6.5%), and palmitic (1.8%).[448]

Pharmacological action — Alterative, stimulant, and bacteriostatic. Seeds are detergent. Various hydnocarpates have checked the growth of acid-fast *Mycobacterium leprae* in culture media. The oil is active against acid-fast bacteria. Sodium salts of chaulmoogric and hydnocarpic acids in 1:100,000 dilution are reported to be bactericidal against *M. tuberculosis.*[178]

Medicinal properties and uses — Hydnocarpus oil is mainly used in the treatment of lepromatous leprosy and is effective in early cases, in decreasing the size of nodules, anesthetic patches, and skin lesions. It is administered internally, the dosage being increased gradually to prevent gastric irritation.[1,3] The best results are obtained by intramuscular injections of the ethyl esters or intravenous injection of the sodium salts — chaulmoogric and hydnocarpic acid. Under the Ayurvedic treatment of leprosy and skin diseases, both chaulmoogra oil and cow's urine are prescribed for internal as well as external use.

Doses — Seed powder — 1 to 3 gr; oil — 1 to 4 ml.

HYDROCOTYLE ASIATICA Linn.
(Syn. *CENTELLA ASIATICA* [Linn.] Urban)

Family: UMBELLIFERAE

Vernacular names — Sanskrit, Mandukaparni; Hindi, Brahmamanduki; English, Indian pennywort; Bengali, Tholkuri; Nepalese, Kholachaghya; Sinhalese, Hingolukola; Japanese, Tsubokura; German, Asiotischer Wassernabel; Unani, Brahmi; Arabian, Zarnab; Persian, Sard Turkistan; Tibetan, Sin-mnar; Burmese, Minkhuabin; Malaysian, Dawoopungah-gah; Tamil, Vallarai.

Habitat — This plant is common all over India, growing plentifully in moist localities, viz., marshy banks of rivers, streams, and ponds and in irrigated lawns and fields.

Parts used — The whole plant-leaves, fruits, roots and seeds.

Pharmacognostical characteristics — The herb trails on the ground and its creeping stems bear roots on their nodes. Leaves small, 2 to 4 cm diameter, rounded or broad kidney shaped, with toothed margins. Flowers minute, pinkish red, 3 to 6 in a cluster. Fruits small, like a grain of barley, 7 to 9 ridged.[435]

Ayurvedic description — *Rasa — tikta, anu rasa-kasaya, madur; Guna — lagu, sar; Veerya — sheeta; Vipak — madur.*

Action and uses — *Tri dosha samak, madhaya, asmiriti wardhak, stamvan, hirdya, bayah sthapak, agni dipak.*

Chemical constituents — Being an important plant in Ayurveda, it has been investigated by various workers since 1893. The presence of alkaloids, glycosides, sterols, tannins, sugars, and inorganic salts has been detected from various extracts of the powdered material. The plant also contains amino acids, viz., aspartic acid, glycine, glutamic acid, α-alanine, and phenylalanine. The total ash contains chloride, sulfate, phosphate, iron, calcium, magnesium, sodium, and potassium.[451]

Rastogi et al.[452] isolated six constituents in the pure state, viz., two saponins, brahmoside, mp 242°C (decomp.), 0.37% and brahminoside, mp 223°C (decomp.), 0.16%; two triterpene acids, brahmic acid, mp 293°C, 0.097% and isobrahmic acid, mp 263°C, 0.9%; betulic acid, mp 308°C, 0.11%, and stigmasterol, mp 170°C, 0.004%. Dutta and Basu[453] isolated two new glycosides and one of these was named thankuniside, $C_{30}H_{47}O_6$. The identity of the other glycoside was not confirmed. Both of them contained glucose and rhamnose as the sugar residues. Thankuniside on acid or alkaline hydrolysis gave a new triterpene acid, thankunic acid, $C_{30}H_{48}O_6$. Singh and Rastogi[454] reported the isolation of asiatic acid, asiaticoside, mesoinosital, and a new oligosaccharide, centellose, from the Indian variety of the plant.

Pharmacological action — Alterative, tonic, diuretic, and local stimulant. The alcoholic extract produced a tranquilizing effect in rats[455] and the glycosidal fraction showed a sedative effect.[451] Ramaswamy et al.[456] reported that Brahmoside possessed sedative action in rats equivalent to that of a minor tranquilizer. Dhar et al.[51] reported antiprotozoal activity of alcoholic extract of the entire plant against *Entamoeba histolytica*. Clinical trials conducted on normal adults showed that the drug increased the mean level of RBC, blood sugar, serum cholesterol, vital capacity, and total protein. The increase in the hemoglobin percentage was quite high. The drug also decreased the mean blood urea level and a moderate decrease in serum phosphatase was observed.[457,458]

To study the effect on mentally retarded children, a clinical trial was conducted and indicated a significant improvement in both general ability and behavioral pattern when the drug was administered for a short period of 12 weeks.[459]

Medicinal properties and uses — The drug is indicated for treatment of ulcerations, chronic and callous, scrofulous and syphilitic with gummatous infiltration in chronic and obstinate eczema, psoriasis, leprosy, epilepsy, and insanity. It is successfully employed in enlargement of glands, in abscess, and in chronic rheumatism either as an ointment or as a dusting powder. Its efficacy has been valued as a stimulant to healthy mucous secretion in infantile diarrhea and ozena. It stimulates fast growth of skin, hairs, and nails. Prepared *ghirtha* improves the color of the body, youth, memory, and longevity.

Doses — Infusion — 12 to 20 ml; powder — 0.5 to 1.5 gr.

HYGROPHILA AURICULATA (Schumach.) Heine.
(Syn. ASTERACANTHA LONGIFOLIA Nees)

Family: ACANTHACEAE

Vernacular names — Sanskrit, Kokilaksha; Hindi, Talmakhana; Bengali, Kulia Khara; Nepalese, Talmakhna; Sinhalese, Ikkirir; German, Langblathriger Sterndorn; Unani, Talmakhana; Tamil, Nirmulli.

Habitat — Common in moist places throughout India and Sri Lanka.

Parts used — The whole plant, seeds, root, leaves, and ashes of the plant.

Morphological characteristics — Leaves 7 to 17 cm long, subsessile. opposite, and lanceolate, six thorns at the node; flowers in a whorl of eight at each node, bright blue, purple corolla, and roots creamy yellow in color.

Pharmacognostical characteristics — Mehrotra and Kundu[129] reported that the stem is characterized by opposite, lanceolate leaves, and six straight or slightly curved thorns at the node. Two kinds of trichomes, uniseriate, 1- to 5- celled, nonglandular trichomes, and glandular trichomes with a stalk and eight head cells are present. Many lenticles are formed sporadically. The epidermis becomes two layered at some places by periclinal divisions; some collenchyma of the hypodermal region develops lignified secondary walls with slit-shaped pits. The inner layer of the cortex forms a regular endodermis with a casparian band on the radial walls of the cells, containing starch grains. The primary phloem fibers along the intervening parenchyma cells form a layer of pericycle. Some wood fibers in the secondary wood are septate. The stomata are mainly caryophyllaceous and are present on both surfaces. Cystoliths are present in large numbers in the epidermis of stem and leaf. Crystals of calcium oxalate are observed in the powdered drug of the leaf.

Ayurvedic description — *Rasa — madhura, tikta; Guna — snigdha, pichil; Veerya — sheeta; Vipak — madhur.*

Action and uses — *Vat-pitta samak, nadi balya, anuloman, sonita sthapan, sotha har, balya, brighan, sukra bardhak, mutral.*

Chemical constituents — The seeds contain 23% of yellow, semidrying oil. The component fatty acids of oil contain the following percentages of acids: linoleic, 71; oleic, 10; stearic, 12; palmitic and myristic, 6%.[130] The acetone, alcohol, and water extract of the aerial part showed the presence of alkaloids. The oil of the root was light reddish brown, but the oil of the aerial part was pale yellow.[131] The unsaponifiable fraction of oil from the seeds yielded a sterol-like compound which was probably a new sterol.[132]

Pharmacological action — Root is cooling, bitter tonic, diuretic, demulcent, and refrigerant. Seeds are diuretic and aphrodisiac.[3]

Medicinal properties and uses — The roots are considered as cooling, diuretic, stimulating, and especially efficacious in dropsical conditions and in cases of stone or gravel in the kidney and other diseases of the genitourinary tract. Leaves and seeds are also useful in jaundice and anasarca. A confection of the seeds containing a large number of aphrodisiac, demulcent, nutritious, and aromatic stimulant substances has been in use for impotence, and seminal and other debilities. For asthmatic complaints, a powder of the *talmakhana* seeds is recommended to be given in a mixture of honey and "ghee".

Doses — Decoction of the root — 14 to 40 ml; infusion of the plant — 14 to 40 ml; seeds — $1/2$ gr.

HYOSCYAMUS NIGER Linn.   Family: SOLANACEAE

Vernacular names — Sanskrit, Yavani; Hindi, Khurasani ajvayan; English, Henbane; Bengali, Korasani ajowan; Nepalese, Emu; Sinhalese, Korasani; Japanese, Hiyosu; Chinese, Lao lang hoa; German, Bilsenkraut; Unani, Ajwan-e-khurasani; Arabian, Bazrul banj; Persian, Bang; Tamil, Korusanai.

Habitat — Grows wild throughout the Western Himalayan range at an, altitude of 5,000 to 11,000 ft from Kashmir to Garhwal.

Parts used — Dried or fresh leaves and flowering tops.

Pharmacognostical characteristics — An annual or biennial bad smelling herb, densely covered with glandular hairs; stem up to 3 ft high. Lower basal leaves 15 to 20 cm, margins toothed. Upper leaves smaller and divided into many segments. Flowers 2 to 3 cm diameter, pale green, streaked with purple, some borne solitary at the place of branching of the stem, others in long terminal, scorpiod spikes. Fruit 1 to 3 cm diameter, globose.[28,435]

A transection of the leaf shows a bifacial structure. Both surfaces exhibit smooth cuticle, epidermal cells with wavy walls, stomata of anisocytic type, and large number of hairs. Calcium oxalate crystals are present in the spongy mesophyll. The calyx exhibits the presence of trichomes and stomata, but the corolla shows trichomes on the outer side only.[69]

Ayurvedic description — *Rasa* — *tikta, kata, kasaya; Guna* — *guru, raksha; Veerya* — *ushna; Vipak* — *katu.*

Action and uses — *Kapha vat samak, pittawardhak, sotha har, badana nasak, pachan, krimi nasak, hirdya awsadak swashar, madak, kam awsadak grahi, nidrajank.*

Chemical constituents — Leaves of *H. niger* contain hyoscyamine, hyoscine, scopolamine, hyosciprin, cholin, fatty oil, mucilage, albumen, and potassium nitrate. Seeds contain hyoscyamine, a fixed or fatty oil (25%); an empyreumatic oil and ash (4 to 5%). The roots contain atropamine, tetramethyldiamenobutane, and tropine. The leaves and roots also contain an unnamed alkaloid.[10,69,50]

Pharmacological action — Seeds are intoxicating, narcotic, anodyne, digestive, astringent, and anthelmintic. Leaves are sedative, anodyne, antispasmodic, stimulant, and mydriatic in effect.

Medicinal properties and uses — Hyoscyamus is largely prescribed in mental and maniacal excitement, epileptic mania, chronic dementia with insomnia, paralysis agitans, convulsions, neuralgia, hypochondriasis, functional palpitations, spasmodic cough, asthma, hiccup, in urinary affections as irritation of the kidneys, uterus, and bladder, and hysteria. It has a sedative effect particulary beneficial in irritable affections of the lungs, bowels, and genitourinary organs such as cystitis.[1,3]

Doses — Leaf powder — 5 to 10 gr; fresh juice — 2 to 4 ml; tincture — 2 to 4 ml; extract of the fresh plant — 1 to 3 ml.

HYSSOPUS OFFICINALIS Linn.   Family: LABIATEAE

Vernacular names — Sanskrit, Zupha; Hindi, Zupha; English, Hyssop, Nepalese, Japapuspa; Japanese, Yanagihakuga; German, Kleinblatt-Rigerysop; Unani, Zufa; Arabian, Zufau-e-yabis; Persian, Zufa-e-Khushk.

Habitat — Found on the hills of western Himalaya from Kashmir to Kamaon, 8,000 to 11,000 ft.

Parts used — Leaves and whole herb.

Morphological characteristics — Perennial herb. Leaves sessile, linear to oblong, usually narrow at both ends, up to 2 in. long. Flowers bluish purple; calyx persistent. The leaves and flowering tops have an agreeable odor and pungent taste.

Ayurvedic description — *Rasa — tikta, katu; Guna — laghu, rooksha, teekshna; Veerya — ushna; Vipak — katu.*

Action and uses — *Kapha vatta samak, pitta sarak, lakhan, sotha-har, anuloman, pitta sark, krimighan, sheleshmher.*

Chemical constituents — The leaves contain glucoside and an essential oil, hyssop oil, in a yield of 0.5, 0.8, and 1 to 1.5% from white, red, and blue variety of the herb. Besides the volatile oil, the herb contains fat, sugar, choline, tannin, carotin (108.1 mg/100 g), and xanthophyll (355.6 mg/100 g). The essential oil contains l-pinocamphene and pinene.[178]

Pharmacological actions — Leaves are stimulant, stomachic, expectorant, diaphoretic and emmenagogue, carminative.[1]

Medicinal properties and uses — Leaves in the form of syrup or infusion are useful in hysteria and colic, cough, asthma, sore throat, and chronic bronchitis; also in uterine affections such as amenorrhea and in indurations of the liver and spleen. A paste of the herb is applied as a salve in catarrhal ophthalmia and as a resolvent and vulnerary. The leaves are anthelmintic, their juice is given with honey for expelling roundworms. Extracts of stalks and leaves have no antibacterial activity.[178]

Doses — 1 to 3 gr.

IPOMOEA NIL (L.) Roth   *Family:* CONVOLVULACEAE
(Syn. *I. HEDERACEA* auct. non Jacq.;
   PHARBITIS NIL Choisy)

Vernacular names — Sanskrit, Krishnavijani; Hindi, Kaladana; English, Pharbitis seeds; Bengali, Kaladana; Sinhalese, Kaladana; German, Kaladana Harz; Unani, Kaladana; Arabian, Hab-un-nil; Persian, Tukhm-e-Ishqpechan.

Habitat — Found throughout India, both cultivated and apparently wild, up to an altitude of 6000 ft in the Himalayas.

Parts used — Dried seeds.

Morphological characteristics — A slender twiner, stem retrorsely hirsute. Leaves 2 to 5 in., alternate ovate-cordate, 3 lobed; lobes ovate-acuminate. Peduncles 1 to 5 flowered, shorter than the petioles; flowers bright blue or pink, $1\frac{1}{2}$ to 2 in. long; funnel-shaped, sepals linear lanceolate, $\frac{3}{4}$ to 1 in. long; more or less hairy below; ovary 3 celled. Capsule $\frac{1}{3}$ in., subglobose, smooth. Seeds 6 or 4, glabrous. The leaves are sometimes quite entire, ovate, or orbicular-cordate.[28]

Ayurvedic description — *Rasa — katu, madhur; Guna — laghu, rooksha, teekshna; Veerya — ushna; Vipak — katu.*

Action and uses — *Kapha pitta har, lakhan, rakt sodhak, sothahar, mutral, artwajanan.*

Chemical constituents — The seeds contain a resin from which a resin glucoside, pharbitin, has been isolated. It has an unnamed alkaloid in leaf and flowers.[10]

Pharmacological action — Cathartic and anthelmintic, purgative.

Medicinal properties and uses — In constipation, seeds and its powders are given as purgative, either alone or in combination with other drugs. It is administered as a substitute for Jalap.

Doses — Infusion — 5 to 15 ml; powder — 10 to 12 gr.

JASMINUM SAMBAC Ait.

Family: OLEACEAE

Vernacular names — Sanskrit, Mallika; Hindi, Belphul Motia, Mugra; English, Arabian jasmine, Tuscan jasmine; Bengali, Mugra; Nepalese, Bela; Sinhalese, Geta Pichcha; Japanese, Matsurika; Chinese, Moli; German, Arabischer Jasmin; Unani, Motiya; Arabian, Sumana; Persian, Gule Supada; Burmese, Mali; Tamil, Isuvatshi.

Habitat — Cultivated throughout India.

Parts used — Whole plant, leaves, and flowers.

Morphological characteristics — A small shrub, leaves opposite, simple, variable in shape, membranous, 1.5 to 3.5 in. long with prominent lateral nerves. Inflorescence terminal, bearing solitary or 3-flowered cymes; flowers white, fragrant; calyx pubescent, lobes 5 to 9, tapering to a fine point; corolla lobes as many as calyx lobes, often numerous, spreading; tube narrow, elongate. Fruit black when ripe, surrounded by persistent calyx.

Ayurvedic description — *Rasa — tikta, katu; Guna — laghu, rooksha; Veerya — ushna; Vipak — katu.*

Action and uses — *Tridosha samak, sothahar, bran ropan, kustangan.*

Chemical constituents — Flowers yield 0.020 to 0.025% attar on water distillation. It contains esters (as benzyl acetate) (32.45 to 35.20%), alcohols (as linalool) (30.73 to 35.58%), methyl anthanilate (2.88 to 3.51%), and indole (2.75 to 2.82%).

Pharmacological action — Anthelmintic, deobstruent, diuretic, and emmenagogue.

Medicinal properties and uses — A poultice of the dry leaves is applied to indolent ulcers and other skin diseases. Root, leaves, and flowers are a powerful lactifuge; the bruised flowers are applied to the breasts of nursing mothers to check secretion of milk if an abscess is being developed. The essential oil extracted from the flowers is deodorant; it is used when discharges from the nose and ear are foul smelling. Plant cooling, used in cases of insanity, weakness of sight, and affections of mouth.[50]

Doses — Decoction — 56 to 112 ml.

JUNIPERUS COMMUNIS Linn. Family: CUPRESSACEAE

Vernacular names — Sanskrit, Hapusa; Hindi, Aaraar; English, Juniper berry; Bengali, Hapusa; Nepalese, Sizar; Sinhalese, Hapusa; German, Wacholder Beere; Unani, Abhal; Arabian, Habhhul araar; Persian, Tukhm erahal; French, Baie de genévrier.

Habitat — A dense more or less procumbent shrub found in the western Himalayas from Kumaon westward at altitudes of 6,000 to 13,000 ft. Common on the alpine valleys of Kashmir and Himachal Pradesh above the tree limit.[460]

Parts used — Fresh ripe berries and the volatile oil.

Morphological characteristics — An evergreen shrub, stem bark reddish brown, peeling off in papery shreds; leaves in whorls of 3, linear subulate, 0.2 to 0.6 in. long; sharply pointed, upper surface concave, glaucous, bluish white, lower surface bluntly keeled; flowers usually dioecious, axillary, fruit subglobose; bluish black when ripe, 0.4 to 0.5 in. in diameter, covered with a waxy bloom; seeds usually 3, elongated, ovoid.[28]

A transverse section of the fruit shows a thin outer epicarp, a yellowish-brown pulpy mesocarp, and three seeds which are hard and woody. Testa of each seed contains large oleoresin glands. The fruits have a pleasant, somewhat terebinthinate odor and sweetish taste.[69]

Ayurvedic description — *Rasa — katu, tikta; Guna — guru, rooksha, teekshna; Veerya — ushna; Vipak — katu.*

Action and uses — *Kapha vata samak, lakhan sotha har, bran ropan, dipan, anuloman, grahi, krimighan, kafa nisark, artawa janan, mutra janan.*

Chemical constituents — The berries contain volatile oil (0.7 to 1.4%), invert sugar (about 30 to 40%), and resin. Of the many compounds present in the oil, the terpene, α-pinnene, camphene, the sesquiterpene cadinene, alcohols, and esters are the most prominent.[69] Geigerone, a C_{12} terpenic ketone, which was isolated from the essential oil of *J. communis* and shown to be *trans*-3-isopropenyl-4-methyl-4-vinylcyclohexanone by Thomas,[461] has now been synthesized.[462]

Pharmacological action — Aromatic, carminative, stimulant, emmenagogue, digestive, diuretic, and antiseptic. Juniper berries were found to be active against parasitic sarcoptic and psoroptic mange in sheep and fungal *(Trichophyton* and *Microsporum)* infections in cattle.[463] Ether extract of *J. communis* berries in linseed oil was reported to be effective against *Sarcoptes scabiei* infection in sheep.[464] The linseed oil extract of powdered berries was highly effective in psoroptic-infected sheep.[465] Rao and Gupta[466,467] reported that ether extract of berries exhibited antifungal activity *in vitro* and *in vivo* against *Trichophyton mentagrophytes* and *T. simi* in experimentally infected goats and rabbits, respectively.

Medicinal properties and uses — Fruit and oil of *J. communis* are useful in scanty urine, hepatic dropsy, cough and pectoral affections, in chronic gonorrhea, and leukorrhea. Locally, powder of berries is rubbed on rheumatic and painful swellings. Ash of the bark is applied in some skin affections.[3,468]

Doses — Powder berries — 1 to 3 gr; oil — 1 to 2 minims (as stimulant and carminative), 4 to 6 minims (as diuretic); infusion — 56 to 74 ml.

JUSTICIA ADHATODA Linn.
(Syn. *ADHATODA ZELANICA* Medic;
ADHATODA VASICA Nees)

Family: ACANTHACEAE

Verancular names — Sanskrit, Vasaka; Hindi, Arusha; English, Malabar nut; Bengali, Bakash; Nepalese, Aeleha; Sinhalese, Adhatoda; Japanese, Adotada; German, Malabar Nuss; Unani, Arusa; Arabian, Hashisa tussoat; Persian, Bansa; Tibetan, Ba-sa; Tamil, Adathodai.

Habitat — This small evergreen subherbaceous bush grows all over the plains of India and in the lower Himalayan ranges, ascending to a height of about 4000 ft above sea level.

Parts used — Root, leaves, and flowers.

Pharmacognostical characteristics — Leaves 10 to 30 cm long and 3 to 10 cm broad, lanceolate to ovate-lanceolate, slightly crenate to entire, acuminate, base tapering, exstipulate; glabrescent, 8 to 10 pairs of lateral veins, bearing a few hairs. Leaves dorsiventral, transcurrent, with 2 layers of palisade; stomata on both surfaces, large in number on the lower surface, clothing trichomes few, 1- to 3-, rarely up to 5-celled, thin-walled, uniseriate up to 50, and glandular trichomes with unicellular stalk and 4-celled head measuring 25 to 36 μm in diameter in surface view; cystoliths in mesophyll layers, elongated, cigar shaped; calcium oxalate crystals in acicular and prismatic forms in mesophyll.[35]

Ayurvedic description — *Rasa — tikta, kasaya; Guna — laghu, rooksha; Veerya — sheeta; Vipaka — katu.*

Action and uses — *Kapha pitta samak, sothahar, rakt sodhak, rakt sdamvak, swash har, kashar, kantya.*

Chemical constituents — Chemical analysis of the leaves shows the presence of two active principles: (1) an alkaloid, vasicine, $C_{11}H_{12}N_2O$, mp 190 to 191°C; (2) traces of a volatile principle of the nature of the essential oil. The leaves contain 0.2 to 0.4%, bark 0.35%, and roots only traces of alkaloids.[36] Gupta et al.[37] obtained a golden yellowish, fragrant essential oil from leaves, flowers, and roots with a yield of 0.075%. Lately investigations of different workers have revealed the presence of new alkaloids, viz., hydroxypeganine, oscine, paganine, vasicinone, and one unnamed from the leaves; vasicine from the flowers; vasicinal, and an unnamed from roots of the plant.[10]

Pharmacological action — Expectorant, diuretic, antispasmodic, and alterative. It produces slight but persistent bronchodilation in experimental animals, and this effect is considerably increased after administration of atropine. The essential oil present in the leaves appears to be chiefly responsible for the expectorant action of the drugs.[1]

Medicinal properties and uses — It is a well-known drug in Ayurvedic medicine and is recommended in bronchitis, asthma, fever, jaundice, and consumption. The leaves and roots are antispasmodic and efficacious in cough. The flowers are bitter, aromatic and antispasmodic. Root is given in malarial fevers, diseases of the respiratory system, diphtheria, and gonorrhea.

Doses — Leaf powder — 2 gr; leaf juice — 8 to 16 ml; bark decoction — 28 to 56 ml; powder of root bark — 10 to 30 gr.

LINUM USITATISSIMUM Linn. *Family:* LINACEAE

Vernacular names — Sanskrit, Uma; Hindi, Tisi, Alsi; English, Linseed; Bengali, Aalas; Sinhalese, Hana; Japanese, Ana; Nepalese, Aalas; Chinese, Hou ma tse; German, Flachs; French, Lin; Unani, Alsi; Arabian, Bazr-e-katan; Persian, Tukhm-e-katan; Tamil, Alisivirai.

Habitat — Extensively cultivated throughout India, up to an altitude of 6000 ft for its oil-yielding seeds.

Parts used — Seeds, oil, and flowers.

Morphological characteristics — An erect annual, 60 to 120 cm high, usually corymbosely branched. Leaves linear or lanceolate; flowers in broad cymes, small, blue or white, 1 in. across, in terminal panicles; fruits capsular, with five cells, each containing two seeds; seeds yellowish or blackish brown, small, flattened, oval, with smooth shining coat.[28]

The testa is glossy and finely pitted with no odor; taste mucilaginous and oily. A transverse section shows a narrow endosperm and two large planoconvex cotyledons. Microscopical examination of the testa has been described by Trease,[69] showing mucilage-containing outer epidermis; one or two layers of collenchyma, the hyaline layers or "cross cells" composed in the ripe seed of partially or completely obliterated parenchymatous cells with their long axis at right angles to those of the sclerenchymatous layer; and an innermost layer of pigment cells. The outer epidermis is composed of cells rectangular or five sided in surface view, which swell up in water and become mucilaginous. The outer cell walls when swollen in water show an outer solid stratified layer and an inner part yielding mucilage. The radial layer or round cells show distinct triangular intercellular air spaces. The pigment layer is composed of cells with thickened pitted walls and containing amorphous reddish-brown contents. The cells of endosperm and cotyledons are polygonal with somewhat thickened walls and contain numerous aleurone grains and globules of fixed oil.

Ayurvedic description — *Rasa* — *madhur, tikta; Guna* — *guru, snigdha, pichil; Veerya* — *usha; Vipak* — *katu*.

Action and uses — *Vat samak, kapha pitta wardhak, bran sothahar, grahi, anuloman, kaf nisark, mutral.*

Chemical constituents — Seeds contain 37 to 44% of a fixed oil which consists of glyceryl combined with 30 to 40% linoleic acid, mucilage (15%) (6% in the testa); proteins, amygdalin, resin, wax, sugar, and ash (3 to 5%). Ash contains sulfates and chloride of potassium, calcium, and magnesium. Linseed oil contains 10 to 15% of mineral substances, chiefly phosphates of potassium, calcium, and magnesium and about 25% of protein substances. Pure fresh oil is colorless; commercial oil is dark yellow; on exposure to the air, oil dries to a transparent varnish consisting chiefly of linoxyn. The seeds, leaves, stem, and roots contain a cyanogenetic glucoside, phaseolunatin (linamarin).[44] The oil has a high iodine value since it contains considerable quantities of glycerides of unsaturated acids. Other constituents such as phylin, lecithin, etc. are reported to be present. β-Carotene forms 22 to 30% of the total carotenoids.[37]

Pharmacological actions — Dumulcent, diuretic, and emollient. Seeds are "aphrodisiac, hot and dry" and roasted seeds are astringent. Flowers are cordial. Poultices dilate the local blood vessels, relax the tissue, and thereby relieve the tension and pain.[1]

Medicinal properties and uses — Infusion of the seeds known as linseed tea given internally as a demulcent and expectorant drink in cold, cough, bronchial affections, irritation of the urinary organs, cystitis, gonorrhea, diarrhea, dysentery, etc. Poultice made from linseed meal is a valuable soothing application to ulcerated and inflamed surfaces, boils, and carbuncles.[50]

Doses — Powder — 3 to 6 gr; oil — $1/2$ to 1 ml.

LITSEA GLUTINOSA (Lour.) C. B. Robins.   *Family:* LAURACEAE
(Syn. *L. CHINENSIS* Lam.;
    L. SEBIFERA Pers.)

Vernacular names — Sanskrit, Maidasak; Hindi, Maidalakri; Bengali, Ratum; Sinhalese, Bombi; Unani, Meda-lakri; Arabian, Maghos Hindi; Persian, Khilza.

Habitat — An evergreen tree up to 75 ft in height, found throughout India, ascending up to an altitude of 4000 ft in the Himalayas.

Parts used — Bark.

Morphological characteristics — Aromatic tree, bark thick, externally gray and internally reddish in color. Leaves 3 to 6 in. long, alternate, entire, scented gland dotted, flowers small, whitish yellow, 2 to 1 sexual, in axillary cymes, bracts deciduous. Perianth inferior, tube short, enlarging in fruit. Lobes 6 in two series. Stamens usually a multiple of the perianth lobes, in 2 to 4 series on the tube, 1 whorl often reduced to staminodes; filaments flattened; often two-glandular at the base. Anthers 2 to 4 celled. Ovary 1 celled with solitary penduluous ovule, style simple. Fruit a fleshy berry. Seeds exalbuminous.[28]

Ayurvedic description — *Rasa — katu, tikta, kasaya; Guna — snigdha; Veerya — ushna; Vipak — katu.*

Action and uses — *Kapha vata samak, dipan, grahi, sotha har, rakt sthamvan, kapha nisark, kamotejok, vata samak.*

Chemical constituents — The bark of *L. glutinosa* has yielded actinodaphnine and β-sitosterol.[469] The aqueous extract of leaves yielded kaempferol-7-glucoside, pelargonidin-5-glucoside, and a new flavonoid glycoside naringerin, identified as naringenin-7-monorhamnoside.[470] The seeds yielded an essential oil comprising ocimene, α-pinene, *d*-limonene, terpinolene, α-terpinene, carvone, caryophyllene, β-amyrin acetate, fenchyle alcohol, α-thujone, and linalool oxide.[471] Upreti et al.[472] isolated an alkaloidal mixture, boldine, laurotetanine, *N*-methyllaurotetanine, and actinodaphnine from the alcoholic extract of leaves and stems of *L. glutinosa*. The defatted trunk bark on ethanolic extraction yielded two new alkaloids, sebiferine and litseferine.[473]

Pharmacological action — Demulcent, astringent, emollient, and aphrodisiac. The essential oil of the plant showed antibacterial and antifungal activity but Bhakuni et al.[11] did not find such activity with 50% ethanolic extract of the whole plant; it, however, showed antispasmodic action on isolated guinea pig ileum. The essential oil from berries exerted a prolonged hypotensive effect in anesthetized dogs.[474] It also caused a reduction in spontaneous motor activity without any concomitant muscle weakness in rats.[475]

Medicinal properties and uses — Due to its balsamic and mucilaginous nature, decoction of bark is demulcent and used in diarrhea and dysentery. Externally, freshly ground bark is used as an emollient application (hemostatic) to bruises, sprains, rheumatic and gouty joints, and to scorpion sting. The wood is stated to be durable and resistant to insect attacks. The paste of the stem bark mixed with common salt is applied externally in bone fractures, inflammation, and sprains.[50,476]

Doses — Bark powder — 1 to 3 gr.

LOBELIA NICOTIANAEFOLIA Heyne. ex Roth *Family:* LOBELIACEAE

Vernacular names — Sanskrit, Nala; Hindi, Nal, Narasal; Bengali, Nal; Nepalese, Tee; Sinhalese, Rasin, German, Lobelic; Unani, Post nila.

Habitat — Found on the Western Ghat from Bombay to Travancore at altitudes of 2000 to 6000 ft.

Parts used — Leaves, seeds, and root.

Morphological characteristics — Biennial or perennial herb, 5 to 12 ft high. Stem stout, hollow, branched at top; leaves oblong or oblong-lanceolate; lower leaves large, upper ones gradually smaller; margins not entire; the main nerve of the leaves whitish, alternate, nearly sessile, flowers large, white tinged with lilac, in long terminal racemes or spikes; capsule sub-oblong, 8 mm in diameter; seeds very small, numerous, ellipsoid, compressed, yellowish-brown.[28,435]

In transection, the stem is characterized by the presence of few trichomes, covered with epidermis, collenchymatous outer cortex, distinct endodermis, laticiferous vessels, and lignified pith cells. Pith is hollow in old stems.[477]

Ayurvedic description — *Rasa* — *madhur, kasaya, tikta; Guna* — *laghu, snigdha; Veerya* — *sheeta; Vipak* — *madhur.*

Action and uses — *Tridosha har, daah prasaman, bran ropan, rakt pitta samak, rakt sodhak, mutral, daah prasaman.*

Chemical constituents — Leaves contain total alkaloid, about 0.8 to 1.14% lobeline.[478] Handa and Nazir[479] reported isolation of norlobelanine also from the plant. The seeds contain an acronarcotic poison.

Pharmacological action — Antiseptic.

Medicinal properties and uses — Leaves and aerial parts of the plant exude a white latex which causes dermatitis. The plant is poisonous and used as an antiseptic.[50] Ethanolic extract of *Lobelia* fruits has been reported to show antifungal activity.[480] The tincture, when given in a mixture of potassium iodide and tincture of belladonna, had beneficial effect as an antispasmodic agent.[481]

Dose — 50 to 125 mg.

LUFFA ACUTANGULA (Linn.) Roxb.   Family: CUCURBITACEAE

Vernacular names — Sanskrit, Koshataki; Hindi, Torai; English, Ribbed gourd; Bengali, Ghosa lata; Nepalese, Ramatoria; Sinhalese, Wetakolu; Japanese, Tokadochechima; Chinese, Szukua; German, Scharfeckige Gurke; French, Pipangua; Unani, Turai; Arabian, Qisha hindi; Persian, Shah turai; Tamil, Paypeerkku.

Habitat — Cultivated throughout greater parts of India.

Parts used — Fruit, fruit juice, seeds, root, and leaves.

Morphological characteristics — An extensive climber. Leaves 4 to 8 in. long, orbicular-reniform; 5 lobed, flower yellow, stamens 3, strongly ribbed ovary. Fruit 6 to 12 in. long, clavate-oblong, obtuse, smooth, longitudinally ribbed, with ten sharp angles. Seed $1/_4$ to $1/_3$ in., ovoid-oblong, much compressed, black, not winged.[28]

Ayurvedic description — *Rasa — tikta, katu; Guna — laghu, rooksha, teekshna; Veerya — ushna; Vipak — katu.*

Action and uses — *Kapha pitta sonsodhan, vamak, rachan, rakt sodhak, sotha har, kafa nisark, kustagan.*

Chemical constituents — The edible fruit contains free amino acids: arginine, glycine, threonine, glutamic acid, leucines. The ripe seeds contain bitter principles — cucurbitacins — B, D, G, H. The roots contain cucurbitacin B and traces of G. Chemical examination of seeds showed the presence of the crystalline bitter principle cucurbitacin B and oleanolic acid. Ripe seeds yield 19.9% oil.[178]

Gopalan[482] reported the presence of calcium (18 mg), phosphorus (0.5 mg), iron (33 mg), riboflavin (0.01 mg), niacin (0 mg), vitamin C (5 mg), carotene (2.6 mg), per 100 g in the plant *L. acutangula*.

Willaman and Li[10] reported that the seed contains an unnamed alkaloid. The alcoholic extract of seeds yielded a crude saponin, which on hydrolysis gave oleanolic acid.[483] The ether extract of fruit mesocarp yielded cucurbitacin B and E, whereas the ethanolic extract gave oleanolic acid.[484]

The fruits are reported to contain oxalate (12.2%),[485] fluoride (0.3 to 0.6 ppm),[486] oxalic acid (84.6 mg%), calcium (572.7 mg%), phosphorus (565.0 mg%),[487] iodine (0.30 ppm), and fluorine (4.4 ppm).[488]

Pharmacological action — Demulcent, diuretic, bitter tonic, nutritive, and expectorant. Aswal et al.[489] reported that 50% of ethanolic extract of the whole plant above ground of *L. acutangula* showed hypoglycemic action in rats. It also exhibited abortifacient activity in female rats.[490] The leaves of the plant support the growth and development of the larvae of the pest *Diacrisia obliqua*.[491] It has been reported that the seeds contain saponin and an enzyme capable of hydrolyzing the saponin which causes salivation, vomiting, and purging in dogs.[492]

Medicinal properties and uses — The plant is a bitter tonic and diuretic; the seeds are emetic and purgative; it is useful in the enlargement of the spleen. Leaves are a good substitute for ipecacuanaha in dysentery. The leaves or its juice are used as a dressing for sores, inflamed spleen, ringworms, piles, leprosy, and bites of insects and snakes. The oil extracted from the seeds is very effective in skin diseases. The root is laxative and is also used in dropsy.[1,3,37]

Doses — Seeds — 1.5 to 2 gr; seeds as demulcent — 5 to 10 gr; fruit juice — 5 to 10 gr; infusion — 28 to 56 ml.

LUFFA CYLINDRICA (Linn.) M. Roem   *Family:* CUCURBITACEAE
(Syn. *L. AEGYPTIACA* Mill. ex Hook. f.)

Vernacular names — Sanskrit, Dhamargava; Hindi, Ghiatarui; English, Sponge gourd, Vegetable sponge; Bengali, Dhundul; Nepalese, Patatosiya; Sinhalese, Neyanwela-kolu; Japanese, Mechima; Chinese, Szu skua; German, Luffa Schwammlobelic; Burmese, Thalhewot.

Habitat — Cultivated in several parts of India.

Parts used — Seeds, fruit, leaves.

Morphological characteristics — A large climber with palmately 5- to 7-angled or -lobed leaves, 4 to 8 in. long. Plants monoecious; male flowers with 5 stamens, in 10- to 20-flowered raceme, female flowers solitary in same axil as males, petals yellow, fruit smooth, cylindrical, usually 5 to 12 in. long; seeds narrowly winged, black or gray.

Ayurvedic description — *Rasa — tikta, katu; Guna — laghu, rooksha, teekshna; Veerya — ushna; Vipaka — katu.*

Action and uses — *Kapha pitta sansodhan, bavan vadan, rakt sodhak, kafanisarak, visaghan.*

Chemical constituents — The seeds were reported to contain aspartic acid, glutamic acid, alanine, valine E, isoleucine, leucine E, lysine E, arginine E, phenylalanine E, tryptophan, cystine, and α-amino butyric acid as free amino acids.[493] Petroleum ether extract of the seeds yielded amarin.[494] The seeds were reported to contain saponins which on hydrolysis yielded sapogenins: oleanolic acid, gypsogenin, and gypsogenin acetate.[495] Varshney and Beg later isolated two saponins, aegyptinin A and aegyptinin B, from ethanolic extract of the defatted seeds.[496]

The fruit contains choline (choline chloride, 10 to 45 mg/g) and phytin.[178]

Pharmacological action — Emetic, cathartic, and diuretic. The aqueous extract of the seeds showed anticancer activity against the ascitic form of Schwartz leukemia in transplanted tumor, at a dose of 8 and 4 mg/kg.[497] In the form of 50% ethanolic extract of the whole plant, *L. cylindrica* did not show any anticancer activity against human epidermoid carcinoma of the nasopharynx in tissue culture or P_{388} lymphocytic leukemia in mice.[489]

Medicinal properties and uses — Fruits are used as vegetables. It is described as cool, demulcent, enhancing appetite, and facilitating production of bile and phlegm. The tender fruit is considered diuretic and lactagogue. The oil obtained from the seeds is used as a substitute for olive oil; it is useful for skin affections. Seeds are emetic and cathartic.[3,37]

MADHUCA LONGIFOLIA (Linn.) Macbride *Family:* SAPOTACEAE
(Syn. *BASSIA LONGIFOLIA* Linn.)

Vernacular names — Sanskrit, Madhuka; Hindi, Mahua; English, Indian butter tree, Mahua; Bengali, Maua; Nepalese, Mahua; Sinhalese, Mipup; Unani, Mohua; Persian, Gul-e-chakan; Tamil, Guupai.

Habitat — Found in the forests of western India from Konkan southward to Kerala. Common in Kanara, Malabar, Mysore, and Anumalais at low elevations.

Parts used — Flowers, fruit, oil of the seeds, leaves, and bark.

Morphological characteristics — A large tree; bark gray to dark brown, scaly, leaves clustered near ends of branches, linear-lanceolate, 3 to 5 × 1 to 2 in.; tapering towards base, glabrous when mature; flowers in dense clusters near ends of branches, many, small; calyx, rusty pubescent; corolla tube fleshy, pale yellow, glabrous, aromatic, caducous; berries ovoid, yellow when ripe; seeds usually one to two, compressed, shining.[28]

Ayurvedic description — *Rasa* — *guru, snigdha; Guna* — *madhur, kasaya; Veerya* — *sheeta; Vipak* — *madhur.*

Action and uses — *Vat pitta samak, badana sthapan, kustaghan, vat samak, sanhen, anuloman, sthamvan, rakt pitta samak, kafanisark, virisya, artawjanan, mutrul, daha prasaman, balya, vringhan.*

Chemical constituents — Flowers contain sugar, cellulose, albuminous substances, ash, and water. Dried flowers contain 50 to 60% sugar; seed contains 50 to 60% of fatty oil, called bassia oil, consisting of olein and palmitin, linolein, and stearin; a bitter principle, probably saponin, albumen, gum, starch, mucilage, and ash. Ash contains silicic, phosphoric, and sulfuric acid, lime, iron, potash, and traces of soda. Juice contains caoutchouc, tannin, starch, calcium oxalate, gum, resins, and formic and acetic acid. Oil is a mixture of 80% of stearin (separated crystals of stearic acid) and 20% of olein. Leaves contain a glucosidic saponin different from that obtained from seeds. Traces of an alkaloid have also been found. Flowers contain a fairly good quality of sugar, enzymes, and yeast and are commercially used for production of fuel alcohol. The fruit contains saccharose and maltose, tannin, and enzymes. It yields 0.03% of an essential oil containing 22.7% of ethyl cinnamate.[50,178]

Alcoholic extract of the kernels of the nuts of the plant yield sucrose, β-sitosterol, and an unidentified sterol glucoside.[498] Triterpenoids as well as free β-sitosterol have been reported from the alcoholic extractive of the pulpy mesocarp. The extract of the nutshell yielded β-D-glucoside of β-sitosterol, quercetin, and dihydroquercetin.[499] Defatted seeds yielded two saponins[500] and the seed oil gave a sterol from its unsaponifiable matter.[501] The leaves have yielded flavonals, myricetin, quercetin, and tannin (4.86%).[502,503] The bark yielded lupeol acetate β-amyrin acetate, α-spinasterol, etc.[504] Nanda[505] reported 0.2 ppm fluoride content. Calcium, phosphorus, vitamin C, iron, and carotene were also found.[506]

Pharmacological action — Fresh juice is alterative and the spirit distilled from the flowers is a powerfully diffusible stimulant and astringent, tonic, and appetizer. Flowers are cooling, demulcent, expectorant, tonic, and nutritive; leaves are astringent and bark is astringent and tonic. Seeds are galactagogue. The 50% of alcoholic extract of stem bark showed hypotensive activity in cats and dogs.[507] The total saponin isolated from the seed revealed spermicidal activity.[508] The flowers, leaves, stem, and stem bark are reported to possess antibacterial activity.[509]

Medicinal properties and uses — An ointment made of the ashes of the leaves is useful for burns and scalds. The bark is astringent and tonic; its decoction is given in diabetes mellitus and rheumatic disease. A lotion of the bark is an excellent gargle for bleeding and spongy gums, acute and chronic tonsillitis, and pharyngitis; this lotion is made by mixing 5 ml of the liquid extract with 280 ml of water; a paste of the bark powder is locally used to cure itch. Flowers are cooling, tonic, nutritive, expectorant, and demulcent; they are given in bronchitis, cough, and are considered useful in piles. The oil extracted from the seeds is locally applied in skin diseases, chronic rheumatism, and headache. As a laxative, it is given in piles and habitual constipation.[1,3]

Doses — Flowers — 25 to 60 gr; bark decoction — 56 to 112 ml.

MALLOTUS PHILIPPENSIS (Lam.) Muell.-Arg.   *Family:* EUPHORBIACEAE
(Syn. ***CROTON PHILIPPENSE*** Lam.)

Vernacular names — Sanskrit, Kampillaka; Hindi, Kamila; English, Rottlera; Bengali, Kamila; Nepalese, Kamila; Sinhalese, Hamperilla; German, Kamala; French, Kamola; Unani, Kamila; Arabian, Kanbil; Persian, Kanbilah.

Habitat — Throughout India and Sri Lanka.

Parts used — Glands and hairs from the fruits.

Morphological characteristics — A small- or middle-sized evergreen tree, young parts densely covered with minute red hairs. Bark gray or pale brown irregularly cracked, wood gray to light red, smooth close grained. Leaves alternate, borne on long stalks; variable in shape; 7 to 20 cm long, lower side dotted with reddish glands, prominently veined. Flowers minute, male and female on separate plants. Female flowers in erect 5 to 8-cm-long spikes, male flowers yellow in 8 to 15-cm-long drooping bunches. Fruit 8 to 13 mm, roundish, 3 lobed, densely covered with reddish-brown powdery substance and minute hairs which are easily rubbed out.[510]

Under the microscope, the powder is seen to consist of characteristic globular glands containing red resin and radiating groups of unicellular curved trichomes.[69]

Ayurvedic description — *Rasa* — *katu*; *Guna* — *laghu, rooksha, teekshna*; *Veerya* — *ushna*; *Vipak* — *katu*.

Action and uses — *Kapha vat samak, pittakark, kustaghan, branropan, bran sodhan, anuloman, rechan, krimighan, asmarivadan, vajkarn, kustaghan.*

Chemical constituents — *M. philippensis* contains rottlerin and the red compound as the main constituent, with small amounts of iso-allorottlerin, the yellow compound, and two more compounds, kamalin I and kamalin II.[511] The fruit is reported to contain terpenes and saponins.[512] Kapoor et al.[513] reported the absence of saponins and alkaloids in dehisced empty fruit. The presence of flavone and chalcone in the fruit has been reported.[515]

The heartwood of the tree yielded betul-3-acetate along with lupeol, lupeol acetate, and sitosterol. Alcoholic extract of heartwood, bark, and leaves yielded bergenin.[515] Stem bark of the plant showed presence of flavonoids and absence of saponins and alkaloids.[516,517]

Pharmacological action — Cathartic, anthelmintic, and aphrodisiac. The crude extract of *M. philippensis* (hair of the fruit) was found to significantly reduce fertility in rats and guinea pigs due to rottlerin. Alcoholic and ethereal extract of fruit showed taenicidal action both *in vitro* and *in vivo*. The 50% ethanolic extract of fruit showed hypoglycemic activity in rats and anticancer activity against human epidermoid carcinoma of nasopharynx in tissue culture and sarcoma 180 in the mouse. It also exhibited antibacterial action and antispasmodic activity on isolated guinea pig ileum.[51] The fruit and resin from the capsules showed significant purgative action in rats.[518,519]

Medicinal properties and uses — *Kamala*, the hairy glands on the fruit, is astringent and anthelmintic; it is considered specific for tapeworms. An ointment made of *kamala* with some bland oil is used for ringworm, scabies, herpes, and other parasitic skin diseases, and its powder is given internally to relieve leprous eruptions. *Kamala* powder alone is applied over syphilitic ulcers.[1,3]

Doses — 1 to 3 gr (children), 10 gr (for adults).

MELIA AZEDARACH Linn.

Family: MELIACEAE

Vernacular names — Sanskrit, Nimba; Hindi, Mahanimba, Bakain, Drek; English, Persian lilac; Bengali, Ghoranim; Nepalese, Neeha; Sinhalese, Kohamba; German, Gemeiner Zedrach; French, Margosier; Unani, Nim; Arabian, Harbit; Persian, Neeb; Tibetan, Ni-mba; Burmese, Kamakak; Pustu, Bakyana; Malaysian, Bay pay; Tamil, Malai.

Habitat — A large evergreen tree found throughout India, often planted.

Parts used — Whole plant, bark, root bark, young fruits, nut or seed, flowers, leaves.

Pharmacognostical characteristics — A tall evergreen tree with brownish bark and delicate foliage. Leaves pinnate, crowded toward the end of branches. Leaflets 9 to 15, opposite or alternate, lanceolate; flowers white, scented, and in short axilliary panicles. Fruit a drupe, greenish yellow when fresh and after drying becomes dark in color, 1.2 to 1.8 cm long. Seeds are dirty brown, 1 cm in length and 4 to 5 mm in width, taste bitter, odor characteristic, one in each fruit.

Bark is channeled, tough, fibrous, brownish gray with a rough scaly surface. Internal bark is yellowish. Under the microscope, the bark of *M. azedarach* shows the rhytidome present along with alternate zones of cork and secondary phloem. The cork cells are rectangular without any intercellular spaces and with indistinct phellogen. The cortex is divided into many layers by newly originated cork layer.[520] The sieve tube elements have compound sieve plates. The phloem fibers have fiber pits. Sclereids are absent. Rosettes and prismatic crystals are present in the ray cells. Stellate and peltate trichomes are present on leaf.[521] The costal cells are elongated, straight walled, and arranged in rows. The stomata are oval. The nonglandular hairs are common on the base of the petiole, and sparse on the margin and veins, unicellular, thick walled, having a rounded or angular foot. The glandular hairs are spherical, short stalked, and multicellular.[522]

Ayurvedic description — *Rasa* — *tikta, kasaya; Guna* — *laghu; Veerya* — *sheeta; Vipak* — *katu*.

Action and uses — *Kapha pitta samak, brana pachan, brana sodhan, daha prasaman, kandughana, brana ropan, badana nasak, rochan, grahi kaphaghan, daha prasaman, rakt sodhak, khustaghana, jawarghna*.

Chemical constituents — The leaves of this plant yield two flavonoids, quercitrin and rutin.[523] The fruits and heartwood contain a bitter principle and a lactone, named bakayanin and bakalactone.[524,525] Petroleum ether extract of the bark led to isolation of 6β-hydroxy-4-stigmasten-3-one and 6β-hydroxy-4-campesten-3-one.[526] Misra and Srivastava[527] isolated from stem bark a new flavone glycoside characterized as 4,5-dihydroxy flavone 7-O-α-L-rhamnopyranosyl-(1-4)-β-D-glucopyranoside. Srivastava et al.[528] isolated 19.25% of protein from seeds which on acid hydrolysis showed the presence of cystine, serine, arginine, glycine, glutamic acid, threonine, methionine, leucine, lycine, and proline.

Pharmacological action — Root bark and fruit are astringent, tonic, and antiperiodic. Bark is bitter, tonic, astringent, antiperiodic, and also vermifuge. Fruit is purgative, emollient, and anthelmintic. Oil from seed and leaves is a local stimulant, insecticide, and antiseptic. Flowers are stomachic.[1,3]

The fine powder of seeds shows insecticidal activity, and unsaponifiable matter obtained from fixed oil of seeds showed antibacterial action against some bacteria.[528-530] The leaves showed significant anthelmintic activity against tapeworms and hookworms.[531] The 50% ethanolic stem bark extract revealed anticancer activity against Walker carcinosarcoma 256 in rats. The extract has antispasmodic action on isolated guinea pig ileum. It showed antiviral activity against Ranikhet disease virus.[11]

Medicinal properties and uses — Bark is used in the form of powder or fluid extract or decoction in cases of intermittent and other paroxysmal fevers to relieve thirst, nausea, vomiting, and general debility, loss of appetite, and skin diseases. A decoction made of this bark and *babula*

(*Acacia* sp.) bark in equal parts is useful in leukorrhea. The oil given in 10-drop doses with milk once a day in combination with other Ayurvedic remedies has been effective in early stages of leprosy. Oil may be used like carbolic acid as a dressing for foul ulcers, and as a linament to rheumatic affections. It is a favorite application in tetanus, leprosy, urticaria, eczema, scrofula, and skin diseases like ringworm, scabies, and pemphigus. In doses of 4 to 16 ml, neem oil is very effective in expelling intestinal parasites.

Doses — Powder root bark — 1 to 2 gr; powder fruits — 1 to 2 gr; decoction of the bark — 56 to 112 ml; decoction of the root bark — 56 to 112 ml; decoction of the leaves — 56 to 112 ml; tincture — 4 to 12 ml; infusion of the flower — 28 to 84 ml; oil — 4 to 10 minims.

MENTHA ARVENSIS Linn.   Family: LABIATAE

Vernacular names — Sanskrit, Puthea; Hindi, Pudina, podina; English, Corn mint; Bengali, Pudina; Nepalese, Nawaghya; Sinhalese, Odutalan; Japanese, Midorihakka; German, Minze; Unani, Pudinah; Arabian, Putnaj; Persian, Pudinah; Burmese, Bhudina; Tamil, Puthina.

Habitat — Cosmopolitan, found chiefly in warm, dry, temperate regions. Occurs in Kashmir at an altitude of about 4000 to 9000 ft.

Parts used — Whole plant and oil with creeping rhizomes and square stems.

Morphological characteristics — Erect branched herb, leaves 2.5 to 5 cm long, shortly petioled or sessile, oblong-ovate or lanceolate, obtusely or acutely serrate, cuneate at the base, sparsely hairy or almost glabrous; flowers lilac, arranged in verticillasters, borne on axils of leaves on upper stem.[510]

Diacytic stomata are present on the lower surface of the leaf. Under the microscope, the leaf also shows 3- to 8-celled clothing trichomes with a striated cuticle. Two types of glandular trichomes, one with a unicellular base and a small single-celled head and the other with a multicellular head characteristic of the family, are present. Calcium oxalate is not present.[69]

Ayurvedic description — *Rasa* — *katu; Guna* — *laghu, rooksha, teekshna; Veerya* — *ushna; Vipak* — *katu.*

Action and uses — *Badana sthapan, dipan, rochan, vata anuloman, krimighan, hiridoutejek, jawarghana.*

Chemical constituents — An essential oil is obtained by steam distillation from the leaves, flowering tops and stems similar to peppermint oil.

The oil of *M. arvensis* growing in Kashmir contains 46% of α-menthol, besides *d*-menthone, menthylacetate, carvomenthone, limonine, β-phellandrene, and pepritone.[532] Jammu-grown *M. arvensis* contained 0.04% menthofuran. The natural oil of *M. arvensis* var. *piperascens* with a menthol content of 70 to 80% yielded only 40 to 50% menthol and 50 to 60% dementholized oil.[533] Arora and Singh[534] found a number of amino acids, viz., aspartic acid, glutamic acid, serine, glycine, threonine, alanine, lysine, etc., in the ethanolic extract of the leaves.

Pharmacological action — Aromatic, carminative, stimulant, antispasmodic, and emmenagogue. Antifertility activity of *M. arvensis* has been observed by several workers, but the results are varying from each other. Alcoholic and aqueous extracts of leaves could inhibit implantation in 60 to 80% in female rats. Petroleum ether extract showed 44% antifertility effect; it seemed to exert some abortifacient activity.[535] Alcoholic extract showed some antiovulatory effect in rabbits.[536] The plant tissue culture exhibited antibacterial activity, also.[537] The oil showed antifungal activity as well.[538]

Medicinal properties and uses — Dried plant is aromatic; given in the form of "sharbat", or syrup for its cooling and diuretic effect. It is used as an antiseptic, carminative, stomachic, and refrigerant.[1] It is considered to be stimulant, emmenagogue, and diuretic.[50] Juice of the leaves is given in diarrhea and dysentery. Infusion of leaves is used in rheumatism and indigestion.[37]

MESUA FERREA Linn. *Family:* GUTTIFERAE

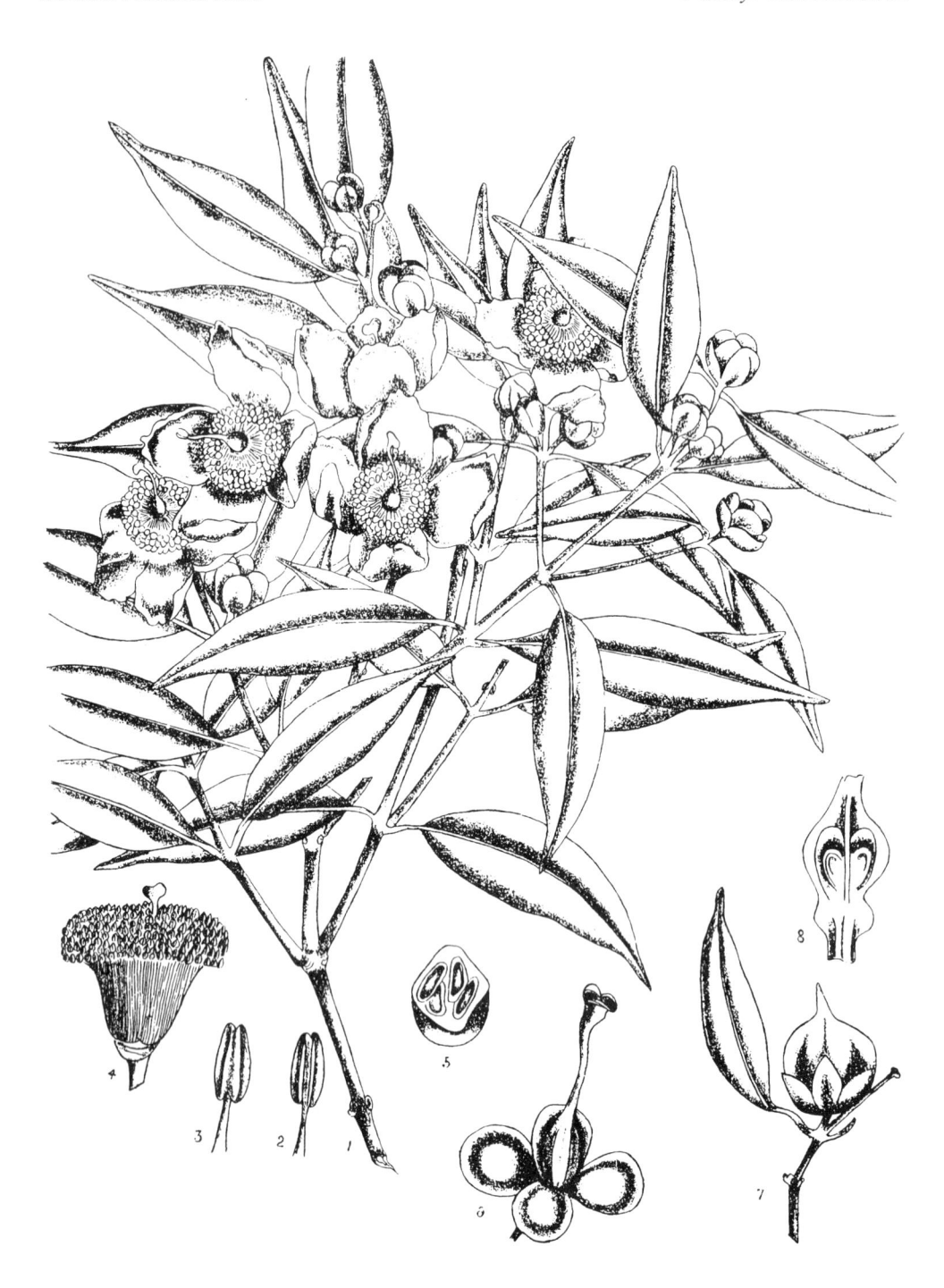

Vernacular names — Sanskrit, Nagkeshara; Hindi, Nagakeshar; English, Cobra's saffron; Bengali, Negesar; Nepalese, Nagakashar; Sinhalese, Namalranu; Japanese, Tagayasan; Chinese, Thiet lucmoc; German, Nagassamen; Unani, Nagkesar; Arabian, Nar-e-misk; Persian, Naz Mushk; Tibetan, Na-ga-ge-sar; Burmese, Gangau; Malaysian, Malopus; Tamil, Chirungappu.

Habitat — Found throughout the eastern Himalayas, Assam, Andemans; Burma, and Bangledesh.[460]

Parts used — Buds, flowers, fruit, seeds, root, and bark.

Pharmacognostical characteristics — A medium-sized evergreen tree, bark grayish or reddish brown, exfoliating in large thin flakes, wood extremely hard; leaves lanceolate, coriaceous, generally covered with a waxy bloom underneath, red when young; flowers large, solitary or in clusters of 2 to 3, white, fragrant; fruits ovoid, nearly woody, 2.5 to 5 cm long, with persistent calyx; seeds 1 to 4, dark brown, up to 2.5 cm diameter.[510]

The presence in a transection of cortical fibers, numerous resin canals, and calcium oxalate crystals in the cortex and pith of the pedicel; anamocytic, anisocytic, or paracytic stomata in sepals and petals; and 3 to 4 zonocolporate pollen grains with reticulate exine surface are the distinguishing characteristics. The stamens are numerous with short filaments and thick, elonged anther lobes. Some fused stamens along with their vascular supply are also seen. In a commercial sample, *nagakeshar* consists of stamens and flowers of *M. ferrea*, appearing purplish brown, and blackish brown, respectively.[539]

Ayurvedic description — *Rasa — kasaya, tikta; Guna — laghu, rooksha; Veerya — ushna; Vipak — katu.*

Action and uses — *Kapha pitta samak, badana nasak, dipan, pachan, arsoghana, grahi, krimighan, shonita sathpan, mutra jank, vajikarna.*

Chemical constituents — Young fruits contain an oleoresin which yields an essential oil. Seeds contain a fixed oil. Pericarp of fruit contains tannin. Essential oil is fragrant, like that of its flowers, and pale yellow in color. The leaves are reported to contain unnamed alkaloids.[10] Seed oil of *M. ferrea* yielded several 4-phenylcoumarin, viz., (1) mesuol, (2) mesuone, (3) mammeigin, and (4) mesaugin. Another coumarin named mammeisin was also reported in a sample from another region.[460] The stamen extract gave α-amyrin, β-amryin, and β-sitosterol, two new biflavonoids, viz., mesuaferrone A and mesuaferrone B, and mesuanic acid.[540-542]

The heartwood of the plant on extraction with acetone yielded three pigments, euxanthone and two new pigments, mesuaxanthone A and mesuaxanthone B.[534,544] Petroleum ether extract of the trunk bark yielded a mixture of ferruol A and ferruol B. The extract also yielded a lupeol type of triterpenoid named guttiferol.[545]

The seed oil was rich in oleic, stearic, and palmitic acids and also linolenic acid.[546-548]

Pharmacological action — Dried blossoms, root and bark are bitter, aromatic, and sudorific; bark is mildly astringent; unripe fruits are aromatic, acrid, and purgative. Dried flowers are astringent and stomachic; also stimulant and carminative.[1,3]

Chakraborty et al.[549] reported that mesuol and mesuone obtained from *M. ferrea* showed antibacterial activity against some bacteria. The essential oil from stamens exhibited anthelmintic activity against tapeworms and hookworms.[550] Phenolic constituents of the seed oil revealed a potent antiasthmatic effect.[551] Gopala Krishnan et al.[552] reported that xanthones exhibited significant anti-inflammatory activity against carrageenin-induced hind paw edema, cotton pellet test, as well as against the granuloma pouch technique in normal and adrenalectomized rats.

Medicinal properties and uses — Leaves are used in the form of a poultice which is applied to the head in severe cold. Bark in the form of decoction, tincture, or infusion is a bitter tonic and is very useful in gastritis and bronchitis. Fixed oil is used as an application for cutaneous affections, such as sores, scabies, and wounds; and as an embrocation in rheumatism. The dried flowers are given in vomiting, dysentery, cough, thirst, irritability of the stomach, excessive perspiration, and bleeding piles. The fruit is stimulating and alterative; it is given in disease of the genitourinary organs as a substitute for cubebs. Chopra et al.[1] reported that the plant possesses antidysenteric properties and is used against snakebite and scorpion sting.

Doses — Decoction of root — 8 to 16 ml; powder — 0.5 to 1 gr.

MICHELIA CHAMPACA Linn.   Family: MAGNOLIACEAE

Vernacular names — Sanskrit, Champaka; Hindi, Champa; English, Yellow champa; Bengali, Champa; Nepalese, Champa; Sinhalese, Sapu; Japanese, Kinkoboku; German, Wohlriechen-de Michele; Unani, Champa; Burmese, Saga; Malaysian, Mang liet; French, Champac; Tamil, Sembagam.

Habitat — Tree growing wild in Assam, Bengal, Burma, and Nepal. Frequently cultivated for its yellow, sweet-scented flowers used in perfumery.

Parts used — Bark, root, root bark, leaves, flowers, fruits.

Morphological characteristics — A tall evergreen tree. Leaves petioled, ovate-lanceolate, acuminate, entire; flowers yellow, scented, axillary and terminal, solitary. Sepals and petals similar, arranged in whorls of 3, soon falling. Stamens indefinite, anther cells adnate, carpels indefinite, free, in one whorl. Style short. Seeds solitary or few, albumen granular, oily. Chaudhuri[553] reported the diagnostic characters of stem bark of *M. champaca* as longitudinal fissures on the outermost layer of the bark which separates on drying, the presence of stone cells with large lumen, phloem fibers with small lumen, secretory cells with mucilage, and prismatic and octahedral crystal of calcium oxalate.

Sharma[554] studied the distribution of sclereids in stem, petiole, and floral parts. The astrosclereids with short, pointed arms are very characteristic of this plant. Nair[555] reported that pollen is unicolpate and bilateral, having a thick exine and granulate sexine with prominent thick granules.

Ayurvedic description — *Rasa* — *tikta, katu, kasaya, madhur; Guna* — *laghu, rooksha, teekshna; Veerya* — *sheeta; Vipak* — *katu.*

Action and uses — *Kapha-pitta samak, daaha prasaman, bran sodhan* and *ropan, ruchivardhak, dipan, ama pachan* and *anuloman, krimighan, virachar, hirdya, rakt sodhak, artwa janan.*

Chemical constituents — The plant contains an essential oil, fixed oil, resin, tannin, mucilage, starch, and sugar. Different parts of the tree yielded the following active principles with different solvents.

1. Stem bark — lirodenine, macheline A.[556,557]
2. Leaves — polyisoprenoid, β-sitosterol and β-sitosterol-3-*O*-β-D-glucoside, and liriodenine.[558]
3. Roots — parthenolide, liriodenine,[559] a new sesquiterpene lactone (mp 147 to 148°C, $C_{15}H_{20}O_3$).[560]
4. Root bark — four sesquiterpene lactones identified as costunolide, parthenolide, dihydroparthenolide, and micheliolide.[561]
5. Flowers — essential oil, along with a new sesquiterpene hydrocarbon champacene.[562]
6. Fruit rinds yield essential oil comprised of 1:8 cineola, phenyl ethyl alcohol, pinacamphene, linalool, ester of pinocampheol, pinocampheol, phenyl ethyl acetate, and α-phellandrene.[563]

Essential oil from leaves contained linalool and linalyl acetate along with methyl heptanone and geraniol as minor components.[564] Seed fats comprised of myristic, palmitic, stearic, hexadecadienoic, oleic, linoleic, arachidic, eicosenoic, and hexadecanoic acid.[565]

Pharmacological action — Deobstruent, alterative, stomachic, tonic, purgative, and carminative. Ethanolic extract of the stem bark showed hypoglycemic action in rats and hypertensive effect in cats and dogs.[51]

Medicinal properties and uses — Decoction of the bark is useful in mild cases of gastritis. Bark is also very effective in chronic rheumatism. The root is bitter, demulcent, and emmenagogue. A paste of the root bark made with curdled milk is efficacious dressing for abscesses and inflammation. Bark is used as febrifuge, stimulant, astringent, and expectorant. Dried root and root bark are purgative; in the form of infusion, useful emmenagogue. Flowers and fruits — stimulant, antiseptic, tonic, stomachic, carminative, bitter, and cooling, used in dyspepsia, nausea, and fever. Useful as diuretic in renal diseases as in gonorrhea; mixed with sesamum oil, forms an external application in vertigo. Oil from flowers, useful application in cephalalgia, ophthalmia, and gout. Juice of leaves given with honey in colic. Seeds and fruit used for healing cracks in feet.[50]

Doses — Bark powder — 1 to 2 gr; infusion — 56 to 75 ml; decoction of the bark — 14 to 28 ml; flower infusion — 14 to 56 ml.

MIMUSOPS ELENGI Linn. *Family:* SAPOTACEAE

Vernacular names — Sanskrit, Bakula; Hindi, Maulsari; Bengali, Bakul; Nepalese, Bakulapuspa; Sinhalese, Munemal; German, Affengesict; French, Karanicim; Unani, Moolsari; Burmese, Kaya; Malaysian, Enengi; Tamil, Magizh.

Habitat — Grows wild and is also cultivated for its ornamental appearance and fragrant flowers throughout India; west peninsula southward from Khandla Ghat on the west and the North Circars on the east side and Andamans. Widely cultivated in Deccan.[460]

Parts used — Bark, flowers, and fruit.

Morphological and pharmacognostical characteristics — An ornamental tree. Bark dark gray, scaly, rough, deeply furrowed, lenticles vertical, blaze pink with red streaks; wood dark red, hard, close grained; leaves shortly accuminate, elliptic, oblong, glabrous; flowers white, fragrant, in fascicles of 2 to 6, calyx 8 lobed; corolla lobed with 24 appendages in 2 rows; stamens 8, resembling petals, clothed on back and margins with white hair; ovary 8 celled; fruit a berry, yellow, ovoid, about 2.5 cm long.

In transection, the root is of the normal dicot pattern, the primary root tetrarch with pith in center. The pith cells are polygonal in outline. The secondary growth produces annual rings. The wood is composed of xylem vessels with broad lumen and large number of fibers, interspersed in a ground tissue composed of parenchyma. From the protoxylem groups emerge one-celled wide, medullary rays, radially outward and extending up to the end of xylem region. Similar uniseriate, secondary xylem rays are abundantly present in the wood. Bordering secondary xylem is the cambium ring, followed by a zone of secondary phloem. Within cortex are found lignified fibrous cells. Crystals absent. Cambium 5 to 8 layered. Cork cells suberized. Stems have the usual dicot structure, showing formation of secondary growth. Pith broad. Presence of rectangular stone cells and 2- to 4-celled phloem rays. The bark liberates brown colored pigments in water.[566]

Ayurvedic description — *Rasa* — *katu, kasaya; Guna* — *guru; Veerya* — *ushna; Vipak* — *katu.*

Action and uses — *Kapha pitta samak, mastiskya balya, grahi, krimighan, hiridya, dant rakshak, rakt sthamvak, jwarghana.*

Chemical constituents — Bark contains tannin, some caoutchouc, wax, coloring matter, starch, and ash. Flowers contain a volatile oil, and seeds a fixed fatty oil. Pulp of the fruit contains a large proportion of sugar and saponin. Seeds of *M. elengi* yielded quercitol, dihydroquercetin, quercetin, β-D-glucoside of β-sitosterol, and α-spinasterol and the flowers gave quercitol, ursolic acid, and lupeol.[567,568] The fatty oil from seeds consisted of capric, lauric, myristic, palmitic, stearic, arachidic, oleic, and linoleic acids; the unsaponifiable matters from the seed fat consisted of β- and γ-sitosterol.[569] The bark and wood of stem of the plant yielded teraxerone, taraxerol, α-spinasterol, sodium ursolate, betulinic acid, β-D-glucoside of β-sitosterol and quercitol,[570] and also *meso*-inositol. Chopra and Kapoor[572] reported the presence of saponins which on hydrolysis yielded β-amyrin and basic acid.[573] Quercitol, hentriacontane, β-carotene, and glucose were isolated from leaves[568] and D-mannitol, β-sitosterol, β-sitosterol-β-D-glucoside, and quercetin were also recovered from the leaves.[571,574] The root yielded lupeol acetate, taraxerol, a spinasterol, and β-D-glucoside of β-sitosterol.[568]

Pharmacological action — Anthelmintic, antihistaminic, astringent, aromatic, cardiotonic, and tonic. The leaf extract of *M. elengi* showed antibacterial activity[575] and that of fruits and leaves exhibited a hypotensive effect in dogs.[576] The saponin from the seeds is reported to be spermicidal in human semen.[577] It showed spasmolytic action on isolated guinea pig ileum. It was most active against histamine.[578]

Medicinal properties and uses — The bark is astringent and tonic, antipyretic, and increases fertility in women; its decoction is given in catarrh of the bladder and urethra as an astringent; in fevers it is given as a febrifuge and tonic. The decoction is a useful mouthwash in diseases of the gums and teeth and excessive salivation. It is also given in leukoderma and heart diseases. Leaves are given in snakebite.[3,50] Pulp of ripe fruit used as astringent for curing chronic dysentery. Seeds bruised and locally applied within the anus of children in cases of constipation. A paste of the roots made with vinegar is applied to swellings on the face; a paste made with water is applied to pustular eruptions of the skin. As an astringent, lotion prepared from the flowers is used for washing wounds and ulcers. A powder of the dried flowers is used as snuff to relieve headache and to induce a copious discharge from the nose. The unripe fruit is astringent. The plant is considered useful by tribals (Santals) in ulcerated tongue, sores, dropsy and anasarca, smallpox, syphilis, sores and carbuncle, consumptive cough, bronchitis, and menorrhagia.[579]

Doses — Bark decoction — 56 to 112 ml; flower powder — 1 to 2 gr.

MORINGA OLEIFERA Lam.
(Syn. *M. PTERYGOSPERMA* C. F. Gaertn.)

Family: MORINGACEAE

Vernacular names — Sanskrit, Sigru; Hindi, Sahinjan; English, Horseradish tree; Bengali, Sojna; Nepalese, Sobhanjan; Sinhalese, Murunga; Japanese, Wasabinoki; French, Moronguier; Unani, Sahinjan; Tamil, Murungai.

Habitat — A beautiful tree growing wild in the sub-Himalayas from Chenab to Uttar-Pradesh. It is commonly cultivated in India and Burma.

Parts used — Bark, roots, fruit, flowers, leaves, seeds, and gum.

Pharmacognostical characteristics — A fairly large and pretty tree. Leaves 30 to 60 cm, usually 3-pinnate; petioles slender, sheathing at the base; leaflets 6 to 9 pairs; 12 to 18 mm long, opposite, pale beneath, leaflets elliptic-ovate or ovate. Flowers white, in large puberulent panicles. Bark corky; wood soft, externally light brown, and inner color white. Cortex is thick and slightly mucilaginous.

Ayurvedic description — *Rasa — katu, khara; Guna — lagu, rooksha; Veerya — ushna; Vipak — katu.*

Action and uses — *Kapha vat samak, bidahi, sothahar, shirovirachan, badana sthapan, sotha har, dipan, pachan, lakhan, graahi, shul prasaman, krimighan, jawaraghana.*

Chemical constituents — Bark contains a white crystalline alkaloid, two resins (one soluble and the other insoluble in ammonia), an inorganic acid, mucilage (gum), and ash (8%). The husked seeds yield, on simple pressure, a clear, limpid, almost colorless fixed oil known as *beni* or *moringa* oil. It contains 60% of liquid oil and 40% of white solid fat. Constituents of moringa oil are myristic acid (7.3%), palmitic acid (4.2%), oleic acid (65.8%), stearic acid (10.8%), behenic acid (8.9%), and lignoceric acid (3.0%). The oil is a good source of a behenic acid in nature. Rao and Georg[580] found that alcoholic extract of fresh roots exhibited strong antibiotic activity due to "ptergospermin" which is a reddish-brown, very active oil. Rangaswami and Subramanian[581] obtained a wax from flowers with mp 69 to 72°C, acid number 10.5, saponification number 29.8, unsaponifiable matter 75.5%. The bark revealed the presence of sterols and terpenes,[582] triterpenoid-bayrenol.[583] The leaves yielded amino acids such as aspartic acid, glutamic acid, serine, glycine, threonine, α-alanine, valine, leucine, isoleucine, histidine, lysine, arginine, tryptophan, cystine, and methionine and also α- and β- carotene.[584] Later, nine amino acids in the flowers, eight in the fruit, and seven each in protein hydrolysate of flowers and fruits were identified. The flowers contained both sucrose and *d*-glucose, but the fruits showed the presence of sucrose only.[585] The stem yielded 4-hydroxymellein, vanillin, β-sitosterol, β-sitosterone, and oclacosanoic acid.[586]

The fatty acid distribution in the seed fat and their glyceride composition have been reported.[587] Pods and leaves were found to be a rich source of vitamin C.[588]

Pharmacological action — Antispasmodic, stimulant, expectorant, and diuretic. Fresh root is acrid and vesicant; internally stimulant, diuretic, and antilithic. Bark is emmenagogue and even abortifacient. Flowers are stimulant, tonic, and diuretic and useful to increase the flow of bile.[1,3]

Juice from leaves and stem bark of *M. oleifera* exhibited antibacterial and antitubercular activity.[589,590] The root bark showed antiviral effect against vaccinia virus.[51] Ethanolic extract of the whole plant showed anticancer activity against human epidermoid carcinoma of nasopharynx in tissue culture and P_{388} lymphocytic leukemia in mice.[591] The root bark showed anti-inflammatory and analgesic activity.[592]

Medicinal properties and uses — The tender leaves are given in scurvy and catarrhal diseases; leaf juice is used as an emetic; it is given with salt to children suffering from flatulence. A poultice of the fresh leaves is applied to wounds, boils, and swellings.[1]

The bark of the stem is diuretic and antiscorbutic. It is used as a cardiac stimulant in asthma, cough, and similar disorders. The fresh root of a young tree is given in intermittent fevers, epilepsy, hysteria, chronic rheumatism, gout, dropsy, dyspepsia, and enlargement of the liver and spleen. The gum that exudes from the stem is given with milk for relief of headache. The gum is an efficacious dressing for glandular swelling and boils. The flowers boiled with milk are aphrodisiac. The tender pods are anthelmintic; also given in diseases of the liver, spleen, and articular pains. The oil is an aperient; locally, it is applied to painful gouty and rheumatic joints.

Doses — Infusion — 1 to 3 ml; decoction — 28 to 56 ml; seed powder — 10 to 16 gr.

MUCUNA PRURIENS (Stickm.) DC. Family: PAPILIONACEAE

Vernacular names — Sanskrit, Kapikachchha; Hindi, Kavach; English, Cow-itch plant; Bengali, Alkusi; Nepalese, Kauchho; Sinhalese, Wandurme; Japanese, Hatsushomame; German, Jackbohne; Unani, Konch; Persian, Hub-ul-kulai; Burmese, Khuele; Malaysian, Nayikuruma.

Habitat — An annual climbing shrub common in the tropics, under cultivation in some parts for its golden brown velvety legumes, which are eaten as a vegetable.

Parts used — Seeds, root, and legumes.

Pharmacognostical characteristics — A climbing green shrub, leaves trifoliolate; leaflets broadly ovate, elliptic or rhomboid ovate, unequal at base; flowers in axillary, pendulous racemose, purple, pods curved, 5 to 10 × 1.5 to 1.8 cm, longitudinally ribbed, turgid, densely clothed with persistent pale brown or gray, irritant bristles; seeds black, pod 4 to 5 in., ovoid with funicular hilum.

A transverse section of the root shows a central porous woody region which forms the major part of the root and an outer thin bark portion. The outermost tissue of the root is the cork which is narrow and light black in color. It consists of four to six rows of tangentially elongated cells with thin dark-brown walls. Exfoliating strips of crumpled cork tissue are present external to this layer at certain regions. On treating with ferric chloride solution, the cork layer turns intensely dark brown.

The seed coat is hard, thick, and glossy. The embryo completely fills the seed and is made up of two large fleshy cotyledons. Transverse section shows an outer testa with a palisade epidermis made up of rod-shaped macrosclereids with thickened anticlinal walls. The hypodermis comprises bone-shaped osteosclereids, broad at the top and base and narrow in the middle. The furrow region at the hilum shows double epidermis of macrosclereids, hypodermis of osteosclereids and a cortical region made up of thin-walled cells with a group of trachoid abutting at the end of raphe. The cotyledons contain oval-shaped starch grains. Microchemical study of the plant shows the presence of tannin, resin, anthraquinone, fat, oil, and saponins.[593,594]

Ayurvedic description — *Rasa — madhur, tikta; Guna — guru, snigdha; Veerya — ushna; Vipak — madhur.*

Action and uses — *Tridosha samak, krimighan, balya, veeresya, bajkarak, yoni sonkochak, mutral.*

Chemical constituents — *M. pruriens* has been reported to be a good source of 3,4-dihydroxyphenylalanine (L-DOPA). It also contains a few amino acids.[595,596] The seeds gave four alkaloids, namely, mucunine, mucunadine, prurienine, and prurieninine.[597,598] Four indole-3-alkylamines and choline have been isolated from different parts of the plant.[599,600] The seed oil contains stearic, palmitic, myristic, arachidic, oleic, and linoleic acids and a sterol.[601]

Pharmacological action — Seeds are astringent, anthelmintic, nervine tonic, and aphrodisiac. Roots are also tonic, stimulant, diuretic, purgative, and emmenagogue. When tested on frogs, prurieninine slows down the heart, dilates the blood vessels, depresses blood pressure, and increases the peristaltic action of the intestine.[1]

Ayurvedic physicians claim that *M. pruriens* is very effective in the treatment of Parkinsonism, but the latest studies on the powdered seed extract showed that it was devoid of anticholinergic activity. The extract showed hypotensive action in dogs and spasmodic action in guineapig preparation. The extract had no effect on frog rectus but revealed a histaminergic activity. Ramaswamy et al.[602] opined that the seed powder may act by some mechanism other than through anticholinergic property in Parkinsonism. Nath et al.[603] in their studies, indicated that the extracts of seed had no anticholinergic activity but had a potent anti-Parkinsonian effect.

M. pruriens seed diet showed a hypocholesterolemic effect in rats.[604] In another study, the protein isolate of the seed in the diet led to a reduction in the cholesterol content of the liver and blood in rats.[605] Bhattacharya and Sanyal[606] reported that the total alkaloids of seeds showed a weak neuromuscular blocking effect on frog rectus abdominus. Bufotenine showed anticholinesterase activity *in vivo* and *in vitro* similar to but 20 to 30 times weaker than that of Physostigmine.[607]

Medicinal properties and uses — The plant is useful for diseases of the nervous system, kidney, and dropsy. An ointment prepared from the roots is applied for elephantiasis. The seeds are astringent and tonic. They possess slight insecticidal activity. The leaves of the plant are applied to ulcers. The hairs were formerly used as vermifuge and were official in some pharmacopoeias. An infusion of hair is used in diseases of the liver and gallbladder and applied externally as a local stimulant and mild vesicant. Seeds are also very useful in the form of powder in leukorrhea, profuse menstruation, paralysis, spermatorrhea, and in cases requiring aphrodisiac action.

Doses — Powder of seeds — 1.5 to 2.5 gr; hairs — 0.625 to 1.25 gr.

MYRISTICA FRAGRANS Houtt.   *Family:* MYRISTICACEAE

Vernacular names — Sanskrit, Jatiphalam; Hindi, Jaiphal; English, Nutmeg; Bengali, Jaiphal; Nepalese, Jaipo; Sinhalese, Sadikka; Japanese, Nikuzuku; German, Echtermuscatnussbaum; French, Muscadier, Macis; Unani, Jaiphal; Arabian, Janzu tbeeb; Persian, Jauzboa; Burmese, Zadi-phu; Malaysian, Bunga pola; Tamil, Jadhikkai.

Habitat — The nutmeg tree is indigenous to Molucca Island and Penang. It has been introduced for cultivation in southern India, Sri Lanka, and West Indies.

Parts used — Dried seed, nutmeg, aril surrounding the seeds, mace, wood, and oil.

Pharmacognostical characteristics — A dioecious evergreen, aromatic tree, usually 25 to 40 ft high, but sometimes reaches a height of about 60 ft. Bark grayish black, longitudinally fissured in old trees; leaves elliptic or oblong-lanceolate, coriaceous; flowers in umbellate cymes, creamy yellow, fragrant; fruits yellow, broadly pyriform or globose, 6 to 9 cm long, glabrous, often droping; pericarp fleshy. Seed broadly ovoid, arillate, albuminous, with a shell-like purplish-brown testa; aril red, fleshy, and laciniate.

Nutmegs, the dried kernels of the seeds of *M. fragrans,* are broadly oval in outline, 2 to 3 × 2 cm. At one end is a lighter colored patch with brown lines radiating from the hilum which is surrounded by a raised ring. From this an ill-defined furrow (imprint of raphe) runs to the chalaza at the opposite end of the kernel, where there is a small, dark depression. Odor strong and aromatic; taste pungent and slightly bitter.[69]

Ayurvedic description — *Rasa* — *katu, teeshana, kasaya; Guna* — *laghu, snigdha, teekshna; Veerya* — *ushna; Vipak* — *katu.*

Action and uses — *Kaf bat nasak, sothhar, badana sthopan, khutaghan, batonuloman, dipan, bat samak, grohi, jawranasak, vajekaran.*

Chemical constituents — Nutmegs yield 5 to 15% of a volatile oil and also 30 to 40% of fat, photosterin, starch, amylodextrin, coloring matter, and a saponin.[69] They yield about 3% of total ash and about 0.2% of acid-insoluble ash. Essential oil of mace is of a yellowish color with the odor of mace, and consists of macene. Mace (arillus) contains a volatile oil (8 to 17%), a fixed oil, resin, fat, sugar, destrin, and mucilage.

The volatile oil (*oleum myristicae,* British Pharmacopoeia) contains pinene and camphene (80%), dipentene (8%), alcohols (about 6%), myristicin (about 4%), safrole (0.6%), and eugenol and isoeugenol (0.2%). By expression or by means of solvents, nutmegs yield a product known as "nutmeg butter" or expressed oil of nutmegs. This consists of 12.5% of volatile oil, 73% of trimyristicin (the glyceride of myristic acid), small quantities of oleic, linoleic, and other acids, and about 8.5% of unsaponifiable matter.[1] Varshney and Sharma[608] found triterpenic saponin and 15% of free myristic acid in the seeds of *M. fragrans.* The arils yield a new neolignan, characterized as *dl*-dehydro-di-isoeugenol, and five other neolignans.[609,610] Raw nutmeg contained 1.5% of total polyphenols and 0.6% tannins. The extract of nutmeg revealed the presence of epicatechin and cyanidin.[611] Nutmeg is also reported to contain calcium, phosphorus, iron, thiamin, riboflavin, and niacin.[612]

Pharmacological action — Nutmeg is aromatic, stimulant, and carminative; in large doses, narcotic. Mace is carminative and aphrodisiac. The extract of the seed and essential oil showed antibacterial activity.[613,614] The seed oil exhibited a depressant effect on isolated frog rectus and a direct relaxant effect on rat ileum.[615]

Medicinal properties and uses — Essential oil is administered in atonic diarrhea and dysentery to relieve pain and is used in combination without stimulating oils for a stimulant action and in plasters for chronic rheumatism. A compound powder called *jatiphaladi churna* made of nutmeg, Indian hemp, camphor, cardamon, cloves, bamboo manna, and *Plumbago zeylanica* is used as a sedative, anodyne antispasmodic in asthma, colic, neuralgia, menorrhagia, dysmenorrhea, spasmodic cough, and lumbago, in doses of 1 to 1.5 gr twice daily with honey. Mace is useful in low stages of fever, in consumptive complaints, and humoral asthma. Roasted nutmeg is useful in obstructions of the liver and spleen.

Doses — Powder nutmeg or mace — 0.5 to 1 gr; oil — 1 to 3 drops.

NARDOSTACHYS JATAMANSI DC.   *Family:* VALERIANACEAE

Vernacular names — Sanskrit, Jatamansi; Hindi, Jatamansi; English, Musk root; Bengali, Jatamansi; Nepalese, Naswaa; Sinhalese, Jatamansi; Chinese, Kan Sung; German, Indische Narde; Unani, Balchar; Arabian, Sunbulatteb; Persian, Sambul-e-hindi; Tibetan, Span-pos; Malaysian, Jata-manchi; French, Nard Indien.

Habitat — The plant is found in alpine Himalaya (11,000 to 15,000 ft), extending eastward from Kumaon to Sikkim (17,000 ft), Bhutan, and Nepal.

Parts used — Rhizomes and roots.

Pharmacognostical characteristics — The plant bears a stem 10 to 60 cm long which is more or less pubescent upward and often glabrate below. Rhizomes stout, long, woody, covered with fibers from the petioles of withered leaves. Leaves usually in pairs, 2.5 to 7.5 cm long, sessile, oblong, or subovate. Flower heads usually 1, 3 or 5, with pubescent bracts. Corolla 5-limbed, somewhat hairy within, corolla tube 6 mm long. Filaments hairy below. Fruit covered by ovate, acute, often dentate calyx teeth, with ascending white hairs. The fibers are produced by an accumulation of the skeletons of the leaves and are matted together forming a kind of network. Internal color is reddish brown. Odor is highly aromatic.

A transverse section of the rhizome shows a brown bark separated by a cambium from the porous wood. The periderm consists of two to eight layers of cork cells. These layers occur in the outer cortex. Phellogen is not distinguishable. The secondary cortex is characterized by the presence of prominent oleoresin cells containing oil and resinous matter. The phloem consists of a number of cells having a diameter of about 10 µm. Cambium ring is distinct. The wood is characterized by the presence of numerous vessels scattered almost uniformly. Medullary rays are not prominent. The rest of the wood is composed of tracheids and a few fibers. Vessels are mostly with scalariform thickening and a few with spiral thickening. Vessels with irregular forms are also present. Average dimensions of vessels are 40 to 50 µm. Mehra et al.[616] described the distinguishing characteristics under microscope as the presence of interxylary and medullary cork, schizogenous cavities in the young, rhizome in the inner cortex, and parenchyma of phloem and xylem, but without any epithelial cells and oil globules in cork cells.

Ayurvedic description — *Rasa* — *tikta, kasay, madhur; Guna* — *laghu, teekshna, snigdha; Veerya* — *sheeta; Vipak* — *katu.*

Action and uses — *Kaf pit samak, daha nasak, badana nasak, baryna, sangya sthapak, acha pya nasak* (antispasmodic), *kesya.*

Chemical constituents — The rhizomes and roots contain volatile essential oil (0.5%) *(oleum jatamansi)*, resin, sugar, starch, bitter extractive matter, and gum. Govindachari et al.[617] isolated a ketonic principle called jatamansone from the rhizomes. The rhizome also yielded sesquiterpene, viz., seychellene and seychelane, and β-sitosterol.[618,619] The roots gave a good number of compounds, viz., valeranone, valeranal, nardol, calarenol, nardostechone, *n*-hexacosanyl arachidate, *n*-hexaconsanol, calarene, *n*-hexacosane, *n*-hexacosanyl isovalerate, and β-sitosterol.[620-622] The root also yielded norseychelanone, seychellen, patchouli alcohol, and also α- and β-patchoulenese.[623] The oil of roots gave terpenic coumarins, oroselol, and a new one named jatamansin.[624] The oil also yielded several hydrocarbons, and a new oxide together with β-eudesmol, elemol, β-sitosterol, angelicin, and jatamansinol.[625]

Pharmacological action — Root is of somewhat bitter taste, aromatic, antispasmodic, diuretic, emmenagogue, nerve sedative, nerve stimulant, tonic, carminative, deobstruent; sedative to the spinal cord, promotes appetite and digestion.[1,3] Various extracts of *N. jatamansi* root showed sedative action in rats and hypotensive activity in cats.[626] Jatamansone exerted tranquillizing hypothermic activity in mice and monkeys, and showed an antiemetic effect in dogs.[627] The alkaloidal fraction from root and rhizome showed a significant and sustained hypotensive action in dogs.[628] Arora et al.[629] observed that essential oil obtained from rhizome exerted prolonged and significant effect in dogs. The extracts from the roots showed antibacterial activity against a number of pathogens.[630] Antifungal activity was also observed by the extract of rhizome and its essential oil. Amin et al.,[632] in a clinical study, showed that root powder has a sedative action. In another clinical study, jatamansone could reduce aggressiveness, restlessness, stubbornness as well as insomnia.[633]

Arora and Madan[634] compared the effect of active principle of *N. jatamansi* with that of quinidine and found that *N. jatamansi* produced less prolongation of refractory period and less slowing of conduction than quinidine. The later property is of distinct advantage over quinidine. In addition, the acute intravenous toxity of *N. jatamansi* in mice was determined and found to be less than that of quinidine. The drug was said to be promising to be of possible therapeutic usefulness.

Medicinal properties and uses — Roots of *jatamansi* are used as an aromatic adjunct in the preparation of medicinal oil and in perfumery. It is a good substitute for the official valerian. Its infusion is employed in the treatment of spasmodic hysterical affections, mainly palpitation of the heart, nervous headache, and flatulence in doses of 28 to 56 ml three times daily. It is used in diseases of the digestive and respiratory organs and in jaundice. It is said to be useful also in leprosy. Mixed with sesamum oil, it is rubbed on the head as a nerve sedative. It also promotes growth and blackness of hair. The drug is prescribed as an antidote in scorpion sting.[1,3]

Doses — Oil — 2 to 6 minims; fluid extract — 14 to 56 ml; infusion — 28 to 56 ml; decoction — 28 to 56 ml; powder — 1 to 1.5 gr.

NELUMBO NUCIFERA Gaertn. *Family:* NYMPHEAECEAE
(Syn. *NELUMBIUM NELUMBO* [Linn.] Druce;
NYMPHAEA NELUMBO Linn.;
NELUMBIUM SPECIOSUM Willd.)

Vernacular names — Sanskrit, Kamal; Hindi, Kanwal; English, Sacred lotus; Bengali, Padma; Nepalese, Pales waa; Sinhalese, Nelum; Japanese, Hasu; Chinese, Lienou; German, Indische Lotosblume; French, Lotus sacre, Nelumbo; Unani, Kanwal; Arabian and Persian, Nilufer.

Habitat — An aquatic herb, found throughout India and Pakistan.

Parts used — Flowers, filaments, anthers, stalks, seeds, leaves, and root.

Pharmacognostical characteristics — Leaves peltate, 60 to 90 cm or more in diameter, orbicular, glaucous, petioles very long; flowers solitary, large, white or rosy; fruit — torus large, top shaped, 5 to 10 cm diameter; spongy with 10 to 30 uni-ovulate carpels sunk separately into cavities on the upper side; carpels maturing into ovoid nut-like achenes.

Mitra et al.[635] reported that two types of leaves, aerial and floating, are borne by *N. nucifera*. Both are orbicular, large peltate, entire, glaucous, and leathery. On drying, they become membranous and brittle. Petioles are smooth, with small brown distinct prickles with fibrous facture. The petiole of the floating leaf is characterized by hydathodes traversed by the trabeculae, the presence of four canals in the center, extreme reduction of xylem, rosette crystals of calcium oxalate on the walls of air canals, and starch grains in the parenchyma. The petiole of the aerial leaf shows similar structure except for thick-walled and heavy cuticularized lower epidermis, subepidermal sclerenchyma, and vascular bundle capped by a group of sclerenchymatous fibers.

The lamina of the floating leaf can be distinguished by the dorsiventral structure, papillate epidermis, thick cuticle, ranunculaceous- or cruciferous-type stomata present on the upper surface. Air chambers are present in spongy parenchyma surrounded by epithelial cells, rosette calcium crystals on the walls of air cavities, and fibrous lignified bundle sheaths capping the phloem. The lamina of the aerial leaf is similar except for the presence of a thick-walled epidermis with a thick cuticle and vascular bundles surrounded by a sheath of parenchyma cells containing starch grains. The leaf powder fluoresced pale green under ultraviolet light, but the powder mounted in nitrocellulose showed dark-green fluorescence under ultraviolet light.

The rhizome, commonly known as *Kamal-Kakri*, is 2 to 5 ft long, yellowish white to yellowish brown in color, smooth with distinct nodes and internodes. Fracture is fibrous and tough. A transverse section of rhizome shows characteristic features, viz., presence of three cortical regions, closed vascular bundles, and two types of starch grains (simple and compound). The three cortical regions are (1) the exocortex, consisting of 5 to 6 layers of collenchyma and 12 to 14 layers of parenchymatous thin-walled cells, containing abundant starch grains; (2) the medicortex, composed of air canals traversed by anastomosing chains of thin-walled, small, single rows of cells; and (3) the endocortex, consisting of spherical cells enclosing large intercellular spaces.[636]

Ayurvedic description — *Rasa* — *madhur, kasaya, tikta; Guna* — *laghu, snigdha, pichil; Veerya* — *sheeta; Vipak* — *madhura*.

Action and uses — *Kapha pitta samak, daah samak, varanya, trisanigraha, hirdya, hirdya sanraksha, sonita sthapan, praja sthapan, jarwarghana, balya*.

Chemical constituents — Willaman and Li[10] have reported the presence of the following alkaloids from different parts of the plant: armepavine, nuciferine, and roemerine from leaves; isoliensinine, liensinine, lotusine, neferine, nuciferine, and pronuciferine from embryo; liensinine from seeds; and nornuciferine from whole plant.

The ether extract of the petals and stamen showed the presence of quercetin and luteolin and that of receptacle only, quercetin. The aqueous extract of petals and stamens yielded isoquercetrin and glucoluteolin; and that of receptacle of the flower yielded isoquercitrin. The leaves contained quercetin, isoquercitrin, leukocyanidin, and leukodelphinidin.[637] The seeds of *N. nucifera* contained 2.11% oil which is comprised of myristic, palmatic, oleic, linoleic, and linolenic acid.[638]

Pharmacological action — Seeds are demulcent and nutritive; flowers are cooling, sedative, astringent, cholagogue, diuretic, bitter, refrigerant, and expectorant; root is demulcent. Dhawan et al.[639] reported that 50% ethanolic extract of the rhizomes showed CNS-depressant effect in mice and diuretic activity in rats.

Medicinal properties and uses — The leaf juice is given in diarrhea; a paste of the leaves mixed with sandalwood is applied to the body in high fevers and in the irritated condition of the skin and mucous membranes; the young tender leaves are used in piles, strangury, and leprosy. The leaves are refrigerant, astringent, cardiac tonic, and diuretic. It is given in the form of a syrup in cough, menorrhagia, and bleeding piles. The seeds are given to check vomiting and in the irritated condition of the intestine. A paste of the seeds is locally applied in skin diseases. Root is mucilaginous and is given in piles. Petals are pounded and administered for syphilis. It is sedative in the uterus, good in thirst, piles, inflammation, and poisoning.[178]

Doses — Seed powder — 4 to 6 gr; pollen — 1 to 2 gr; root infusion — 3 to 5 drams.

NERIUM INDICUM Mill.   *Family:* APOCYNACEAE
(Syn. *N. ODORUM* Soland.)

Vernacular names — Sanskrit, Karavira; Hindi, Kaner; English, Roseberry spurge; Bengali, Karabi; Nepalese, Kaniha-swaa; Sinhalese, Kenera; Japanese, Kyochikuto; Chinese, Kia tchou tao; German, Wahlriechender Oleandes; Unani, Kaner; Arabian, Dafla; Persian, Khurjahrah; Tamil, Alari.

Habitat — Found in the Himalayas from Kashmir to Nepal up to an altitude of 6500 ft and also in Uttar Pradesh and Central India. Cultivated in gardens throughout India with single or double flowers; apparently wild in South India and in Maharashtra along the banks of streams.[28]

Pharmacognostical characteristics — A large, glabrous evergreen shrub. Leaves in threes, shortly stalked, 4 to 6 in. long, linear-lanceolate; acuminate; dark green and shining above. Flowers 1$^1/_2$ in. diameter, red, rose, or white, fragrant. Follicles 6 to 9 in. long. Closely allied to European oleander, *N. oleander* Linn.[28]

There are two varieties of *N. indicum,* viz., the white flowered and the red flowered. A transverse section of leaf, stem, and root shows the presence of lactifers and calcium oxalate crystals. The root has well developed vascular cylinder in the center occupying about two thirds of mature root. The vessels are arranged usually in rows and medullary rays are single-celled and thick in transection. Vessels and xylem fibers are quite frequent in the powder. The stem has a group of fibers present just outside the phloem. Intraxylary phloem is also present. The powder shows abundance of fibers, vessels, and xylem fibers. The leaves show pits having stomata and simple nonglandular trichomes on the lower surface. The pallisade tissue is on both surfaces and the venation consists of a midvein from which several secondary veins arise and run almost parallel to each other. The vein islets are usually quadrangular or pentangular and the vein endings are not frequent.[640]

Ayurvedic description — *Rasa — katu, tikta; Guna — lagu, rooksha, teekshna; Veerya — ushna; Vipak — katu.*

Action and uses — *Kaf-bat samak, kusta ghna, brana ropan, soth har, dipan, bidahi, bhadan.*

Chemical constituents — Root, bark, and seeds contain toxic principles: neriodorin, neriodorein, and karabin. Like neriodorin, karabin is a powerful cardiac poison and acts on the heart, and like ditigalin, acts on the spinal cord, in more or less the same way as strychnine. The toxic properties of neriodorein are of much milder character than those of either karabin or neriodorin.[1]

The petroleum extract of the roots yielded a lactonic constituent, viz., plumericin.[641] The alcoholic extract of root bark showed the presence of α-amyrin and β-sitosterol in the petroleum ether fraction, kaempferol in the ether fraction, and odoroside in the chloroform fraction.[642] Chopra et al.[178] reported the isolation of oleandrin, neriodin, ursolic acid, and a dynerin from the leaves. A new pentacyclic triterpene, oleanderol, and the known betulin, betulinic acid, ursolic acid, and oleanolic acid, have been isolated from the fresh leaves of *N. oleander* by Siddiqui et al.[643]

Pharmacological action — All parts of the plant are poisonous. Root and the root bark are powerful diuretic and cardiac tonics like strophanthus and digitaline. Abortifacient. The 50% ethanolic extract of the roots showed spasmogenic activity on the isolated guinea pig ileum and CNS-depressant effect on mice.[51] Plumieride showed antipyretic effect as well as anti-inflammatory and analgesic activity.[644] A glycoside obtained from the roots increased adaptability of rats and mice against stressful conditions. Dried leaf or its alcoholic extract taken orally or by injection is an effective cardiac stimulant with broader action and less toxicity.[178]

Medicinal properties and uses — Paste of the root is used as an external application in hemorrhoids, chancres, and ulcers on the penis. An oil prepared from the root bark is used in skin diseases of a scaly nature and in leprosy. Root is a powerful resolvent and attenuant. Fresh juice of leaves is dropped into the eyes for inducing lachrymation in ophthalmia. Powder of the root is rubbed to the head in headache. Decoction of leaves is applied externally to reduce swellings. Leaf juice is given in very small dose in snakebites and other powerful venomous bites. The antidote is "ghee".

Doses — Powder — 0.2 gr.

NIGELLA SATIVA Linn. *Family:* RANUNCULACEAE

Vernacular names—Sanskrit, Upakunchika; Hindi, Kalanji, Kalajira; English, Black cumin; Bengali, Mugrela; Nepalese, Mugrelo; Sinhalese, Kaluduroo; Japanese, Nigera; German, Schwarzkümmel; Unani, Gandana; Arabian, Kurras; Persian, Gandana; Burmese, Samonne; French, Cumin noir.

Habitat—This plant is cultivated in Punjab, Himachal Pradesh, Bihar, and Assam; also found as an occasional weed of cultivation.

Parts used—Dried fruit and seeds.

Morphological characteristics—An annual herb. Leaves 2 to 3, pinnatisect, 2.5 to 5 cm long, cut into linear lanceolate segments; flowers pale blue, 2 to 2.5 cm across; seeds trigonous, black.

Ayurvedic description—*Rasa—katu, tikta; Guna—laghu, rooksha, teekshna; Veerya—ushna; Vipak—katu.*

Action and uses—*Kapha vat samak, pitta-wardhak, lakhan, sotha har, badana sthapan, vtejek, rochan, dipan, pachan, anuloman, grahi, krimighan, kaphanisark, jwaraghana.*

Chemical constituents—Black cumin contains volatile oil (0.5 to 1.6%), ether extract (fatty oil; 35.6 to 41.6%), and oleic acid (3.4 to 6.3%). Oil contains carvone (45 to 60%), *d*-limonene, and cymene. Besides the volatile and fatty oil, seeds contain a bitter principle, tannins, resins, proteins, reducing sugar, glycosidal saponin, melanthin, and also 1.0% melanthigenin.[1]

Bose et al.[646] reported that *N. sativa* seeds yield esters of unsaturated fatty acids, dehydrostearic acid, and linoleic acid with C_{15} and higher terpenoids and aliphatic alcohols and α-β-unsaturated hydroxy ketone. Nigellone, the crystalline active principle, is the single constituent of the carbonyl fraction of the oil.[178]

Pharmacological action—Seeds are aromatic, diuretic, diaphoretic, antibilious, stomachic, stimulant, and carminative; also anthelmintic, emmenagogue, galactagogue.[3]

Medicinal properties and uses—The oil is applied externally for eruptions of skin. Seeds are useful in the form of tincture in indigestion, loss of appetite, fever, diarrhea, dropsy, and puerperal fevers. Decoction of the seeds is given to recently delivered females in combination with a few other medicines; it stimulates uterine contraction. In doses of 10 to 20 gr they are useful in ammenorrhea and dysmenorrhea and in large doses cause abortions. Roasted seeds also have antibilious property and are administered internally in intermittent fevers and to arrest vomiting. Alcoholic extracts of the seeds show antibacterial activity against *Micrococcus pyogenes* var. *aureus* and *Escherichia coli.*

Doses—1 to 3 gr.

NYCTANTHES ARBOR-TRISTIS Linn.

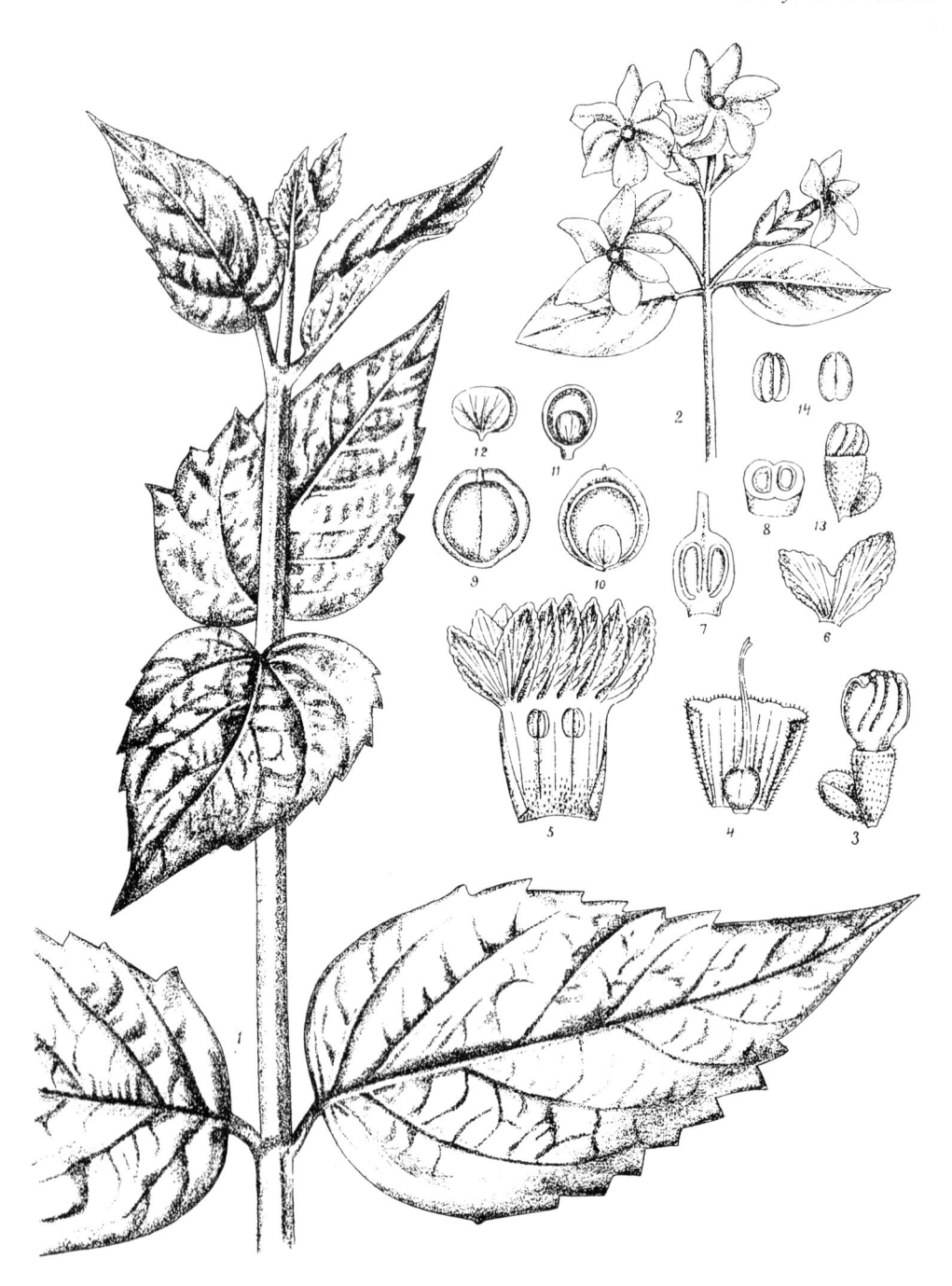

Family: OLEACEAE

Vernacular names — Sanskrit, Parijata; Hindi, Harsinghar; English, Night jasmine; Bengali, Sedi, Singhar; Nepalese, Palija swaa; Sinhalese, Sepalika; Unani, Harsinghar; Burmese, Seikpalu.

Habitat — Found growing wild in the forests of sub-Himalayan regions, from Chenab to Nepal, Assam, Bengal, Madhya Bharat, and southwards to the Godavari. Cultivated in many parts of India.

Parts used — Leaves, seeds, and bark.

Morphological characteristics — Bark greenish white and rough. Leaves ovate, acuminate, entire or with a few large distant teeth, rough and scabrous above, densely pubescent beneath; flowers small, 3 to 7 in head, arranged in trichotomous cymes; corolla fragrant, white, 4 to 8 lobed, with bright orange tubes; capsules suborbicular, compressed into two flat 1-seeded compartments.

Ayurvedic description — *Rasa* — *tikta; Guna* — *laghu, rooksha; Veerya* — *ushna; Vipak* — *katu.*

Action and uses — *Kapha vat har, pitta sansodhan, dipan, anuloman, pitta sark, krimighan, rakt sodhak, kaphaghan, mutral, jwaraghana.*

Chemical constituents — Flowers yield an essential oil and crystalline nyctanthin. Leaves contain an alkaloid, an astringent principle, a resinous substance, an amorphous glucoside, sugar, and a trace of peppermint-like oily substance. The seed kernels yield 12 to 16% of a pale yellow-brown fixed oil.[50]

Detailed studies of the oil obtained from flowers showed the presence of α-pinene, p-cymene, 1-hexanol, methyl heptanone, phenylacetaldehyde, 1-decanol, and anisaldehyde.[647] The acetone extract of the corolla tubes yielded the β-monogentiobioside ester of α-crocetin as major component and the β-digentiobioside ester of α-crocetin as minor components.[648]

The alcoholic extract of the stem yielded β-sitosterol and a new glycoside, naringenin-4-O-β-glucopyranosyl-α-xylopyranoside. The plant also contains oleanolic acid.[649]

The leaves of *N. arbor-tristis* showed the presence of β-amyrin, β-sitosterol, hentriacontane, benzoic acid with free glucose, and fructose.[650] The ethanolic extract of the leaves yielded two flavonal glycosides, astragelin and nicotiflorin,[651] and an unidentified alkaloid.[652] Kapoor et al.[653] reported the presence of traces of alkaloids and the absence of saponins and flavonoids.

Pharmacological action — Cholagogue, anthelmintic, and laxative. The ethanolic extract (50%) of the whole plant showed CNS-depressant effect and produced hypothermia in mice.[11] Alcoholic extract of the leaves showed hypotensive and respiratory stimulant action in dogs.[654] An alkaloid isolated from leaves showed a relaxant action and also a nonspecific antispasmodic action on isolated rabbit ileum. The alkaloid had marked hypotensive action in the anesthetized dog.[652] Lately, it was observed that alcoholic extract of leaves exerted a significant tranquilizing and antipyretic effect and histamine-antagonistic and purgative activities. Significant anti-inflammatory activity against acute, subacute, and chronic models of inflammation in rats was observed.[655,656]

Medicinal properties and uses — Fresh leaf juice is a mild cholagogue and a safe purgative for infants. It is given with honey in chronic and bilious fevers. Expressed juice of leaves is cholagogue, laxative, and is a mild bitter tonic given with a little sugar to children as a remedy for intestinal worms. In the form of infusion, it is useful in fever, rheumatism, and obstinate sciatica. Leaves are used as an antidote to reptile venom. Seeds in the form of powder are employed as a paste to cure scurvy and affections of the scalp.[3,50]

Doses — Infusion — 28 to 56 ml.

OCHROCARPUS LONGIFOLIUS Benth. & Hook. f.   Family: GUTTIFERAE

Vernacular names — Sanskrit, Punnaga; Hindi, Nag kesar; English, Alexandrian laurel; Bengali, Nagkeshar; Nepalese, Hyangu noga; Unani, Sarpan; Persian, Naramushka.

Habitat — Found in the forests of the west coast of India from Kanara to Konkan.

Parts used — Fruit and flowers, flower buds.

Morphological characteristics — A large tree with greenish, resinous juice. Bark reddish gray. Leaves opposite, evergreen, and coriaceous, 5 to 6 in. long and 2 to 2.5 in. broad, bluntly pointed, glabrous. Flowers showy, numerous, bisexual, scented, yellowish red in color. Sepals 2 to 6, imbricate; petals 2 to 6, contorted, white streaked with red. Stamens usually indefinite, sterile in female flower, ovary 1- to many-celled. Style usually short or absent; ovule 1, fruit about 1 in. long with usually one seed.[108]

Ayurvedic description — *Rasa* — kasaya, tikta, amla; *Guna* — laghu, rooksha; *Veerya* — ushna; *Vipak* — katu.

Action and uses — *Dipan, pachan, vatta pitta samak, arsoghan, grahi, krimighan, hiridya, sonitya sthapan, vajikaran, mutral, kustaghan, jwaraghana.*

Chemical constituents — It contains an essential oil.

Pharmacological action — Astringent, aromatic, stimulant, and carminative.

Medicinal properties and uses — Fruit is edible. Dried flower buds are stimulant, aromatic, stomachic, and astringent; used as a fragrant adjunct to decoction and medicated oil. They are used to quench thirst, and cure irritability of the stomach; also given in dysentery. A paste made of them is used to fill cavities of carious teeth to relieve toothache. Flowers are useful in some forms of dyspepsia and hemorrhoids. The drug is also used in scorpion sting.[50]

Flower buds are used for dyeing silk.

Doses — 1 to 3 gr.

OCIMUM SANCTUM Linn. *Family:* LABIATAE

Vernacular names — Sanskrit, Tulssi; Hindi, Kala tulasi; English, Holy basil; Bengali, Krishna tulsai; Nepalese, Newari-Tulsi; German, Basttikum; French, Basilic odorant; Unani, Tulsi; Arabian, Rayhan; Persian, Rayhan; Burmese, Lum; Malaysian, Krishna tulsi; Tamil, Thulasi.

Habitat — This is a small herb found throughout India and cultivated near temples and gardens.

Parts used — Leaves, seeds, root, and flower.

Pharmacognostical characteristics — *O. sanctum* is an erect, hairy, annual herb. Leaves ovate or elliptic-oblong, entire, serrate or dentate, softly hairy, stalked, about 2 in. long and 1 in. broad; minute gland dots present. Inflorescence of elongate racemes; flowers closely whorled; calyx 2-lipped; corolla hardly longer than the calyx, pale purple. Fruit forming 4 dry, small, reddish-brown nutlets.

Gupta[657] reported the salient anatomical features which distinguish *O. sanctum* from other species. The stomata are present on both surfaces of the epidermal cells of the leaf. Fibers are nonlignified, acutely pointed. The xylem vessels have pits of annular and spiral type. The sessile glands are abundant. Two types of trichomes are reported: (1) simple, uniseriate, unbranched, and warty; (2) glandular trichomes are sessile or unistalked, small, head being pear or globular shaped, and one to eight celled.

Ayurvedic description — *Rasa* — *katu, tikta; Guna* — *laghu, rooksha; Veerya* — *ushna; Vipak* — *katu.*

Action and uses — *Kaf bat samak, pitwardhak, dipan, pachan, anuloman, rakt sodak, jawarghana.*

Chemical constituents — The leaf contains the highest percentage of essential oil, followed by inflorescence and stem, but roots are devoid of any oil.[658] Seeds contain a large amount of mucilage. Lal et al.[659] reported the presence of 70% eugenol as a major constituent in the essential oil. The other components identified were nerol, eugenol methyl ether, caryophyllene, terpinene-4-ol, decylaldehyde, γ-selinene, α-pinene, β-pinene, camphor, and carvacrol. The leaves also yield ursolic acid, apigenin, luteolin, apigenin 7-*o*-glucuronide, luteolin, orientin, and molludistin.[660] Old leaves of *O. sanctum* contained 3.15% calcium, 0.34% phosphorus, and 4.97% insoluble oxalate.[661]

Pharmacological action — Demulcent, expectorant, and antiperiodic. Root is febrifuge; seeds are mucilaginous and demulcent. Dried plant is stomachic and expectorant. Leaves are anticatarrhal, expectorant, fragrant, and aromatic. *O. sanctum* leaves are reported to show abortifacient and antifertility activity.[662] Ethanolic extract (50%) of leaves showed a hypoglycemic effect in rats and antispasmodic activity against spasmogen-induced spasms in isolated guinea pig ileum.[51] Singh et al.[663] reported that crude watery extract of leaves showed a transient hypotensive effect in anesthetized dogs and cats and a negative inotropic and chronotropic effect on rabbit's heart. The extract inhibited the spasm of smooth muscles induced by acetylcholine, carbachol, and histamine and potentiated hexobarbitone-induced hyposis in mice.

The essential oil showed larvicidal activity against *Culex pipiens fatigans*.[664] The seed mucilage of *O. sanctum* also showed marked larvicidal activity against *C. pipiens fatigans* and *Aedes aegypti*.[665] Grover and Rao[666] reported the antibacterial activity of the essential oil and its major constituent, eugenol. They also showed antifungal activity. The seeds exhibited anticoagulose activity as evidenced by the suppression of coagulose activity and mannitol fermentability of pathogenic staphylococci.[667] The essential oil also showed antimicrobial and antimycotic effects *in vivo*.[666] Ethanolic extract of leaves showed fungitoxic activity against *Rhizoctonia solani*.[668]

Medicinal properties and uses — Infusion of the leaves is given in malaria and as a stomachic in gastric diseases of children and hepatic affections. Juice of the leaves should be taken internally and is very effective in skin diseases such as itches, ringworm, and leprosy, and in impurities of the blood. Dried plant in decoction is a domestic remedy for catarrh, bronchitis, and diarrhea. Fresh leaves also cure chronic fever, hemorrhage, dysentery, and dyspepsia. With honey and ginger juice, it is a good expectorant, useful in cough, bronchitis, and children's fever. Popular as herbal tea.[3,50]

Doses — Seed powder — 1.5 to 2 gr; leaf infusion — 4 to 12 ml; decoction — 28 to 56 ml.

OPERCULINA TURPETHUM (Linn.) Silva, Manso   Family: CONVOLVULACEAE
(Syn. *IPOMOEA TURPETHUM* Linn.;
   CONVOLVULUS TURPETHUM Linn.)

Vernacular names — Sanskrit, Trivrit; Hindi, Nisotha; English, Indian jalap; Bengali, Tauri; Nepalese, Newari-Temal; German, Brast Liauische Jalapa; Unani, Turbud; Arabian, Turbud; Persian, Turbud; Tamil, Sivathai.

Habitat — The plant is a perennial found throughout India, and distributed to Sri Lanka, Southeast Asia, Malaya, and Australia; it is cultivated occasionally in gardens as an ornamental plant.

Parts used — Root bark, root, stem.

Pharmacognostical characteristics — The external surface of the root (which may be black or brown) is deeply furrowed longitudinally, giving a ropelike appearance. The drug root bark occurs in cylindrical pieces, varying in length from 2 to 10 cm and in thickness from 6 to 25 mm. The color of the drug of the white variety is gray or reddish gray and of the black, brown.

In transverse section, root shows a prominent cortex including secondary phloem and a central woody portion. The phellogen cells, as seen in transverse section, appear rectangular and radially flattened, a bit curved toward the periphery at intervals. The phellem or cork cells outside the phellogen layer are for the most part filled with brown granular contents. Below the periderm, the secondary cortex is found to be composed of thin-walled parenchyma, mostly filled with starch grains and crystals of calcium oxalate. The cells of the cortex are of different sizes, tapering below toward the phloem. The cortex includes a number of laticiferous ducts of different dimensions containing resin. The secondary xylem consists of big vessels with simple, pitted tracheids of different dimensions. Wood fibers are simple with pointed end. Ray cells are thin walled; xylem parenchyma has pitted walls. The stem of *O. turpethum* shows patches of collenchyma in the cortex, lignified pericycle, intraxylary phloem, and an amphiphloic siphonostele with several radial arms of delignified parenchymatous tissue in xylem. Resin ducts are abundant in cortex.[669]

Ayurvedic description — *Rasa* — *katu, tikta, madhur, kasaya; Guna* — *iaghu, rooksha, teekshna; Veerya* — *ushna; Vipak* — *katu.*

Action and uses — *Pit kaf samak, bat bardhak, bhadan, lakhan.*

Chemical constituents — Turpeth contains 9 to 13% of resin which is supposed to be a mixture of α- and β-turpethein, glycosides, and turpethin. It is a white amorphous compound of colloid nature, insoluble in water, mixes with traces of alkalis to form collodial solutions. Besides the resin, it also contains coumarin, scopoletin, glucose, rhamnose, and fucose.[670]

Pharmacological action — Root and root bark of white "turpeth" is cathartic and laxative. The extracts of roots showed anti-inflammatory activity against carrageenin-induced rat paw edema, cotton pellet-induced granuloma, and formalin-induced arthritis in rats. Aqueous extract was most potent.[671]

Medicinal properties and uses — The roots are cathartic; they are as good as imported *Ipomoea purga* (jalap) and superior to rhubarb; they are given as a mild hydragogue in chronic constipation, ascites, enlargement of the spleen, and other disorders where a purgative is required; they are also prescribed in rheumatic and paralytic diseases in doses of 3 to 4 gr. The tuberous roots, in doses of 2 to 4 ml mixed with chebulic myrobalan, are particularly efficacious for dropsical affections, melancholia, gout, leprosy, rheumatism, and paralysis.[3,50]

Doses — Powder — 1 to 5 gr.

OROXYLUM INDICUM (Linn.) Vent.　　　　　　　　　　　　　*Family:* BIGNONIACEAE

Vernacular names — Sanskrit, Shyonaka; Hindi, Snapatha; Bengali, Sondala; Nepalese, Balchatasi; Sinhalese, Totilla; Chinese, Ch'len Tseng; Unani, Tenta; Arabian, Abnos; Persian, Abnos; Burmese, Kyoungsha; Malaysian, Bulai.

Habitat — A small, medium-sized deciduous tree growing throughout the greater part of India, up to 3000 ft except in western, drier areas.

Parts used — Root bark, stem bark, fruit, and seeds.

Pharmacognostical characteristics — Leaves opposite, 1 to 3 ft long, pinnate at the apex, bipinnate or tripinnate at the base; leaflets numerous, 3 to 8 in. long, ovate, short stalk. Flowers in stiff, erect, terminal racemes, large, malodorous; flower stalk long, stout; calyx leathery, thick, 1 in. long; corolla bell-shaped, fleshy, 2 to 3 in. long. Pods large, flat, woody, black. Seeds winged, flat, thin; wings broad, silvery. Root bark is 3 to 8 mm thick, rough, grayish to buff color, with cracks and fissures and circular or vertically elongated lenticles on the external surface. Taste initially sweet without any offensive smell. The stem bark is grayish yellow, with slightly mucilaginous brown-colored tissue and transversely elongated or circular lenticles.[672]

In transverse section, the mature root bark shows a zone of cork cells, a group of stone cells in the phelloderm, secondary phloem, and phloem fibers. The stem bark consists of a narrower zone of cork cells. A larger group of stone cells, and narrower tri- to tetraseriate medullary rays in the inner region and multiseriate in the outer region are seen. The phloem fibers are longer, more or less straight, and tapering, with occasional bulging and peg-like outgrowths. The powder of root bark mounted in nitrocellulose fluoresced brown and that of stem bark gave a bright orange color.

Ayurvedic description — *Rasa — tikta, kasaya; Guna — laghu, rooksha; Veerya — sheeta; Vipak — katu.*

Action and uses — *Tridosh samak, sothahar, bran ropan, badanasthapan, dipan, pachan, krimighan, sthamvan, jawraghana, mutral.*

Chemical constituents — The ethanolic extract of *O. indicum* leaves gave baicalein and scutellarein from the ether fraction; a flavone glycoside identified as baicalein-6-glucuronide from the ethyl acetate fraction; and two glucuronides (scutellarein) and (baicalein) from the aqueous mother liquor.[673] The defatted leaves on extraction with chloroform gave a gummy solid-yielding anthraquinone, aloe emodin.[674]

Ethanolic extract of stem bark on fractionation yielded oroxylin A, baicalein, scutellarein, and scutellarein-7-rutinoside, in different solvent extractives. Chrysin and baicalein-7-glucuronide and *p*-coumaric acid were also isolated from the mother liquor.[675] Benzene extract of heartwood of *O. indica* yielded β-sitosterol and prunetin-4′,5-dihydroxy-7-methoxyisoflavone.[676]

The seeds showed the presence of terpenes, alkaloids, and saponins.[677] A new flavone oroxinden, characterized as 5-hydroxy-8-methoxy-7-*O*-β-D-glucopyranuronosyl flavone, was isolated from the enthanolic extract of seeds by Nair and Joshi.[678]

Pharmacological action — Root bark is astringent, bitter tonic, stomachic, anodyne, and sudorific.

Medicinal properties and uses — Root bark is one of the ten ingredients of the Ayurvedic preparation, *dasamula*. Infusion or decoction in doses of 14 to 28 ml is useful in diarrhea and dysentery. Stem bark is used in acute rheumatism and used as diaphoretic. Roots used in dropsy and leaves as emollient.[1,3] The medicated oil made from the root bark is used in otorrhea. The tender fruit is carminative and stomachic. The seeds are purgative.

Doses — Bark powder — 5 to 15 gr; decoction — 28 to 56 ml.

PAEDERIA FOETIDA Linn. *Family:* RUBIACEAE

Vernacular names — Sanskrit, Prasarini; Hindi, Gandhaprasarini; English, Chinese flower plant; Bengali, Gandha-bhadulia; Nepalese, Biree; Sinhalese, Prasarini; Unani, Gandha prasarini; Malaysian, Talanili.

Habitat — Found in the Himalayas from Dehra-Dun eastward up to an altitude of 5500 ft; also in Bihar, Orissa, Bengal, and Assam.

Parts used — Whole plant, root, and leaves.

Pharmacognostical characteristics — It is an extensive climber. Leaves ovate to lanceolate, 5.15 cm long and 2.7 cm broad, entire, membranous with long petioles; flower in scorpioid cymes, purple or violet; fruit ellipsoid, compressed, red or black. The shape of the root is cylindrical or subcylindrical with both sides somewhat compressed. Outer surface is full of root scars. Fracture fibrous. Outer color brownish and internal color light brown; taste bitter.

In a transection, the roots are characterized by the presence of six to eight layers of cork with deep brown contents. Prominent phellogen and phelloderm cells which are arranged in radial order are distinct from outer layer of the cortex. Two types of medullary rays, viz., broad multiseriate rays ending in funnel-shaped dilations and narrow uni- and biseriate rays with intervening xylem and phloem wedges. Needle-shaped raphides and starch grains along with oil globules are found in parenchyma. The secondary phloem tissue is prominent and forms faint phloem wedges. The wood is characterized by the presence of broad ray cells. The secondary xylem is characterized by the presence of thick-walled cells consisting of a longitudinal and a transverse radiating system. The longitudinal system consists of elongated and overlocked overlapping cells, such as tracheids, fibers, and vessels, and longitudinal row of parenchyma. The radial system chiefly consists of parenchyma cells with their long axes at right angles to the longitudinal axis of the central cylinder, and they consist of xylem rays. Starch grains are present in the wood parenchyma and wood fibers. The vessels have a very large lumen with pitted thickenings on the walls. Few spiral elements are present.[680,681]

Ayurvedic description — *Rasa — tikta; Guna — guru, sar; Veerya — ushna; Vipaka — katu.*

Action and uses — *Kaf bat samak, pit sansodak, badana nask, soth har, batanuloman, mridubirachan, virisya.*

Chemical constituents — Basu et al.[682] had earlier reported the presence of a foul-smelling essential oil and two alkaloids, α-paederine and β-paederine. Leaf protein exhibited the presence of amino acids like arginine, histidine, lysine, tyrosine, tryptophan, phenylalanine, cystine, methionine, thereonine, and valine. Leaves are reportedly rich in carotene and vitamin C.

Hexane extract of the whole plant yielded epifriedelinol acetate,[683] whereas the petroleum extract gave sitosterol and friedelin. Shukla et al.[685] reported the presence of iridoid glycosides, viz., asperuloside, paederoside, scandoside; triterpenoids, viz., hentriacontane, cerylalcohol, hentriacontanol, palmitic acid; and the steroids, viz., sitosterol, stigmasterol, campesterol, and ursolic acid. Bose et al.[686] reported the presence of methyl mercaptan responsible for the fetid odor of the plant.

Pharmacological action — It is diuretic and externally effective in rheumatism, contraction, and stiffness of the joints.

Decoction of the whole plant showed significant anti-inflammatory activity.[687] Ethanolic extract of leaves showed antispasmodic activity on the isolated guinea pig ileum. It also showed gross depressant effects and hypothermia in mice and anticancer activity against human epidermoid carcinoma of the nasopharynx in tissue culture.[51] The leaf juice exhibited a potent anthelmintic effect against bovine helminths.[688]

Medicinal properties and uses — Boiled and mashed leaves are applied to the abdomen in cases of retention of urine; a decoction of the leaves is reported to possess diuretic properties and also to dissolve vesical calculi. Roots and bark are employed as emetics and the fruits are used by the hill tribes to blacken the teeth with a view to prevent toothache. The juice of the root is prescribed in piles and inflammation of spleen. Leaf juice is astringent and given to children in diarrhea. All parts of the plant are considered specific for rheumatic affection. A poultice of the leaves is applied to the abdomen to relieve distension due to flatulence. Several oils or liniments for external application are prepared with this plant.[1,3]

Doses — Infusion — 12 to 24 ml; decoction — 56 to 112 ml.

PANDANUS TECTORIUS Soland. ex Parkinson *Family:* PANDANACEAE
(Syn. *P. ODORATISSIMUS* Roxb.)

Vernacular names — Sanskrit, Kataki; Hindi, Keora; English, Fragrant screwpine; Bengali, Keora; Nepalese, Nalu; Sinhalese, Wetakeya; Japanese, Togenashiadan; German, Schrauben Palme; Unani, Kewrah; Arabian, Kazi; Persian, Kader; Tamil, Thazh.

Habitat — Southern India, sea coast of Indian peninsula on both sides, Andamans.

Parts used — Anthers, root, seeds, and oil.

Morphological characteristics — A large shrub with large leaves spirally arranged. Leaves glaucous, green, 0.9 to 1.5 m long, ensiform, caudate, acuminate, coriaceous with spines on the margins and on the midrib; spadix of male flowers 25 to 50 cm long, with numerous subsessile cylindric spikes, 5 to 10 cm long, enclosed in long white, fragrant, caudate acuminate spathes; spadix of female flowers solitary, 5 cm in diameter; fruit an oblong or globose syncarpium, 15 to 25 cm in diameter. Yellow or red.

Ayurvedic description — *Rasa — tikta, madhura, katu; Guna — laghu, snigdha; Veerya — ushna; Vipak — katu.*

Action and uses — *Tridosh samak, varnya badana nask, dipan, pachan, anuloman, virsya, vajikarn, kustagan, jwaraghana.*

Chemical constituents — Floral brackets contain an essential oil in a yield of 0.28 to 0.31%, light yellow, with strong characteristic odor. "Keora oil" contains methyl ether of β-phenylethyl alcohol (65 to 80%).[689] The blossoms yield an essential oil containing benzyl benzoate, benzyl salicylate, benzyl acetate, benzyl alcohol, geraniol, linalool, linalyl acetate, bromostyrene, guaiacol, phenylethyl alcohol, and aldehydes.[50,690]

Dhingra et al.[691] reported that keora oil contains dipentene, *d*-linalool, phenylethyl acetate, citral, phenylethyl alcohol, ester of phthalic acid, fatty acids, and steroptene.

Pharmacological action — Bitter, purgative and aromatic.

Medicinal properties and uses — The leaves are said to be valuable in leprosy, smallpox, scabies, syphilis, and leukoderma. The anthers of the male flowers are given in earache, headache, and diseases of the blood. Oil and the attar are stimulant and antispasmodic and are used in headache and rheumatism.

Doses — Arka (distillate) — 56 to 70 ml; sharbet (syrup) — 28 to 56 ml.

PAPAVER SOMNIFERUM Linn.

Family: PAPAVERACEAE

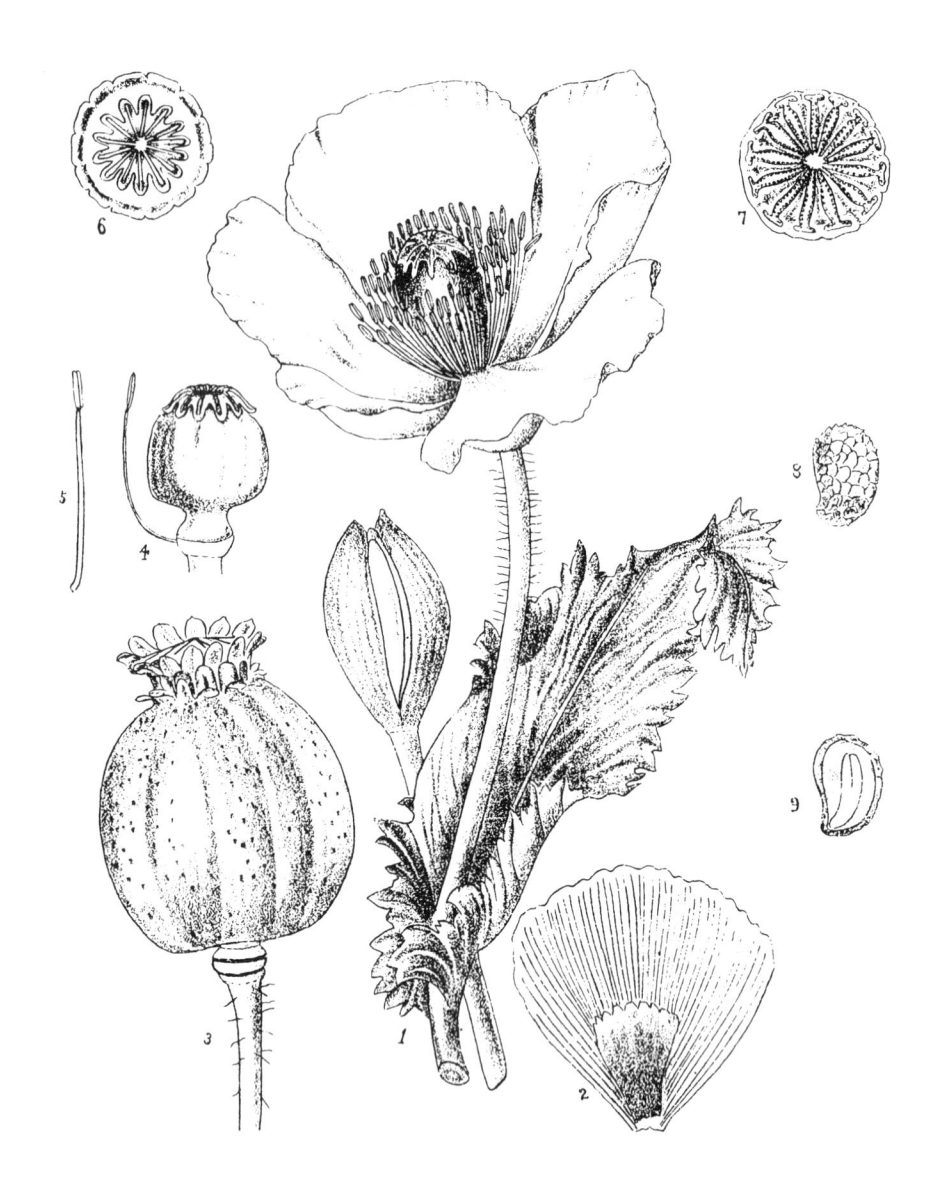

Vernacular names — Sanskrit, Ahiphenam; Hindi, Afim; English, Opium poppy; Bengali, Postodheri; Nepalese, Afim; Sinhalese, Abin; Japanese, Keshi; Chinese, Ya-Pin; German, Mohn; Unani, Afiun; Arabian, Labanul Khashkhash; Persian, Afiun; Burmese, Bhain; Malaysian, Affium; Tamil, Abini.

Habitat — It is cultivated in some districts of Uttar Pradesh, Madhya, Pradesh, and Rajasthan. A variety with dark seeds is cultivated in some parts of Jammu (Kashmir). It is also grown in Nepal, Burma, China, Egypt, and southeastern Europe. White-flowered variety of poppy is largely grown in India.

Parts used — Ripe and dried capsules, seeds, and the latex (opium).

Morphological characteristics — Poppy plant is annual herb about 3 to 4 ft high and bears about 5 to 8 capsules. Leaves ovate-oblong or linear-oblong; flowers large, usually bluish white with a purple base or white, purple; capsule large, 2.5 cm diameter; seeds many, white or black, reniform.

Ayurvedic description — *Rasa — sukshama, rooksha; Guna — jikta, kasaaya; Veerya — ushna; Vipaka — katu.*

Action and uses — *Kapha-bat samaka, pitta prakopak, badanasama ka* (analgesic).

Chemical constituents — Opium is the air-dried concrete milky latex or exudation obtained by incising unripe capsules. Fresh opium is brownish and has a characteristic fruity odor. Opium contains nearly 25 alkaloids; morphine, codeine, thebaine, narcotine, narceine, and papaverine are the chief alkaloids and of these, morphine is the most abundant and most important. Morphine exists in combination with meconic and sulfuric acid in the form of salts readily soluble in water. Other alkaloids occur in opium partly in the free state and partly as salts. The usual range of alkaloids in Indian opium are morphine, 9 to 14%; narcotin, 3 to 10%; codeine, 1.25 to 3.75%; papaverine, 0.5 to 2.75% and thebaine, 1.5 to 3.0%.[1]

Pharmacological action — Hypnotic, narcotic. Tonic to brain and useful in cough, phthisis, weak liver, kidneys, and urinary diseases.

Medicinal properties and uses — Morphine has both analgesic and sedative action, and where both of these actions are required as in severe injury, burns, and pain, opium is recommended. It is used in diarrhea, dysentery, and cough. Syrup of codeine phosphate is popular in cough and in bronchitis and useful in respiratory diseases.

Dwarkanath[692] described the properties of opium from Ayurvedic literature as tonic, antiphlegmatic, aphrodisiac, cleanses bodily impurities, binds the bowels, causes biliousness, nervous excitement, mental confusion, and promotes dryness. Poppy seeds were described as heavy, tonic, aphrodisiac, promotes luster of the body, enhances capacity to perform muscular work, allays nervous excitement, and causes the production of phlegm. Poppy capsules are reported dry, cool in potency, light, bitter and astringent in taste; promotes taste. They cause nervous excitement, garrulousness, intoxication, bind the bowels, cause dryness of the body, mental confusion, impotency, and promote the utilization of nutrition by the tissue.

Opium is used in convulsions and rheumatism.

Doses — 30 to 125 mg.

PEGANUM HARMALA Linn. *Family:* RUTACEAE

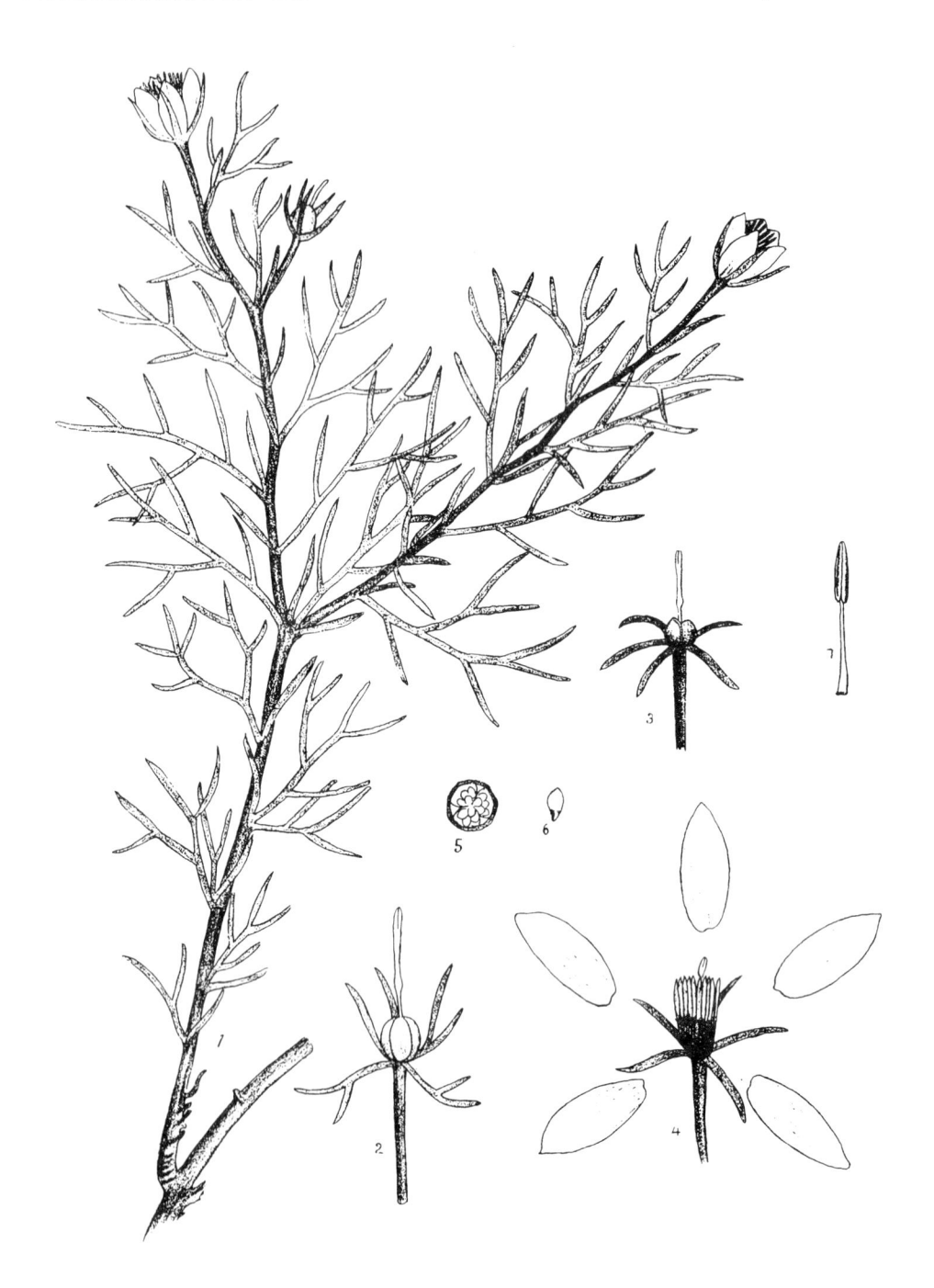

Vernacular names — Sanskrit, Harmal; Hindi, Harmal; English, Syrian rue; Bengali, Isband; Sinhalese, Rata arooda; German, Harmelraute; Unani, Hurmal; Arabian, Isband; Persian, Ispand; Pustu, Spail-anai.

Habitat — Very common in drier waste places and fields of Cutch, the Punjab, Kashmir, Delhi, Uttar Pradesh, Bihar, Konkan, and western Deccan.

Parts used — Seeds.

Morphological characteristics — A glabrous bush, 1 to 3 ft high, dichotomously and corymbosely branched, densely foliaged. Leaves 2 to 3 in., multifid; segments linear, acute. Flowers $1/2$ to $3/4$ in. diameter, white. Calyx lobes linear, exceeding the petals; petals elliptic-oblong. Capsule globose, about $1/3$ in. diameter, deeply lobed. It has three chambers, each containing one angular seed.

Ayurvedic description — *Rasa — tikta; Guna — laghu, rooksha; Veerya — ushna; Vipaka — katu.*

Action and uses — *Kapha vata samak, pitta wardhak, asepa har, badana sthapan, vata anuloman, krimighan, rakta sodhak, kafonisarak, artaw janan, vaji karan, mutral, kustaghan, jwaraghana.*

Chemical constituents — Seeds yield a red dye; roots and seeds contain four alkaloids, harmolol, harmaline, harmine, and peganine, to the extent of 4%, and a soft red-colored resin with aromatic odor resembling that of *Cannabis indica*. Harmaline, when treated with hydrochloric acid, yields harmatol in orange-red crystals sparingly soluble in water. Harmine occurs as colorless crystals. Blossoms and stems yield alkaloid peganine identified with l-peganine (vasicine).[50] Hermine is the most important, being present in all parts of plant. Besides these, the plant contains quinazoline.[178]

Pharmacological action — Alterative, antiperiodic, stimulant, emmenagogue, and abortifacient.

Medicinal properties and uses — Seeds in the form of powder are given as anthelmintic against tapeworms and in the treatment of intermittent and remittent fevers. The drug is useful in chronic malaria. Its decoction is a good anodyne in asthma, colic, and jaundice; it is given in ammenorrhea; it increases the flow of milk and menses. It is used for a gargle in laryngitis and is used for producing abortions. The smoke is believed to have antiseptic properties; the fumigation is applied in palsy and lumbago. A paste of the seeds made with mustard oil is used to kill head lice. The alkaloids are ineffective as contact poison but active in vapor form. It is effective against algae; in higher concentrations it is effective against water animals and lethal to molds, bacteria, and intestinal parasites.[178]

Doses — 1 to 3 gr.

PEUCEDANUM GRAVEOLENS Linn.   *Family:* UMBELLIFERAE
(Syn. *ANETHEUM SOWA* Kurz.)

Vernacular names — Sanskrit, Misroya Satapushpi; Hindi, Sowa; English, Indian dill; Bengali, Soolpha; Sinhalese, Mahaduru; Japanese, Indndo; German, Garten Dill; Unani, Sowa; Arabian, Shibbat; Persian, Shoot; Tibetan, Sa-ta-du-spa; Tamil, Chaiakuppai.

Habitat — Cultivated in Indian gardens.

Parts used — Oil, seeds, leaves.

Pharmacognostical characteristics — Annual herb, leaves 2 to 3 in. long. Flowers in umbels, yellowish in color; fruit small, light blackish, longer (twice as long as broad) and more strongly convex, paler color of dorsal ridges, which render them more conspicuous than those of the European true dill.

Ayurvedic description — *Rasa — katu, tikta; Guna — laghu, rooksha, teekshna; Veerya — ushna; Vipaka — katu.*

Action and uses — *Kapha vat samak, badana nask, soth har, bran pachan, dipan, pachan, anuloman* and *krimighan, hirdya utajak, kaf nasak, artawjanan, astanya, jwarghana, sukra nask.*

Chemical constituents — Dill fruit contains 3 to 4% volatile oil and fixed oil. The volatile oil is composed of anethine, phellandrene and *d*-limonene, and apiol; also carvol (*d*-carvone) and traces of antheole, anisaldehyde, eugenol, and thymol.[96]

The Indian oil shows a higher specific gravity due to the presence of a large amount of dill apiol.[64] The English fruit yields about 4.0%, German 3.8%, and East Indian about 3.19%, of volatile oil. The oil is obtained in two different fractions: a fraction with low specific gravity known as the "light oil" and another with a high specific gravity known as the "heavy oil".[97]

The seeds contain petraselinic acid, triglyceride, and β-sitosterol glucoside, mp 230 to 31°C.[95]

Pharmacological actions — Carminative, stomachic, aromatic, stimulant, diuretic, resolvent, emmenagogue, and galactagogue.[1]

Medicinal properties and uses — Distilled water of the fruit is much used in flatulence, hiccup, colic, and abdominal pain. It is used to diminish the griping of purgatives and the tormina of dysentery. An infusion of bruised fruits or seeds is also a very useful drink to women after confinement. Leaves are moistened with a little oil and warmed and applied to boils and abscesses to hasten suppurations.[1]

Doses — Powder — 1.5 to 3 gr; oil — 5 to 15 minims.

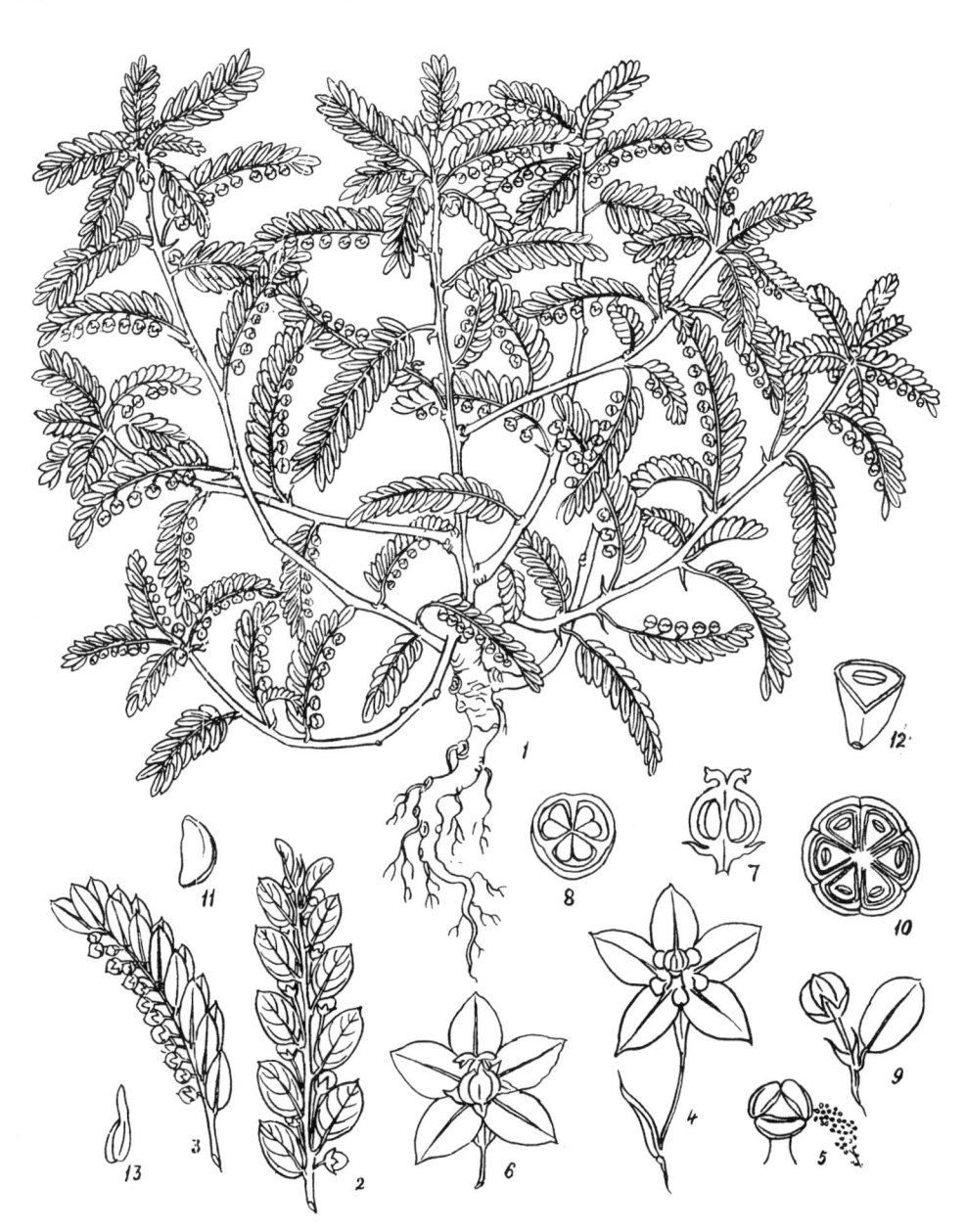

PHYLLANTHUS FRATERNUS Webster *Family:* EUPHORBIACEAE
(Syn. *P. NIRURI* auct. non Linn.)

Vernacular names — Sanskrit, Bhumyaamlaki; Hindi, Jangli amla, Bhuyi-avla; Bengali, Bhuiamla; Nepalese, Bhumyaamlaki; Sinhalese, Pittawakku; Japanese, Kidachimikanso; French, Niruri; Unani, Bhumiamla; Tibetan, Ta-ma-la; Burmese, Miziphiyu; Tamil, Kizharelli.

Habitat — Throughout the hotter parts of India.

Parts used — Whole plant, leaves, and root.

Morphological characteristics — An annual weed about 2 ft high. Leaves small, alternate, arranged in two rows, membranous, usually thin and glaucous under surface, elliptic, narrow at the base; stipules two. Flowers very small, monoecious, in pairs in the axils of leaves. Capsule globose, slightly depressed at the top with six ridges.

In a transection, the stem shows a broad cortex, a circular stellar region, and large prominent pith. The epidermis of a single layer of broad barrel-shaped cells with a thick cutinized wall is followed by continuous ring of collenchyma, chlorenchyma, and parenchyma. Vascular bundles arranged in a ring are many, open collateral, and endarch. Medullary rays are unicellular. Central pith persists and the stem does not form cork or bark tissue. Two rings of xylem are occasionally present.[693] Starch grains, mineral crystals, or latex vessels are not seen either in stem or root.

The leaves consist of ranunculaceous stomata mainly on the lower epidermis; the upper epidermis has a thin cuticle. Reduced vascular elements are seen running beneath the collecting cells which form a row beneath the palisade layer.

Ayurvedic description — *Rasa* — *tikta, kasaya, madhur; Guna* — *laghu, rooksha; Veerya* — *sheeta; Vipak* — *madhur.*

Action and uses — *Kapha pitta samak, bran ropan, sothahar, kustaghan, dipan, pachan, anuloman, sthamvan, rakt sodhak, rakt pitta har, kasha har, swas har, grvasysa sothahar, mutral kustaghan, jwaraghana.*

Chemical constituents — Krishnamurthy and Seshadri[694] isolated a bitter and a nonbitter constituent from leaves, called phyllanthum and hypophyllanthum, which were later identified as lignans by Row et al.[695] The hexane extract of leaves yielded three additional lignans, viz., niranthin, nirtetralin, and phyltetralin.[696] The aerial parts of the plant yielded two alkaloids: 4-methoxy-securinine (phylanthine) and 4-methoxy-norsecurinine.[697] The roots yielded two new glycoflavones and lupa-20(29)-ene-3-β-ol and its acetate.[698] Another new compound, viz., lintetraline, was also isolated by Ward et al.[699]

Pharmacological action — Diuretic, astringent, cooling, laxative, and bitter tonic. Petrol extract of the whole plant of *P. fraternus* showed antifungal activity.[700,701] Ethanolic extract (50%) of the whole plant showed anticancer activity against Freund virus leukemia (solid) in the mouse and antispasmodic activity on isolated guinea pig ileum.[41] The aqueous extract of the leaves was reported to produce hypoglycemic action in normal as well as alloxan-diabetic rabbits. This activity appeared to be higher than that of tolbutamide.[702] Dixit and Achar[703] in a clinical study observed that *P. fraternus* was very effective in the treatment of infective hepatitis without any side effects.

Medicinal properties and uses — Decoction of the herb is given for dropsical disorders, gonorrhea and other genitourinary diseases, jaundice, constipation, stomachache, dyspepsia, and dysentery. The juice of the plant is an efficacious dressing for offensive sores; mixed with some bland oil, the juice is used in ophthalmia. The young leaves are useful in the milder forms of intermittent fevers; boiled in milk, they are given in dropsical disorders and urinary complaints. A poultice of the leaves made with rice water is applied to ulcers or edematous swelling and is used for the treatment of itch and scabies and other skin diseases. The fresh root is given in jaundice; with rice water they are given for menorrhagia; with milk they are given as a galactagogue.[1,3]

Doses — Infusion — 14 to 28 ml; powder — 3 to 6 gr.

PICRORHIZA KURROA Royle ex Benth.   *Family:* SCROPHULARIACEAE

Vernacular names — Sanskrit, Katula; Hindi, Katki, Kuru; Bengali, Kuru; Nepalese, Kutakee; Sinhalese, Katukarosana; Japanese, Kooren; Chinese, Hu Huang Line; Unani, Kutki; Arabian, Khairbaque; Persian, Kharbaq Siyah; Tibetan, Hon-len; Tamil, Kadukurokani.

Habitat — A perennial herb is only found in the higher mountains at 9,000 to 12,000 ft in the northwestern Himalayas from Kashmir to Sikkim.

Parts used — Dried rhizome.

Pharmacognostical characteristics — The herb is more or less hairy, with perennial woody bitter rhizome 15 to 25 cm long clothed with dry leaf bases. Leaves spathulate, serrate, 5 to 10 cm long, rather coriaceous with rounded tips, and the base is narrowed into winged sheathing petioles. Flowers small, in spikes, bracteate, bracts oblong or lanceolate, as long as the calyx. Sepals 5, lanceolate 6 mm long, ciliate; flowers dimorphic. They are of two kinds: some have a 8-mm-long filament, others have 2-cm-long filaments. Fruit 1 to 3 cm long.[510] The rhizomes of the plant are cylindrical with an average diameter of 2 cm.

In a transection, the cortex consists of collenchymatous and parenchymatous cells, cortical bundles, radially arranged xylem which is endarch, and the pith. A large number of root and leaf traces, leaf gaps, and branch connections are visible. The vascular ring is not continuous and consists of 3 to 6 vascular arcs of different sizes.[704,705] The cells of cortex and pith are spongy and filled with starch grains.[706] The roots show only thick parenchymatous cells and four to seven arcs of exarch xylem. Cortical bundles and pith are absent.[704]

Ayurvedic description — *Rasa—tikta; Guna—rooksha; Veerya—sheeta; Vipak—katu.*

Action and uses — *Kaf pit har, pachan, dipan, yakrit utagek, pit sark, krimighan, prame hangn, daha prasamk.*

Chemical constituents — Kapoor et al.[708] in a preliminary study reported the presence of flavonoid, but absence of saponin and alkaloids, in the roots and rhizomes of *P. kurroa*. Roots yielded kutkin, a glucosidal bitter principle, which was reported to be a stable mixture of two glycosides, viz., picroside-1 and a new glycoside, kutkoside.[709,710] The petroleum ether extract also yielded D-mannitol, kutkiol, and kutkisterol, and a ketone which was found to be identical with apocynin. Picrorhizin also reported to be its bitter principle, as glucosidovanilloyl glucose was isolated from the plant.[178]

Pharmacological action — In small doses, it is a bitter stomachic and laxative, and in large doses, a cathartic. It is reputed as an antiperiodic, cholagogue. It is a bitter tonic and is reported to have beneficial action in case of dropsy, and is also used in snakebite and scorpion sting.[1,50] The root is also used in diseases of liver and spleen including jaundice and anemia.[707]

Medicinal properties and uses — Root powder when given with sugar and hot water acts as a mild purgative; 1 to 1.5 gr of the powder with aromatics or pepper, asafetida, triphala, and salts are very useful in constipation due to scanty intestinal secretions. In bilious fever a compound decoction of its root, liquorice, raisins, and nim bark is very curative; in dyspepsia with severe pains, the powder of katuki, *Acorus calamus,* chebulic myrobalan, and plumbago root in equal parts is given in doses of 28 ml with cow's urine. The drug is also used in scorpion sting. Recently it has been tried and found beneficial in many cases of ill-defined fever, such as low fever with constipation, symptomatic fever of elephantiasis, and fever of material origin which resisted other home remedies.

In clinical studies on patients of infective hepatitis with jaundice, *P. kurroa* was reported to have led to a rapid fall in serum bilirubin levels toward normal range and quicker clinical recovery with no untoward effects.[713] Rajaram[714] in another study claimed that *P. kurroa* led to beneficial results in the management of bronchial asthma. The drug is also reported to produce marked reduction in serum cholesterol and coagulation time.[715]

Doses — Powder as a tonic — 1 to 1.5 g; as antiperiodic — 3 to 3.5 g.

PIPER LONGUM Linn.
Family: PIPERACEAE

Vernacular names — Sanskrit, Pipali; Hindi, Pipal; English, Long pepper; Bengali, Pepul; Nepalese, Chihi pipee; Sinhalese, Tippili; Japanese, Hihatsu; Chinese, Pipo; French, Racines de poivre long; Unani, Pipal; Arabian, Dar-e-filfil; Persian, Fil-fil-e-daras; Tibetan, Dro-sman; Burmese, Peikchin; Malaysian, Lada; Tamil, Thippilli.

Habitat — The plant grows in warmer regions of India, viz., Western Ghats, central Himalayas to Assam, Khasi, and Miker Hills, and the lower hills of Bengal.[460]

Parts used — Fruits and root.

Pharmacognostical characteristics — A small aromatic plant trailing on the ground, also climbing with erect and thin branches. Leaves smooth, entire, 7-ribbed; upper leaves ovate or ovate-oblong, narrow pointed, often unequal sided, sessile, base surrounding the stem; lower leaves 6 to 10 cm long, ovate; deeply cordate with big lobes at the base, dark green and shining, stalked. Flowers dioecious, minute; inflorescence a spike; male spike narrow, 3 to 4 in. long, female spike circular. Fruit small, ovoid, sunk in fleshy spike which is 2.5 to 4 cm, ovoid, oblong, blackish green, and shining.

The root in a transection shows thick-walled parenchyma, simple or compound starch grains, lignified and striated stone cells, resinous cells in the cortex, perivascular fibers in the phloem, and radial strips of xylem which meet at the center. Pith is absent. The stem has a secretory cavity in the center. The cortex shows starch grains as well as resinous and some stone cells. The phloem is capped by perivascular fibers and xylem is arranged in V-shaped groups.[716] Mehra and Puri[717] described the fruiting spike of *P. longum* as black, cylindrical, irregular, up to 2 to 5 cm long, and compact. The fruits are one seeded with three-layered pericarp. Endocarp is wavy in outline, which is a distinguishing character. Dasgupta and Datta,[718] while giving details of anatomy of the fruits, described the fruitlet of *P. longum* as thick-walled with heavy brown contents in the outermost layer, mesocarp with thickened cells, endocarp and seed coat fused to form a deep zone with hyaline content in the outer layers, and orange-red pigment.

Ayurvedic description — *Rasa* — katu; *Guna* — laghu, snigdha, teekshna; *Veerya* — sheeta; *Vipak* — madur.

Action and uses — *Kapha vat samak, pitta samak, dipan, vatanuloman, swashar, mutral, jawaraghana, khustaghana, rasayan, balya.*

Chemical constituents — The fruits gave positive tests for the presence of volatile oil, starch, protein and alkaloids, saponins, carbohydrates, and amygdalin, but no tannins.[718]

The alkaloids isolated from roots and stems were identified as piperine and piperlongumine and also methyl-3,4,5-trimethoxycinnamate.[719-721]

Fruit yielded sesamin, a lignan[722] dihydrostimasterol, piperine, and two low-melting unstable compounds, one of which appeared to be isobutylamide of an unsaturated acid. The petroleum ether extract yielded *N*-isobutyl-decatrans-2-*trans*-4-dienamide.[723] Hexane extract of fruit also yielded piperine. Steam distillation of dried fruits of *P. longum* yields an essential oil consisting of *n*-hexadecane, *n*-heptadecane, *n*-octadecane, *n*-nonadecane, *n*-cicosane, *n*-hencosane, α-thujene, terpinolene, zingiberene, *p*-cymene, *p*-methoxyacetophenone, dehydrocarveol, phenylethyl alcohol, and two new monocyclic sesquiterpenes.[725,726] The presence of L-tyrosine, L-cysteine hydrochloride, DL-serine, and L-aspartic acid as free acids has also been reported in the fruits.[727]

The seeds of *P. longum* gave sylvatine, sesamin, and dieudesmin.[728] The component fatty acids of crushed seeds were reported to be palmitic, hexadecenoic, stearic, linoleic, oleic, linolenic, higher saturated acids, arachidic, and behenic acids.[729]

The leaves gave hentriacontane, hentriacontane-16-one, triacontanol, and β-sitosterol.[730]

The content of calcium, phosphorus, and iron was reported to be 1230, 190, and 62.1 mg/100 g, respectively.[731]

Pharmacological action — Infusion is stimulant, carminative, alterative, and tonic; more powerful than black pepper; also aphrodisiac, diuretic, vermifuge, and emmenagogue. Root is stimulant.

P. longum is reported to exhibit significant antitubercular activity.[732,733] The essential oil of fruit showed antibacterial, antifungal, and anthelmintic activity,[734-736] as well as insecticidal and insect-repellant activity.[737] Sharma and Singh[738] observed a marked anti-inflammatory activity of fruit decoction against carrageenin-induced rat paw edema. Kholkute et al.[739] observed that benzene extract of *P. longum* in combination with methanol extract of *Embelia ribes* berries led to inhibition of pregnancy in 80% of the animals.

Dhar et al.[51] reported that ethanolic extract of the whole plant exerted hypoglycemic effect in rats and also counteracted the spasms induced by various spasmogens in isolated guinea pig ileum. Kulshrestha[740,741] observed that petroleum ether extract produced respiratory stimulation in smaller doses, but higher doses caused convulsion in laboratory animals. This may be due to the presence of some medullary stimulant factor in the extract. Banga et al. reported[742] that crude extract of *P. longum* as well as piplartine, one of its alkaloids, suppressed the ciliary movements of the esophagus of frog, which may be due to the suppression of cough reflex.

Medicinal properties and uses — The berries are a cardiac stimulant, carminative, alterative, tonic, laxative, digestive, stomachic, and antiseptic. It is given with honey in doses of 5 to 10 gr for indigestion, dyspepsia, flatulant colic, cough, chronic bronchitis, chest affections, and in asthma. It is also very useful in enlarged spleen, palsy, gout, rheumatism, and lumbago. Fruit is vermifuge and also used after childbirth to check post-partum hemorrhage.

Root is used as stimulant. The drug is also used in snakebite and scorpion sting.[1,3]

Doses — Fruit powder — 0.5 to 1.5 gr.

PIPER NIGRUM Linn.   *Family*: PIPERACEAE

Vernacular names — Sanskrit, Maricha; Hindi, Kalimirich; English, Black pepper; Bengali, Golmarich; Nepalese, Maley; Sinhalese, Gammiris, Chinese, Huchio; German, Schwartzer Pfeffer; French, Poivre noir; Unani, Siyah Mirch, Arabian, Filfil-us-siyah; Persian, Sihay Pipal; Tibetan, Na-le-sam; Burmese, Nayukon; Malaysian, Laddahitam; Tamil, Cheviyan, Molagu.

Habitat — This perennial climbing shrub is cultivated in the hot and damp parts of India.

Parts used — Dried unripe fruit.

Pharmacognostical characteristics — A stout, glabrous, woody creeper, much swollen at the nodes. Leaves broadly ovate, 4 to 9 in. long, 4$^1/_2$ in. broad, leathery, 5 to 9 ribbed, paler beneath. Flowers unisexual and bisexual, in slender, drooping spikes; berries in racemes, rather fleshy, one sided, yellow, turning red when ripe. The fruits of *P. nigrum* are distinguished by the absence of peripheral sclereids of pericarp, dark-brown mesocarp zone, asymmetrically thickened wall of the endocarp, shining seed coat layer, and yellow pigment cell in the kernel.[743]

Ayurvedic description — *Rasa — katu; Guna — laghu, teekshna; Veerya — ushna; Vipak — katu.*

Action and uses — *Vata pitta samak, lakhan, lolanisark, dipan, pachan, vatanuloman, utegek, artabjanan, kaphani sarak, krimighana, jawarghana.*

Chemical constituents — The black pepper contains an alkaloid piperine (5 to 9%), piperidine (5%), a balsamic volatile essential oil (1 to 2%), fat (7%); mesocarp contains chavicine, a balsamic volatile oil, starch, lignin, gum, fat (1%); proteids (7%) and ash containing organic matter (5%). Chavacine is a soluble pungent concrete resin.

The fruits yielded *N*-isobutyl eicosa-*trans*-2-*trans*-4-dienamide in addition to earlier reported piperine, piperetine, piperidine amides, viz., piperlin, piperolein A, and piperolein B.[744] Extraction of the powdered stem with petroleum ether led to the isolation of piperine, hentriacontanone, hentriacontane, hentriacontanol, β-sitosterol, and an unidentified compound.[745] The fatty acid composition of the crushed seeds includes palmitic, linoleic, oleic, and linolenic acids.[746] Nanda[486] reported the presence of 3.6 ppm of fluoride content.

Pharmacological action — *P. nigrum* is acrid, pungent hot, carminative, also used as antiperiodic. Externally, it is rubefacient and stimulant to the skin and resolvent. The extract and essential oil of *P. nigrum* is reported to be antibacterial and antifungal. The fruits exhibited taenicidal activity.[460]

Medicinal properties and uses — A paste of black pepper is a rubefacient and stimulant; it is locally used for boils, relaxed sore throat, piles, paralytic affections, rheumatic pain, headache, prolapsed rectum, and toothache. Black pepper is aromatic, given in dyspepsia, flatulence, debility, diarrhea, cholera, disorders of the urinary system, cough, gonorrhea, and malarial fever.[3,50]

Doses — 5 to 20 gr.

PLANTAGO OVATA Forsk.   *Family:* PLANTAGINACEAE

Vernacular names — Sanskrit, Ashwagolam; Hindi, Issufgul; English, Spogel seeds, Ispaghula; Bengali, Isabgul; Nepalese, Isabgol; Sinhalese, Isphagol Vithai; Japanese, Obeko; Chinese, Ch'-Ch'ientzu; German, Indische Psylli-samen; Unani, Aspaghol; Arabian, Bazarqutuna; Persian, Aspaghol; Tamil, Grappicol.

Habitat — This herb is found in northwestern India, the Punjab, Sind, and Baluchistan; extensively cultivated in Gujerate.

Parts used — Seeds.

Pharmacognostical characteristics — Scapigerous herb; leaves 5 to 12 cm. long; variable in breadth, ovate or oblong-ovate; spike 2.0 to 5 cm long; flowers minute, scattered or crowded; fruit 8 to 12 mm long, the upper half separating like a lid.

The seeds of *P. ovata* are dull, pinkish gray-brown, boat-shaped, having an ovate outline; the dorsal surface is convex with a small elliptical or elongated shining reddish-brown spot, while the ventral surface is concave with a deep furrow, not quite reaching either end of the seeds. The seeds are 2.0 to 3.3 mm in length and 1 to 1.6 mm in breadth.[747]

Ayurvedic description — *Rasa* — *madhur; Guna* — *snigdha, guru, pichil; Veerya* — *sheeta; Vipak* — *madhura.*

Action and uses — *Vata-pitta samak, dah prasman, sanahan, balya, brihan, jawaraghana.*

Chemical constituents — The seeds contain mucilage, fixed oil, and albuminous matter. The presence of a glucoside named aucubin in small quantities is also reported. Atal et al.[748] reported that embryo oil of the seeds is a good source of linoleic acid. Patel et al.[749] found a number of amino acids in the combined form, viz., valine, alanine, glycine, glutamic acid, cystine, lysine, leucine, and tyrosine. Valine, alanine, and glutamic acid were also found in the free form.

Pharmacological action — Seeds are cooling, demulcent, mildly astringent, emollient, laxative, and diuretic.[1,3]

The alcoholic extract of *P. ovata* seeds in small doses produced a transient precipitous fall of blood pressure in cats and dogs, but in higher doses it produced a rise in blood pressure.[750] The oil of the embryo showed a more potent hypocholesterolemic effect than safflower oil in rabbits rendered hypercholerolemic. In normal rabbits it had no effect.[751] Bijlani and Thomas[752] reported that *P. ovata* enhanced the digestive and absorptive functions of the sac of hamster ileum.

Medicinal properties and uses — Ispaghula seeds are used in catarrh, chronic dysentery, intestinal fluxes, diarrhea, blenorrhea, affections of the bladder, urethra, and kidney, and inflammatory and functional derangements of digestive organs in doses of 6 to 12 gr powder. In many affections of the kidneys and bladder, in gonorrhea, attended with pain, local irritation, and scalding or difficulty in passing urine, and in piles, a strained decoction of the bruised seeds is given (56 to 112 ml) three times a day. A poultice of the bruised seeds is very effective in rheumatic and gouty affection, glandular swelling, and for irritable surface of the skin.[1,3] Jain and Tarafder[753] reported that tribals use the plant for pain and bronchitis.

Doses — 6 to 12 gr.

PLUMBAGO ZEYLANICA Linn.   Family: PLUMBAGINACEAE

Vernacular names — Sanskrit, Chitraka, Hindi, Chitra; English, White leadwort; Bengali, Chita; Nepalese, Chita; Sinhalese, Elanitul; Japanese, Indo matsuri; Chinese, Pai Hau; German, Bleiwurz; Unani, Chita; Arabian, Shaytraj Hindi; Persian, Bekh-e-barandah; Burmese, Kanchoppyiju; Malaysian, Vellakotuveri; French, Dentalaire de Ceylon; Tamil, Kodiveli.

Habitat — Largely cultivated in gardens throughout India. Grows wild in South India and Bengal.

Parts used — Roots.

Pharmacognostical characteristics — A perennial undershrub with rambling branches. Leaves $1^1/_2$ to 4 in., ovate, acute, suddenly narrowed into a short amplexicaul petiole with a dilated base. Spikes 4 to 12 in., often branched, the rachis glandular. Flowers bisexual, white; calyx persistent, tubular, with conspicuous stalked viscid glands; 5-ribbed; corolla tube 1 in. long, slender. Base of style glabrous.[28] Capsule included in the persistent calyx, opening transversely near the base. The external color of the roots is light yellow when fresh and reddish brown when dry. Internal color is brown and striated. Fracture short, taste acrid and biting; wood hard and reddish in color.[460,510]

In transverse section, the root shows a very small cork, consisting of 8 to 10 layers of cells and a very wide cortex, and the size of wood is medium. The phelloderm is a wide zone $^1/_2$ to $^2/_5$ of total root diameter; cells thick walled and polyhedral. The phelloderm cells contain starch. The cortical cells are polygonal, parenchymatous, and also contain starch. The important feature of the cortex is the presence of many nonlignified fibers scattered throughout the cortex in groups or chains. The wood is characterized by the presence of numerous vessels of different sizes and is mainly composed of tracheids. The xylem rays are thin to moderately broad and many in number. The vessels and the tracheids have pitted walls, and the vessels generally have end perforations. Wood fibers are also present. Powder of root is olive green but gives a green color when mounted in nitracellulose under fluorescent lights.[754]

While locating the distribution of active principles in plant tissues, Iyenger and Pendse[755] reported that plumbagin is distributed mostly in cells of secondary cortex and medullary ray cells in roots as an amorphous yellow substance. The root powder also showed simple and compound starch grains.

Ayurvedic description — *Rasa — katu; Guna — laghu, rooksha, teeshana; Veerya — ushana; Vipak — katu.*

Action and uses — *Kaf bat samak, pit-bardhak, dipan, pachan, grahi, shotha-har, vagi karan, grabha shyasan kochak* (uterine constrictor).

Chemical constituents — Root of *P. zeylanica* contains an acrid crystalline principle called "plumbagin" about (0.91%); in the form of yellow needles, mp 72°C, slightly soluble in boiling water, freely soluble in alcohol and ether, partly volatizes when heated.[1] Apart from plumbagin, the root yielded new pigments, viz., 3-chloroplumbagin, 3,3-biplumbagin, binaphthoquinone identified as 3′,6′-biplumbagin, and four other pigments identified as isozeylinone, zeylinone, elliptinone, and droserone.[756-758]

Pharmacological action — Alterative, gastric stimulant, and appetizer; in large doses it is acronarcotic poison. It has a specific action on the uterus and is abortifacient.

Pharmacological studies were mainly concentrated on the antifertility activity of *P. zeylanica*. Gupta et al.[759] reported successful anti-implantation activity in 50% alcoholic extract of the roots and fruit, but Bhakuni et al.[11] could not confirm it. Aswal et al[760] found that 50% ethanolic extract was devoid of abortifacient activity in rats and anti-implantation effect in hamsters. Premakumari et al.[761] observed significant anti-implantation and abortifacient activity in albino rats without any teratogenic effect. Plumbagin caused a selective testicular lesion in dogs.[762] Antitumor or anticancer activity was not reported by Kundu and Mondal[763] or Bhakuni et al.[11] The extracts, however, showed antibacterial effect against some organisms.[764,765] However, no antifungal activity was reported. Topical application of plumbagin (isolated from *P. indica*) was reported[11] to be useful in patients with common warts.[766]

Medicinal properties and uses — Powder of root tempered with some bland oil is used as a rubefacient application in rheumatism, paralytic affections, and in enlarged glands. It is also effective in some cases of leukoderma and other skin diseases. When scraped root is introduced into the mouth of the womb to procure abortion, it will expel the fetus from the womb whether dead or alive. A tincture of the root is employed as an antiperiodic.

Dose — 0.5 to 1 gr.

PREMNA OBTUSIFOLIA R. Br.
(Syn. *P. INTEGRIFOLIA* Linn.;
P. CORYMBOSA auct. non Rottl. and Willd.;
CORNUTIA CORYMBOSA Burm. f.)

Family: VERBENACEAE

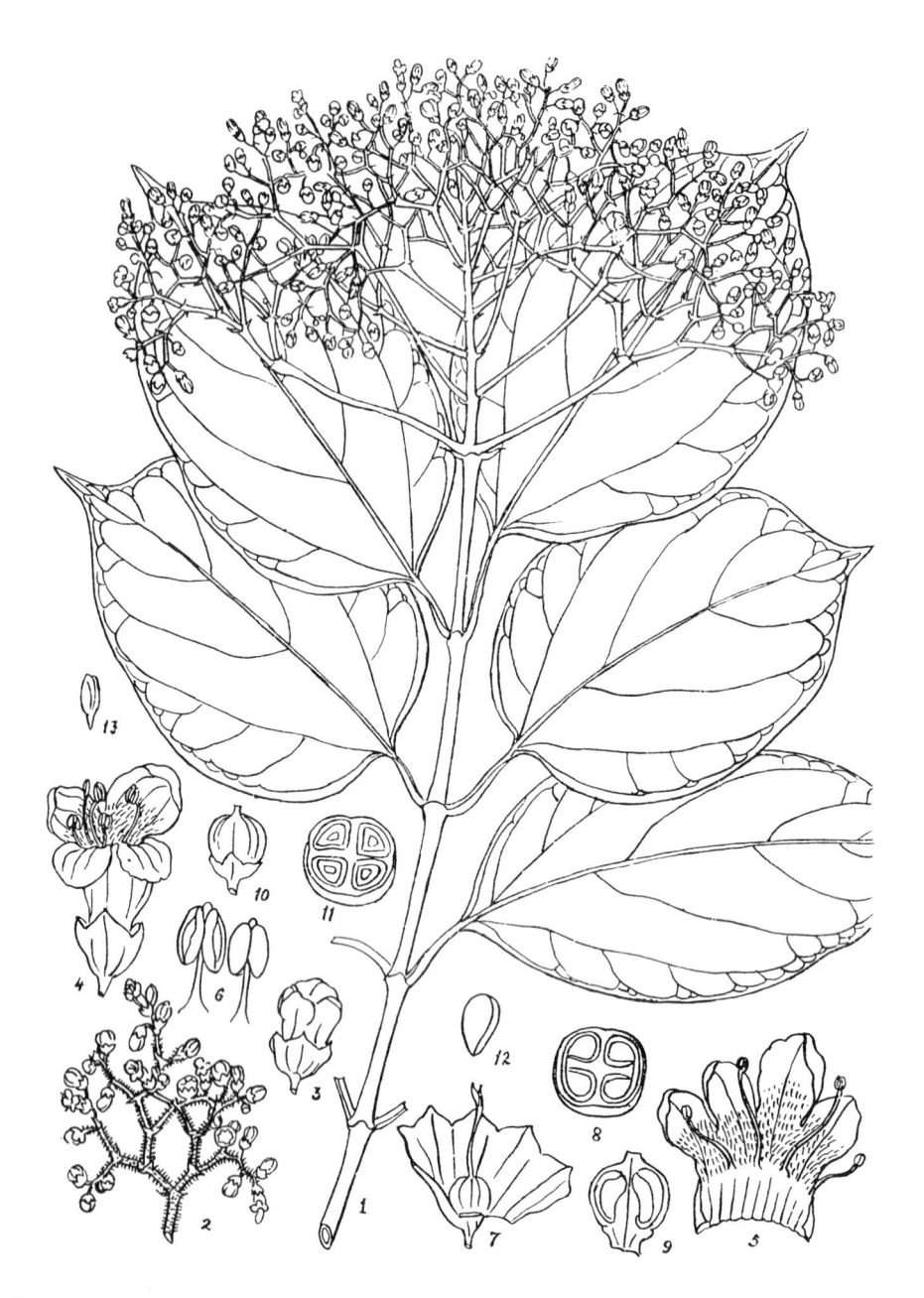

Vernacular names — Sanskrit, Agnimantha; Hindi, Arni; Bengali, Bhut-bhiravi; Nepalese, Gineri; Sinhalese, Midi; Unani, Arni; Burmese, Toung-than-ghee.

Habitat — It is common along the Indian and Andaman coasts. It is also recorded as occurring in the plains of Bombay, Gujerat, Assam, and Khasi Hills where it grows under mesophytic conditions.

Parts used — The roots, root bark, and leaves.

Pharmacognostical characteristics — It is a small-sized tree or large shrub with a comparatively short trunk and numerous branches, bearing simple, opposite, dark-green, broad, elliptic, obtuse leaves, very shortly acuminate and cymose panicles of small inconspicuous flowers. Its older parts are often armed with large opposite, straight, strong, woody spines that represent the basal parts of stunted branches. Bark smooth, thin, dark brown or brownish yellow, and lenticellate. The leaves and flowers emit a very strong offensive or unpleasant odor when bruised. Flowers are small, bracteate, pale greenish white or greenish yellow; bracts are minute, lanceolate. The calyx is about 2.5 mm long, thick, two lipped with one lip toothed and the other subentire so that the calyx appears three lobed; corolla is glabrous, from outside the tube cylindrical and hairy in the throat region. The stamens are oblong and rounded, slightly exserted, while filaments are hairy at the base. The ovary and style are glabrous, stigma has two equal maricate lobes. The fruit is 4 mm long, pear shaped, and four seeded.[767]

The roots are light brownish or yellowish brown, woody, strong, and possess a very agreeably scented bark. In transverse section, the root is circular or oval but the outline is not regular; narrow zones of two to five or more rows of thick-walled cells alternating with broader zones of four to eight or more rows of thin, broadly rectangular cells are seen. The cortical zone is composed of thin-walled cells of various sizes. The bast forms the most prominent part of the bark and shows broad segments of phloem composed of regularly arranged parenchymatous elements and large sieve tubes. Few stone cells, either isolated or arranged in irregular concentric rings, occur in the bast. Medullary rays are many, narrow, and mostly uniseriate with occasional two- and three-seriate rays. Cells in the phloem are thin walled, but in the xylem they have thick pitted walls and are loaded with starch. There is no pith in the center.[768,769]

Ayurvedic description — *Rasa* — *tikta, katu, kasaya, madhura; Guna* — *rooksha, laghu; Veerya* — *ushna; Vipak* — *katu.*

Action and uses — *Kaf bat samak, soth har, badana, sthapak, samak, rakt sodhak, and parmehaghana.*

Chemical constituents — Stem bark of *P. obtusifolia* contains an alkaloid premnine and ganiarine.[770,771] It also contains resin and tannin. Ramaiah et al.[772] obtained betulin and β-sitosterol from the stem and β-sitosterol from the leaves. Bheemasankara Rao et al.[773] isolated polyisoprenoid and β-sitosterol. Dasgupta et al.[774] reported the isolation of first spermine alkaloid, viz., aphelandrine, from the stem bark and lutiolin from the leaves.

Pharmacological action — It is a cardial, stomachic, carminative, alterative, and tonic. Dhar et al.[51] reported that stem bark extract showed hypoglycemic action in rats and hypotensive activity in cats and dogs. The root extract also had hypoglycemic action in rats. The extract did not exhibit any antibacterial activity. Kurup and Kurup[775] found that a phenolic compound showed antibacterial activity against *Staphylococcus aureus, Bacillus subtilis,* and *Streptococcus haemolyticus*. Decoction of the plant exhibited marked anti-inflammatory and antiarthritic activity against acute, subacute, and chronic inflammation induced in both immunological and nonimmunicological experimental models.[776]

Medicinal properties and use — Decoction of root is cardial, stomachic, and good for liver complaints. It is prescribed in rheumatism and neuralgia. Infusion of the leaves is administered with pepper in colds and fevers; in decoction, it is very useful in flatulence; in the form of soup it is used as stomachic and carminative. Root forms an ingredient of "dasamula" and thus is used in a variety of affections. Root rubbed into a paste with water is recommended to be taken with clarified butter in urticaria and roseola for a week. Decoction of the root is also given in doses of 56 to 112 ml in gonorrhea.[3,50]

Doses — Infusion of the leaves — 28 to 56 ml; decoction of the root — 56 to 112 ml.

PRUNUS AMYGDALUS Batsch.
(Syn. *AMYGDALUS COMMUNIS* Linn.)

Family: ROSACEAE

Vernacular names — Sanskrit, Badama; Hindi, Badam; English, Almond; Bengali, Badam; Nepalese, Badam; Sinhalese, Ratakotabansba; German, Mandelbaum; Unani, Badam; Arabian, Louza; Persian, Badam; Burmese, Badam; Tamil, Badampisil.

Habitat — The almond tree is a native of western Asia. It is cultivated in cooler parts of India, in the Punjab, and Kashmir.

Parts used — Sweet almonds, almond shell, ripe seeds. bitter almond, and oil.

Morphological characteristics — Small- to medium-sized tree with oblong-lanceolate, serrulate leaves; petiole glandular, as long or longer than the breadth of the leaf; stipules fimbriate. Flowers copious, peduncled, white tinged with red, appearing before the leaves. Fruit (drupes) have a soft, velvety pericarp, the inner part of which gradually becomes sclerenchymatous as the fruit ripens to form a pitted endocarp or shell, separating into two halves when ripe. Two varieties, viz., amara and dulcis, the bitter and sweet almonds, are recognized.[28]

Sweet almond is 2 to 3 cm long, rounded at one end and pointed at the other. The bitter almond is 1.5 to 2.0 cm in length but of similar breadth to sweet almond. Both varieties have a thin, cinnamon-brown testa. The kernel consists of two large, oily, planoconvex cotyledons and a small plumule and radicle, the latter lying at the pointed end of the seed.

Ayurvedic description — *Rasa — madhur; Guna — guru, snigdha; Veerya — ushna; Vipak — madhur.*

Action and uses — *Vata samak, kaphapitta wardhak, dipan, sanahan, anumolan, mridu rochan, kapha nisarak, sukrajanan, vajekaran, artaw janan, mutral, balya, brihhan.*

Chemical constituents — Sweet almonds, besides the fixed oil, contain an albuminous principle or ferment "emulsin" soluble in water, mucilage (3%), sugar (6%), proteins (18.58%), and ash (3 to 5%), containing potassium, calcium, and magnesium phosphates.

Bitter almonds contain a fixed oil (45%), amygdalin (3%), proteins (25%), emulsin and sugar (3%) mucilage (3%), ash (3 to 5%), hydrocyanic acid (2 to 4%), and amygdalin. The seeds of var. *amara* contain an alkaloid, phenethylamine.

The essential oil of bitter almond is not present as such in the almonds, but is formed by hydrolysis of the glucoside amygdalin by the enzyme emulsion which is found in the seeds. The chief constituents of the essential oil are benzaldehyde hydrocyanic acid and benzaldehydecyanhydrin (or mandelonitrile).[28]

Almonds yield two distinct oils, a fixed or fatty oil and an essential oil. Sweet almonds do not yield any essential oil, but the yield of fixed oil is about the same as in the case of bitter almonds varying from 40 to 55%.[69]

Pharmacological action — Sweet almonds are demulcent, stimulant, nutritive, nervine tonic, and emollient. Bitter almonds are emollient, demulcent, and laxative and are used as sedative and in cough. Root is alterative.[3,50]

Medicinal properties and uses — Expressed oil of sweet almonds is slightly laxative; it is given in constipation. Almond nut cream is recommended for brain workers. A paste of bitter almond made with vinegar is used to relieve neuralgic pains; they are also used for removing obstructions of the liver and spleen. Applied to the head, they kill lice; as a suppository they are a valuable application to irritable sores and skin eruptions. Juice of almonds mixed with sugar is used in cough. Almonds soaked in honey at night and taken in early morning are a very nutritious food for all those who wish to build up a strong and healthy constitution. A confection made of sweet almonds together with several other ingredients is recommen ded as useful in polyuria due to kidney affection, in building up the kidney tissue and nervous tissue, and also increasing and thickening the semen. Seeds of the bitter variety are useful in certain skin affections but not for internal use. Sweet almond meal has been recommended as a suitable diet for diabetic patients as it contains no starch.

Doses — Confection — 7 to 14 ml; seeds — 7 to 10.

PRUNUS PUDDUM Roxb. ex Wall. *Family:* ROSACEAE

Vernacular names — Sanskrit, Padmaka; Hindi, Paddam; English, Bird-cherry; Bengali, Padma Kasta; Nepalese, Padmakha; Sinhalese, Padamkasta; German, Traubenkirsche; Unani, Padmakh; Burmese, Panni.

Habitat — Found wild in temperate Himalaya from Garhwal to Sikkim and Bhuttan at altitudes of 3000 to 6000 ft; also met with in the hills of Kodaikanal and Ootacamund in South India.

Morphological characteristics — A moderate-sized to large tree of brilliant appearance when in flower. Leaves 3 to 5 in., ovate or oblong-lanceolate, long-acuminate, sharply serrate, petiole 2 to 4 glandular. Flowers solitary, fascicled or umbelled, 1 in. in diameter, pink or white, appearing before the leaves. Calyx tube narrowly campanulate. Petals obovate or linear-oblong. Fruit a drupe, $1/2$ to $3/4$ in. long, oblong or ellipsoid, obtuse at both ends; flesh scanty, reddish or yellowish, acid; stone wrinkled and furrowed.[28]

Ayurvedic description — *Rasa — kasaya, tikt; Guna — laghu, snigdha; Veerya — sheeta; Vipak — katu.*

Action and uses — *Tridosha har, kafa pitta samak, daah samak, kustaghan, dipan, pachan, sthamvan, rakt pitta samak, swashar, hicca har, mutral, sukra janan, grapha sthapan, balya.*

Chemical constituents — Bark, leaves, and kernels contain amygdalin which yields hydrocyanic acid.

Pharmacological action — Astringent.

Medicinal properties and uses — Fruit is acidic and somewhat astringent. Kernel used in kidney stone and gravel. Smaller branches used as substitute for hydrocyanic acid.[50]

Doses — 24 to 45 gr.

PSORALEA CORYLIFOLIA Linn.   Family: PAPILIONACEAE

Vernacular names — Sanskrit, Vakuchi; Hindi, Babchi; English, Babchi seeds; Bengali, Bavachi; Nepalese, Bakuchi; Sinhalese, Bodi ata; German, Bawchan; Unani, Babchi; Persian, Ba bakhi; Tamil, Karpokarisi.

Habitat — This plant is found as a weed in waste places all over the plains of India. It is also cultivated in some places.

Parts used — Leaves, seeds.

Pharmacognostical characteristics — It is an erect herb with densely gland-dotted branches. Leaves stalked, simple, roundish, 1 to 3 in. long, 1 to 2 in. broad, firm in texture, covered with numerous black dots, hairs few; stalks up to 1 in. long, hairy, gland dotted. Flowers are dense, 10 to 30 in a bunch arising in axils of leaves; corolla yellow or bluish purple. Fruit small, subglobose, without hairs, slightly compressed, pitted, black, seed one, smooth. Datta and Das[777] described the seeds as kidney shaped, 2 to 4 mm long, 2 to 3 mm broad, and 1 to 1.5 mm thick, smooth, exalbuminous, with straw-colored hard testa. The testa and tegmen are separable and the embryo consists of two elongated kidney-shaped cotyledons. Under the microscope, macrosclereids can be seen in the testa which contains mucilage and I-shaped osteosclereids below the palisade layer, and thick-walled cells of tegmen containing a brown pigment. Microchemical tests showed the presence of lignin, alkaloids, tannin, and saponin in the seed coat; oil, starch, and protein in the kernel; and amygdalin in the palisade cells of the seed coat.

Ayurvedic description — *Rasa — katu, tikta; Guna — laghu, rooksha; Veerya — ushna; Vipak — katu.*

Action and uses — *Kapha bat samak, branaya karak, branropan, kesya, dipan, pachan, anuloman, pramehayn, bajikarn.*

Chemical constituents — Chemical studies by various workers revealed a number of compounds belonging to different chemical groups such as furanocoumarins, coumesterol group, chalcones, and flavones.

The seeds contained an essential oil, a nonvolatile terpenoid oil, a pigment, a resin, fixed-oil,[778,779] raffinose,[780] two coumarin compounds, viz., psoralen and isopsoralen,[781,782] psoralidin,[783] corylifolean,[784] and isopsoralidin.[785,786] The seed oil yielded limonene, β-caryophylenoxide, 4-terpineol, linalool, geranylacetate, angelicin, psoralen and bakuchiol.[787]

From the petroleum extract of *P. corylifolia* roots, daidzein, trilaurin, and coumesterol were isolated for the first time. Angelicin, psoralen, and sitosterol were also isolated from the extract.[788]

Pharmacological action — Seeds are laxative, anthelmintic, diuretic, and diaphoretic. Oil has a powerful effect against skin streptocci.

Rashid Ali and Agarwala[789] made extensive studies and showed that psoralen significantly accelerates the photooxidation of DOPA (diydroxyphenylalanine) under sunlight as well as photo flood lamplight. The alcoholic and aqueous extracts of seeds exerted antibacterial activity.[790,791] Volatile oil of seeds inhibited the growth of staphylococci. Ether and chloroform extracts also showed similar activity.[792] Essential oil of *P. corylifolia* showed a moderate selective antifungal activity.

Petroleum ether extract of seeds exhibited anthelmintic activity against earthworms.[795] The extract also produced a rise in blood pressure on anesthetized dogs and caused a stimulation of the intestinal smooth muscle.[792] Chandhoke and Ray Ghatak[796] reported that isopsoralene exhibited tranquillosedative, anticonvulsant, and central muscle-relaxant properties in rats, mice, and rabbits. Bavachinine, a flavonoid isolated from *P. coryliflora,* showed anti-inflammatory activity against carrageenin-induced edema in rats.[797] Mukerjee[798] reported that oleoresin extract of the seeds (containing most of the essential oil) was found to be a most effective preparation when applied locally on the patches of leukoderma.

Medicinal properties and uses — Seed powder is useful in leprosy and leukoderma internally; and is also applied in the form of paste or ointment externally. As a laxative it is particularly used in bilious disorders.

Doses — Powder — 1 to 3 gr; oil — externally.

PTEROCARPUS MARSUPIUM Roxb. *Family:* PAPILIONACEAE

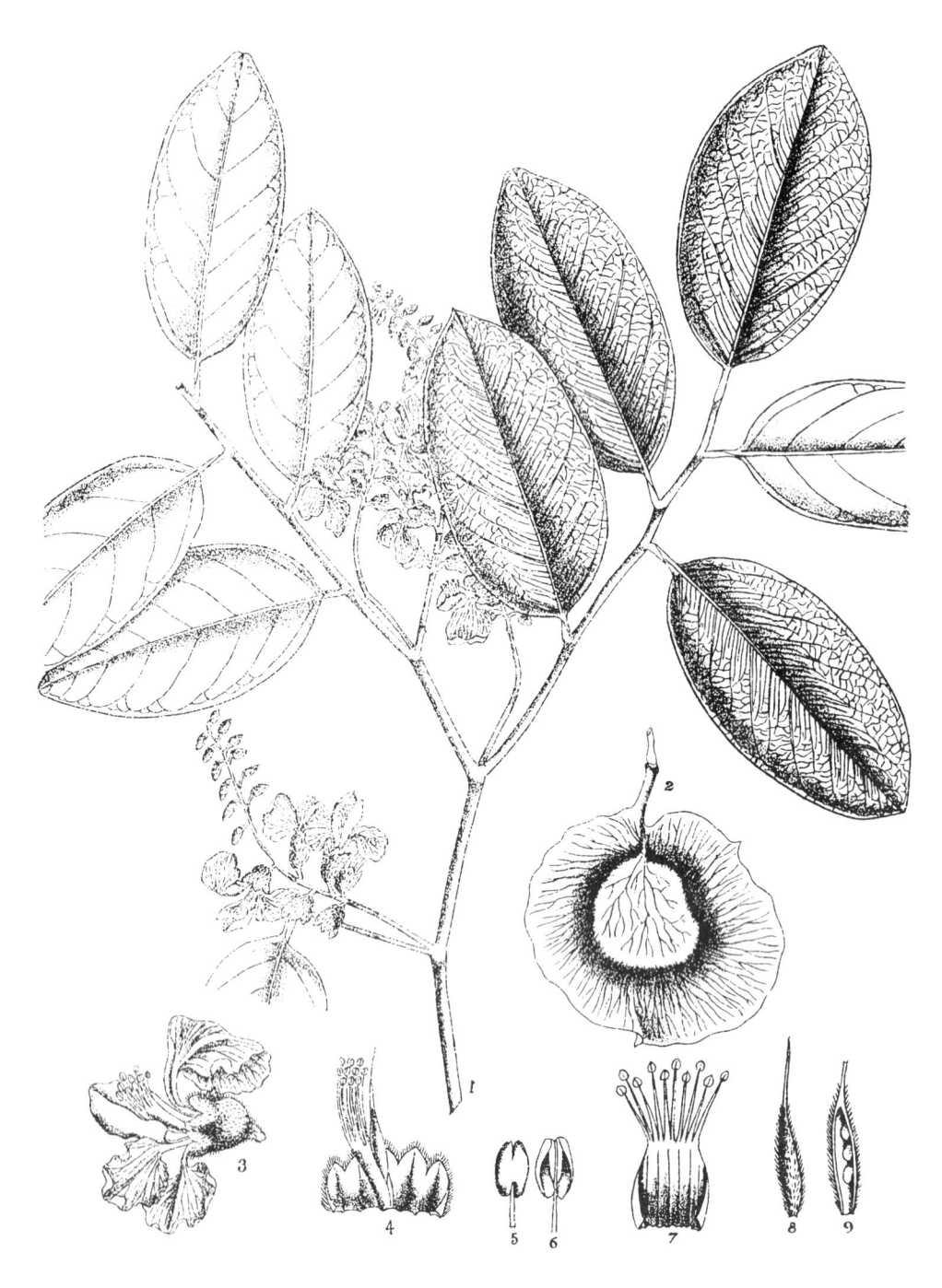

Vernacular names — Sanskrit, Pitasala; Hindi, Bijasal; English, Malabarkino; Bengali, Pitsal; Nepalese, Bijasar; Sinhalese, Gammalu; German, Malabarkino; French, Pterocarp; Unani, Dammul-akhajan; Arabian, Dammul Akhwayn; Persian, Khoon-e-siyaun-shan; Tamil, Vengai.

Habitat — A moderate to large deciduous tree about 90 ft or more high, common in central and peninsular India; found at 3000 ft in Gujarat, Madhya Pradesh, and sub-Himalayan tracts.[460]

Parts used — Kino (gum), bark, leaves.

Pharmacognostical characteristics — A large handsome tree. Bark gray, longitudinally cracked. Leaves compound; with 5 to 7 leaflets, 3 to 5 in. long, oblong or elliptic, margin wavy. Flower about 1.5 cm long, yellow, in very large, dense bunches. Fruit 2 to 5 cm long, roundish, winged, with one seed.

The heartwood of this tree is golden yellowish brown with dark streaks, staining yellow when damp and turning darker on exposure, strong and tough. The wood consists of vessels, tracheids, fiber tracheids, and wood parenchyma, all the elements being lignified and filled with tannin. Vessels are medium sized, drum shaped, scattered, leading to semiring-porous condition, tyloses present. Tracheids are long, abundant, thick walled, with tapering ends and simple pits on the side walls. Xylem parenchyma is small, thick walled with blunt ends, rectangular, simple pitted surrounding the vessel. Medullary rays are uni- or biseriate with ray cells 6 to 13 cells high and 1 to 2 cells wide. A few crystal fibers are observed in tangential section of the wood.[799]

Tree bark yields a reddish gum known as kino gum, which becomes brittle on hardening and is very astringent.[800] Kino contains kino-tannic acid.

Ayurvedic description — *Rasa—kasaya; Guna—laghu, rooksha; Veerya—sheeta; Vipak — katu.*

Action and uses — *Kapha pitta samak, sothahar, sandhaniya, kustaghan, kesya, sthamvan, krimighan, rakt sodhak, rakt pitta samak, rasayan, pramehagan.*

Chemical constituents — *P. marsupium* roots have yielded a new C-glycosyl-β-hydroxy-dihydrochalkone pterosupin, pseudobaptigenin, liquiritigenin, isoliquiritigenin, garbanzol, 5-deoxykaempferol, and *p*-hydroxybenzalaldehyde.[810] On extraction with petrol, the root wood yielded a new sesquiterpene alcohol selin-4(15)en-1b,11-diol, besides β-eudesmol, erythrodiol-3-monoacetate, and pterostilbene.[802]

The heartwood, which was soluble in alkali, gave isoliquiritigenin, liquiritigenin, and pterostilbine, but the sapwood showed the presence of pterostilbene.[803] The bark contained *l*-epicatechin and pterostilbene. Various workers obtained a novel isoflavonoid glycol marsupol, carpusin, propterol, and propterol B from the heartwood of the tree.[804-806]

Pharmacological action — Astringent, alterative, hypoglycemic. Pandey and Sharma[807] observed hypoglycemic action in rats rendered diabetic by alloxan when a decoction of *P. marsupium* was administered orally. Trivedi[808] reported that aqueous extract of the plant exhibited hypoglycemic effect in both acute and chronic experiments on normal rabbits.

In another study by Pandey and Sharma,[809] decoction of the bark showed hypocholesterolemic effect in rabbits. Trivedi[810] reported that marsupinol-α-flavonoid isolated from the heartwood stimulated cardiac contraction without affecting the heart rate. In a preliminary clinical trial on the extract of *P. marsupium*, heartwood showed encouraging hypoglycemic effects in 14 diabetic patients.[811] Water stored overnight in a tumbler made of *P. marsupium* heartwood is believed to have an antidiabetic effect.[812]

Medicinal properties and uses — Kino is a simple astringent, administered in diarrhea; it is also used for toothache. Bark is used in the form of powder or decoction in diarrhea. Bruised leaves are applied as paste to boils, sores, and skin diseases. Decoction of the bark is very useful for diabetic patients.

Doses — Decoction — 56 to 112 ml; powder — 3 to 6 gr; extracted juice — 125 mg.

PTEROCARPUS SANTALINUS Linn.

Family: PAPILIONACEAE

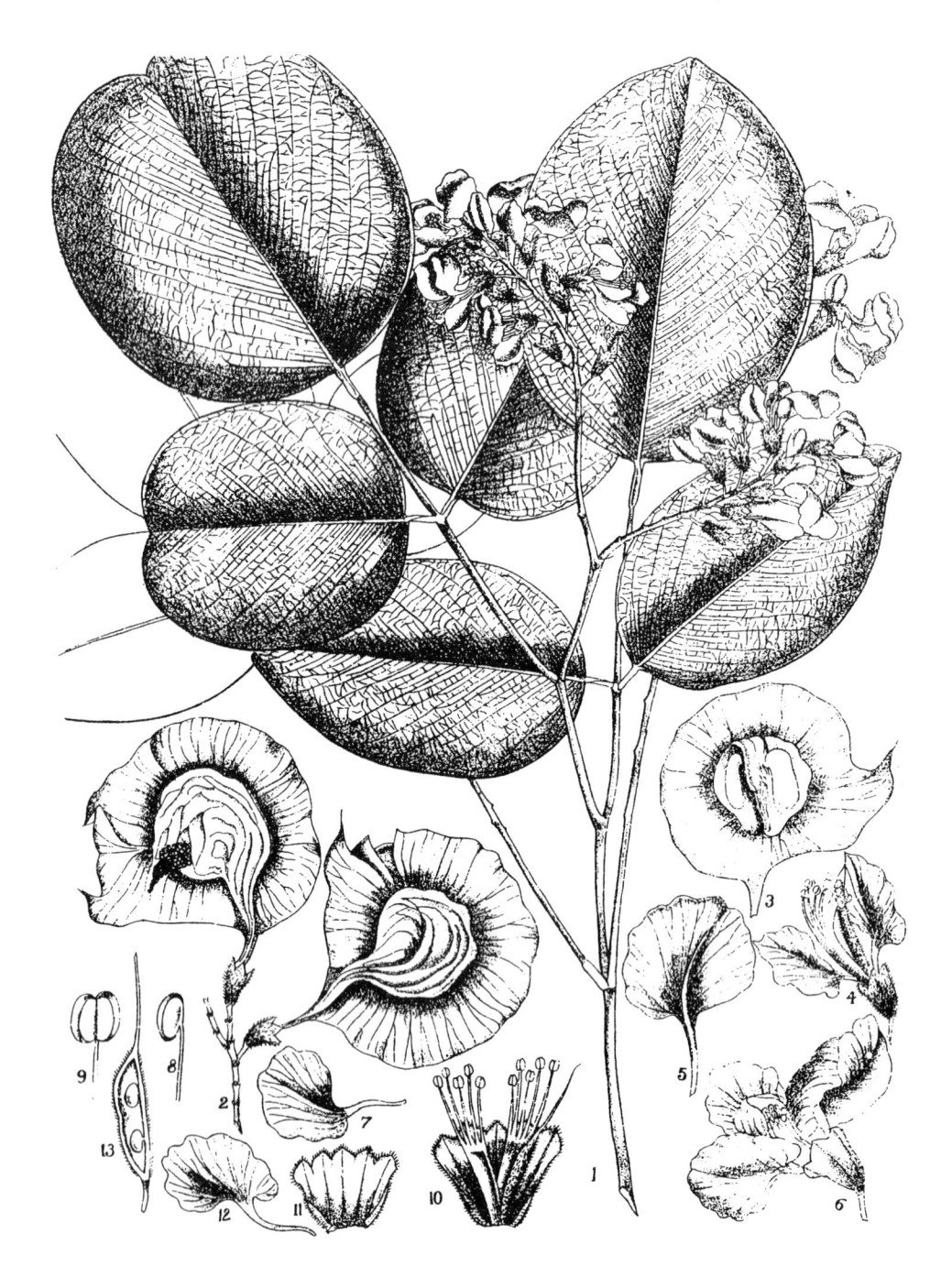

Vernacular names — Sanskrit, Rakta chandana; Hindi, Lal chandan; English, Red sandalwood; Bengali, Rakta chandana; Nepalese, Raktachandana; Sinhalese, Rath handunn; Chinese, Tan hasiang; German, Dunkelrothe Flugel-Frucht; French, Santal rouge; Unani, Sandal surkh; Arabian, Sandal-e-Ahmar; Persian, Sandal-e-surkh; Burmese, Nasani.

Habitat — This tree is generally met with in forests of southern India.

Parts used — Wood.

Morphological characteristics — Bark blackish brown, deeply cut into elevated rectangular patches. Bark, when injured, exudes red juice; wood dark red or maroon, externally hard; when rubbed against a hard surface it gives a red-colored aromatic paste. Leaves alternate, leaflets 5 to 10 cm long, rounded at both ends; flowers yellowish, in copious panicled racemes.

Ayurvedic description — *Rasa — tikta, madhur; Guna — guru, rooksha; Veerya — sheeta; Vipak — katu.*

Action and uses — *Kapha pitta samak, daah samak, sthamvan, sothahar, rakt pitta samak, kustaghan, jwaraghana, visaghan.*

Chemical constituents — Ravindernath and Seshadri[813,814] isolated from heartwood red pigment and two major compounds, viz., santalin A and santalin B, both of which gave permethyl ether on methylation. Kumar et al.[815] isolated a new sesquiterpene, viz., isopterocarpolone, ptercarptriol, and isopterocarpone, along with known sesquiterpenes, viz., pterocarpdiolone, pterocarpol, and cryptomeridiol, and triterpenes, viz., acetyl oleanolic aldehyde, acetyl oleanolic acid, and pterostilbine. The sapwood extracts yielded acetyl oleanolic acid and erythrodiol as well as santalin A and santalin B.[815] The bark extract exhibited the presence of β-amyrone, lupeonone, epilupiol, and lupeol, a new lupene diol and a new triterpene betulin.[816]

The leaves of *P. santalinus* yielded lupenone, β-amyrone, epilupeol, β-amyrin, lupeol, sitosterol, and stigmasterol, as well as erythrodiol, butlin, leukoanthocyanin, and anthocyanidin.[817,818]

Pharmacological action — Antiperiodic, antipyretic, aphrodisiac, astringent, bitter, cooling, diaphoretic, and vermifuge. Dhawan et al.[819] reported that the extract of the stem showed antispasmodic activity in isolated guinea pig ileum and stem bark exhibited semen coagulant activity. The seed extract showed hypoglycemic activity in rats.

Medicinal properties and uses — The wood of the tree is astringent and tonic; its decoction is given in chronic dysentery; its powder is given with milk in bleeding piles; the wood is administered in bilious disorders, skin diseases, and as a diaphoretic. A paste of the wood is a cooling application for boils, inflammatory diseases of the skin, swollen limbs, ophthalmia, sore eyes, and headache. The wood is an ingredient of many medicinal oils and pharmaceutical preparations.

Doses — Powder — 1.5 to 3 gr; decoction — 28 to 56 ml.

PUNICA GRANATUM Linn.

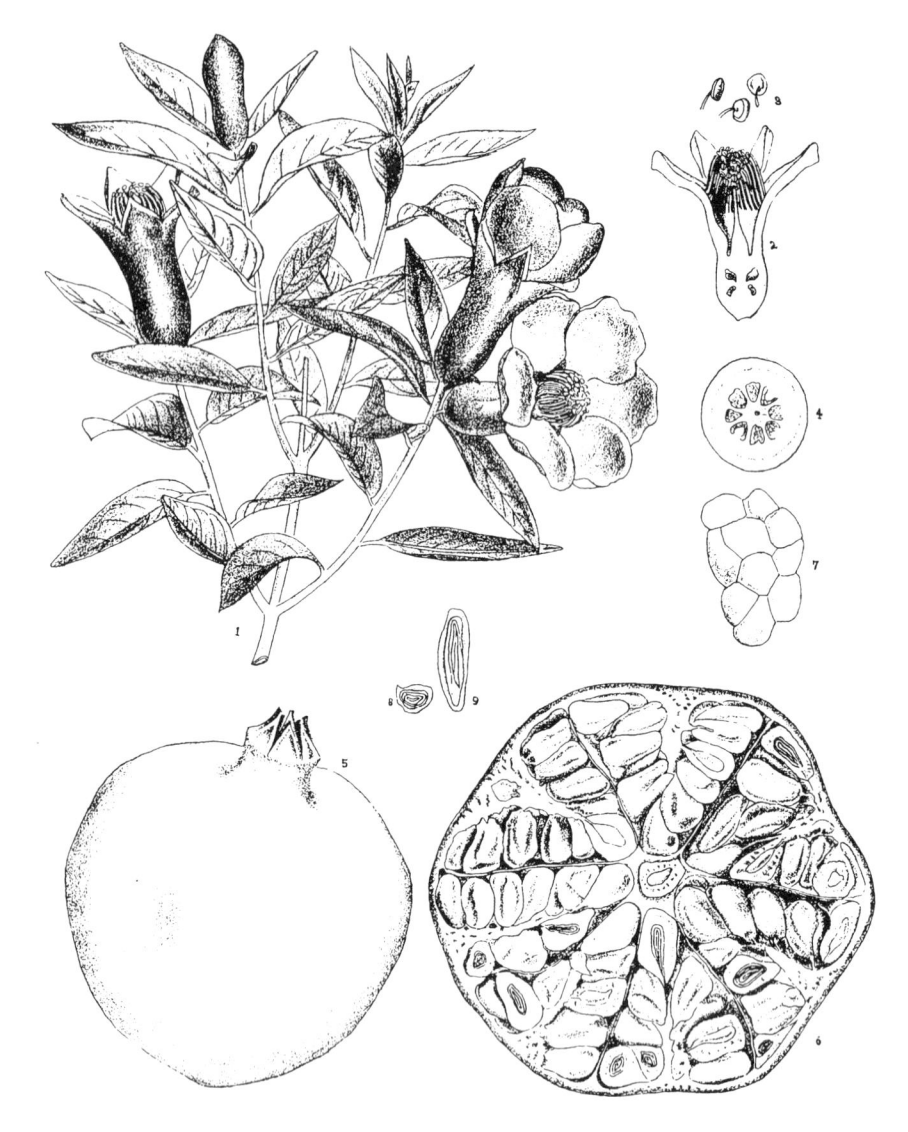

Family: PUNICACEAE

Vernacular names — Sanskrit, Dadima; Hindi, Anar; English, Pomegranate; Bengali, Dalim; Nepalese, Dhale; Sinhalese, Delum; Japanese, Zakuro; Chinese, An-shih-liu; German, Granatbaum; French, Grenadier; Unani, Anar shirin; Arabian, Rumman-ul-halw; Persian, Anar-e-shirin; Tibetan, Bal-po'1-se'u; Burmese, Salebin; Tamil, Mathul.

Habitat — Afghanistan and Baluchistan. This plant grows wild in warm valleys and outer hills of the Himalayas between 2500 and 6600 ft and is cultivated in many parts of India.

Parts used — Flowers, rind of the fruit, fresh fruit juice, dried bark of the stem, and root.

Pharmacognostical characteristics — A large deciduous shrub or a small tree, often armed. Leaves 1 to 2.5 in. long, minutely pellucid punctate, oblong or obovate, base narrowed into a short petiole. Flowers 1.5 to 2 in. long and as much across, bell shaped. Fruit 1.5 to 3 in. diameter, the interior septate with membranous walls and containing numerous seeds, angular from mutual pressure.[28]

According to Chaudhuri,[820] the stem bark of *P. granatum* is grayish brown with easily removable abraded patches of cork. The inner surface is light yellow or yellowish brown. Under microscopic examination, the phellem consists of thin-walled loosely arranged cells containing clustered and prismatic calcium oxalate crystals and starch grains. The phloem as usual consists of sieve tubes, companion cells, phloem parenchyma, medullary ray cells, and stone cells. Stone cells are larger in number and present in the cortical and in the phloem region. The medullary ray cells and phloem parenchyma are very prominent.

The root bark occurs in thin fragments, curved or channeled pieces 5 to 10 cm long and 1 to 3 mm wide. The outer surface of the bark is yellowish gray and the inner one is grayish yellow. Under the microscope, the phellem of the root bark has almost the same structure as that of stem bark. The phloem parenchyma contains tannins and alkaloids. The sieve tubes and companion cells are quite prominent in the secondary phloem. Stone cells are comparatively large in size.

Dey and Das[821] reported that under the microscope the leaf showed a thick cuticle, a single-layered palisade with large solitary crystals of calcium oxalate, compact spongy cells, and a distinct crystal layer surrounding the vascular bundle of the petiole. The epicarp of the fruit has thick cuticularized epidermis, patches of sclereids, specialized cells containing starch grains, and secretory canals.

Ayurvedic description — *Rasa* — *kasaya, madur, amola; Guna* — *lagu, snigdha; Veerya* — *ushana; Vipak* — *madur.*

Action and uses — *Tridosha ghan, dipan, grahi krimighan, hirdya, sonita sthapan.*

Chemical constituents — Bark and the rind of the fruit contain tannin (22 to 25%) and the root bark contains punico tannic acid, mannite, sugar, gum, pectin, an active liquid alkaloid, "pelletierine" and "isopelletierine", and two inactive alkaloids methyl-pelletieriene and pseudo-pelletieriene.[78]

Ponniah and Seshadri[822] reported the presence of a pigment pelargonidin 3,5-diglucoside in the flowers of *P. granatum*. Sharma and Seshadri[823] reported the presence of malvidine pentose glycoside in the seeds and a mixture of pentose glycosides (malvidin and pentunidin) in the rind. The petroleum ether and chloroform extracts of flowers yielded sitosterol and ursolic acid apart from maslinic acid, asiatic acid, and sitosterol-β-D-glucoside as the major components. The alcoholic extract gave D-mannitol, ellagic acid, and gallic acid.[824] The fruit rind yielded ellagic acid.[825] *P. granatum* has been reported to contain fluoride (0.2 to 0.3 ppm); and calcium (11.3), magnesium (3.6), phosphate (70.9), and vitamin C (3.8) content mg%.[486] Gopolan et al.[482] revealed the composition (per 100 g) as calcium (10.0%) phosphorus (70.0%) iron (0.30%) thiamin (0.06%) riboflavin (0.2%) niacin (0.30%) and vitamin C (16.0 mg).

Pharmacological action — The root and stem bark and their alkaloids are astringent, anthelmintic, and taenifuge. Flowers and rind of the fruit are also astringent and stomachic, juice of the fruit is cooling and refrigerant. Alkaloid "pelletierin" is anthelmintic and taenicide.

Trivedi and Kazmi[826] reported that the extract of flowers and rind showed good inhibitory action against most bacteria, the stem bark extract being least active in this respect. Janardhanan et al.[827] showed that acetone extract of *P. granatum* was fungitoxic to *Pyricularia oryzae* and *Colletotrichum falcatum*. The extracts of bark, fruit, pulp, flowers, and leaves completely inhibited the spore germination of fungi, viz., *Dreschlera rostrata* and *Curvularia lunata*.[828] The alcoholic extract showed a positive anthelmintic activity and so did chloroform extract of stem and root of *P. granatum*.[829]

Medicinal properties and uses — The juice of the fresh leaves and young fruits is given in dysentery. Paste of the leaves is locally used in conjunctivitis. The bark of stem and roots is a well-known anthelmintic; the root bark is preferred, as it contains a larger quantity of alkaloids than the stem bark, which is highly toxic to tapeworms. The dried flowers, known as "gulnar", are used in hematuria, hemorrhoids, hemoptysis, dysentery, chronic diarrhea, and bronchitis. The infusion or decoction of the rind of fruit is used in relaxation of the gums and throat, mucous discharge, prolapse of the rectum, uterus, and for leukorrhea.[1,3]

Doses — Fruit powder — 4 to 8 gr; flower powder — 4 to 5 gr; root and bark powder — 1.5 to 3 gr; bark decoction — 100 to 200 ml; bark decoction — 28 to 56 ml (child).

RANDIA DUMETORUM Lam. Family: RUBIACEAE

Vernacular names — Sanskrit, Madana; Hindi, Mainphal; English, Emetic nut; Bengali, Menphal; Nepalese, Menphal; Sinhalese, Kokuruman; Japanese, Harizakuro; Chinese, Gang tu hu; German, Chelafruchte; Unani, Mainphal; Arabian, Teuzal Qay; Persian, Tauzul kehu; Tibetan, Po-son-cha; Burmese, Thaminsa; Tamil, Karai.

Habitat — Found in the sub-Himalayan tract from the Punjab eastward, ascending in Sikkim up to an altitude of 4000 ft. Southward it extends to Chittagong in Bangladesh and peninsular India.[28]

Parts used — Rind of the fruits, bark, and seeds.

Pharmacognostical characteristics — A small tree or a rigid shrub; spines axillary stout, straight, horizontal. Leaves 1 to 2 in. long, obovate, obtuse or subacute, glabrous or pubescent. Flowers solitary, rarely 2 to 3 on peduncle, subsessile, greenish yellow or white, fragrant. Calyx tube strigose, corolla $1/2$ to $3/4$ in. diameter, hairy outside. Berry $3/4$ to $1 1/2$ in. long, globose or ovoid, glabrous or pubescent, smooth or obscurely ribbed, yellow, crowned with the persistent calyx limb; pericarp thick, contains four lobed light-black seeds.

Ayurvedic description — *Rasa — madur, tikta, kasaya, katu; Guna — laghu, rooksha; Veerya — ushna; Vipak — katu.*

Action and uses — *Kaph vat sodhak, pitta nisark, bamak, sothahar, badana nasak, krimighan, rakt sodk, lakhan.*

Chemical constituents — The fruit contains saponin in the pericarp, a glucosidic saponin in the pulp, and seeds are said to contain traces of alkaloids.[26] An essential oil is also present.[1] From the powdered root of the plant scopoletin was isolated. Dried fruits yielded a microcrystalline saponin named urso saponin as well as β-sitosterol and a new triterpine. The drug has a high hemolytic index.[178]

Pharmacological action — Rind of the fruit is diaphoretic and antispasmodic. Bark is sedative and nervine calmative.

Medicinal properties and uses — Bark is given internally and applied externally as anodyne in rheumatism and to relieve pain of bruises and bone aches during fevers. It also acts as an astringent and is useful in diarrhea and dysentery. Fruit is irritating, emetic. It is useful in cases of acute bronchitis and asthma. The drug is also used in scorpion sting.

Doses — 3 to 5 gr (for emesis), 1 to 30 gr (for other purposes); bark decoction — 56 to 112 ml.

RAPHANUS SATIVUS Linn. *Family:* CRUCIFERAE

Vernacular names — Sanskrit, Moolaka; Hindi, Muli; English, Radish; Bengali, Moola; Nepalese, Laiey-Mula; Sinhalese, Rabu; Japanese, Daikon; Chinese, Lai fu; German, Riibenrettig; Unani, Muli; Arabian, Phujul; Persian, Turb; Burmese Moula; French, Raifort cultivé; Tamil, Mullangi.

Habitat — Cultivated throughout India up to 10,000 ft.

Parts used — Seeds, root, and leaves.

Morphological characteristics — An annually cultivated plant, leaves very deeply lobed, end lobe very broad, leaf stalks arising directly from the thick taproot, spreading in rosette. Taproot long, coarse, thick, white. Inflorescence a terminal raceme; stalk short at first, much elongated when in fruit; flowers white or lilac, with purple veins. Pods indehiscent, $1/2$ in. long, columnar, tapering at the apex to a pointed beak, constricted between the seeds.

Ayurvedic description — *Rasa — katu; Guna — laghu, guru, teekshna; Veerya — ushna; Vipak — katu.*

Action and uses — *Tridosha har, rochan, dipan, pachan, anuloman, bhadan, kasha har, mutral, asmari bhadan, jwarghana, artawjanan.*

Chemical constituents — Fresh vegetable contains 79 to 91.0% moisture; and the completely dried material contains ether extract (4.0%), albuminoids (18.0%) (containing nitrogen, 2.88%), soluble carbohydrates (52.66%), woody fibers (9.34%), and ash (16.0%) (containing sand, 0.33%). Seeds and root contain a fixed oil, essential oil, a sulfuretted volatile oil, containing sulfur and phosphoric acid. The root is reported to contain arsenic, 0.01 mg in 100 g. It also contains glucoside, enzyme, and methyl mercaptan; recently, cyanin and pelargonin have been isolated from violet-red and yellowish-red varieties of radish. The pigment in the purple-colored edible part has been identified as malvin chloride.[178]

Pharmacological action — Seeds and leaves are diuretic, laxative, and lithnotriptic. Juice is antiscorbutic.

Medicinal properties and uses — Leaf juice is prescribed in dysuria, stranguary, calculi, and as a stimulant. The taproot is prescribed in urinary and syphilitic diseases, piles, stranguary, and stomachache; it is also strongly antiscorbutic. Root in the form of syrup is given in hoarseness, whooping cough, bronchial disorders, and other chest complaints. The seeds are peptic, expectorant, diuretic, laxative, lithontriptic, carminative, emmenagogue and stimulant.[1,3,50]

Doses — Leaf juice — 2.5 to 5 ml; taproot juice — 5 to 10 ml; root powder — 1 to 2 gr; root decoction — 28 to 56 ml.

RAUWOLFIA SERPENTINA (L.) Benth. ex Kurz. *Family:* APOCYNACEAE

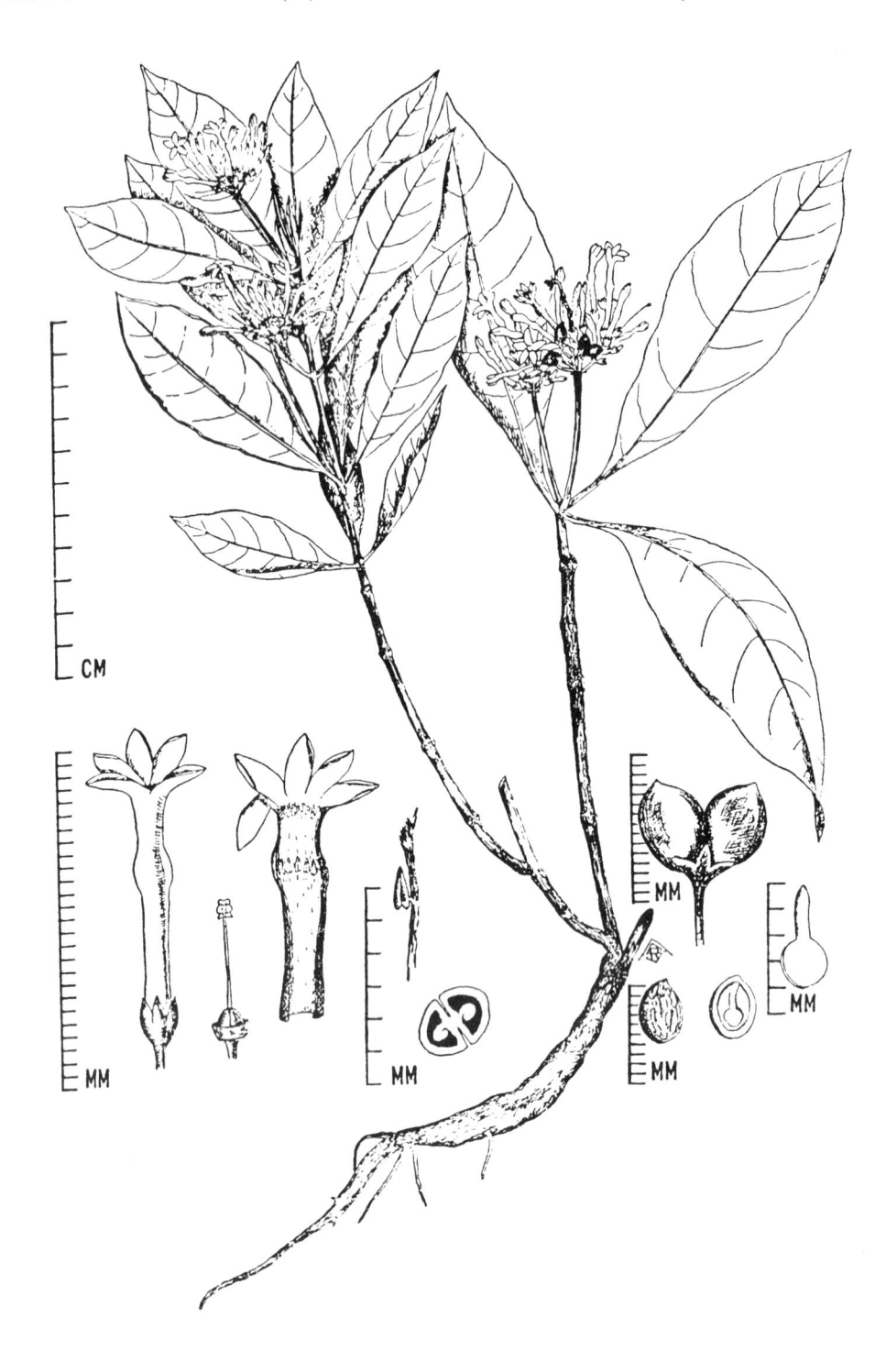

Vernacular names — Sanskrit, Sarpagandha; Hindi, Chota chand; English, Serpentina; Bengali, Chandra; Nepalese, Sharpagandha; Sinhalese, Ekaweriya; Japanese, Indojaboku; German, Rauwalfia; Unani, Asrol; Burmese, Bongmaiza.

Habitat — A small herbaceous plant found in the sub-Himalayan tracts from Sirhand eastward to Assam, especially in Dehra Dun, Siwalk range, Rohilkhand, Gorkhpur, ascending to 4000 ft. Also in Konkon, Kanara and Maharashtra, Eastern and Western Ghat of South India, Bihar, and Bengal.

Parts used — Roots.

Pharmacognostical characteristics — An erect, glabrous shrub 1 to 3 ft. high, leaves whorled, 3 to 7 in. long, lanceolate or oblanceolate, acute or acuminate tapering gradually into the petiole, thin. Flowers white or pinkish, peduncle 2 to 5 in. long, pedicels and calyx red. Calyx lobes $1/_{10}$ in. long, lanceolate. Corolla about $1/_2$ in. long, tube slender, inflated a little above the middle. Lobes much shorter than the tube, obtuse. Drupes $1/_4$ in. diameter, single or didymous and more or less connate, purplish black when ripe.[28]

R. serpentina root shows the following microscopical characteristics: (1) a stratified cork of from usually two to eight alternating tangential zones of smaller and radially narrower and larger and radially broader phellem cells, the former consisting of two to five layers of cells with suberized walls, the latter of one to five or more layers of cells with lignified walls; (2) a phellogen consisting of a layer of tangentially elongated rectangular meristematic cells; (3) a secondary cortex (phelloderm) of tangentially elongated to rounded cortical parenchyma cells, mostly containing starch. Resin cells with brownish oleoresin contents occur interspersed through this region and are most abundant in the Dehra Dun variety of R. serpentina. They have been stated by some authors to represent short latex cells; (4) secondary phloem of starch- and crystal-bearing phloem parenchyma and sieve tissue traversed by phloem rays which curve outward; (5) cambium composed of a line of thin-walled cells; (6) secondary xylem consisting of numerous, radially elongated wood wedges separated by xylem rays, the latter from 1 to 8, occasionally up to about 14 layers of cells in width, as observed in tangential longitudinal sections. The wood wedges contain numerous vessels, tracheids, wood fibers, and wood parenchyma cells. The vessels are cylindrical and from about 180 to 432 μm in length and up to about 57 μm in diameter. They possess lignified walls with mostly simple pits, but those adjacent to the xylem rays have bordered pits. Their end walls are transverse to oblique and marked by openings surrounded by prominent perforation rims. However, a number of them also occur with closed end walls and lateral perforation plates. Some of them may show tyloses and gummy lignin deposits in their cavities. The tracheids are of the pitted type with moderately thick lignified beaded walls and broad lumina. The wood parenchyma cells possess moderately thick pitted walls, and are usually loaded with starch. The wood fibers possess thick, greatly lignified walls with transverse to oblique pits and pointed to bifurcate ends. They range from about 200 to 750 μm in length; (7) a tetrarch primary xylem in the center. The rhizome exhibits a cork somewhat similar to that of the root, a phellogen of rectangular cells, a phelloderm of up to about six layers of parenchyma which contain scattered latex cells with resinous contents, a narrow pericycle which contains pericyclic fibers arranged singly or in small groups, a broad circle of bicollateral bundles separated by vascular rays, and a small central pith. The pericyclic fibers possess thick nonlignified walls, tapering and frequently lobed ends, and subterminal enlargements. The parenchymatous cells of cortex, phloem, pericycle, xylem, and pith contain starch grains. The wood wedges of the xylem are composed chiefly of starch-bearing parenchyma through which course radially arranged pitted vessels, tracheids, and fibers. The vessel elements measure up to 485 μm in length and are mostly up to 45 μm in width. The wood fibers are up to about 750 μm in length and up to about 32 μm in width and they are generally less wavy than those of the root. Beneath the xylem and embedded in the outer region of the pith are a variable number of internal phloem strands. The pith is composed largely of starch-bearing parenchyma through which are interspersed short latex cells with yellowish contents which stain brown with iodine test solution. The phloem of both rhizome and root shows scattered crystal cells containing tabular to small angular calcium oxalate crystals from 3 to 20 μm in diameter.[830]

Powdered *R. serpentina* — This is pale yellow to pale yellowish brown in color and possesses an indistinct odor and a bitter taste. Under the microscope, it exhibits numerous single and usually two to three compound, occasionally four-compound, starch grains, the individual grains being spheroidal, ovate, muller-shaped, and plano- to angular-convex with a Y-shaped, stellate or irregularly cleft hilum, the unaltered grains usually up to 34 μm in diameter, the altered grains up to 50 μm in diameter; less numerous crystals of calcium oxalate of the monoclinic type up to about 20 μm in diameter; numerous fragments of cork tissue, its cells of polygonal and rectangular outline and with either lignified or suberized walls; numerous fragments of pitted tracheids and fewer vessels with lignified walls often associated with thick-walled wood fibers and crossed by xylem ray cells; numerous fragments of starch-bearing parenchyma; a number of resin masses of yellowish to brown color and latex cells. If either rhizome or aerial stem is present in the powder, a number of narrow, colorless, nonlignified pericyclic fibers, single and in small groups, will also be found.[831]

Ayurvedic description — *Rasa—tikta; Guna—rooksha; Veerya—ushna; Vipak—katu.*

Action and uses — *Kapha vat samak, nidra kar, badna nasak, hirdya awsadak, rakt-wahini prasark, rakt varnasak, kamawsadak.*

Chemical constituents — The following alkaloids occur in the root of *R. serpentina*: reserpine, rescinnamine, reserpinine, serpentine, serpentinine, sarpagine, raubasine, yohimbine, ajmaline, ajmalinine, ajmalicine, isoajmaline, neoajmaline, rauwolfinine, rauhimbine, thebaine, and papaverine. The drug also contains an oleoresin, a sterol starch, and calcium oxalate.[831]

The yield of total alkaloids in *R. serpentina* ranges from about 0.8 to 1.3%.[832]

Pharmacological action — Root is bitter tonic and possesses well-marked sedative properties. It acts also as febrifuge.

The ajamaline group acts as a general depressant to the heart, respiration, and nerves, and the serpentine group paralyzes respiration and depresses the nervous system, but stimulates the heart.

Root is hypnotic, sedative, specific for insanity, reduces blood pressure, remedy in painful affections of bowels, etc. Reserpine is a CNS depressant. Sedative, hypotensive with bradycardia. Used as tranquilizer. Deserpidine modifies the hypophysiogenital system of the rat. In the female, it suppresses vaginal keralinization, induces permanent estrus, and stimulates the mammary gland; in the male it dissociates the germinal and endocrine functions of testicles. Reserpinine has hypotensive, sedative, and bradycardiac activity similar to that of reserpine. Sarpajmaline an alkaloid complex, is a much more potent hypotensive than reserpine without the sedative and CNS depressant action of the latter.

Ajmalicine is a CNS depressant and adrenergic blocking agent.

Serpentine is the major yellow anhydronium base of *R. serpentina;* it is twice as hypotensive as ajmaline, and shows synergistic activity when administered with resperine.[178]

Medicinal properties and uses — Decoction of the root is employed to increase uterine contraction and promote expulsion of the fetus. Root is a valuable remedy in dysentery, painful affection of the bowels, and recently it has attained prominence as a remedy for insomnia, hypochondria, and irritative condition of the central nervous system. Powdered root in doses of 20 to 30 gr given twice daily produces not only sedative effects, but also a reduction of the blood pressure.[1,50]

The hypnotic action of the drug appears to have been known to the poorer classes in Bihar and the practice of putting children to sleep by this drug is stated to be still present in certain parts of that province. In Uttar Pradesh and Bihar the drug is sold as *pagal-ka-dawa* (insanity specific) and its use is common among practioners of indigenous medicine.[50,178]

Doses — Powder — 20 to 30 gr (for lowering the blood pressure), 1 to 3 gr (as hypnotic), 1 to 3 gr (in insanity).

RHEUM EMODI Wall. *Family:* POLYGONACEAE

Vernacular names — Sanskrit, Amlavetasa; Hindi, Ravand chini; English, Indian rhubarb; Bengali, Rheuchini; Nepalese, Amalwed; Sinhalese, Gorak; Japanese, Chosendaio; Chinese, Yunn-anta-huang; German, Rhabarber; Unani, Revand chini; Arabian, Rewand; Persian, Bekh-e-riwas; Tibetan, Star-bu; French, Rhubrabde; Tamil, Gr. Valchini.

Habitat — Subalpine and alpine Himalayas, viz., Himachal, Pradesh, Kashmir, and Nepal, Sikim at an altitude of 7,000 to 12,000 ft.

Parts used — Root.

Morphological and pharmacognostical characteristics — Stout herb with woody large roots, cylindrical, barrel shaped, conical, mostly 2 to 20 cm in length and 1.5 to 8 cm in diameter. Outer surface irregularly longitudinally wrinkled, covered with brownish cortex.

Transection of the root shows a brown bark composed of 10 to 12 layers of cells. Cortex is composed of a few layers of cells, mostly irregular, rounded, thin walled, and parenchymatous. The cortical cells contain starch grains, tannins, and large clusters of crystals of calcium oxalate. The cambium is much compressed. Ray cells are very prominent and linear rectangular in shape and generally consist of one to two layers of cells. Vascular bundles from one- to two- layer chains in the radial direction and the central cylinder of wood are found to consist of many radiating rays formed in this way. The rest of the wood is composed of tracheids and xylem parenchyma.[833,834]

Ayurvedic description — *Rasa — amala; Guna — laghu, rooksha, tikshna; Veerya — ushna; Vipak — amla.*

Action and uses — *Kapha vat samak, pittawardhak dipan, pachan, anuloman, bedhan, hridotajek swas har, hicaahar.*

Chemical constituents — Root contains a large amount of chrysophanic acid, emodin, a glucoside rhaponticin, a tannin known as rheo-tannic acid, resins, mucilage, tannin and gallic acid, sugar, starch, pectin, lignin, calcium oxalate, and various inorganic salts. Leaves contain oxalic acid. Roots and rhizomes contain certain anthraquinone derivatives as their chief constituents.[1] Roots contain rhein, emodin. Rhizomes are reported to yield 0.05 essential oil containing eugenol, a terpene alcohol, and methylheptyl ketone.[50] Fresh rhubarb roots yielded hetrodianthrones sennidin C, reidin B, and reidin C.[178]

Pharmacological action — Rhubarb is stomachic, bitter, tonic, cathartic, purgative, and is also considered emmenagogue and diuretic.

Medicinal properties and uses — The tuber is very useful in biliousness, lumbago, heating of the brain, sore eyes, piles, chronic bronchitis, chronic fever, asthma, and coryza. Rhubarb forms an important ingredient of a large variety of compounds.[3] It is given for irritation of the bowels common among children when teething, and in chronic dysentery, duodenal cattarrh of the biliary ducts, with jaundice, and in certain skin diseases.

Doses — Root powder — 5 to 20 gr.

RHUS SUCCEDANEA Linn.

Family: ANACARDIACEAE

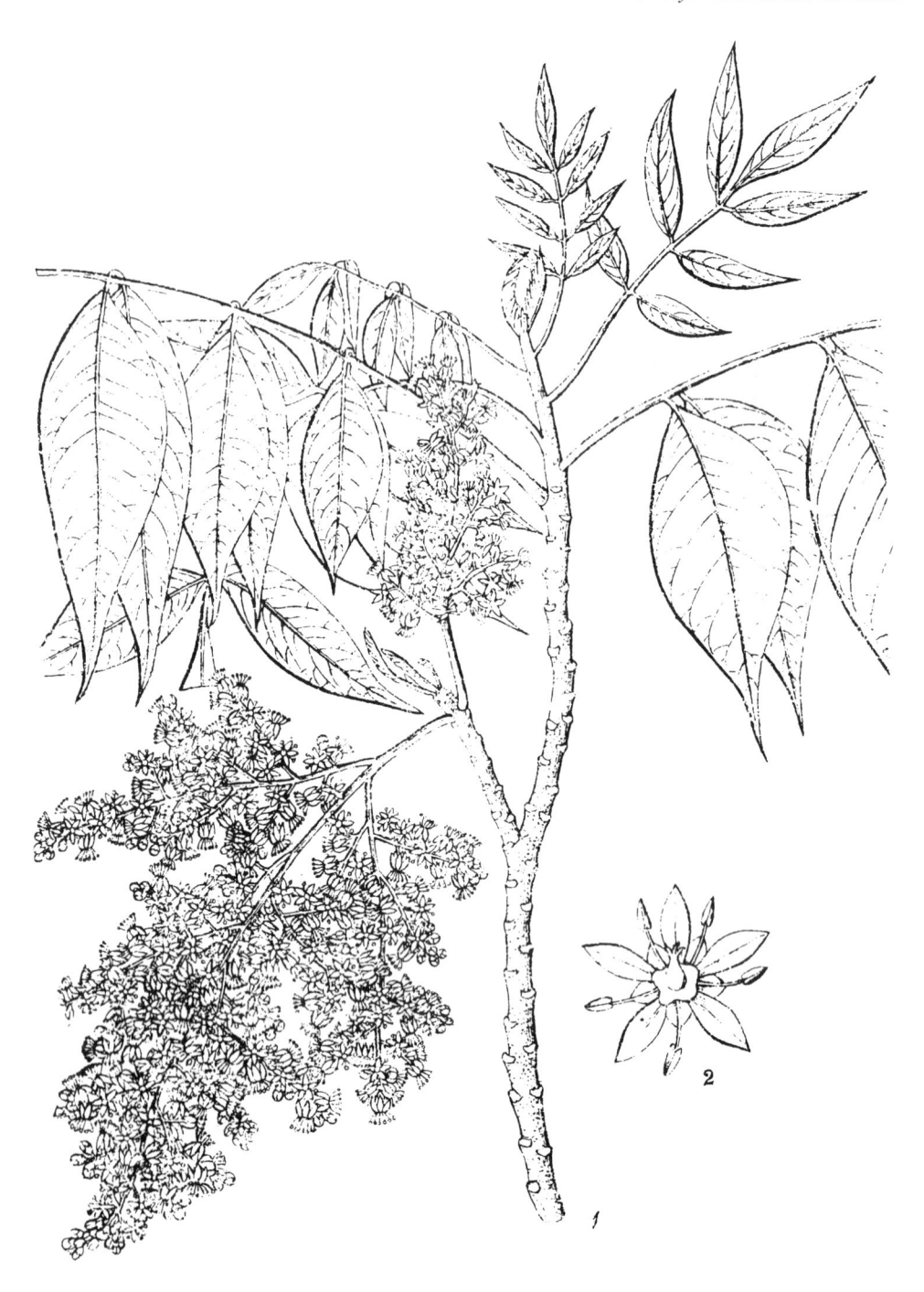

Vernacular names — Sanskrit, Karkatashringi; Hindi, Kakra Singi; English, Galls; Bengali, Kakrasringi; Nepalese, Karkatasharngi; Sinhalese, Karkatashringi; Japanese, Hazenoki; Chinese, Lu; German, Sumach; Unani, Kakrasinghi; Tamil, Karkoda gasinghi.

Habitat — Found in the temperate Himalayas from Kashmir to Sikkim and Bhutan at altitudes of 3000 to 8000 ft; also in Khasia Hills between 2000 and 6000 ft.

Parts used — Galls.

Pharmacognostical characteristics — A medium-sized deciduous tree. Leaves pinnate, 12 to 24 in. long, rachis glabrous, acuminate, entire, glabrous, oblique; petiole $1/5$ to $2/5$ in. Flower green-yellow; $1/7$ in. across, in slender, drooping, axillary panicles half as long as the leaves; pedicels $1/10$ to $1/5$ in. long. Drupes nearly $1/3$ in. diameter, compressed, glabrous, shining, epicarp dehiscing irregularly; mesocarp fibrous, waxy.[28]

Ayurvedic description — *Rasa — kasaya, tikata; Guna — lagu, rooksha; Vipak — katu; Veerya — ushana.*

Action and uses — *Kaf bat samak, soth har, rakt rodhak, dipan, bata anuloman, grahi kaf nisark, hicca nasak* and used in the inflammatory condition of uterus with profuse discharges.

Chemical constituents — Fruits yield Japan wax. Leaves contain 20% of tannin. Milky juice yields laccol which is identical with urushiol and a toxic phenol.[1,50] Leaves give a new glycoside apigenin.[178]

Pharmacological action — Astringent, tonic, expectorant, and stimulant. Gall is also a cholagogue.

Medicinal properties and uses — Galls are useful in cough, phthisis, asthma, fever, want of appetite, irritability of the stomach, and conditions of the respiratory tract. It is much used in combination with other astringents in diarrhea, since the drug by itself contains a large amount of tannin. This is also useful in infantile diarrhea and gastrointestinal troubles during teething. Externally a paste of the galls is recommended as application in psoriasis. In the form of decoction or lotion, the galls are used as gargle to suppress hemmorrhage from gum; also used to suppress bleeding from the nose, discharges from mucous membranes such as gleet and leukorrhea. Galls are also used as an antidote to snake venom and scorpion sting.[1]

Doses — Gall powder — 0.5 to 1.5 gr.

RICINUS COMMUNIS Linn. *Family:* EUPHORBIACEAE

Vernacular names — Sanskrit, Eranda; Hindi, Endi; English, Castor oil plant; Bengali, Aranda; Nepalese, Alama; Sinhalese, Endaru; Japanese, Togoma; Chinese, Peima; German, Rhizinus; French, Ricin; Unani, Arand; Arabian, Kharva; Persian, Bed-e-ynjeer; Tibetan, E-ran; Burmese, Kesusi; Malaysian, Jarak; Pustu, Arhand; Tamil, Amanakk.

Habitat — Originally, probably from Africa, it is extensively cultivated for its oil-bearing seeds and has also become naturalized near habitation in many parts of India.[28]

Parts used — Oil, leaves, seeds, and root.

Pharmacognostical characteristics — A tall, stout, glabrous and glaucous annual or perennial and subarboreous. Leaves alternate, large, palmately lobed, lobes 7 or more, serrate. Flowers monoecious, in terminal subpaniculate racemes, the upper male, crowded, the lower female. Calyx in male membranous, splitting into $3/5$ valvate segments; in female spathaceous, caducous. Petals 0. Stamens many, connate in several branched columns, anther clusters on the final branches. Ovary 3-celled, style entire, bifid or two-partite; ovules solitary in each cell. Fruit a prickly capsule of three 2-valved cocci. Seeds oblong with large caruncle and crustaceous testa; albumen fleshy. The fruits when ripe dehisce explosively and scatter the seed.

A transverse section of the root 12 mm in diameter is circular and the outline regular except for an occasional lenticel opening. There is a very broad cream-colored woody region in the center having a diameter of 8.5 mm surrounded by bark. Cork in the outer tissue is thin and composed of 6 to 12 rows of rectangular, elongated, thin-walled cells. Phellogen composed of a single row of narrow thin-walled, tangentially elongated cells. A phelloderm formed of three rows of cells is present next within. The cortex is comparatively narrow. Some of these cells contain large oil globules of a light yellow color. Most of the cells contain rounded starch grains. A few cortical cells contain rosette crystals of calcium oxalate. Phloem is well developed and is seen as wedge-shaped strips with their apices projecting into the cortex and alternating with medullary rays. The thin walled elements of phloem consist of sieve tubes with their companion cells and a phloem parenchyma. Sieve tubes are wider and more prominent at the inner region of the phloem.

A cambium zone is present just within the phloem consisting of three to five rows of rectangular, thin-walled cells. The wood forms the major parts of the root. There are so many vessels uniformly distributed throughout the wood. They occur in small groups of two to three or four that are arranged in radial rows. Solitary vessels are also common. The wood parenchyma cells are much less in number when compared with the wood fibers and are mostly seen adjacent to the vessels. The wood fibers form a considerable part of the wood. They are long and thick walled.[835,836]

Ayurvedic description — *Rasa — madhur, katu, kasaya; Guna — guru, snigdha, teeshana, suskshama; Vipak — maduar; Veerya — ushana.*

Action and uses — *Bat samak, sansharan, balya, badana samak, sothahar, asthana dosha har, mutra bisodhan, kriminisarak.*

Chemical constituents — Seeds contain fixed oil (45 to 52%) soluble in alcohol, proteids (20%), starch, mucilage, sugar, and ash (10%). The oil chiefly consists of glycerides of ricinoleic, isoricinoleic, stearic, and dihydroxystearic acids. The purgative action of the oil is said to be due to free ricinoleic and its stereoisomer, which are produced by hydrolysis in the duodenum. Apart from the oil which is contained in the kernels, a very toxic substance "ricin", an albuminoid poisonous body, is present in the seed, but not present in the oil to any extent.

Castor oil congeals to a gel mass when the alcoholic solution is distilled in the presence of sodium salts of higher fatty acids. This gel is useful in dermatosis and is a good protective in occupational eczemas and dermatitis.

After expression of oil, the residue contains an extremely poisonous toxin known as ricin which makes it unfit for cattle feed. The ricin is reported to have antitumor properties. The seeds also contain liposes and a crystalline alkaloid ricinine which is not so toxic; its structure is related to nicotinamide.[69]

Pharmacological action — Oil is nonirritant purgative; when it reaches the duodenum it is decomposed by the pancreatic juice in ricinoleic acid which irritates the bowels, stimulates the intestinal glands and the muscular coat, and causes purgation.

Medicinal properties and uses — The leaf juice is given as an emetic in narcotic poisoning; a decoction of the leaves is a purgative, lactagogue, and emmenagogue. A poultice of the leaves is applied to boils and swellings; coated with some bland oil the hot leaves are applied over the abdomen of children to relieve flatulence; over the pubic region of women to promote the menstrual flow; and over the breasts of nursing mothers as a lactagogue.

The decoction of the roots with the addition of potassium carbonate is very useful in lumbago, rheumatism, and sciatica. The root bark is a strong purgative. In intestinal colic of children, oil is given in small doses repeatedly. Oil is usually given in constipation during pregnancy and before and after childbirth to the mother. The oil is locally applied in conjunctivitis. Castor oil is much praised for its efficacy in chronic articular rheumatism in which it is used in various combinations.

Doses — Oil — 14 to 28 ml (adults — 12 to 28 ml, children — 5 ml); leaf paste — 9 to 15 gr; root paste — 3 to 6 gr; seeds — 2 to 6.

RUBIA CORDIFOLIA Linn.   Family: RUBIACEAE

Vernacular names — Sanskrit, Manjista; Hindi, Manjit; English, Indian madder; Bengali, Manjit; Nepalese, Manu; Sinhalese, Velmadata; Japanese, Akane; Chinese, Ch'ien-ts'oa; German, Farberwurzel; French, Garance; Unani, Manjit; Tibetan, Brtsod; Tamil, Manji.

Habitat — Climbing plant growing in the northwestern Himalayas, Nilgiris, and other hilly districts of India.

Parts used — Roots.

Pharmacognostical characteristics — A perennial climber. Leaves heart shaped, 2 to 4 in. long. Flowers small, yellow, scaly. Fruits $1/_5$ in. long, round, violet color; seeds light black.

Ayurvedic description — *Rasa — kasaya, tikta, madhura; Guna — guru, rooksha; Veerya — ushna; Vipak — katu.*

Action and uses — *Kapha pita samak, sothahar, bran ropan, khutaghana, madak, dipan, pachan, sthamvan* and *krimighan, sonit sthapan, garvasaya utajak, artavjanan, sthanya sodhan, pramehghana jawarghana, balaya, rasayan.*

Chemical constituents — Roots of *R. cordifolia* contain resinous and extractive matter, gum, sugar, coloring matter, and salts of lime. Coloring matter consists of a red crystalline principle, purpurin; a yellow principle glucoside, manjistin, garancin, alizarin (orange-red in color); and xanthine (yellow in color).[37,50]

Pharmacological action — Dried root is emmenagogue, astringent, and diuretic.

Medicinal properties and uses — The root is astringent, alterative, deobstruent, and tonic. It is given either as a decoction or infusion in paralysis, jaundice, obstruction in urinary passages, menstrual disorders, inflammatory condition of the chest, and scanty lochial discharge after childbirth. Paste made of the root with honey is applied over swelling, inflammations, skin diseases such as ulcers, leukoderma, freckles, and discoloration of the skin. Stems used in cobra bite and scorpion sting.[1] In tuberculous and intestinal ulceration, it acted as an anodyne.

Doses — Decoction — 56 to 112 ml; powder — 1 to 3 gr.

RUTA GRAVEOLENS Linn.   *Family:* RUTACEAE

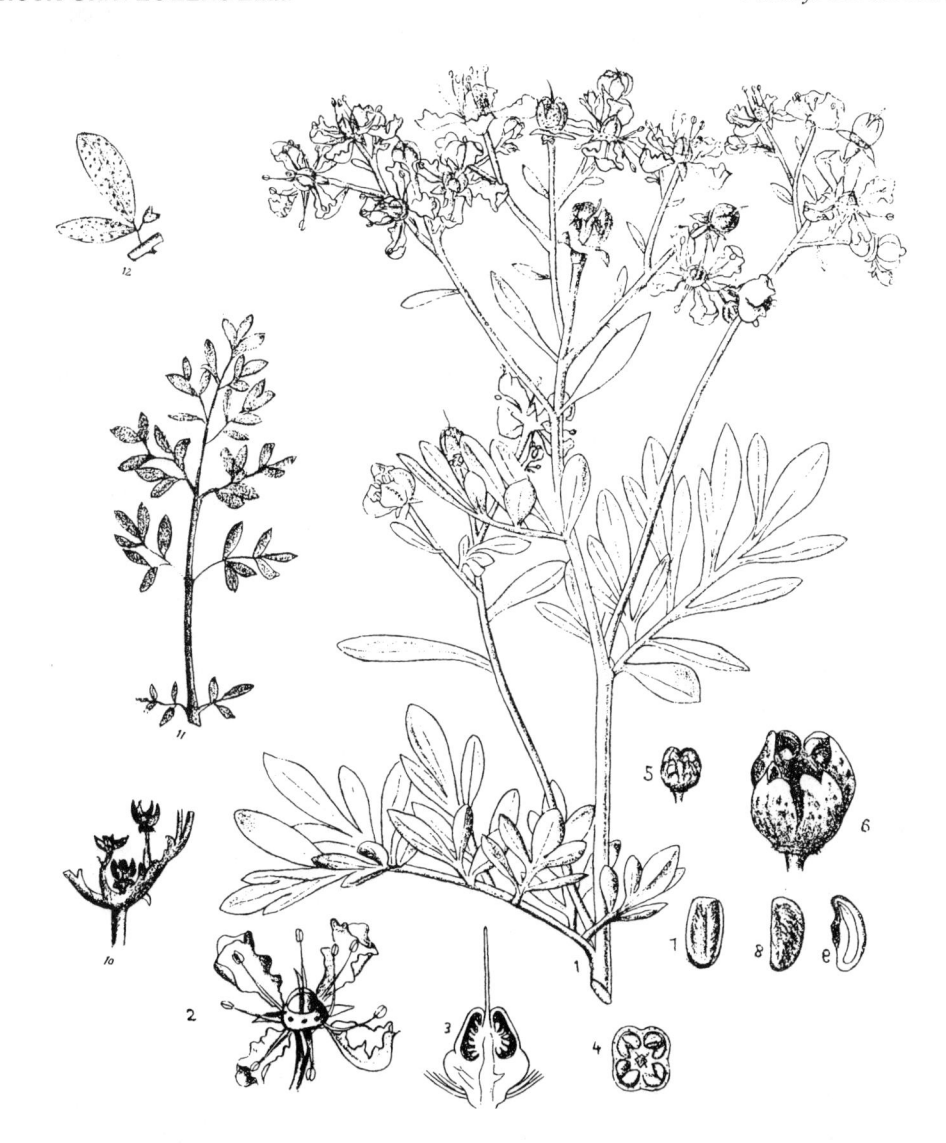

Vernacular names — Sanskrit, Sadapaha; Hindi, Sadap; English, Garden rue; Bengali, Ispand; Sinhalese, Aruda; Japanese, Matskareso; Chinese, T'sao; German, Raute; Unani, Sudab; Arabian, Sudab; Persian, Sudab; Tamil, Aruvada.

Habitat — Commonly cultivated in gardens throughout India for medicinal properties of its leaves and seeds.

Parts used — Whole plant, leaves, and oil.

Morphological characteristics — A small branching undershrub 2 to 3 ft high. Leaves petioled, decompound; segments various: flowers in divaricately spreading corymbs; bracts lanceolate; sepals triangular acute; petals oblong-obovate, pectinate, abruptly clawed. Capsule obtuse, shortly pedicelled.[28]

Ayurvedic description — *Rasa* — *tikta; Guna* — *laghu, rooksha, teekshna; Veerya* — *ushna; Vipak* — *katu.*

Action and uses — *Kapha vat samak, badana sthapan, naribalya, dipan, anuloman, krimighan, artaw janan, mutral, jwaraghana.*

Chemical constituents — From the fresh herb, a volatile oil (oil of rue) is obtained in 0.06% yield by steam distillation. It consists mainly of ketone, methyl-*n*-nonyl ketone, with about 5% of methyl-*n*-heptyl-ketone. Strong disagreeable odor and an acrid and nauseous taste. A glucoside rutin and a coumarin-like odoriferous principle have also been isolated from the plant.[837]

Pharmacological action — Antiseptic, stimulant, emmenagogue, and abortifacient. The glucoside rutin is reported to restore capillary fragility to normal, thus preventing capillary hemorrhage; and therefore it is employed in cases of hypertension and radiation injury.[37,69]

Medicinal properties and uses — Juice of the plant is stimulant, expectorant, and antispasmodic, also anthelmintic to children; in large doses a narcotic poison. Internally, it is useful in hysteria and in flatulent colic. Fresh leaves bruised and mixed with brandy are used as an external application in the first stage of paralysis. Powdered leaves combined with some aromatics are given in dyspepsia. Volatile oil is a valuable resolvent, diuretic, and emmenagogue. It is found to be a powerful aphrodisiac and is abortifacient to pregnant women. Externally, it acts as a rubefacient and also as a vermicide.[3,50]

Doses — Leaf juice — 2 ml; leaf powder — 0.5 to 1.5 gr; oil — 1 to 5 minims.

SALVADORA PERSICA Linn.   *Family:* SALVADORACEAE

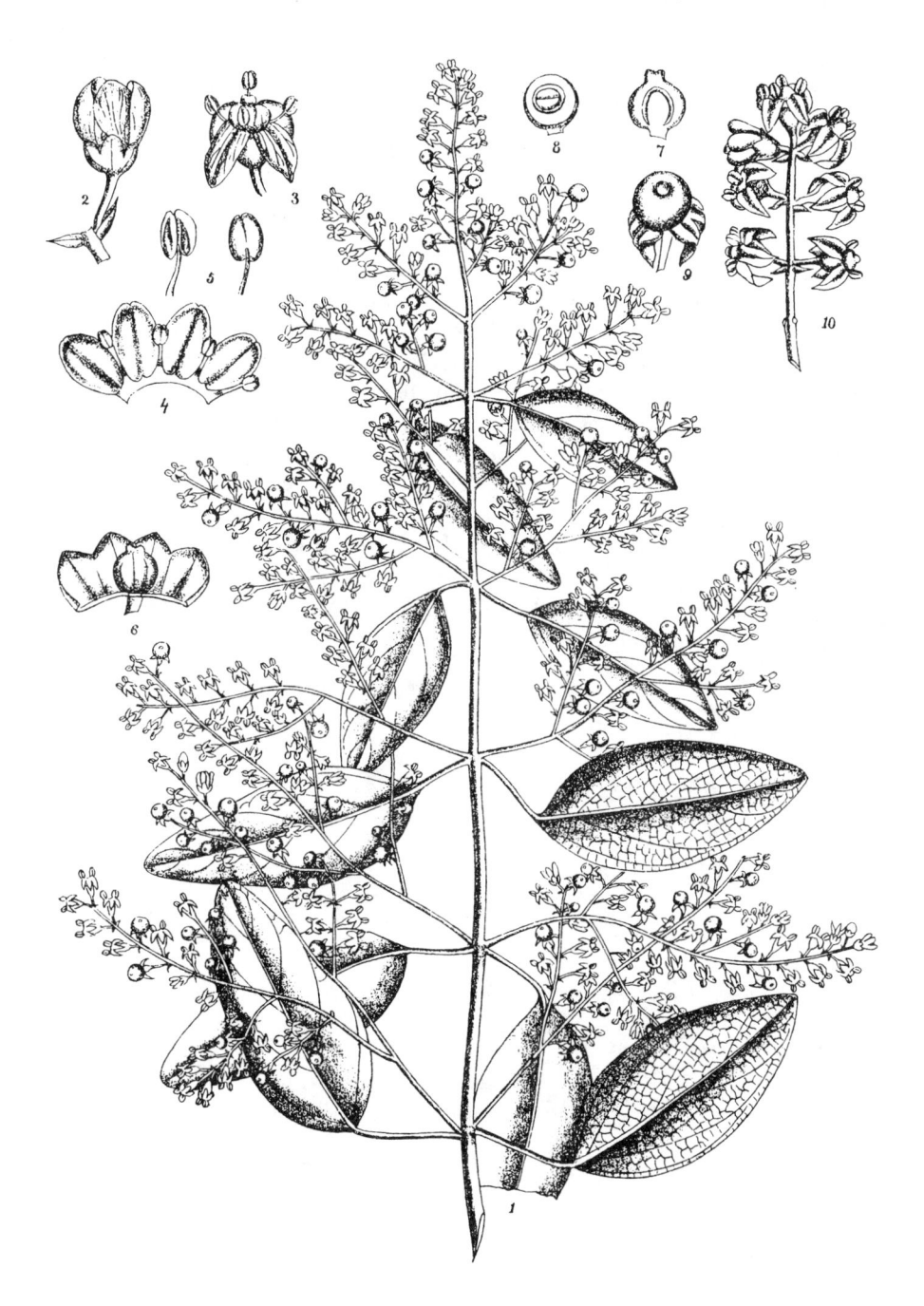

Vernacular names — Sanskrit, Pilu; Hindi, Chota pilu; English, Tooth brush tree; Bengali, Chota pilu; Nepalese, Data okhar; German, Persische; Unani, Pilu; Arab, Erak; Persian, Darkhat-e-misbak; French, Salvadore de persa; Pustu, Plewan; Tamil, Uka.

Habitat — Found in the drier parts of India, especially on saline lands and often on the black cotton soil, viz., Rajasthan, Punjab, Bihar, Konkan, Deccan, and Karnatic; often planted near graveyards.

Parts used — Root, bark, leaves, flowers, and fruits.

Morphological characteristics — An evergreen shrub or small tree with usually a short and crooked trunk, branches drooping, leaves 1 to 2 in. × $7/_{10}$ to $1^1/_3$ in., elliptic-lanceolate or ovate, obtuse, often mucronate, base cuneate or rounded. Flowers greenish yellow, in axillary and terminal lax, panicles 2 to 5 in. long, numerous toward the tips of branches, pedicels $1/_{16}$ to $1/_8$ in. long. Corolla twice as long as the calyx. Drupe globose, $1/_8$ in. diameter. Red when ripe, one seeded.[28]

Ayurvedic description — *Rasa* — *katu, tikta, madhur; Guna* — *laghu, snigdha, tikshna; Veerya* — *ushna; Vipak* — *katu.*

Action and uses — *Kapha vat samak, tridosha har, sothahar, dant sodhak, rakt sodhak, rakt pitta prasaman, sirovirachan, mutral, jawarghana.*

Chemical constituents — Root bark contains resin, coloring matter, traces of an alkaloid, trimethylamine, and ash containing large amounts of chlorine.[50] Fruits contain sugar, fat coloring matter, and the alkaloid. Seeds contain a white fat and yellow coloring matter. Oil cake from the seeds contains nitrogen (4.8%), potash (2.8%), and phosphoric anhydride (1.05%).

Pharmacological action — Carminative, diuretic, and deobstruent.

Medicinal properties and uses — Leaves are used as an external application in rheumatism and their juice is given in scurvy. A poultice of the leaves is a useful application to painful tumors and piles. In the form of a decoction, it is given in asthma and cough. Fruits are carminative, diuretic, deobstruent. Flowers yield an oil, which is stimulant and laxative and is beneficial in wind, phlegm, worms, leprosy, gonorrhea, and headache. It is applied to painful rheumatic affections.[3,37,50]

Doses — Fruit infusion — 7 to 14 ml; seed powder — 1 to 3 gr; leaf powder — 1 to 3 gr. Root bark decoction — 28 to 56 ml.

SANTALUM ALBUM Linn. Family: SANTALACEAE

Vernacular names — Sanskrit, Chandanam; Hindi, Safed Chandan; English, White sandalwood; Bengali, Chandan; Nepalese, Shrikhanda; Sinhalese, Sudu handun; Japanese, Byakudan; Chinese, Tan hsiang; German, Weisses Sandelholz; French, Santal; Unani, Chandan safed; Arabian, Sandal-e-Abyaz; Persian, Sandan safed; Tibetan, Tsa-ndna; Burmese-Sanduku; Tamil-Chandanam.

Habitat — *S. album* is a small evergreen parasitic tree that grows wild and is systematically cultivated in Karnatica and also in Coimbatore, Salam, and other southern parts of Madras State.

Parts used — Wood and volatile oil.

Pharmacognostical characteristics — A middle-sized evergreen tree, branches almost drooping; bark dark, rough, with vertical cracks; mature wood scented. Leaves 4 to 7 cm long, opposite, shining on upper surface. Flowers small, dull, purplish in small bunches. Fruits roundish, 6 mm diameter, purple black, succulent.[37,510] The sandalwood tree is of a parasitic nature. A few months after germination, haustoria from the roots penetrate into the roots of grasses, small shrubs and eventually of large trees. Trees growing on hard, rocky, ferruginous soils are richer in oil than those growing on fertile tracts.

The volatile oil is contained in all the elements of the wood — namely, medullary ray cells, vessels, wood fibers, and wood parenchyma.[3,7]

Ayurvedic description — *Rasa* — *tikta, madhura*; *Guna* — *laghu, rooksha*; *Veerya* — *sheeta*; *Vipak* — *katu*.

Action and uses — *Kapha pitta samak, daah prasaman, branaya, trisanigraha, grahi, krimighan, hirdya, rakt sodhak, rakt pitta samak, mutrala, mutra daah samak, jwarghana.*

Chemical constituents — The essential oil of sandalwood is distilled from the small chips and billets cut out of the heartwood. Oil is extremely viscid, of a light yellow color, and possesses a characteristic roseate and penetrating odor and a bitterish, slightly acrid taste. It is soluble in from 3 to 6 volumes of 70% alcohol at 20°C and has the following characteristics: sp gr 0.973 to 0.985; optical rotation 14 to 21°; refractive index 1.5040 to 1.5100; acid value 0.5 to 6; ester value 3 to 17; sesquiterpene alcohol 90 to 96%. Heartwood contains a volatile oil (2.5 to 6%), a dark resin and tannic acid. Constituents of oil are (1) santalol, a body or a mixture of isomerics or sesquiterpene alcohol with different boiling points; it is the principal constituent of the oil, occurring therein to the extent of 90% or more. It is a mixture of two isomers known as α-santalol (bp 300 to 301°C) and β-santalol (bp 170 to 171°C). The rest is composed of aldehydes, santalol and ketones, isovaleric aldehyde, santonone, santalone, esters, and free acid. Fruits yield betulic acid, β-sitosterol, and a fatty oil.[178]

Pharmacological actions — Wood is bitter, cooling, sedative, and astringent. Oil is astringent and disinfectant to the mucous membranes of the genitourinary and bronchial tracts; also diuretic, expectorant, and stimulant.

Medicinal properties and uses — Emulsion or a paste of the wood is a cooling dressing in inflammatory and eruptive skin diseases such as erysipelas, prurigo, and prickly heat. It has also been used as a diaphoretic and as an aphrodisiac. In cases of morbid thirst, wood powder is taken in coconut water; with addition of sugar, honey, and rice water it is given to check gastric irritability and dysentery and to relieve thirst and heat of body.

Sandalwood is a popular remedy in gonorrhea, cystitis, gleet, and urethral hemorrhage. It is valuable also in bronchial catarrh. Externally, the oil is an excellent application in scabies in every stage and form.[50]

Doses — Powder — 3 to 6 gr; oil — 5 to 25 minims.

SARACA INDICA Linn.

Family: LEGUMINOSAE

Vernacular names — Sanskrit, Asoka; Hindi, Angana priya; English, Asoka tree; Bengali, Anganapriya; Nepalese, Aasalaswaa; Sinhalese, Diyaratmal; Japanese, Muyuju; Unani, Ashok; Arabian, Bags; Persian, Shamshad; Tibetan, Mya-nan-med; Burmese, Thawka; Malaysian, Asogam; Tamil, Asogu.

Habitat — Asoka is distributed throughout India, particularly in central and eastern Himalayas and extending up to Burma and Sri Lanka. It is considered one of the sacred trees in India. Its bark is collected by making transverse and longitudinal incisions. It is also cultivated for beauty as well as for medicinal and religious purposes.

Parts used — Stem bark, flower, and seeds.

Pharmacogostical characteristics — It is a deciduous tree. Bark from the old stem is dark green in color, but this is often marked by bluish and ash white patches of lichens. The outer surface is rough due to the presence of rounded and oval lenticels, the formation of short vertical narrow fissures of varying lengths, and partial exfoliation of the outer rind in thin flakes. The thickness of the whole bark varies from 5 mm to 1 cm, depending upon the age of the tree. The entire cut surface turns reddish on exposure to air.

In transverse section, the outermost tissue of cork has a thickness of 1 to 1.5 mm and consists of 30 rows of elongated cells with adhering portions of the old cortical tissue composed of parenchyma cells and stone cells. Most of the cells contain reddish-brown contents. The entire two to three rows of the cork are nearly rectangular and thin walled. In two distinct rings, the phelloderm contains stone cells; crystal fibers are also present. Medullary rays are funnel-shaped and uniseriate in the inner bark.[835,838]

Ayurvedic description — *Rasa* — *kasaaya, tikta; Guna* — *laghu, rooksha; Veerya* — *sheeta; Vipa* — *katu.*

Action and uses — *Garahi, rakta sangrahak.*

Chemical constituents — Bark of the asoka tree contains tannin, catachin, an organic compound of calcium, hemotoxyline, and a ketosterol. The alcoholic extract, which was mostly soluble in hot water, showed the presence of a fair amount of tannin and probably an organic substance containing iron.[839] Chopra et al.[178] reported the isolation of ketosterol, a glycosidal fraction, a saponin, and an organic calcium compound from the whole plant.

Pharmacological action — Bark is strongly astringent and is a uterine sedative. It acts directly on the muscular fibers of the uterus. It has a stimulating effect on the edometrium and ovarian tissue. Ketosterol and the calcium salt in the extract of the plant are considered important in the treatment of menorrhagia.[178]

Medicinal properties and uses — Asoka bark is very useful in uterine affections, especially in menorrhagia. Bark is useful in internal bleeding, hemorrhoids, and also hemorrhagic dysentery.

Doses — Decoction — 28 to 112 ml; powder — 1 to 3 gr.

SAUSSUREA LAPPA C.B. Clarke *Family:* COMPOSITAE

Vernacular names — Sanskrit, Kushtha; Hindi, Kust; English, Costus; Bengali, Kur; Nepalese, Kut; Sinhalese, Kottam; German, Kostwurz; Unani, Kut; Tibetan, Ru-rta; Arabian, Qust; Persian, Kushnah; Tamil, Kottam.

Habitat — These herbs grow abundantly on the Himalayas and valley of Kashmir.

Parts used — Roots.

Pharmacognostical characteristics — It is a tall perennial herb with erect stem, 3 to 6 ft high. Leaves membranous, irregularly toothed; basal ones very large, 1.5 to 3.0 ft long, with a long winged stalk; upper leaves smaller, sometimes with stalks; two small lobes at the base of these leaves almost clasping the stem; end lobe often 30 cm in diameter. Stem leaves are smaller. Flowers about 2 cm long, bluish purple or almost black, borne on round flower heads, few flower heads clustered together in axils of leaves or at tops of stems. The hairs on fruits (pappus) about 1 to 7 cm long, feathery, giving a curious fluffy appearance to the fruiting flower heads.

The color of the outer surface of the root in comparatively fresh samples is dirty gray to light yellow, while the internal color is white and is often visible through the thin gray cork wherever the outer surface is bruised during the transit of the drug. In older samples, the external surface is yellowish brown and the internal surface light brown to cinnamon color. It also has a pleasant odor which is characteristic and aromatic.

The cork is composed of about three to five layers of radially arranged but tangentially elongated cells with somewhat thick, brown, suberized walls. The phellogen or cork cambium consists of one to two layers of thin-walled, rectangular, and tangentially elongated cells. The phelloderm consists of three to six layers of thin-walled polygonal cells. The secondary phloem consists of 60 to 80 layers of cells, decreasing in older roots to 30 to 50 layers. It is mostly composed of storage parenchyma, a small group of sieve tubes with companion cells, and small bundles of phloem fibers. The cells of the phloem parenchyma are thin walled, and not easily distinguishable from those of phloem rays, excepting that they are similar and mostly polygonal without prominent intercellular spaces. Phloem rays are much larger and often tangentially elongated, with large intercellular spaces.

The sieve tubes, which are usually accompanied by a few companion cells, are 128 to 142 μm long, and 6 to 12 μm in diameter. Phloem fibers are thick walled, lignified, and very much elongated with tapering ends. A thin-walled cambium of two to three layers, with rectangular and tangentially elongated cells in transverse section, follows. The secondary xylem is composed of strands of one to ten vessels associated with tracheids, xylem fibers, fiber tracheids, and xylem parenchyma, separated by wide xylem rays. In young roots, the vessels are arranged mostly in one radial file. In older roots, the vessels occupy more than one radial file and the fibers are more strongly developed and arranged, in a more or less concentric manner. The xylem vessels are mostly polygonal or circular in transverse view. In longitudinal section, they are cylindrical with oblique articulation and well-marked perforation rims. Resin canals, which appear as subspherical or ellipsoidal cavities in transverse section, occur abundantly almost in any part of the root. They are to be found in the phloem, phloem rays, xylem, and xylem rays. When the roots become older, they are the only structures that are distinct among the disintegrating mass of parenchyma. They are bounded by a layer or two of elongated cells, the long dimension of which are extended parallel to the long dimension of the canals. The storage parenchyma of both the xylem and the phloem regions contains inulin, which gives a violet coloration with α-naphthol and sulfuric acid.[833,840]

Ayurvedic description — *Rasa* — *tikta, katu, madur; Guna* — *lagu, rooksha, teekshna; Veerya* — *ushna; Vipak* — *katu.*

Action and uses — *Kaf bat samak, badana nasak, shurka sodhak, bajkarn, branya, durgandh nasak, rasayan, sawashar, ashep har, takt sodhak, dipan, pachan, anuloman, grahi.*

Chemical constituents — Roots of *S. lappa* contain an odorous principle composed of two liquid resins, an alkaloid, a solid resin, salt of valeric acid, an astringent principle, and ash which contains manganese. The oil of the root was observed to have camphene (0.04%), phellandrene (0.4%), terpene alcohol (0.2%), d-costen (6.0%), β-costen (6.0%), aplotaxene (20.0%), costol (7.0%), di-hydrocostus lactone (15.0%), costos lactone (10.0%), costic acid (14.0%). Later on, reinvestigation of the active principles of the root were found to have (1) an essential oil of a strong aromatic, penetrating, and fragrant odor (1.5%); (2) a glucoside; and (3) an alkaloid saussurine (0.05%) (in leaves, 0.025% only), resin (6.0%), traces of a bitter substance, small quantitites of tannin, inulin (about 18.0%), a fixed oil, potassium nitrate, and sugar. Leaves do not contain the essential oil.[841] Chopra et al.[178] reported the presence of a new lactone, saussurea lactone, in 20 to 25% yield.

Pharmacological action — It is carminative; strongly antiseptic and disinfectant against streptococcus and staphylococcus; an expectorant and a diuretic; and relaxes the involuntary muscle tissue and is a cardiac stimulant.[1]

Medicinal properties and uses — In India, the root is used as an aphrodisiac and as a tonic. Ayurvedic physicians describe the drug as a bitter, acrid stimulant and alleviative of wind, phlegm, fever, phthisis, cough, and dyspepsia. Root also used as a carminative, antiseptic, prophylactic, anthelmintic, astringent, sedative, alterative, antispasmodic, and aphrodisiac and as an aromatic stimulant. Infusion with cardamons is used in cough, asthma, chronic rheumatism, skin diseases, and dyspepsia. *Hakims* and *vaidyas* use the roots in the treatment of quartan malaria, leprosy, persistent hiccup and rheumatism. Dried root powder is a useful hair wash and an astringent stimulant. Its ointment is applied in wounds and severe ulcerations and for resolving tumors.[3,50]

Doses — Powder — 0.5 to 1.5 gr.

SEMECARPUS ANACARDIUM Linn. f. *Family:* ANACARDIACEAE

Vernacular names — Sanskrit, Bhallataka; Hindi, Bhilawa; English, Marking nut tree, Bengali, Bhela; Nepalese, Bhala; Sinhalese, Senkottam; Japanese, Sumiurushinoki; German, Ostindis Chertintenbaum; French, Noix a marquer; Unani, Bhilawa; Arabian, Habbul Qalb; Persian, Beladur; Tibetan, Bsre-sin; Burmese, Che; Tamil, Cherangettai.

Habitat — A deciduous tree of sub-Himalayan tract from the Sutlij eastward, ascending to 3500 ft, and is found growing in tropical parts of India as far east as Assam.

Parts used — Fruit, gum, and oil.

Pharmacognostical characteristics — A moderate-sized tree, exuding dark acrid juice. Leaves 7 to 24 × 4 to 12 in. Crowded at the ends of branches, obovate-oblong, rounded at apex, sometimes shortly auricled at base, more or less pubescent beneath. Flowers $1/5$ to $1/3$ in. across, greenish yellow, in fascicles arranged in large pubescent terminal panicles. Petals 5, ovary densely pilose. Drupe 1 in. long, obliquely ovoid or oblong, smooth, shining, black when ripe. The unripe fruit yields milk-like juice which turns black on contact with air.[28]

Ayurvedic description — *Rasa* — *madhur, kasaya; Guna* — *laghu, snigdha, tiksna; Veerya* — *ushana; Vipak* — *madhura.*

Action and uses — *Kapha vat samak, pitta sansodhan, dipan, pachan, vadan, krimighan, vajekarn, jawarghana, rasayan.*

Chemical constituents — The pericarp of the fruit contains a bitter and powerful astringent principle. Siddiqui[847] investigated the juice of pericarp and isolated the following constituents:

1. A monohydroxyphenol, which forms 0.1% of the extract. This has been named semecarpol (bp 185 to 190°C) congealing below 25°C to a fatty mass.
2. An *o*-dihydroxy compound forming 46% of the extract (15% of the nut). This has been called *bhilawanol*. (This distills at 225 to 226°C and congeals below 5°C.)
3. A tarry, nonvolatile, corrosive residue forming about 18% of the nut.

It has been further shown that the pericarp contains 20% of oil which can be distilled out of nuts by slow heating to 350°C in large retorts. It is a dark, viscous, highly vesicant liquid which contains *bhilawanol* and other compounds.

Root bark contains an acrid, viscid juice similar to that found in the pericarp of the fruit.

Pharmacological action — Ripe fruit is an acrid, stimulant, digestive, sedative, antispasmodic, alterative nervine tonic and escharotic. It is a good cardiac tonic.[1]

Medicinal properties and uses — The juice of the shell of the nut is a powerful escharotic; it is given in small doses with some bland oil or butter in leprous and scrofulous affections, syphilis, skin diseases, epilepsy, nervous debility, neuralgia, asthma, dyspepsia, and piles. The bruised nut is used as an abortifacient by placing it in the mouth of the uterus; also given as vermifuge.[50]

A brownish gum exuding from the bark is regarded as valuable in scrofulous, venereal and leprous affections, and nervous debility.

Equal parts of marking nut and chebulic myrobalams and sesamum seeds are made into a confection *(kshir pak)* and administered in doses of 3 to 4 gr. It is used with advantage in simple chronic enlargement of the spleen without any hepatic complication or fever. It is useful in many neurotic, cardiac troubles; the rate of the heartbeat is usually increased under its influence. It is also useful in cases of pneumonia.

Doses — Oil — 1 to 2 drops; kshir pak — 12 to 24 gr.

SESAMUM INDICUM Linn.
(Syn. *S. ORIENTALE* Linn.)

Family: PEDALIACEAE

Vernacular names — Sanskrit, Tila; Hindi, Til; English, Sesamum; Bengali, Tel; Nepalese, Hamo; Sinhalese, Tala; Japanese, Goma; Chinese, Hu ma; German, Sesam; French, Sesame; Unani, Til; Arabian, Simsim; Persian, Kunjud; Burmese, Hnan; Malaysian, Bijan; Tamil, Ellu.

Habitat — Extensively cultivated throughout India, being grown as a winter crop in the warmer parts of the country and as summer crop in the colder areas.

Parts used — Seeds, leaves, and oil.

Morphological characteristics — An erect, pubescent annual, 1 to 3 ft high, more or less fetid, branching from the base. Leaves 3 to 5 in. long, upper often narrowly oblong and subentire, middle ovate, and toothed, the lower ones lobed or often pedatisect. Flowers 1 to $1\frac{1}{4}$ in. long, purple or whitish with yellow or purple marks, pubescent. Capsule 1 in. long, erect, bluntly quadrangular, shortly beaked, dehiscent from above to about halfway down. Seeds black or white.[28]

Ayurvedic description — *Rasa — madhur; Guna — guru, snigdha; Veerya — ushna; Vipak — madhur.*

Action and uses — *Tridosha samak, sanehan, sandha niya, bran sodhan, ropan, dipan, grahi, shula prasaman, rakt srawnasak, vaj karn, artabjann.*

Chemical constituents — Two well-marked varieties of the seeds, black or white, yield 50 to 60% of fixed sesame oil (til, sesame, or gingelly oil) which is clear and limpid. It contains about 1% of lignan sesamin and the related sesomolin. The prominent components of the oil are glycerides of oleic and linoleic acids with small proportions of palmitic, stearic, and arachidic acids. Refined sesame oil resembles olive oil in its properties. It is a pale-yellow bland oil, which on cooling to about –4°C solidifies to a buttery mass; it has a saponification value the same as that of olive oil and a somewhat higher iodine value (104 to 120). It is being used as a substitute for olive oil.[69] Seeds yield a fixed oil (50 to 60%); leaves contain gummy matter.

Pharmacological action — Demulcent, emollient, diuretic, emmenagogue, lactagogue and laxative.

Medicinal properties and uses — The leaves, being highly mucilaginous, are used as a demulcent in respiratory affections, infantile cholera, diarrhea, dysentery, and other bowel affections and bladder diseases such as catarrh, cystitis, and stranguary. Decoction or confection of the seeds given in bleeding piles; their powder is useful in amenorrhea and dysmenorrhea. A poultice of the seeds is applied over ulcers, burns, and scalds. The oil extracted from the seeds is rubbed on the head for relief of migraine and vertigo. The oil is used as a substitute for olive oil in pharmaceutical preparations for external uses.

Doses — Seed powder — 1 to 2 gr; decoction — 28 to 56 ml.

SIDA CORDIFOLIA Linn.   *Family:* MALVACEAE

Vernacular names — Sanskrit, Bala; Hindi, Khareti; English, Country mallow; Bengali, Bala; Nepalese, Balu; Sinhalese, Kotikanbevila; Japanese, Marubakingojikuwa; Chinese, Kedong; Unani, Kharenti; Tibetan, Ba-la; Malaysian, Kelulut pulih; Tamil, Arival.

Habitat — Occurs throughout the tropical and subtropical plains all over India, Pakistan, and Sri Lanka as a weed, usually in waste places and open scrub forests.

Parts used — Roots, leaves, and seeds.

Pharmacognostical characteristics — A small much branched shrub, minute star-shaped hairs present all over the plant; leaves 2 to 5 cm long, ovate or roundish, thick, margins toothed, petioles shorter than leaves. Flowers yellow, small, one or a few together; fruits 6 to 8 mm in diameter, divided into 7 to 10 parts, each strongly reticulated and with two awns (spiny projections) on tip.[435]

Ayurvedic description — *Rasa* — *madhura; Guna* — *guru, snigdha, pichil; Veerya* — *sheeta; Vipak* — *madhura.*

Action and uses — *Vat pitta samak, badana sthapan, sotha har, balya, bat har, sanehan, anuloman, grahi, hirdya, rakt pit samak, sukral, balya brighan, oja bardhak.*

Chemical constituents — Ghosh and Dutt[848] in a systematic chemical examination of the plant showed the presence of (1) fatty oil, phytosterol, mucins, potassium nitrate, resin, resin acids, but no tannins or glucoside; and (2) alkaloid to the extent of 0.085%. The hydrochloride of alkaloid occurs in colorless needles, freely soluble in water, but sparingly soluble in absolute alcohol. The main portion of the alkaloid was identified to be ephedrine. The seeds contained about four times as much alkaloid than either the stem, root, or leaves.[37]

Pharmacological action — Roots are cooling, astringent, stomachic, and tonic; aromatic, bitter, febrifuge, demulcent, and diuretic. Chopra and De[849] have shown the presence of a sympathomimetic alkaloid whose pharmacological action closely resembled that of ephedrine, so its use as a cardiac stimulant is justified in Ayurvedic medicine.

Medicinal properties and uses — The juice of the plant is given in rheumatism, gonorrhea, and spermatorrhea; its infusion is given in gonorrhea, rheumatism and febrile diseases. The leaves are mucilaginous; their infusion is given as a demulcent in fever; they are given as a vegetable to patients suffering from bleeding piles.

In the form of infusion, roots are given in urinary diseases and disorders of the blood and bile; it is also useful in bleeding piles, hematuria, in gonorrhea, cystitis, leukorrhea, chronic dysentery, nervous diseases as insanity, facial paralysis, and in asthma as a cardiac tonic.[50] As an alterative, root bark is useful in rheumatism; in elephantiasis, a paste of the root made with the juice of palmyra palm is locally used; the juice of the roots is applied as a sedative over wounds and ulcers.[1]

Doses — Root infusion — 2 to 20 ml; powder — 1 to 3 gr.

SOLANUM INDICUM Linn.

Family: SOLANACEAE

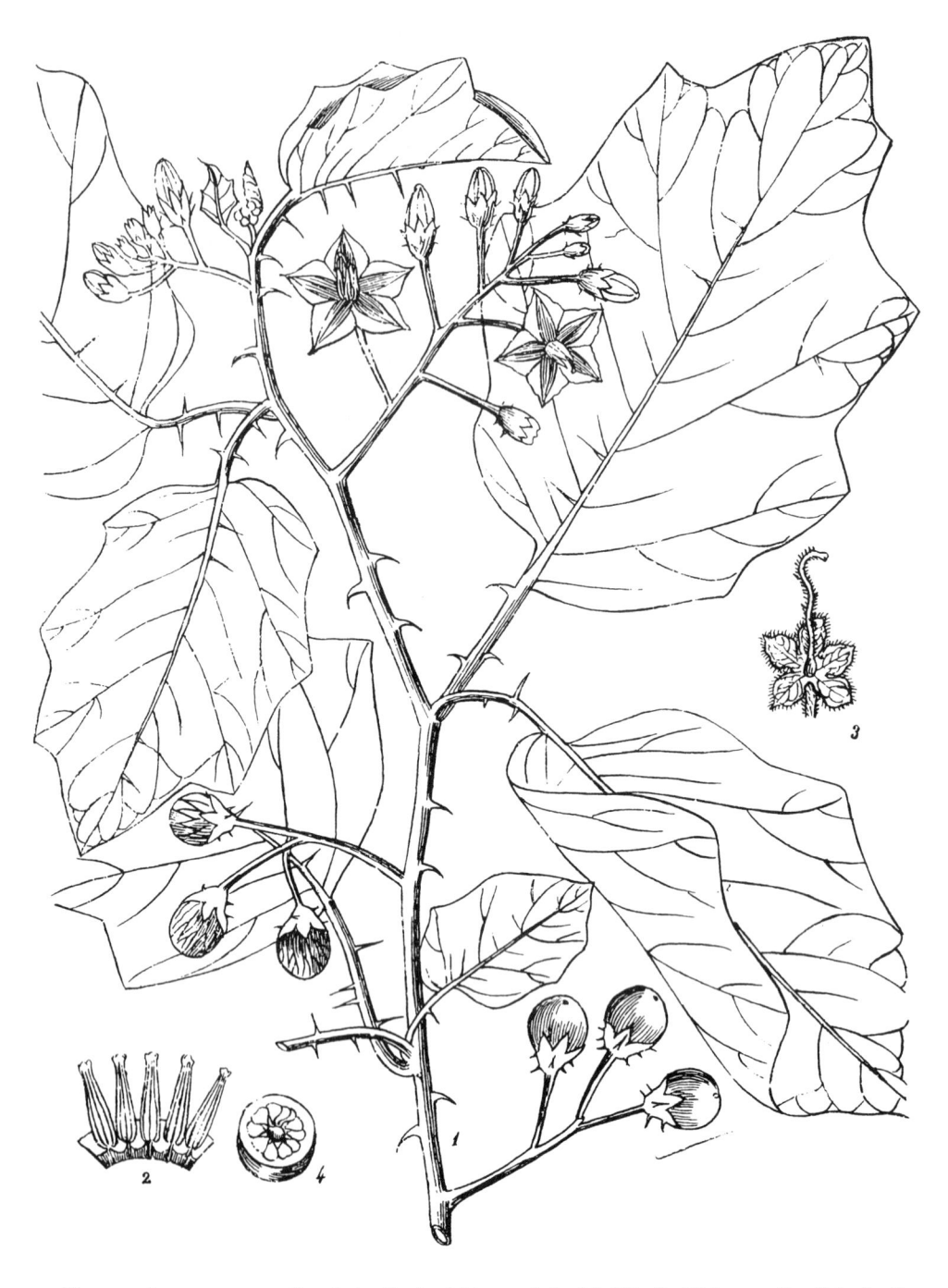

Vernacular names — Sanskrit, Brahati Vanavrinktaki; Hindi, Birhatta; English, Indian nightshade; Bengali, Byakura, Tita, bhekuri; Nepalese, Bihee; Sinhalese, Elabatu; Japanese, Shirosuzume-nasubi; Chinese, Housang kiue; German, Indisch Nachtschatten; Unani, Katai; Arabian, Shaukat-ul-aqrab; Tibetan, Brl-ha-ti; Tamil, Kari.

Habitat — This plant is commonly found throughout tropical India.

Parts used — Whole plant.

Pharmacognostical characteristics — A stiff, prickly herb; prickles stout, recurved; leaves 3 to 6 in. long, 1 to 4 in. broad, alternate, lobed, entire; spines present on petiole and midrib. Flowers blue in lateral racemes. Calyx 5 to 7 lobed, somewhat enlarged in the fruit. Corolla rotate, 5 lobed. Stamen attached to corolla throat, filament short, anthers large. Berry globose, green, white lining when ripe becomes yellow. Seeds many, discoid.[28]

Ayurvedic description — *Rasa — katu, tikta; Guna — laghu, rooksha, teekshna; Veerya — ushna; Vipak — katu.*

Action and uses — *Kapha vat samak, badana sthapan, kesya, dipan, pachan, grahi, krimighan, vajikarn, swashar, kashar, sothhar, mutral, jawrahar.*

Chemical constituents — Fruit and root contain wax, fatty acids, and the alkaloids solanine and solanidine occur in roots and leaves. Fruits are reported to contain enzymes.[1,850]

Pharmacological action — Plant is reported as aphrodisiac, astringent, carminative, cardiac tonic, and resolvent.

Medicinal properties and uses — *S. indicum* is useful in asthma, dry cough, difficult parturition, and chronic febrile affection; colic with flatulence, worms, scorpion sting, in dysuria, and inchuria. Root forms one of the ingredients of *dasamula kvatha* in Ayurvedic medicine. It is regarded as diuretic; useful in dropsy, expectorant, and catarrhal affections; also diaphoretic and stimulant. Juice of leaves with fresh juice of ginger stops vomiting. Leaves and fruit rubbed up with sugar used as external application for itch.[3,50]

Doses — Decoction — 48 to 96 ml; powder — 1 to 2 gr.

SOLANUM NIGRUM Linn.   *Family:* SOLANACEAE

Vernacular names — Sanskrit, Kakamachi; Hindi, Makoi; English, Black nightshade; Bengali, Kakmachi; Nepalese, Kakmachi; Sinhalese, Kalukemmeriya; Japanese, Inubozuki; Chinese, Ti'en kui tse; German, Alpkraut; Unani, Makoff; Arabian, Anab-us-salab; Persian, Anqur-e-shefa; Malaysian, Trong parachichit; Pustu, Karosgi, Tamil, Manalthakali.

Habitat — Throughout India, up to 9000 ft in the western Himalayas.

Parts used — Whole plant and fruits.

Morphological characteristics — An erect, nearly glabrous, much branched annual herb, 1 to $1^1/_2$ ft high. Leaves 1 to 4 in. long, ovate or oblong, entire or sinuate or sometimes toothed or lobed. Flowers white, $^1/_4$ to $^1/_2$ in., drooping-subumbellate, and on extra-axillary peduncles. Calyx $^1/_8$ in. long, teeth small, obtuse. Berry $^1/_4$ in. diameter, yellow, red, or black when ripe, smooth, and shining.[28,37]

Ayurvedic description — *Rasa — tikta; Guna — laghu, snigdha; Veerya — ushna; Vipak — katu.*

Action and uses — *Tridosha nasak, sothahar, badana sthapan, dipan, pittasark, rakt sodhak, swashar, mutral, kustaghan, jwarghana.*

Chemical constituents — The plant contains solanine and tropine alkaloid with mydriatic action. Solanine is contained in the berries. It was from this species that the alkaloid solanine was first isolated in 1820 by Defosses.[28] The plants growing in Azerbaijan yield alkaloids (0.04%), and glycosides (1.8%). Isolation of alkaloid solamargine and oil (4.4%) in berries and 18 to 22% in seeds has been reported.[178]

Pharmacological action — Alterative, sedative, diaphoretic, diuretic, and expectorant. Berries are tonic and diuretic.[1]

Medicinal properties and uses — It is used as a vegetable or as a decoction in dropsy, chronic enlargement of the liver, and jaundice. The plant juice is used in chronic skin diseases, blood spitting, and piles; as an anodyne, a decoction of the plant is used for washing inflamed, irritated, and painful parts of the body. The leaf juice is given in inflammation of the kidneys and bladder and in gonorrhea.[50]

The berries contain the toxic alkaloid solanine; they are alterative, diuretic, and tonic; used in eye diseases; hydrophobia.[50]

Doses — Infusion — 15 to 25 ml; fruit powder — 1 to 2 gr; decoction — 28 to 56 ml.

SOLANUM SURATTENSE Burm. f. *Family:* SOLANACEAE
(Syn. *S. XANTHOCARPUM* Schrad. and Wendl.)

Vernacular names — Sanskrit, Kantakari; Hindi, Kateli; English, Kantakari; Bengali, Kantikari; Nepalese, Kantakari; Sinhalese, Katuvelbattu; Japanese, Kinginnasubi; German, Nacutsebattin; Unani, Katai khurd; Arabian, Hadaka; Persian, Katai khurd; Tibetan, Ka-nta-ka-ri, Malaysian, Velvottu-valutina; Tamil, Kandamgkathri.

Habitat — It is found in all dry districts in the plains as well as low hills throughout India from Punjab and Assam to Kanya Kumari. In South India, it is found abundantly along the Coromandel Coast and in the district of Tinnevelly.

Parts used — Whole plant.

Pharmacognostical characteristics — A prickly, much branched herb, spreading or diffuse. Leaves ovate, elliptic, or subpinnatifid, glabrescent, very prickly. Leaves 10 to 12.5 × 5 to 15 cm, spines 1.2 cm, straight, petiole 2.5 cm. Inflorescence — few flowered lateral cyme, peduncles short, mostly extraaxillary; flowers few, all perfect; pedicels and calyx stellately pubescent or at length glabrous. Lobes ovate-oblong, usually prickly, hardly enlarged in fruit; corolla blue, 2.5 cm diameter, pubescent outside, lobes shallow. Fruit — berry, 12 to 18 mm diameter, globose, glabrous, much exceeding the calyx lobes, yellow or whitish in color, seed 52 mm diameter, glabrous.

A transverse section of the root shows a periderm tissue, as most of these are crushed and fall out during growth, a small cortex, and a large central xylem region. Cortical cells are more or less elongated and thin walled; the upper cortical cells are irregular. The vessels are present in large numbers and vary in their arrangement and size of lumen. The medullary rays are usually small, generally one to two cell layers. The wood also consists of tracheids with pitted walls and in various shapes and sizes. Wood fibers are of various sizes, generally with irregular walls, the ends are irregularly tapering, and with somewhat wide lumen. The wood parenchyma is somewhat scantily developed. Starch grains are present in cortical cells.[833,851]

Ayurvedic description — *Rasa* — *tikta, katu; Guna* — *lagu, rooksha, teeshna; Veerya* — *ushna; Vipak* — *katu*.

Action and uses — *Kaf bat samak, sothahar, krimighna, rachan, kafa nisarak, hicca har, vage karak, kanthya.*

Chemical constituents — Fruit of the plant yields carpesteral and 1.3% glucoalkaloid solanocarpine and solanine-S, which on hydrolysis yields alkaloid solanidine-S. Solanocarpine obtained from the seeds is believed to be identical with solanine-S.[852-854] Chopra et al.[178] reported the presence of 1% of alkaloid which could form a source material for cortisone and sex hormone preparations.

Pharmacological action — It is carminative, an effective diuretic, expectorant, and febrifuge. It is also pungent, bitter, digestive, alterative, and astringent.[50]

Medicinal properties and uses — Roots are one of the constituents of dashmul asava. The plant is useful in fever, cough, and asthma. Fumigation with the vapors of the burning seeds is administered to cure toothache. The juice of the berries is useful in sore throat. A decoction of the plant is used in gonorrhea and rheumatism. It also promotes conception in the female. A fine powder of the fruits of this plant mixed with honey is very useful for chronic cough in children. A decoction of the root with that of *Tinospora cordifolia* is used as a tonic in fever and cough.[1,3]

Doses — Decoction — 56 to 112 ml; powder — 1 to 2 gr.

SPHAERANTHUS INDICUS Linn. *Family:* COMPOSITAE

Vernacular names — Sanskrit, Munditika; Hindi, Gorakmundi; English, East Indian globe thistle; Bengali, Nurmuriya; Ghork-mundi; Nepalese, Gorakhamundi; Sinhalese, Mudamahan; Unani, Mundi; Arabian, Kamazarus; Persian, Zakhmi-e-Hayat; Tamil, Kottaikaranthai.

Habitat — Throughout India, ascending the Himalayas up to 5000 ft from Kumaon to Sikkim.

Parts used — Root, root bark, flowers, and leaves.

Pharmacognostical characteristics — A much-branched, strongly scented, glandular-hairy herb, about 1 to 2 ft high, branches ascending with toothed wings. Leaves 1 to 2 in. long, sessile, decurrent, obvate-oblong, narrowed to the base, dentate or serrate, the teeth often bristle pointed. Clusters of head $^3/_8$ to $^5/_8$ in. diameter, globose or shortly oblong on winged peduncles. Involucral bracts ciliate at apex. Flowers pink or purple; achenes stalked.[28]

Ayurvedic description — *Rasa — tikta, katu, madhur; Guna — laghu, rooksha; Veerya — ushna; Vipak — katu.*

Action and uses — *Tridosha samak, sothahar kaphanghan, virisya; mutral, swedajanan; kustaghanan, jawaraghana; daah prasaman, bran sodhan, rakt sthamban, dipan anuloman, artwa janan, lakhan.*

Chemical constituents — The fresh flowering herbs yield 0.01% deep cherry-colored essential oil. Stems, leaves, and flowers contain a bitter alkaloid, sphaeranthine.[50,178]

Pharmacological action — Bitter, tonic, stomachic, stimulant, alterative, and externally emollient.[1]

Medicinal properties and uses — Distilled water prepared like rosewater from the herb is recommended for bilious affections and for the dispersion of various types of tumors. Root and seeds are used as a stomachic and anthelmintic. Flowers are alterative and tonic, useful as blood purifiers in skin diseases.[50] Root bark is a valuable remedy in bleeding piles. Dried leaves are used in doses of 20 gr twice a day in chronic skin diseases as antisyphilitic and nervine tonic. Decoction of the whole plant is used as diuretic in urethral discharges. Flowers are swallowed to cure conjunctivitis.[178]

Doses — Root powder — 1 to 3 gr; dried leaves — 0.5 to 1.5 gr.

STRYCHNOS NUX-VOMICA Linn. Family: LOGANIACEAE

Vernacular names — Sanskrit, Kupilu; Hindi, Kuchla; English, Nux vomica; Bengali, Kuchila; Nepalese, Kuchila; Sinhalese, Godakaduru; Japanese, Machin; Chinese, Fan Mu Pieh; German, Gemeinerbrech Nussbaum; Unani, Kuchla; Arabian, Ezaaraqee; Persian, Kuchlah; Tibetan, Ka-ka-nda; Tamil, Etti; Burmese, Kabaung; Malaysian, Kanjiram; French, Noizvomique, vomiquier.

Habitat — The tree is found growing in the state of nature throughout tropical India, in the forests of Gorakhpur, Bihar, Orissa, Konkan, Karnataka, Madras, Coromandel Coasts, Travancore, etc.

Parts used — Dried, ripe seeds, stem bark.

Pharmacognostical characteristics — A deciduous tree often with short, axillary spines. Leaves 3 to 6 × 1 1/2 to 3 in., broadly elliptic, acute, obtuse, or shortly acuminate, glabrous and shining, 5-nerved. Flowers many, greenish white, in terminal short-peduncled pubescent compound cymes; calyx 1/10 in. long. Corolla 1/2 in. long, 5 lobed, tube hairy inside below, the throat glabrous, lobes 1/6 in. long. Ovary and style glabrous. Berry globose, 1 to 3 in. diameter, orange-red when ripe. Seeds several, satiny, surrounded by white pulp. Seeds gray or greenish gray, disk shaped, nearly flat, umbonate, but sometimes irregularly bent; 10 to 30 mm in diameter and 4 to 6 mm thick; margin rounded or somewhat acute; hilum raised and connected to the micropyle by a radial ridge. Surface silky, densely covered with radiately arranged, closely appressed, outwardly directly lignified trichomes; endosperm translucent, horny, having a central, disk-shaped cavity in which, adjacent to the micropyle, lies the embryo, with 2 small, thin cordate, 5- to 7-nerved cotyledons, and terete radicle.[28]

Testa of one integument; outer epidermis of lignified thick-walled cells with sinuous polygonal outlines in surface view, small-branched lumina, and oblique linear pits, each cell prolonged externally into a closely appressed trichome up to 1 mm long, the wall having about ten strongly lignified longitudinal ribs of the thickening; remainder of testa consisting of flattened parenchyma appearing in section as a brown band. In the region of the hilum, some small spiral vessels occur as components of a short vascular strand. Endosperm, with very thick-walled hemicellulosic polyhedral cells with no obvious pits, but connected by plasmodesmal strands and containing oil, plasma, and aleurone grains up to about 30 μm in diameter; color reactions for strychnine and brucine given by the contents of the defatted endosperm. Embryo of small parenchymatous cells with oil and small aleurone grains.[35]

Ayurvedic description — *Rasa — tikta, katu; Guna — lagu, rooksha, teekshana; Veerya — ushana; Vipak — katu.*

Action and uses — *Kaf bat samak, badanahar, dipan, pachan, grahi, shula prasaman, vajikaran, katu postic.*

Chemical constituents — Seeds of Indian nux vomica contain 2.6 to 3% total alkaloids, of which 1.25 to 1.5% is strychnine; 1.7% brucine and vomicine, igasurine, or impure brucine in combination with igasuric or strychnic acid; loganin, a glucoside, proteids (11%), yellow coloring matter, a concrete oil or fat, gum, starch, sugar (6%), wax, phosphates, and ash (2%); wood, bark, and leaves contain brucine, but do not contain strychnine. Young fresh bark contains the largest percentage of brucine, i.e., 3.1%.

Chopra et al.[1,50] reported the presence of three new alkaloids, viz., α-and β-colubrine and pseudostrychnine, vomicine. The fruit pulp contains the glycoside loganin. Leaves contain the alkaloids brucine, strychnine, and strychnicine. The bark contains chiefly brucine and traces of strychnine, but no strychnicine. The younger bark contains 3.1% and the older bark shows 1.68% brucine.

Pharmacological action — Dried seeds are nervine, stomachic, tonic, aphrodisiac, a spinal stimulant; also respiratory and cardiac stimulant. In excessive doses, it is a virulent poison producing tetanic convulsions.[1]

Medicinal properties and uses — Powdered nux vomica seeds are used in the treatment of dyspepsia and diseases of the nervous system. It is used as a remedy in chronic dysentery, atonic diarrhea, paralytic and neuralgic affections, mental emotion, epilepsy, chronic constipation from atony of bowels, prolapsus of the rectum, gout, chronic rheumatism, and hydrophobia. In neuralgia of the face and gastralgia, in sexual impotence, spasmodic diseases as nausea of pregnancy, and epilepsy, its effects are well marked. Nux vomica is also useful in functional paralysis due to anemia of the cord, general exhaustion, spermatorrhea, and incontinence of urine in children. Leaves of nux vomica are applied as poultice to sloughing wounds or ulcers infested with maggots. Root bark ground into a fine paste with lime juice and made into pills is effectual in cholera. Seeds with some aromatics are prescribed in colic. Wood is used for dysentery, fevers, and dyspepsia.[50]

Doses — Seed powder — 0.5 to 3 gr; extract — 0.5 to 1 gr; tincture — 0.5 to 1 ml.

SWERTIA CHIRATA Buch-Ham   *Family:* GENTIANACEAE

Vernacular names — Sanskrit, Kirata tikta; Hindi, Charayatah; English, Chireta; Bengali, Chirata; Nepalese, Khatu; Sinhalese, Keratha; Japanese, Senburi; Chinese, Toyaku; German, Chirata-kraut; Unani, Chirata; Arabian, Aasbuzzarirah; Persian, Naynahawandi; Burmese, Sekhage; Malaysian, Kiriyat.

Habitat — The plant occurs in the temperate Himalayas between 3500 to 4000 ft altitude from Kashmir to Assam.

Parts used — Whole parts.

Pharmacognostical characteristics — An annual herb up to 5 ft high, stem branching, robust. Leaves in opposite pairs, lanceolate, about 4 in. long, without stalks and pointed tips. Flowers numerous, in large leafy panicles, very small, green-yellow, lined with purple, white, or pink hairs, each petal lobe has a pair of green glands. Capsules minute, 6 mm or more long, ovoid.[435]

Ayurvedic description — *Rasa — tikta; Guna — lagu, rooksha; Veerya — sheeta; Vipak — katu.*

Action and uses — *Kaf pit samak, bransodhan, dipan, aam pachan, pitsarak, anuloman, krimighan, rakatsodhak, sothhar, dahahar, katu postik.*

Chemical constituent — Chireta plant is reported to contain a bitter glycoside chiratin which is precipitated by tannin and yields on hydrolysis two bitter principles, ophelic acid and chiratogenin, the latter being insoluble in water. Ophelic acid is a brown hygroscopic substance which is readily soluble in water and in alcohol. The drug also contains resin, tannin, and 4 to 8% of ash.[50,855]

Pharmacological action — Bitter tonic, stomachic, anthelmintic, laxative, alterative, antidiarrheic, and antiperiodic.[1,3]

Medicinal properties and uses — It is a well-reputed drug for intermittent fevers, skin diseases, and intestinal worms. It is given in the form of infusion or tincture in bronchial asthma; the infusion made in hot water with cloves, cinnamon, etc. is given in doses of 14 to 28 ml as a tonic. To check hiccup and vomiting, the decoction of root is taken in doses of 0.5 to 2 gr with honey. Chireta is used in scorpion sting, also. It is given as tonic to a gouty person.[1]

Doses — Root powder — 0.5 to 2 gr.

SYMPLOCOS RACEMOSA Roxb.　　　　　　　　　　　*Family:* SYMPLOCACEAE

Vernacular names — Sanskrit, Lodhra; Hindi, Lodh; English, Lodh tree; Bengali, Lodhar; Nepalese, Lodha; Sinhalese, Loth; Japanese, Hainoki; Unani, Lodh pathani; Tibetan, Senphrom; Burmese, Daukyat; Tamil, Velli.

Habitat — A small tree found in the plains and lower hills of Bengal, Assam, and Burma, and forests of Chota Nagpur.

Parts used — Bark.

Pharmacognostical characteristics — A small tree 15 to 20 ft high. Leaves 3 to 8 in. long, 1 to 2$^1/_2$ in. broad, oblong-elliptic, leathery, slightly toothed or entire, dark green above, midrib slightly hairy. Flowers in small axillary clusters, small, about 1 to 2 cm diameter, each flower having 3 pubescent bracts, small, white at first, later yellowish. Fruit l to l.3 cm long, purplish black. Bark has a pinkish color.[857]

Ayurvedic description — *Rasa — kasaya; Guna — lagu, rooksha; Veerya — sheeta; Vipak — katu.*

Action and uses — *Kapha pitta samak, soth har, rakt stamvan, bran ropan, khustaghan, gravasya srab nask, jawaraghana.*

Chemical constituents — Bark is reported to contain three alkaloids: (1) loturine (0.24%), (2) colloturine (0.02%), and (3) loturidine (0.06%); and quinovin or kinovin, ash which contains carbonate of soda, large amount of red coloring matter, but no tannin.[50] Later on, Spath[858] showed that loturine was identical with abrine and harman. Bark also contains an amorphous lactone.[178,856]

Pharmacological action — Bark is cooling and mild astringent.

Medicinal properties and uses — Bark is commonly used in bowel complaints, menorrhagia, discharges due to the relaxed condition of mucous membranes, and dropsy. It is also given in the form of powders in liver complaints, fevers, and skin diseases. Decoction of the bark is a mouthwash for bleeding gums. It is also used as an eye bath in eye disease and as a wash for ulcers.[50] Alcoholic and water extracts of lodh are frequently used as astringents for looseness of the bowels.

In cases of chyluria and elephantiasis due to filarial infection, lodh has been a favorite remedy with many physicians in the country.

Doses — Powder — 1 to 3 gr; decoction — 56 to 112 ml.

TAMARINDUS INDICA Linn.

Family: CAESALPINIACEAE

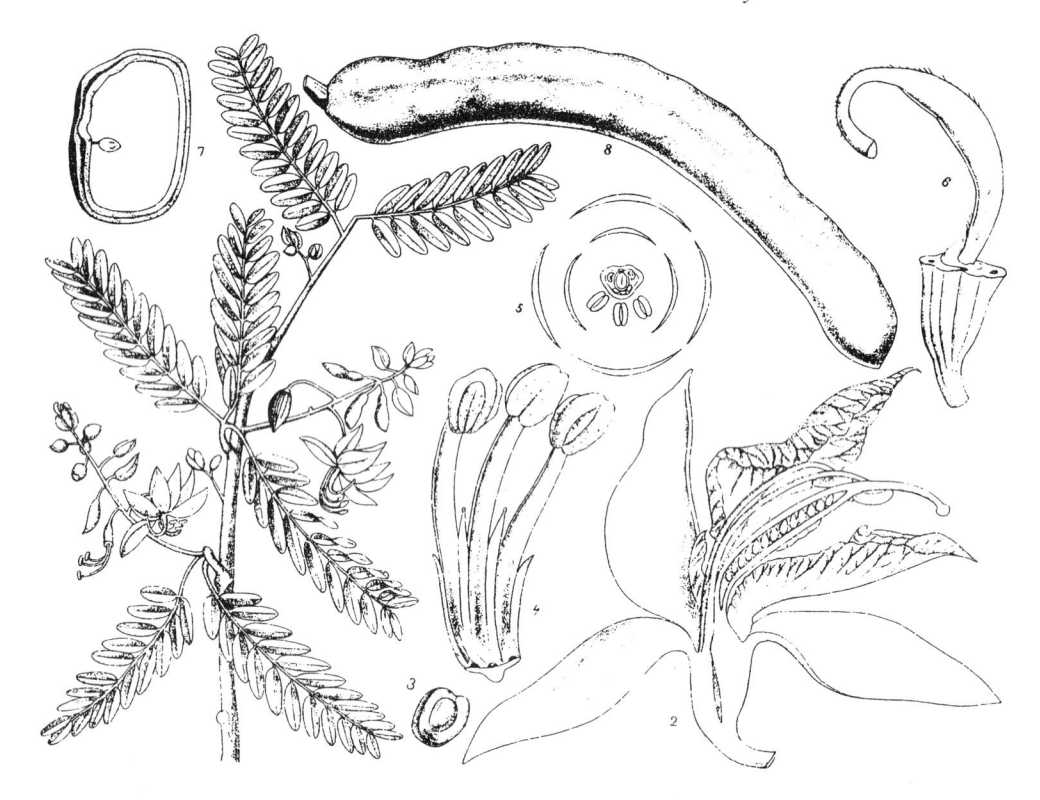

Vernacular names — Sanskrit, Amlika; Hindi, Imli; English, Tamarind; Bengali, Tentul; Nepalese, Imali; Sinhalese, Siyambula; Japanese, Tomarindo; Chinese, Makkham; German, Tamarindenbaum; French, Tamarinier; Unani, Imli; Arabian, Subaru; Persian, Tamar-e-hindi; Malaysian, Assamjava; Tibetan, Bse-yab; Tamil, Pulladi.

Habitat — An evergreen tree which occurs commonly in the central and southern regions of India. It is also planted throughout India on roadsides and in gardens.

Parts used — Pulp of the fruits, seeds, leaves, flowers, and bark.

Morphological characteristics — A large tree, about 70 to 90 ft high; leaves 7.5 to 15 cm long, leaflets 10 to 20 pairs, about 1 cm long; oblong, obtuse; flowers yellow striped with red, in small clusters among the leaves. Fruit a pod, 7.5 to 15 × 2.5 to 3.0 cm, ligulate, fleshy, pendulous, brown in color, seeds 3 to 12, dark brown, shining, embedded in the fleshy fibrous mass which is well-known acid pulp of tamarind.[435]

Ayurvedic description — *Rasa — amla; Guna — guru, rooksha; Veerya — ushna; Vipak — amala.*

Action and uses — *Vat samak, kapha pitta wardhak, rakta pittakarak, sotha har, badana sthapan, dipan, shul prasaman, daah prasaman, jawaraghana.*

Chemical constituents — Tamarind pulp contains tartaric acid (5%), citric acid (4%), malic and acetic acid, tartaric of potassium (8%), invert sugar (25 to 40%), gum, and pectin. Seed testa contains a fixed oil and insoluble matter. Seeds contain albuminoids, fat, carbohydrate (63.32%), fiber, and ash containing phosphorus and nitrogen. Fruits contain a small amount of oxalic acid. Tamarind kernel contains polysaccharides. The leaves are reported to contain glycosides.[69,178]

Pharmacological action — Ripe fruit is sweet, acidic, refrigerant, carminative, digestive and laxative, a valuable drug in febrile diseases, antiscorbulic, and antibilious. Tender leaves and flowers are cooling and antibilious. Bark is astringent and tonic.

Medicinal properties and uses — Pulp of the ripe fruits as well as a poultice of the leaves is applied externally to inflammatory swelling to relieve pain. Pulp is also very useful for checking bilious vomiting, for purging the system of bile, and to adjust "humors." Poultice of flowers is useful in inflammatory affection of conjunctivitis. Juice expressed from the flowers is given orally for bleeding piles. Decoction of the leaves is used as a gargle in throat affections and also used as a wash for indolent ulcers; promotes healthy action. The bark is used topically for loss of sensation in paralysis. The ash is given for urinary discharges and gonorrhea.

The ripe fruit is appetizing, laxative, tonic to the heart, anthelmintic, and heals wounds and fractures. The seeds are useful in vaginal discharges and ulcers. The drug is used in scorpion sting, also.[1,108]

Doses — Fruits — 1 to 3 gr; seed powder — 1 to 2 gr.

TARAXACUM OFFICINALE Weber. *Family:* COMPOSITAE

Vernacular names — Sanskrit, Dugdha feni; Hindi, Dudal; English, Dandelion; Japanese, Seiyotanpope; Chinese, P'u kung ying; German, Lowenzahn; Unani, Kanphul.

Habitat — Found in the temperate Himalayas, from 5,000 to 12,00 ft, and to some extent in Nilgiris. Also found in Tibet.

Parts used — Root and leaves.

Morphological characteristics — Perennial herb; root cylindrical, 3.5 to 9.5 cm in length and 3 to 5 mm in thickness, externally yellowish brown, longitudinally wrinkled, internally whitish; leaves entire radical, sessile, 5 to 20 cm, oblanceolate, acute; heads 8 to 50 cm in diameter, bracteate, inner white, outer short; flowers yellow. Achenes glabrous, flattened, ribbed, narrowed to the base, minutely spiny on the upper half, abruptly contracted into a long, slender beak crowned by the pappus.[108] White latex oozes from leaves and roots when injured. Generally, the plant flowers soon after the snow melts.

Ayurvedic description — *Rasa — tikta, katu; Guna — laghu, rooksha, teekshna; Veerya — ushna; Vipak — katu.*

Action and uses — *Kapha pitta har, bran sodhan, dipan, pitta sarak, rochan, rakkt sodhak, sothahar, mutral, jawaraghana, vishaghan.*

Chemical constituents — The milky juice contains a bitter, amorphous principle — taraxacin, a crystalline substance — taraxacerin; also potassium and calcium salts. Roots contain inulin (25%), pectin, sugar, levulin, and ash (5 to 7%). The presence of phytosterols, taraxasterol and homotaraxasterol, and the saponin in the drug have been reported.[50] The seeds contain an unnamed alkaloid.[10]

Pharmacological action — Stimulant, diuretic, diaphoretic, and tonic.

Medicinal properties and uses — Root of *T. officinale* is a valuable hepatic stimulant and very useful in obstructions of the liver, chronic disorders of the kidney, and visceral diseases. Decoction of the root with podophylum is useful in jaundice, hepatitis, and in indigestion. Root is also given in dropsy and chronic skin diseases. Leaves used for fomentation.

Doses — Root powder — 0.5 to 1 gr; decoction — 28 to 56 ml.

TAXUS BACCATA Linn. *Family:* CONIFERAE

Vernacular names — Sanskrit, Talispatra; Hindi, Thuneer, Biemi; English, Himalayan fir; Bengali, Sugandh; Nepalese, Tariswaa, Dhengresala; Sinhalese, Talispaturu; Unani, Birmi; Pusto, Kharoa; Tamil, Thalizopathari; German, Eilec.

Habitat — This tree is met with in the temperate Himalayas at altitudes of 6000 to 11,000 ft and in Khasi Hills at altitudes of 5000 ft.

Parts used — Leaf.

Morphological characteristics — A small, medium-sized evergreen tree, stem fluted, branches horizontal. Leaves 1 to 1$^1/_2$ in. long, linear, flattened, distichous, acute, narrowed into short petiole which is decurrent along the twig, dark green and shining above, pale yellowish, brown, or rusty red below. Flowers usually dioecious. Male flowers in solitary, axillary, subglobose catkins, stamens about 10, pollen sacs 5 to 9, globose, arranged around the filament beneath the peltate tip of the stamen. Female flowers solitary, axillary, resembling leaf buds, each consisting of a few imbricate scales around a single erect ovule, which is surrounded at the base by a membranous disk. As the young fruit matures, the disk enlarges, becomes succulent, and finally forms a bright red fleshy cup about $^1/_3$ in. long which surrounds the olive green seed of which only the tip is exposed.[28]

Leaf has a single-layered cutinized epidermis; cuticle of the lower epidermis is with globular projection. Dumbbell-shaped sunken stomata present on the under surface are arranged in a row and are overarched by subsidiary and other cells which appear as a multicellular structure in surface view. Cells contain starch grains and calcium oxalate crystals.

Ayurvedic description — *Rasa* — *tikta; Guna* — *laghu, teekshna; Veerya* — *ushna; Vipaka* — *madhur.*

Action and uses — *Kapha vat samak, badana sthapan, rochan, dipan, vatta anuloman, tikta, katu, teekshna, ushna, shelemhar, jawarghana, bal wardhak.*

Chemical constituents — Leaves, shoots, and seeds contain an alkaloid known as taxine, a toxic principle which is vigorously active heart poison. Alkaloid taxine is a complex mixture consisting of pure alkaloid taxine A and taxine B. Taxine B is the main alkaloid.[178] Leaves contain alkaloid taxine, taxinine, and traces of ephedrine. They also yield a glucoside taxicatin. Leaves also contain a volatile oil, tannic and gallic acid, and resinous substance called toxim.[1] In addition to alkaloids, a cyanogenetic glycoside and antitumor agent have been reported in the genus.[69]

Pharmacological action — Carminative, expectorant, stomachic, and tonic. Leaves and fruits emmenagogue, sedative, and antispasmodic.

Medicinal properties and uses — Leaves are somewhat similar in property to digitalis. In the form of tincture, it is used as antispasmodic and given in asthma, hiccup, for indigestion, hemoptysis, epilepsy, and other spasmodic affections. Fruits are given for their emmenagogue, sedative, and antispasmodic effects. They act as antilithic in calculus and nervousness.

Doses — Tincture — 2 to 5 ml; infusion — 14 to 28 ml; powder — 450 to 900 gr.

TEPHROSIA PURPUREA (Linn.) Pers.   *Family:* LEGUMINOSAE
(Syn. *T. MAXIMA* Pers. or
    T. PURPUREA var. *MAXIMA*;
    T. LANCEOLATA R. Grah.)

Vernacular names — Sanskrit, Sarapunkha; Hindi, Sarphankha; English, Purple tephrosia; Bengali, Bonnil; Nepalese, Sarapungkha; Sinhalese, Pila; Japanese, Nabankusafuji; Chinese, Nah troi; Unani, Sarphoka; Persian-Barg-e-sofar; Tamil, Kat Kolingi.

Habitat — Found throughout India, ascending in the Himalayas up to an altitude of 6000 ft.

Parts used — Leaves, seeds, and root.

Pharmacognostical characteristics — A copiously branched, suberect, herbaceous perennial 1 to 2 ft high, branches glabrous. Leaves 3 to 6 in. long, leaflets 13 to 21, $3/4$ to 1 in. long, oblanceolate, obtuse or retuse, mucronate, glabrous above, obscurely silky beneath. Flowers $1/4$ in. long; purple, in leaf-opposed lax racemes 3-6 in. long, the lower flowers fascicled. The calyx teeth linear-subulate, as long as the tube. Style flattened glabrescent; stigma penicillate. Pods $1 1/2$ to 2 in. long. Finally glabrescent, slightly curved, 6 to 10 seeded.[864,865]

Ayurvedic description — *Rasa — tikta, kashaya; Guna — lagu, rooksha, teekshna; Veerya — ushna; Vipak — katu.*

Action and uses — *Kaph vat samak, soth har, bran ropan, jawaraghana, dipan, anuloman, pita sark, krimighan, rakt pitta samak, raktsodhak, kaphanisark, rasayan, piliha har.*

Chemical constituents — The roots of *T. purpurea* are reported to contain tephrosin, deguelin, isotephrosin, rotenone.[859,860] The leaves contain about 2.0% of glucoside, osyritin.[861] Rangaswami and Sastry[862] isolated "lanceolatine A" from the pods. Rutin has been obtained from the leaves in 1 to 4% yield by Rangaswami and Rao.[863] Chopra et al.[178] reported that roots yield "maxima substance A", 0.8%, "maxima substance B", 0.1%, and "maxima substance C", 0.8%, three crystalline compounds. Maxima substance C showed isoflavone structure chemically related to rotenone. The pods give purpurin A, purpurin B, and maximin. Leaves yielded 1.4% rutin.

Pharmacological action — Febrifuge, cholagogue, diuretic, deobstruent, tonic, and laxative.

Medicinal properties and uses — It is a tonic, laxative, cordial, blood purifier, deobstruent, diuretic, anthelmintic, antipyretic, and alterative. It is also useful in diseases of the liver and spleen.[1] The root is diaphoratic and bitter; it is useful in dyspepsia, tympanitis, chronic diarrhea, bronchitis, asthma, and inflammation.[50] Extracted oil from the seeds is an efficacious local application for eczema.

Doses — Powder — 3 to 6 gr; infusion — 12 to 20 ml.

TERMINALIA ARJUNA Wight & Arn.   *Family:* COMBRETACEAE

Vernacular names — Sanskrit, Arjuna; Hindi, Arjuna; English, Arjuna myrobalan; Bengali, Arjun; Nepalese, Keshi; Singalese, Kumbuk; Unani, Arjun; Burmese, Toukhyan; Tamil, Marutham.

Habitat — Throughout the greater part of India; in the sub-Himalayan tract, Madhya Pradesh, parts of Bombay and South India.

Parts used — Bark.

Pharmacognostical characteristics — Arjuna is a large deciduous tree with a height of about 60 to 80 ft and trunk 10 to 12 ft in circumference. Leaves subopposite or alternate, sometimes clustered at the ends of the twigs and often bearing large glands on the petiole or near the base of the midrib beneath. Flowers small, green or white in simple or panicled spikes; hermaphrodite or the upper flowers of the spike sometimes male. Calyx tube produced above the ovary with a companulate mouth; limb of 5 short, valvate, triangular lobes, deciduous. Petals 0. Stamens 10 in two series; epigynous disk within them densely hairy. Ovules two to three. Fruit ovoid, very various in size, smooth or angular or winged with two to five wings, indehiscent, coriaceous. Bark flat or slightly curved. Outer surface smooth and white. Inner surface soft and reddish.

Ayurvedic description — *Rasa — kasaya; Guna — lagu, rooksha; Veerya — sheeta; Vipak — katu.*

Action and uses — *Sheeta veerya, kapha batsamak, hiridya, raktsangrahik, sothaghana, sandhaniya, bran ropan.*

Chemical constituents — Bark contains a large amount of calcium salts but smaller amounts of aluminum and magnesium salts. The watery extract contains as much as 23% of calcium salts and 16% tannins. Bark contains a crystalline compound arjunine, a lactone, arjunetin, essential oil, tannin, reducing sugars, and coloring matter.[50] Arjuna also contains a saponin-like substance which is responsible for the diuretic property. Chopra et al.[178] reported the isolation of arjunalic acid, saponin, and (+)-leukodelphinidin.

Pharmacological action — Its bark is recommended in heart diseases. It is a tonic, astringent, and lithontriptic.[1]

Medicinal properties and uses — Arjuna is popularly used as a cardiac tonic.[866] In doses of 14 to 28 ml in the form of decoction it is very useful in hemorrhages and other fluxes; also in diarrhea, dysentery, and sprue. It is also useful in bilious affections and as an antidote to poisons. The decoction of thick bark of arjuna (3 g), cane sugar (24 g), and boiled cow's milk (200 ml) is highly recommended in heart disease complicated with endocarditis, pericarditis, and angina. Ayurvedic physicians recommend the use of arjuna bark in derangement of all the three humors, *kapha, pitta,* and *vayu,* and all sorts of conditions of cardiac failure and dropsy. Powdered bark relieves hypertension, and has a diuretic and a general tonic effect in case of cirrhosis of the liver. It is considered a cardiac stimulant.[178] Bark is prescribed for bilious affections, for sores, and as an antidote to poisons. The ash of bark is prescibed in scorpion sting. Fruit is tonic and deobstruent.[1,3]

Doses — Swarasa — 12 to 24 gr; decoction — 60 to 120 ml; powder — 1 to 3 gr.

TERMINALIA BELLERICA (Gaertn.) Roxb.   *Family:* COMBRETACEAE

Vernacular names — Sanskrit, Vibhitaka; Hindi, Bhaira; English, Belleric myrobalan; Bengali, Bohera; Nepalese, Bihala; Sinhalese, Bulu; Japanese, Bererikamiro baran; Chinese, Bang nut; German, Myrobalane; Unani, Bahera; Arabian, Balilaj; Persian, Balilah; Tibetan, Ba-ru-ka; Burmese, Bankha; Malaysian, Jilawei; Tamil, Thantri, Tani.

Habitat — Found throughout India in the plains and lower hills with the exception of arid tracts in the west.

Parts used — Fruits.

Pharmacognostical characteristics — A large deciduous tree 60 to 80 ft high, leaves 3 to 8 in., broadly elliptic, cuneate at base, clustered at the ends of branchlets, generally punctate on the upper surface; petiole 1 to 4 in. Spikes solitary, axillary, simple. Flowers pale greenish yellow, with an offensive odor, upper flowers of the spike male, lower hermaphrodite. Drupe 1 to $1^1/_2$ in. diameter, subglobose, gray-tomentose, obscurely 5 angled when dried; stone hard and pentagonal and contains a sweet oily kernel.[28]

Ayurvedic description — *Rasa—kasaya; Guna—rooksha, laghu; Veerya—ushna; Vipak —madhur.*

Action and uses — *Tridosha har, sothahar, rakt stamvan, dipan, anuloman, krimighan, kaf nasak, vajikarn, dhatu bardhak, cekachusya.*

Chemical constituents — *T. belerica* fruit contains about 17% tannin substances. Heartwood, bark, and fruits contain ellagic acid and the seed coat of the fruit contains a gallic acid. Seeds contain about 20 to 25% yellow oil.[867]

Pharmacological action — Astringent, tonic, expectorant, and laxative.

Medicinal properties and uses — Fruit is useful in cough, hoarseness, and eye diseases. It is one of the constituents of "triphala", which is prescribed in diseases of the liver and gastrointestinal tracts.

Oil expressed from the kernel is used as a dressing for the hair; also externally it is applied in rheumatism. Unripe fruit is purgative. Dried ripe fruit is astringent and employed in dropsy, piles, and diarrhea.[1,3]

Doses — Powder — 1 to 3 gr.

TERMINALIA CHEBULA Retz.   *Family:* COMBRETACEAE

Vernacular names — Sanskrit, Haritaki; Hindi, Harara; English, Chebulic myrobalan; Bengali, Haritaki; Nepalese, Halla; Sinhalese, Aralu; Japanese, Shirobarannoki; Chinese, He lile; German, Rispiger Myrobalanenbaum; Unani, Halaila zard; Arabian, Ahlilaj asfar; Persian, Halaila-e-zard; Tibetan, A-ru-ra; Burmese, Pangah; Malaysian, Buah kaduka; Tamil, Kadukkai.

Habitat — This tree is abundantly found in the forests of northern India, central provinces, and Bengal; common in Madras, Mysore, and in the southern parts of Bombay.

Parts used — Dried fruits, immature fruits; mostly the outer skin of the fruits.

Pharmacognostical characteristics — A moderate-sized or large deciduous tree. Leaves not clustered, often subopposite, 4 to 7 in., ovate or elliptic, usually acute, rounded at the base, petiole 1 in., often panicled. Flowers dull white with an offensive smell, all hermaphrodite. Drupe $1/3$ to $1 1/4$ in. long, glabrous, ellipsoidal or obovoid, yellowish green, five to six ribbed when dry due to the five-ribbed endocarp.[28]

Ayurvedic description — *Rasa — kasaya; Guna — lagu, rooksha; Veerya — ushna; Vipak — madhur.*

Action and uses — *Tridoshahar, soth har, bran sodhan, dipan, pachan, anuloman, balya, medhya, sonitasthapan, verisya, praja sthapan, chachu bal bardhak, mutral, rasayan.*

Chemical constituents — Fruits contain about 30% of an astringent substance which is due to the characteristic principle chebulinic acid; also contains tannic acid (20 to 40%), gallic acid, resin, and some purgative principle of the nature of anthraquinone.[50]

Pharmacological action — It is a safe and effective purgative, astringent, and alterative. Unripe fruits are more purgative and the ripe ones are astringent.

Medicinal properties and uses — Myrobalans are used in fevers, cough, asthma, urinary diseases, piles, and worms. It is also useful in chronic diarrhea and dysentery, flatulence, vomiting, colic, and enlarged spleen and liver. Infusion is used as a gargle in sore mouth and stomatitis, spongy and ulcerated gums. Chebulic myrobalans are extensively used in combination with belleric and embolic myrobalans under the name of "triphala" and also as adjuncts to other medicines in numerous diseases.

Doses — Powder — 1 to 1.5 gr; powder — 3 to 6 gr (laxative); decoction — 56 to 112 ml.

THALICTRUM FOLIOLOSUM DC.   *Family:* RANUNCULACEAE

Vernacular names — Sanskrit, Tryamana; Hindi, Pilijari, Mamira; English, Gold thread; Bengali, Trayamana; Nepalese, Tamal; Sinhalese, Tra yamana; Unani, Mamiran-e-chini; Arabian, Mamiran-e-seeni; Persian, Mamiran-e-chini.

Habitat — Throughout temperate Himalayas, at 5000 to 8000 ft, Khasia Hills, and in Nilgiris.

Parts used — Root.

Morphological characteristics — It is a tall, rigid, perennial, glabrous herb: Leaves pinnately compound; leaflets orbicular, crenate; flowers blue, polygamous, in much-branched panicles; achenes oblong, ribbed.

Ayurvedic description — *Rasa — tikta; Guna — rooksha; Veerya — ushna; Vipak — katu.*

Action and uses — *Lakhan, sothahar, dipan, pachan, anuloman, mutral, visuchica har, sarpa, visha har.*

Chemical constituents — Roots of this plant are reported to contain two alkaloids, berberine and thalictrine, each about 0.2% yield. Chatterjee et al.[868] investigated the rhizome which yielded berberine (0.35%), palmatine (0.03%), jatrorrhizine (0.02%), but no thalictrine. It was presumed that thalictrine is a mixture of palmitine and a phelonic base jatrorrhizine. Gopinath et al.[869] identified thalictrine.

Pharmacological action — Tonic, aperient, purgative, diuretic, and febrifuge.

Medicinal properties and uses — Root is a good remedy for chronic dyspepsia; useful in convalescence after acute diseases and as application for ophthalmia It is useful in jaundice, flatulence, and visceral obstructions. Root also possesses aperient and diuretic properties; a good substitute for rhubarb. A snuff prepared from it clears the brain; used in coryza; and relieves toothache.[50]

This is one of the few berberine-containing plants which is used as household treatment for conjunctivitis, inflammation of cornea, and generally in the form of a collyrium .[1]

Doses — Infusion — 14 to 28 ml; tincture — 1 to 2 ml; root powder — 0.5 to 1 gr.

TINOSPORA CORDIFOLIA (Willd.) Miers   Family: MENISPERMACEAE

Vernacular names — Sanskrit, Guduchi; Hindi, Gilo; Bengali, Gurach; Nepalese, Gurgau; Sinhalese, Raskinda; Japanese, Ibonashitsu zurabuji; Chinese, Kuan chu hisng; Unani, Gilo; Arabian, Julu; Persian, Gulue; Tibetan, Sle-tres; Burmese, Singomone; Tamil, Sindil.

Habitat — The plant occurs throughout tropical regions of India.

Parts used — Stem, root, and whole plant.

Pharmacognostical characteristics — A large climber with succulent stems, throwing aerial roots like the common banyan (ficus) tree. Stems and branches specked with white glands. Leaves 5 to 10 cm, ovate or roundish, 7 to 9 nerved, petioles slightly shorter than leaves. Flower minute, male and female separate; male flowers grouped in axils of bracts; female solitary. Fruits size of a pea, red.[435]

Ayurvedic description — *Rasa — tikta, kasaya; Guna — guru, snigdha; Veerya — ushana; Vipak — madhur.*

Action and uses — *Tridosh samak, khutaghan, trisa nigraha, dipan pachan, rakt sodhak, rakt wardhak, kafaghan, veersya, prameha har, daha samak, rasayan.*

Chemical constituents — Stem contains 0.1% of a bitter substance, another bitter principle, and a neutral substance.[50] Kidwai et al.[870] obtained the crystalline substance, viz., (1) giloin, $C_{23}H_{32}O_{10}H_{20}$, a glycoside, mp 226 to 228°C after drying to a constant weight over P_2O_5, yield 0.2%, and bitter at a dilution of 1:10,000; (b) gilenin, $C_{17}H_{18}O_5$, a nonglycoside bitter, mp 210 to 212°C, yield 0.001%, and bitter in dilution of 1:1000; (c) gilosterol, $C_{28}H_{48}O$ mp 192 to 193°C. Chatterjee and Ghosh[871] reported the presence of tinosporine, the furanoid bitter principle. Hanuman et al.[872] isolated a new clerodane furano-diterpene 2 with the molecular formula $C_{20}H_{22}O_8$ from the stems of *T. cordifolia*. Its spectral characteristics are very similar to those of known furano-diterpenes.

Pharmacological action — Alterative, diuretic, aphrodisiac, tonic.[1]

Medicinal properties and uses — The starch obtained from the roots and stems of the plant is useful in diarrhea and dysentery; it is also a nutrient. The fresh plant is more efficacious than the dried plant. Its watery extract, known as Indian quinine, is very effective in fevers due to cold or indigestion; the plant is commonly used in rheumatism, urinary diseases, dyspepsia, general debility, syphilis, skin diseases, bronchitis, spermatorrhea, and impotence. In gonorrhea, cough, and chronic fever the juice of the fresh plant is administered in doses of 56 to 112 ml with long pepper and honey. Antidote for snakebite.[1]

Doses — Decoction — 56 to 112 ml.; powder — 1 to 3 gr; extract — 1 to 2 gr.

TRIBULUS TERRESTRIS Linn.*   Family: ZYGOPHYLLACEAE

Vernacular names — Sanskrit, Gokshura; Hindi, Chota gokhru; English, Small caltrops; Bengali, Gokhuri; Nepalese, (Newari), Gokhur; Sinhalese, Nerenchi; Japanese, Hamabishi; Chinese, Chili; French, Croix de chevalier; Unani, Gokshuru khurd; Arabian, Khara-khusk; Persian, Khasak; Tibetan, Gze-ma; Burmese, Charatte, Malaysian, Nerungil, Pushtu, Malkundai; Tamil, Nerunjil.

Habitat — This trailing plant is common in sandy soil throughout India, Pakistan, and Sri Lanka.

Parts used — Fruit and root; the entire plant is also used.

Pharmacognostical characteristics — A prostrate spreading herb, densely covered with minute hairs. Leaves in opposite pairs, 5 to 8 cm long, compound; leaflets 4 to 7 pairs, 8 to 12 mm long. Flowers pale yellow, 1 to 1.5 cm diameter, growing solitary opposite to leaves or in axils of leaves on peduncles. Fruit usually hairy, cocci each with two very sharp, rigid spines and two shorter ones.[28]

Ayurvedic description — *Rasa—madhur; Guna—guru, snigdha; Veerya—sheeta; Vipak—madhur.*

Action and uses — *Bat pit samak, mutral, rakt pit nasak, sothhar, anuloman, grahi, veerisya, balya.*

Chemical constituents — Extract of the powdered fruit contains an alkaloid, a resin, fat, and mineral matter (14%); fruit also contains an aromatic substance,[50] when burnt it gives off a fragrant odor. The fruit contains an alkaloid in traces of 0.001%, a fixed oil (3.5%) consisting mainly of unsaturated acids, an essential oil in very small quantities, resins, and fair amounts of nitrates. An aqueous solution of the tartrate of the alkaloid, after removal of the alkaloid, was found to contain sugar, but no physiologically active substance.[50] Chopra et al.[178] reported the presence of harman in the herb and harmine in the seeds.

Pharmacological action — Plant and spiny fruit are cooling, demulcent, diuretic, tonic, and aphrodisiac. The diuretic properties of the plant are due to the presence of a large amount of nitrates as well as the essential oil which occurs in the seeds.[1,3]

Medicinal properties and uses — Plant and dried spiny fruits are used in decoctions or infusions in cases of spermatorrhea, phosphaturia, and diseases of the genitourinary system such as dysuria, gonorrhea, gleet, chronic cystitis, calculous affections, urinary disorders, gout, and impotence; also in uterine disorders after parturition, kidney diseases, and gravel. It is used in northern India in cough, diseases of the heart, and suppression of urine. The fruits are also an ingredient of the *dasamula kvatha*. The action of the drug on the mucous membrane of the urinary tract is stated to resemble that of *bachu* leaves and *Uva ursi* flowers. The drug undoubtedly has diuretic properties.[28] Decoction of the fruits with addition of carbonate of potash is given in painful micturition.

Doses — Infusion — 4 to 8 ml; decoction — 56 to 112 ml; powder — 0.5 to 1 gr.

---

* Illustration is on page 326.

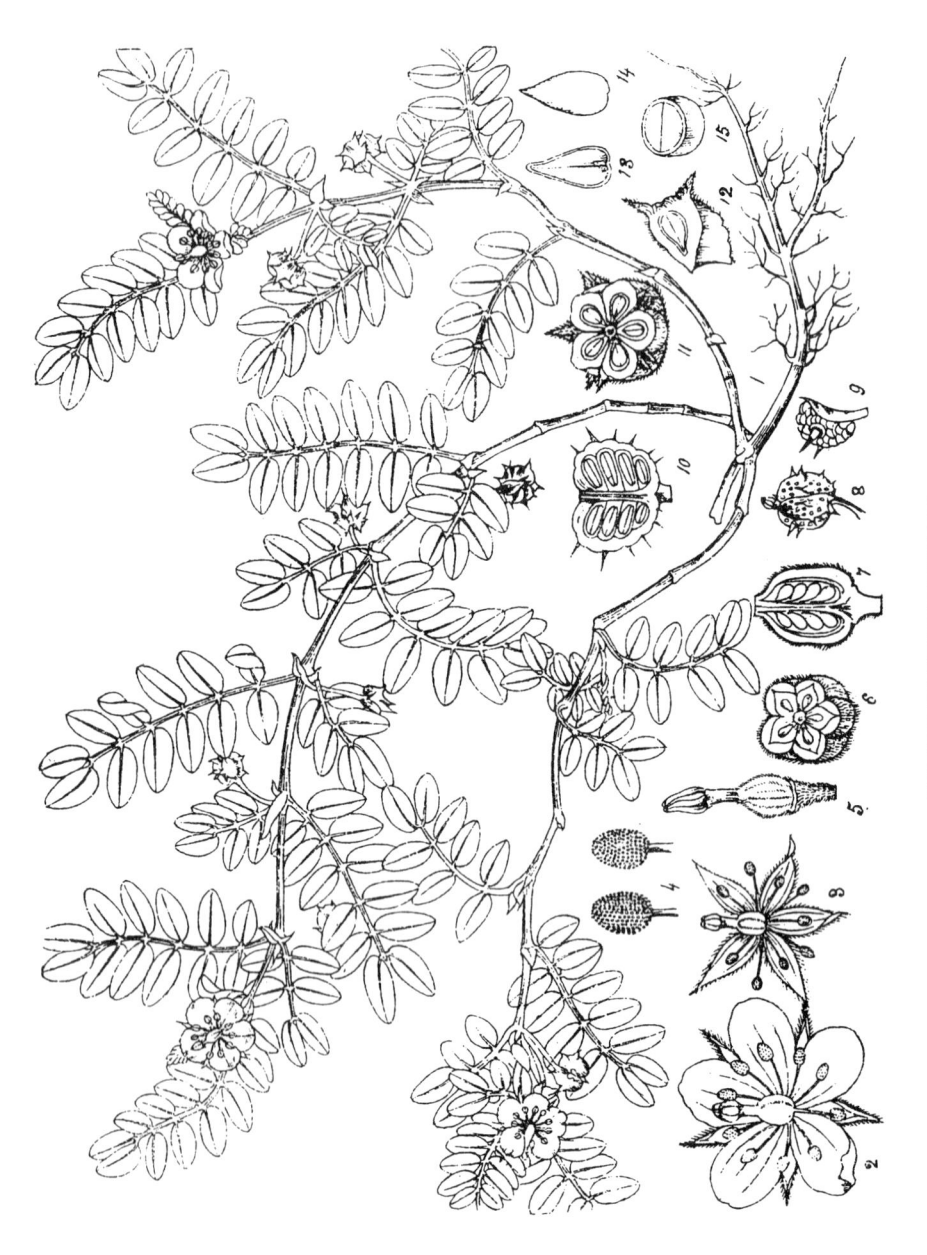

326   *CRC Handbook of Ayurvedic Medicinal Plants*

TRIBULUS TERRESTRIS Linn.

TRIGONELLA FOENUM-GRAECUM Linn. Family: LEGUMINOSAE

Vernacular names — Sanskrit, Medhika; Hindi, Methi; English, Fenugreek; Bengali, Methi; Nepalese, Mee; Sinhalese, Uluwa; Japanese, Koroha; French, Fenugre; Unani, Methi; Arabian, Hulba; Persian, Shamlit; Burmese, Penanlazi; Tamil, Vendaya.

Habitat — Found growing wild in Punjab and Kashmir, but now cultivated all over India.

Parts used — Whole plant, leaves, and seeds.

Morphological characteristics — An annual herb, raised from seed. Leaves compound 3 to 4 in. long; stipules entire; leaflets 3, lanceolate or obovate, toothed. Flowers axillary, 1 to 2 sessile, yellow. Pods 2 to 3 in. long, thin, pointed, 10 to 20 seeds, emitting a peculiar aroma, brownish yellow, having an oblique furrow along a part of their length.

Trease and Evans[69] observed that different commercial samples vary according to their geographic origin. The seeds may be irregularly rhomboidal, oblong, or square in outline and yellow, olive green, or yellowish brown to dark brown in color.

Ayurvedic description — *Rasa — madhur, tikta; Guna — laghu; Veerya — ushna; Vipak — madhur.*

Action and uses — *Tridosha samak, dipan, anuloman, grahi, sotha har, jawraghana, kapha nisarak, verisya, sandhaniya, visachana.*

Chemical constituents — Seeds contain alkaloid trigonelline and choline, saponin, essential oil, fixed oil, prolamin, mucilage, bitter extractive, and a coloring substance. Air-dried seeds contain 0.38% trigonelline and 3 mg% nicotinic acid.[1] Chopra et al.[178] reported the presence of 1.0% saponin in the seeds; and lactation-promoting factor in fenugreek oil. Seeds also contain an oil (7%) used as a galactagogue. The defatted meal contains indigestible hydrophilic mucilage (50%). Fazli and Hardman[873] reported the presence of 0.8 to 2.2% steroidal sapogenins, particularly diosgenin, which is contained in the oily embryo of the seeds. Fenugreek seed is attracting great interest as a source material for extraction of diosgenin.[874,875]

Pharmacological action — Demulcent, emmenagogue, aromatic, diuretic, nutritive, tonic, lactagogue, emollient, astringent, carminative, and aphrodisiac.[1]

Medicinal properties and uses — A poultice of the herb is applied to reduce swellings; a passary made of the herb or of lint saturated with its decoction is used for the treatment of leukorrhea. The leaves are aperient; they relieve indigestion and bilious disorders. Paste of the leaves is applied over swelling and burns; it is a hair tonic and is applied to the head for premature loss of hair.[1,3] Leaves are used both internally and externally for their cooling properties.

Seeds are eaten, boiled or roasted, as a vegetable in dyspepsia, diarrhea, dysentery, colic, flatulence, rheumatism, enlargement of the liver and spleen, and chronic cough. As a lactagogue, a gruel of the seeds is given with milk and sugar. An infusion of seeds is given to smallpox patients as a cooling drink. Toasted and then infused, it is administered for dysentery. A paste of the seeds is used as a cosmetic to keep the skin smooth and clean; poultice of the seeds is applied to reduce swellings and inflammatory affections; the mucilage produced when they are soaked in water is locally used over boils, carbuncles, and abscesses.

Doses — Powder — 3 to 6 gr.

URGINEA INDICA Kunth *Family:* LILIACEAE

Vernacular names — Sanskrit, Vana palandam; Hindi, Jangli piyaz, Kanda; English, Indian squill; Bengali, Jangli piyaz; Nepalese, Banapayaj; Sinhalese, Vallum; German, Indische Meerzwiebel; Unani, Ansal; Arabian, Basl-ul-barri; Persian, Piyaz-e-dasht; Tamil, Kattuvvengayam.

Habitat — Found in drier sub-Himalayan tracts of western Himalayas; also in Bihar, Konkan, and on the Coromandel Coast.

Parts used — Bulb.

Morphological characteristics — Scapigerous herbs with tunicate pale bulbs, 2 to 4 in. long, ovoid, thick. Leaves appearing after the flowers, 6 to 18 × $1/2$ to 1 in., subbifarious, linear, acute, flat. Scape erect, 12 to 18 in. long, brittle. Flowers drooping or spreading, dingy brown or greenish white, very distant; bracts minute, evanescent; pedicels 1 to $1 1/2$ in. long. Perianth segments $3/8$ in. long, 3-nerved in the middle. Filaments flattened. Capsule $1/2$ to $3/4$ in. long, oblong, the cells 6 to 9 seeded.[876]

Ayurvedic description — *Rasa — katu, tikta; Guna — teekshna, laghu; Veerya — ushna; Vipak — katu.*

Action and uses — *Vatta kapha samak, pitta wardhak, bran kark, balya, mutral, artwojanan, vamak, vachan, krimighan.*

Chemical constituents — Seshadri and Subramanium[878] observed that *U. indica* as available in the market yields two glycosidal fractions: (1) the water-insoluble part (A) is found to be a mixture of glycosides, which on hydrolysis with acid yields, besides glucose and rhamnose, scillaren A, as predominantly the major component — a new aglucone melting at 265 to 267°C with molecular formula $C_{26}H_{32}O_4$; and (2) the water-soluble part (B) resembling scillaren B which gives on hydrolysis a crystalline aglucone, mp 227 to 230°C, and glucose. Later studies revealed that glycoside from *U. indica* is identical with scillaren A.[877]

Pharmacological action — In small doses, it is expectorant, cardiac, stimulant, diuretic, deobstruent, and emmenagogue; in large doses it is emetic and cathartic.[1]

Medicinal properties and uses — Bulb is rubbed on the soles of the feet to relieve burning; in small doses it acts like an expectorant. Bulb mixed with some other ingredients is given in the form of syrup in acute bronchitis, where the sputa are tenacious and scanty, in chronic bronchitis associated with emphysema, and in spasmodic croup. Physiologically, the drug slows heartbeat and increases the flow of urine. It is excreted by the bronchial, genitourinary, and gastrointestinal secretions. In excessive doses, it is a narcotic acrid poison causing nausea, stranguary, and bloody urine. In the form of tincture, it is given in cardiac and renal dropsy, ascites; also in asthma, rheumatism, calculous and paralytic affections, leprosy, and skin diseases. A powder of it is locally applied to remove warts.[1,3]

Doses — Powder — 125 mg; tincture — 0.5 to 2 ml; syrup — 2 to 4 ml.

VALERIANA JATAMANSI Jones
(Syn. *V. WALLICHII* DC.)

Family: VALERIANACEAE

Vernacular names — Sanskrit, Tagara; Hindi, Tagar, Mushkbala; English, Indian valerian; Bengali, Tagar; Nepalese, Sugandhawala; Sinhalese, Thuwarala; Japanese, Himarayakan okoso; German, Indische Baldrian; Unani, Tagar; Arabian, Asarum; Persian, Asaroon; Tibetan, Span-spos; Tamil, Kiranthitakaram.

Parts used — Whole plant and root.

Habitat — Indigenous to temperate Himalayas and found from Kashmir to Bhutan.

Morphological characteristics — Perennial, erect herb; 2 to 3 ft high. Root yellowish brown, 1.5 to 7 cm long, 1 to 2 mm thick; rhizome yellowish to brownish, 4.0 to 7.0 cm long and 1 cm thick, subcylindrical; leaves radical, persistent, stalked, cordate-ovate, acute, toothed; flowers white or tinged with pink in a terminal corymb 2.5 to 8 cm across, often unisexual, and the male and female on different plants. Fruit small, smooth, without hairs. The market samples of the rhizomes are unbranched and somewhat flattened dorsiventrally. The upper surface bears leaf scars and the lower surface roots or root scars. The rhizome breaks with a short fracture and the horny interior shows a thin, dark bark, a well-marked cambium, about 12 to 15 light-colored xylem bundles, a dark pith, and medullary rays. The odor is valerianous and taste bitter and camphoraceous.[69]

Ayurvedic description — *Rasa—tikta, katu, madhur, kasaya; Guna—laghu, snigdha, sar; Veerya—ushna; Vipak—katu.*

Action and uses — *Tridosha har, bran ropan, mastiskya blaya, dipan, shul prasaman, sarak, hiridyotegek, swashar, vajikaran, pitta samak, jawarghana, chachyusya.*

Chemical constituents — Rhizomes contain a large proportion of volatile oil (ethereal valerianic oil) (1%), containing esters of iso-valerianic acid and formic acid. The rhizome is reported to contain tartaric, citric, malic, and succinic acids. An active compound from roots and rhizomes has been isolated. Maliol has been isolated from the oil distilled from the roots of rhizomes of *V. wallichii*.[178]

Pharmacological action — Stomachic, stimulant, carminative, and sedative.

Medicinal properties and uses — Valerian is given for nervous debility and failing reflexes; also as a hypnotic, and spastic disorders like chorea and gastrospasms. Root is useful in hysteria, epilepsy, shell shock, and neurosis.[1,3]

Doses — 1 to 3 grs.

VANDA ROXBURGHII R. Br.
(Syn. *V. TESSELLATA* Hook. ex G. Don.)

Family: ORCHIDACEAE

Vernacular names — Sanskrit, Rasna; Hindi, Rasna; Bengali, Rasna; Nepalese, Rasna; Sinhalese, Arattha; Unani, Rasha; Arabian, Zanjabeel-e-shami; Persian, Zanjabeel-e-shami.
Habitat — An epiphytic herb which grows throughout the hotter parts of India.
Parts used — Whole plant.

Morphological and pharmacognostical characteristics — Stem 30 to 60 cm long, stout, simple or branching roots. Leaves thickly coriaceous, 15 to 20 × 1.3 to 3 cm, recurved, obtusely keeled with usually two unequal rounded lobes and an acute interposed one. Flowers 6 to 10, in racemes 15 to 25 cm long; bracts scarious, 3 mm long, ovate, acute; pedicels with ovary 3.5 to 5 cm long. Sepals yellow, tessellated, with brown lines and white margins. Petals yellow with brown lines and white margins, 13 mm wide; shorter than sepals, lip 16 mm long, bluish dotted with purple. Column very short; pollinia ellipsoid or subglobose, caudicle short, broad; gland large. Capsules 7.5 to 9 cm long, narrowly clavate-oblong with acute ribs and short pedicel.

Presence of velamen and extremely thick-walled endodermis with passage cells in the root is seen. The leaves are characterized by the presence of a fibrous thick-walled hypodermis. Stomata are present on both surfaces. Raphides and volatile oil are present in the parenchymatous tissue of the leaf, stem, and root. The ground tissue of the stem consists of large, thick-walled lignified cells and small, thin-walled nonlignified cells with scattered vascular bundles.[879,880]

Ayurvedic description — *Rasa — tikta; Guna — guru; Veerya — ushna; Vipak — katu.*

Action and uses — *Kapha vat nasak, sotha nasak, swashar, uteder roga nasak, jawarghana, vishaghan, vatta vikara.*

Chemical constituents — It contains an unnamed alkaloid and a substance of glycosidal nature, tannin, saponins, sterols, fatty oil, resins, and coloring matter.

Pharmacological action — Anodyne, carminative, expectorant, and nervine tonic. Stimulates cholinergic nerve endings, lowers blood pressure.

Medicinal properties and uses — It enters into the composition of several medicated oils for external application in rheumatism and allied disorders and also diseases of the nervous system. It is a remedy for secondary syphilis and scorpion sting. Leaves made into a paste applied to the body during fever; juice is introduced into aural meatus as a remedy for otitis media.[50]

The root is bitter, heating, alexiteric, antipyretic; useful in dyspepsia, bronchitis, inflammations, rheumatic pains, diseases of the abdomen, hiccup and tremors.[108]

VERNONIA CINEREA Less.   *Family:* COMPOSITAE

Vernacular names — Sanskrit, Sahadevi; Hindi, Sahadevi; English, Ash-colored fleabane; Bengali, Kukseem; Nepalese, Sahadevi; Sinhalese, Monorkudumbia; Japanese, Yanbaruhikodai; Unani, Sahdai; Malaysian, Ekor kudah; Tamil, Neichatti.

Habitat — Indigenous to Bengal, east and west coasts of India, and South India.

Parts used — Whole plant, seed, leaves, and root.

Morphological characteristics — Annual, erect, 15 to 75 cm high; stem stiff, cylindric, striate, more or less pubescent, slightly branched. Leaves petioled, 2.5 to 5 × 2 to 3.8 cm (the upper leaves the smallest), variable in shape, broadly elliptic or lanceolate, obtuse or acute, shortly mucronate, more or less pubescent on both sides, irregularly toothed or shallowly crenate-serrate; petioles variable, 6 to 13 mm long. Heads small, about 20-flowered, 6 mm diameter, in lax divaricate terminal corymbs, with a minute linear bract beneath each head of flowers and with small bracts in the forks of the peduncles; flowers pinkish violet. Involucral bracts linear-lanceolate, awned, silky on the back. Pappus white, the exterior row short, about 0.5 mm long. Achenes 1.25 mm long, oblong, terete (not ribbed), slightly narrowed at the base, clothed with appressed white hairs.[108]

Ayurvedic description — *Rasa—katu; Guna—laghu, rooksh; Veerya—ushna; Vipak—katu.*

Action and uses — *Kapha vatta samak, sothahar, badana sthapan, jawaraghana, anuloman, krimighan, rakt sodhak, asmarivedan, mutral.*

Pharmacological action — Febrifuge, diaphoretic, alterative.

Chemical constituents — The whole plant gives the chemical constituents: amyrin acetate, lupeol acetate, β-amyrin, lupeol, β-sitosterol, stigmasterol, α-spinasterol and KCl.[178]

Medicinal properties and uses — Decoction of the plant is used to promote perspiration in febrile conditions. Seeds are employed as an alexipharmic and anthelmintic; also as alterative in leprosy and chronic skin diseases. Whole plant is a good remedy for spasm of the bladder and stranguary. Poultice of the leaves is a useful application in guineaworms. Flowers are administered for conjunctivitis. Root is given for dropsy. It is also useful in scorpion sting. Juice of plant given for piles.[50]

Doses — Infusion — 14 to 28 ml.; decoction — 56 to 112 ml; seed powder — 0.5 to 1 gr.

VIOLA ODORATA Linn. *Family:* VIOLACEAE

Vernacular names — Sanskrit, Banaphsha; Hindi, Banaphsha; English, Wild violet; Bengali, Banosa; Japanese, Nioisumaire; German, Wildnechendes Veilchen; Unani, Banafsha; Arabian, Banafsaj; Persian, Banafshah.

Habitat — A herb found in Kashmir and in the temperate western Himalayas above 5000 ft.

Parts used — Leaves, stem, and flowers.

Morphological characteristics — Perennial stock short, but sometimes branched, knotted with the remains of the old leafstalks and stipules, and usually producing creeping runners or scions. Leaves in radical (or rather terminal) tufts, broadly cordate, rounded at the top, and crenate, downy, or shortly hairy with rather long stalks. Stipules narrow-lanceolate or linear, and entire. Peduncles about as long as the leafstalks, with a pair of small bracts about half way up. Flowers nodding, of the bluish-purple color named after them or white, more or less scented. Sepals obtuse. Spur of the lower petal short. Stigma pointed, horizontal, or turned downward.[881,882]

Ayurvedic description — *Rasa—madhur, tikta; Guna—laghu, snigdha; Veerya—sheeta; Vipak—madhur.*

Action and uses — *Vatta pitta samak, kapha nisark, daah samak, rakt rodhak, sothahar, anuloman, virachan, rakt pitta samak, jawarghana.*

Chemical constituents — Flowers and root contain an alkaloid "violine", a glycoside violaquercitrin, which is probably identical with rutin, and a saponin. Roots, leaves, and blossoms contain methyl salicylate in the form of glucoside.[50] Flowers also contain traces of a volatile oil. The roots and rhizomes of the plant are also reported to contain an alkaloid odoratine.[10]

Pharmacological action — Whole plant is emollient, demulcent, diaphoretic, diuretic, aperient, antipyretic, and febrifuge.[1]

Medicinal properties and uses — The fresh leaf of *V. odorata* is a reputed drug for the treatment of cancer; leaves relieve pain of cancerous growth, especially in the throat; 2.5 oz of the fresh leaves are infused in a pint of boiling water in a covered stone jar for 12 h; the strained liquid is taken in the course of a day, in doses of a full glass at a time; for the treatment of cancer of the tongue, only half the quantity is taken in a day; the other half is used to foment the tongue. The leaves are emollient and laxative.[1,3]

The underground root is emetic and purgative; an infusion made with 2 oz of the roots acts as a purge and emetic; its juice causes nausea and vomiting.

The flowers are given in bilious affection, epilepsy, nervous disorders, prolapse of the rectum and uterus, and inflammatory swelling. The flowers are popularly used as a diaphoretic and for the treatment of coughs, sore throat, kidney diseases, and liver disorders. It is used either as an infusion or as a syrup.[50]

Doses — Flower infusion — 28 to 56 ml.

VITIS VINIFERA Linn.   Family: VITACEAE

Vernacular names — Sanskrit, Draksha; Hindi, Angur; English, Grapes; Bengali, Angur; Nepalese, Dhakha; Sinhalese, Muddarppalam; Japanese, Budo; Chinese, P'u t'ao; German, Rosinen; French, Raisin, Vignecultive; Unani, Angur; Arabian, Angwi; Persian, Angur; Tibetan, R Gun-brum; Burmese, Sabisi; Pustu, Kwar; Tamil, Thrashai.

Habitat — Cultivated in some parts of India, in the Deccan, Mysore, Andhra, Punjab, Uttar Pradesh, and in Kashmir.

Parts used — Fruit and leaves.

Morphological characteristics — Shrubs climbing by means of tendrils, stems and branches nodose, juice copious, watery. Leaves alternate, simple or compound, stipulate, glossy. Flowers bisexual, usually small and greenish, in compound usually leaf-opposed inflorescences; peduncles often transformed into simple tendrils. Calyx smooth, entire or 4 to 5 toothed or lobed. Petals 4 to 5, valvate, free, usually caducous. Stamens 4 to 5, opposite the petals, inserted at the base of disk; anthers free. Ovary usually partially sunk in the disk, 2 to 6 celled; ovules 1 to 2 in each cell. Fruit a berry, often watery, 1 to 6 celled; cells 1 to 2 seeded.

Ayurvedic description — *Rasa — madhur; Guna — snigdha; Veerya — sheeta; Vipak — madhur.*

Action and uses — *Ratak rodhak, kapha nisark, daah samak, anuloman, rakt pitta samak, jawarghana.*

Chemical constituents — Fruit contains malic, tartaric, and racemic acids. Unripe fruit has oxalic acid. Arsenic (up to 0.05 mg in 100 cc fruit juice) has been reported.[50] Seeds contain a fixed oil and tannic acid. Skin of the fruit contains tannin.

Pharmacological action — Grapes are demulcent, laxative, refrigerant, stomachic, diuretic, and cooling.

Medicinal properties and uses — Grapes are useful in bilious dyspepsia, hemorrhages, dysuria, in chronic bronchitis, heart diseases, and gout. Grape juice is given to children for constipation. Dried grapes or raisins are useful in thirst attendant on fevers, in cough, catarrh, jaundice, and in subacute cases of enlarged liver and spleen. Leaves are used in diarrhea. The bland fixed oil of the seeds or the seeds in powder have proved efficacious in many cases of chronic diarrhea.[1]

Doses — According to digestive power.

WITHANIA SOMNIFERA Dunal   Family: SOLANACEAE

Vernacular names — Sanskrit, Ashvagandha; Hindi, Asgandh; English, Winter cherry; Bengali, Ashwagandha; Nepalese, Aasoganda; Sinhalese, Amukkara; Japanese, Aswangandha; Unani, Asgandh valaiti; Arabian, Bahman; Persian, Bahman; Tibetan, Ba-dzi-gandha; Pustu, Kutilad; Tamil, Amurkkuralckizhangu.

Habitat — The plant is distributed throughout the drier regions of India, especially in wastelands ascending to an altitude of 5500 ft in the Himalayas. It is also cultivated.

Parts used — Root and leaves.

Pharmacognostical characteristics — A small or middle-sized undershrub, erect, grayish or hoary, branching perennial about 3 to 5 ft in height. Stem and branches covered with minute star-shaped hairs. One or more fairly long tuberous roots and short stem. It is in flower nearly throughout the year. Leaves simple, up to 10 cm long, ovate hair-like branches, petiolate, and alternate. It bears small, about 1 cm long, greenish or yellow flowers; borne together in short axillary clusters. Fruit 6 mm diameter, globose, smooth, red, enclosed in the inflated and membranous calyx.

A transverse section of the root shows only a narrow cork and cortex enclosing a wide central woody region which is parenchymatous for the most part and storage in function. A transverse section of a fresh root 2 cm in diameter shows a cork, light brown in color, at the outside, 320 µm in width. Cork is composed of 12 to 16 rows of thin-walled cubical to slightly tangentially elongated cells. One or two rows of cork cells just surrounding the phellogen have a yellow color. Most of the cortical cells are fully packed with moderately large starch grains. The starch grains are compound, having two to five components, and almost rounded, with a diameter of 15 to 30 µm.

The different components are also found separately. There are many smaller starch grains 9 µm in diameter. The wood which forms the major part of the root is 12 to 14 mm in diameter. It is composed almost entirely of secondary xylem and woody rays. The peripheral part of the wood is composed of thick-walled fibers, wherein lie small patches of thin-walled parenchyma as well as scattered or radially arranged rows of small groups of vessels and tracheids, all together forming a circular zone or band of thick-walled tissue. The parenchyma cells are all thin walled, fairly large, rectangular, and most of them fully loaded with starch grains similar to those in the cortical cells.[883,884]

Ayurvedic description — *Rasa* — *madhura, kasaya, tikta; Guna* — *lagu, snigdha; Vipaka* — *madhura; Veerya* — *ushna*.

Action and uses — *Kaf bat samak, soth har, badana asthapan, nari doraybalya, batbikara*.

Chemical constituents — Extraction with 45% alcohol yields the highest percentage of alkaloids. Isolation of nicotine, somniferiene, somniferinine, withanine, withananine, and pseudowithanine has been reported.[887] Isolation of sucrose, β-sitosterol an acid, and a neutral compound, as also the occurrence of isopelletierine in the roots, have been reported.[889] Roots contain tropine, pseudotropine, 3α-tigloyloxytropane, choline, cuscohygrine, *dl*-isopelletierine, and new alkaloids anaferine and anhygrine.[890] Leaf contains withanone (empirical formula $C_{24}H_{32}O_5$) and the berries have amino acids.[891,892]

Pharmacological action — Tonic, alterative, astringent, aphrodisiac, and nervine sedative. Seeds possess the property of coagulating milk. Leaves and root are narcotic. Root is also diuretic and deobstruent, tonic, alterative, and aphrodisiac.

Medicinal properties and uses — Roots and leaves of *W. somnifera* are used as a hypnotic in alcoholism and emphysematous dysphonea. Leaves are used as an anthelmintic and as an application to carbuncles. Root is used as an application in obstinate ulcers and rheumatic swelling. It is given in doses of about 2 gr in consumption, emaciation of children, senile debility, rheumatism, in all cases of general debility, nervous exhaustion, brain fag, loss of memory, loss of muscular energy, and spermatorrhea. It infuses fresh energy and vigor in a system worn out owing to any constitutional disease like syphilis, and in rheumatic fever. Powdered root is very useful with equal parts of ghee and honey for impotence or seminal debility. As nutrient and health restorative to the pregnant and old people, a decoction of the root is recommended. For scrofulous and other glandular swelling, fresh green root of *W. somnifera* reduced to paste with cow's urine or with heated water is applied to the affected parts. As a galactagogue, the decoction of the roots of *W. somnifera, Batatas paniculata,* and licorice is recommended to be given in cow's milk. For improving sight, a mixture of *W. somnifera* powder, licorice powder, and juice of emblic myrobalans is recommended.

Doses — Powder — 3 to 6 gr.

ZANTHOXYLUM ALATUM Roxb.   *Family:* RUTACEAE

Vernacular names — Sanskrit, Tejpal, Tumburu; Hindi, Tumru, Tejbal; Bengali, Napalidhania; Nepalese, Timur (Newari-Tebhu); Japanese, Asakurazansho; Chinese, Chiao; German, Gelbholz; Unani, Tejbal; Pustu, Dambre; Persian, Kababe-jahanulsha.

Habitat — Found in the hot valleys of the subtropical Himalayas from the trans-Indus areas to Bhutan up to an altitude of 7000 ft; also in the Khasia Hills between 2000 and 3000 ft; and in the hills of Ganjam and Vizagapatam at about 4500 ft.

Parts used — Bark, seeds, and fruits.

Pharmacognostical characteristics — A shrub or small tree, armed on the branchlets, leaf rachis, and midrib with broad, flattened prickles; those on the older branches usually with a conical corky base. Leaves imparipinnate, rachis usually winged. Leaflets 2 to 6 pairs, opposite, and terminal, $1\frac{1}{2}$ to $3 \times \frac{2}{5}$ to 1 in. Lanceolate, serrate, sessile. Flowers yellow, in dense pubescent lateral panicles 2 to 6 in. long, polygamous, calyx segments 6 to 8; petals 0; stamens 6 to 8. Fruit of 1 to 3 small, red, globose drupes, the size of a peppercorn, ultimately splitting into 2 valves and exposing the solitary black shining seed.[28]

Ayurvedic description — *Rasa — katu, tikta; Guna — laghu, rooksha, teekshna; Veerya — ushana; Vipak — katu.*

Action and uses — *Kaph vat samak, pittawardhak, dant sodhan, dipan, pachan, krimighan, jawarghana.*

Chemical constituents — Bark contains a bitter crystalline principle identical with berberine, a volatile oil, and some resin. The fruit contains about 1.5% of an essential oil consisting chiefly of 1-α-phellandrene with small amounts of linalol etc. Leaves yield an essential oil which has carbonyl compound identified as methyl *n*-nonyl ketone. The ketone-free fraction contained linalyl acetate, sesquiterpene, hydrocarbon, and tricosane. From the stem bark, dictamnine has been isolated.[178] The roots yielded the alkaloids dictamnine, γ-fagarine, magnoflorine, skimmianine, xanthoplanine. The stem bark and wood also yield the alkaloid magnoflorine.[10,893]

Pharmacological action — Tonic, stomachic, and carminative.

Medicinal properties and uses — Seeds and bark are used as aromatic, tonic in fever, dyspepsia, and cholera. Infusion and decoction of the bark is used in doses of 28 to 56 ml. Fruit as well as the branches and thorn are used as a remedy for toothache; also stomachic and carminative.[1,3]

Doses — Infusion or decoction of bark — 28 to 56 ml; fruit powder — 0.6 to 1.2 gr; bark powder — 1 to 3 gr.

ZINGIBER OFFICINALE Rosc.   Family: ZINGIBERACEAE

Vernacular names — Sanskrit, Ardhrakam; Hindi, Adrak; English, Ginger; Bengali, Ada; Nepalese, Palu; Sinhalese, Inguru; Japanese, Shoga; Chinese, Chiang; German, Inguere; French, Gingembre; Unani, Adrak; Arabian, Zanjabil urratab; Persian, Zanjabil ee-e-tar; Tibetan, Sge u-gser; Burmese, Khenseing; Malaysian, Alea; Tamil, Gnji.

Habitat — The plant is widely cultivated all over India; on a large scale in the warm, moist regions of Madras, Cochin (Travancore), and to a somewhat lesser extent in Bengal and the Punjab.

Parts used — Scraped and dried rhizomes as well as the green ones.

Pharmacognostical characteristics — *Z. officinale* is a perennial herb with a stout, horizontal, tuberous, jointed, aromatic rootstock, having several lateral tubers with an annual elongate, erect, leafy shoot 60 cm to 1 m high, bearing simple, alternate, distichous, narrow, oblong-lanceolate leaves 15 to 30 cm long and 2 to 3 cm broad with sheathing bases, the blade gradually tapering to a point, and on lateral slender peduncles 15 to 30 cm. Long, cylindric, oblong spikes of large bracteate greenish yellow, zygomorphic, epigynous sterile flowers with small dark-purple or purplish-black lips.

The dried, scraped drug "ginger" occurs in sympodially branched pieces known as "hands" or "races". These are 7 to 15 cm long 1 to 1.5 cm broad, and laterally compressed. The branches arise obliquely from the rhizome, are about 1 to 3 cm in length, and terminate in depressive scars or in undeveloped buds. The drug breaks with a short fracture, the fibers of the fibrovascular bundles often projecting from the broken surface. It has an agreeable aromatic odor and a pungent taste.

The unpeeled rhizome, in transverse section, shows a zone of cork tissue having an outer zone of irregularly arranged cells produced by suberization of the cortical cells without division and an inner zone of cells arranged in radial rows and produced by tangential division of the cortical cells. No cork cambium is differentiated. Within the cork is a broad cortex, differentiated into an outer zone of flattened parenchyma and an inner zone of normal parenchyma. The cortical cells contain abundant starch grains which are simple, ovoid, or sack shaped. Scattered in the cortex are numerous oil cells, with suberized walls enclosing yellowish-brown olearesin. The inner cortical zone contains about three rings of collateral, closed vascular bundles. Each vascular bundle contains phloem, showing well-marked sieve tubes and a xylem composed of 1 to 14 vessels with annular, spiral, or reticulate thickening.

Axially elongated secretion cells with dark contents occasionally accompany the vessels. The inner limit of the cortex is marked by a single-layer endodermis free from starch. The outermost layer of the steel is marked by a single-layered pericycle. The vascular bundles of the stele resemble those of the cortex and are, except for a ring of small bundles, immediately within the pericycle, scattered, as is typical of monocotyledonous stems. The ground mass of the stele is composed of parenchyma resembling the cortical parenchyma and containing much starch and numerous oil cells. Cork cells are absent from the scraped drug.[37,69,894]

Ayurvedic description — *Rasa* — *katu; Guna* — *lagu, singdha, guru, rooksha, teekshna; Veerya* — *ushna; Vipaka* — *madhura.*

Action and uses — *Kaf-bata samaka, sotha har, badanasamak, dipan, pachan, batanuloman, arsoghna, kafaghana, sawas har.*

Chemical constituents — Indian ginger contains 1 to 2% volatile oil of yellow color, having a characteristic odor and containing camphene, phellandrene, zingiberine, cineol, and borneol; gingerol, a yellow pungent liquid; an oleoresin "gingerin" the active principle, resin, starch, and K oxalate. The pungency of gingerol and ginger is destroyed when boiled with 2% potassium hydroxyde.

Trease and Evans[69] reported that the pungency of ginger is due to gingerol, an oily liquid consisting of homologous phenols. The principal of these is [6] gingerol (i.e., where n = 4). It is formed in the plant from phenylalanine, malonate, and hexanoate. Gingerols with 7, 8, 9, and possibly 16 carbon atoms in the side chain have been reported. New minor constituents of an extract were gingediols, methylgingediols, gingediacetates, and methylgingediacetates.

Pharmacological action — Aromatic, carminative, stimulant to the gastrointestinal tract, and stomachic; also sialogoge and digestive. Externally a local stimulant and rubefacient.

Medicinal properties and uses — Ginger is extremely valuable in dyspepsia, flatulence, colic, vomiting, spas, and other painful affections of the stomach, and the bowels unattended by fevers for cold, cough, asthma, dyspepsia, and indigestion. A paste of ginger is a local stimulant and rubefacient in headache, toothache, and near sightedness due to deficient contractile power of the iris; ginger powder rubbed on the extremities of the limbs checks cold perspiration and improves blood circulation in the collapse stage of cholera. Dry ginger is best given either in powder in doses of 1 to 2 gr, which may be taken with 0.5 gr of sodium carbonate or potash in gout and chronic rheumatism, or in the form of infusion in doses of 28 to 56 ml every hour for indigestion and want of appetite.

Trease and Evans[69] reported that powdered ginger may be a more effective antiemetic than dimenhydrinate (Dramamine). It was suggested that it may ameliorate the effects of motion sickness in the gastrointestinal tract itself, in contrast to antihistamines which act centrally. It is more widely used as a condiment than as a drug.

Doses — Infusion — 7 to 20 ml; sonth powder — 1 to 2 gr.

ZIZYPHUS MARTIANA Lam. *Family:* RHAMNACEAE
(Syn. *Z. JUJUBA* Lam. non Mill.)

Vernacular names — Sanskrit, Badri; Hindi, Baer; English, Jujube fruit; Bengali, Kula; Nepalese, Bayer; Sinhalese, Masan; Japanese, Gnumatsume; Chinese, Hong tsao; German, Stumpfblattriger Judendorn; French, Jujubier; Unani, Ber unnab; Arabian, Banque; Persian, Kanar; Tibetan, Gya-sug; Burmese, Ziben; Pustu, Berra; Tamil, Glanda maram.

Habitat — Found wild and cultivated in many parts of India and in the outer Himalayas up to 4500 ft.

Parts used — Leaves, berries, and bark.

Morphological characteristics — A small subdeciduous tree with dense spreading crown, commonly 1.5 ft girth and 18 to 25 ft high. Bark blackish to gray or brown, rough, regularly and deeply furrowed, the furrows about 1.2 cm apart. Blade 9 to 13 mm, pink, with or without paler streaks, the juice turning purplish black on the blade of a knife. Branches usually armed with spines, mostly in pairs, one straight, the other curved. Young shoots more or less densely pubescent. Leaves 3 to 6.3 × 2.5 to 5 cm oblong or ovate, usually minutely serrulate or apex distinctly toothed, obtuse, base oblique and 3-nerved, nerves depressed on the glabrous shining upper surface, densely clothed beneath, with white or buff tomentum. Petiole 2.5 to 10 mm long. Flowers 3.8 to 5 mm diameter, greenish, in dense axillary tomentose cymes or fascicles 1.2 to 1.9 cm long. Drupe 1.2 to 2.5 cm diameter, globose, first yellow then orange and finally reddish brown, containing a single stone surrounded by fleshy pulp.[108]

Ayurvedic description — *Rasa — amla, madhur, kasaya; Guna — guru, snigdha, pichil; Veerya — sheeta; Vipak — madhur.*

Action and uses — *Vatta pitta samak, daah prasaman, bran sodhan, bran ropan, lakhan, dipan, anuloman, grahi, hirdya, daah prasaman, jawarghana, brinhan.*

Chemical constituents — Fruits contain mucilage and sugar in addition to fruit acids. Bark contains tannin and a crystallizable principle, zizyphic acid. The root bark contains an unnamed alkaloid.[10]

Pharmacological action — Astringent and stomachic.

Medicinal poperties and uses — The leaves boiled in milk are given in virulent gonorrhea. An infusion of the leaves is used as an eye lotion in conjunctivitis; a paste of the tender leaves and twigs is applied to boils, carbuncles, and abscesses to promote suppuration; and the boiled leaves are applied over the navel and the pubic organ in dysuria.

The bark is bitter and astringent; it is a household remedy in diarrhea, dysentery, colic, and inflammation of the gums. The root and bark are tonic. The leaves are anthelmintic; good in stomatitis and gum bleeding; heal wounds, syphilitic ulcers; cure asthma; good in liver complaints. The flowers afford a good collyrium in eye troubles. The unripe fruit increases thirst, lessens expectoration and biliousness. The ripe fruit is sweet, sour, and has flavor; not good for digestion; causes diarrhea in large doses; useful in fevers, and for wounds and ulcers. The seed is astringent; tonic to the heart and brain; allays thirst.[108] The berries are a blood purifier and an aid to digestion. The dried ripe fruit is a mild laxative and expectorant. An ointment made of the seeds with some bland oil is locally used as a liniment in rheumatism. The plant is considered to have antitubercular properties.

Doses — Fruit — 5 to 7; bark decoction — 56 to 112 ml.

ZIZYPHUS VULGARIS Lam.   *Family:* RHAMNACEAE

Vernacular names — Sanskrit, Unnab; Hindi, Unnab; English, Jujube; Bengali, Kul; Nepalese, Bayear; Japanese, Natsume; Chinese, Suan tsao; German, Brustbeeren; Unani, Unnab; Arabian and Persian, Unnab; Pushtu, Karkamber.

Habitat — Found in Punjab and Punjab Himalayas up to 6500 ft and eastward to Bengal.

Parts used — Fruit and bark.

Morphological characteristics — A small deciduous tree, often shrubby, quite glabrous; branches of young plants armed with very short spines, one straight, 2.5 cm long, the other much shorter, recurved; older trees usually unarmed, flowering shoots about 15 to 20 cm. long, often fascicled on dwarf branches. Leaves 2.5 to 5 cm long, ovate-lanceolate, glabrous, crenate-serrate, oblique, 3-nerved; petiole 2.5 to 7.5 mm. long. Flowers in few-flowered, axillary clusters. Petals clawed, tips truncate. Disk obscurely lobed. Styles 2, united to the middle. Drupe ellipsoid, 2 cm long, stone tuberculate.[108]

Ayurvedic description — *Rasa — amla, madhur, kasaya; Guna — guru, snigdha, pichil; Veerya — sheeta; Vipak — madhur.*

Action and uses — *Vat pitta samak, daah samak, dipan, anuloman, hiridya, virnghan, sonita sthapan, hiccahar, mutral.*

Chemical constituents — Bark and leaves contain bitter substances and zizyphic acid and tannin. The root and bark of *Z. vulgaris* is reported to contain an unnamed alkaloid[10] Leaves, when chewed completely, anesthetize the taste for 5 to 20 min. Yields 1.7% of amorphous or microcrystalline substance with high potency and a gummy fraction with lower potency.[895]

Pharmacological action — Diuretic, antidiabetic, emollient, and pectoral.[1]

Medicinal properties and uses — It is good for checking bilious complaints and improving digestion. Bark powder is very effective in ulcer. Juice of the root bark is used as a purgative and externally in gout and rheumatism. Decoction of the root bark is given in fever and delirium. Syrup of dried fruits used for bronchitis. The bark is used to heal ulcers and wounds. The gum is good in eye diseases. The leaves are laxative; used in scabies, throat trouble, burning of the body. The fruit is sweet or sour, expectorant; purifies and enriches the blood; good in chronic bronchitis, fever, enlargement of the liver. The seeds are good in dry cough and for skin eruptions.[1,3] The drupes are emollient and pectoral.[50] A syrup of the dried fruit is used for bronchitis.

Doses — Decoction — 56 to 112 ml; fruit — 5 to 7.

BASIC CONCEPTS OF AYURVEDA

Col. Sir R. N. Chopra, Chairman, in his report, presented a note on the basic concepts of Ayurveda — The Tridosha and the pharmacodynamic aspects.[*]

This may be of interest to readers and is reproduced here.

The *Tridoshas*, viz., *Vaata, Pitta* and *Kapha*:

Vaata is that primal constituents of the living body, whose structure is Akaasha-Vaayu, and whose function is *raajasic,* it being concerned with the production of those physical and mental processes which are predominantly *raajasic (activating or dynamic)* in nature; hence, the presence of *vaata* is to be inferred in such mental phenomena as exhibition of enthusiasm, concentration, etc., as also in such physical phenomena as respiration, circulation, voluntary action of every kind, excretion and so on. It will be seen that many of these physical phenomena are included among those which modern physiologists would assign primarily to the activities of the nervous system (both cerebrospinal and autonomic).

Pitta is that primal constituent of the living body, whose structure is *tejas* and whose function is *saatwic,* it being concerned with the production of those physical and mental processes which are predominantly *saatwic* (balancing or transformative) in nature; hence the presence of *pitta* is to be inferred in such mental phenomena like intellection and clear concentration, as also in such physical phenomena as digestion, assimilation, heat production and so on; it will be seen that many of these physical phenomena are among those which modern physiologists would include under the activities of thermogenetic and nutritional systems (including thermogenesis and the activities of glandular or secretory structures), especially of the endocrine glands whose internal secretions or hormones are now known to be of such vital importance in digestion, assimilation, tissue-building and metabolism generally.

Kapha is that primal constituent of the living body, whose structure is *ap-prithvi* and whose function is *taamasic,* it being concerned with the production of those physical and mental processes which are predominantly *taamasic* (conserving or stabilizing) in nature; hence the presence of *kapha* is to be inferred in such mental phenomena as the exhibition of courage, forbearance, etc., as also in such physical phenomena as the promotion of bodily strength and build, integration of the structural elements of the body into stable structures, the maintenance of smooth working joints, and so on. It will be seen that many of these physical phenomena are among those which modern physiologists would include under the activities of the skeletal and anabolic systems.

Pharmacodynamic Aspects:

The term *rasa* refers among other things to taste, chyle *(annarasa),* circulating fluid (the *rasadhaatu*), mercury, etc. In the context of the pharmacodynamical doctrine of Ayurveda, it refers to the taste of a substance as sensed by the tongue. However, it has particular reference to the molecular structure of organic substances which determine, among other things, their tastes. The sweetness of the substance, for example, is determined for the most part by the preponderance in the molecule (the *anu*) of atoms *(paramaanus)* belonging to the *bhautic* species *prithvi* and *ap*. The principle applies with equal force to the other taste, viz., acid/sour, saline, astringent, acrid and bitter also. Technically, therefore, the term *rasa* refers to the structural composition of molecules and not to tastes as much.

The term *guna* refers to physical qualities of substances as determined by the preponderance of the one or the other of the five species of *bhautic* elements in the structural composition of molecules *(anus).* The classical Indian medicine has described eight primary and twenty secondary qualities.

---

[*] Extracted from the report of the Committee on Indigenous Systems of Medicine (1948), Ministry of Health, Government of India, Volume II, Appendix B, p. 347.

The term *veerya* refers to energy modalities. The *Charakasamhita* has defined this term thus: "*Veerya* is the power by which action is performed. No action is possible without *Veerya*." All actions are due to *Veerya*.[1] This definition has been accepted by all the ancient authorities. *Veerya* is of two kinds, viz., *sheeta* (potential?) and *ushna* (kinetic?). Some ancient authorities have treated the eight primary qualities referred to in *guna*, as *veeryas*. Later in the 13th century A.D., Hemadri described *veerya* as *shakti* or power/energy which, in his opinion, was well known in the world.[2]

Literally rendered into English, the term *"vipaaka"* means special transformation/reactions. In the context of its usage in Ayurvedic pharmacology, it refers to biochemical changes foods and drugs undergo during the gastro-intestinal digestion, in the first instance, and towards the conclusion of metabolism in the second. The latter is stated to complete the action ascribed to foods and drugs.[3]

The term 'prabhaava' described the specific and characteristic actions of drugs which cannot be explained from the points of view of *'rasa'*, *'guna'*, *'veerya'* and *'vipaaka'*. In this concept, drugs that possess identical *rasa, guna, veerya* and *vipaaka* — which are referable to their molecular structure, energy modalities, and changes they undergo during the gastrointestinal digestion and metabolism, can perform different and contrary actions. A close examination of this concept shows that it compares with isomers and isomeric action.

1. "Veeryam tu kriyate yenayaakriyaa. Naa veeryam kurute kinchit sarvaakritaakriyaa" (Charaka Sootra 26, 35)
2. "Lokapiprasiddah": (Hemaadri, Ayurveda Rasaayna commentary on Ast, hrd, Sootra 1, 14).
3. "Vipaaka karmanisthayaa" (Charaka, Sootra 26, 66) and "Karmanishtaa kryaaparisamaapti" Chakrapaani Datta on the above.

INTRODUCTORY NOTES ON THE FUNDAMENTAL PRINCIPLES OF AYURVEDIC PHARMACOLOGY

DEFINITION

Dravyaguna is the science of drugs *(dravya)*, their properties, and actions *(guna)*.

Pharmacognosy *(rupa-vijan)* deals with various names and synonyms (name) of drugs and also their morphological characters *(rupa)*. Most of this information is gained by the help of synonyms.

Pharmacology *(guna-karmvijnana)* discusses the properties *(gunas)* and action *(karma)* of drugs. The action of drugs is interpreted on the basis of their properties.

Panca mahabhutas are the basic foundation on which the anatomy, physiopathology and pharmacology of Ayurveda stands; that is why *Sushruta* has observed that there is no need of any other subject of consideration in medicine than *panca mahabhutas*. Particularly in pharmacology, as we shall see, *panca mahabhutas* are intimately related with the soil, seasons, *dravya, guna, virya, rasa,* and *vipak,* and as such no study can be complete without them. According to Ayurveda, the body of the individual is composed of five *mahabhutas.* Similarly, other things of the world are also composed of five *mahabhutas*, as are explained in terms of *dosha, dhatus,* and *mala,* and in drugs they represent the (1) *rasa* (taste), (2) *guna* (qualities), (3) *virya* or *veerya* (potency), and (4) *vipaka* (the taste arising after digestion and metabolism of a substance).

These five mahabhutas are named (1) *Prithvi mahabhuta,* (2) *jala mahabhuta,* (3) *teja mahabhuta,* (4) *vayu mahabhuta,* and (5) *akasa mahabhuta.*

Prithvi mahabhuta — Substances that are heavy, tough, hard, stable, nonslimy, dense, gross, and abounding in the quality of smell, are dominated by *prithvi*. They promote plumpness, compactness, heaviness, and stability.

Jala mahabhuta — Substances that are liquid unctuous, cold, dull, soft, slimy, and abounding in the qualities of taste, are dominated by *'jala'*. They promote stickiness, unctuousness, compactness, moistness, softness, and happiness.

Teja mahabhuta — Substances that are hot, sharp, subtle, light unctuous, nonslimy and abounding in the qualities of vision are dominated by *tejas*. They promote combustion, metabolism, luster, radiance, and clour.

Vayu mahabhuta — Substances that are soft, light, cold, ununctuous, rough, nonslimy, subtle, and abounding in the qualities of touch are dominated by *vayu*. They promote roughness, aversion, nonsliminess, and lightness.

Akasa mahbuta — Substances that are soft, light, subtle, smooth, and dominated by the qualities of sound are dominated by *akasa*. They promote softness, porosity and lightness.

Rasa — *Rasa* is the object of the gustatory sense organ which is located in the tongue. It is not only perception of taste, but is an indicator of the composition, properties, and probable action of the drug. There are six *rasas,* each composed of predominately two *Mahabhuta,* such as:

Rasa	*Mahabhuta*
1. *Madhura* (sweet)	*Prithvi + jala*
2. *Amla* (sour)	*Prithvi + teja*
3. *Lavan* (saline)	*Jal + tejas*
4. *Katu* (pungent)	*Vayu + tejas*
5. *Tikta* (bitter)	*Vayu + akash*
6. *Kasaya* (astringent)	*Vayu + prithvi*

1. *Madhura rasa* (sweet) — It is pleasing, brain tonic, healing, anti-abortifacient, beneficial for burning sensation, thirst, heart, throat, skin, hairs, galactagogue, antipoison. It is *snigdhasita* and *guru*. It is *kapha* increasing, *pitta* decreasing.

2. *Amla* (sour) — It is pleasing, sialogogue, appetizer, digestive, promotes bleeding (anticoagulant), and *snigdha, ushna, laghu, pitta* increasing.
3. *Lavan* (saline) — Moistening, breaking, appetizer, digestive sialogogue, expectorant, diuretic, vitiates blood. *Snigdha, ushna,* and *guru* increase *kapha* and *pitta*. Increases water content in the body.
4. *Katu* (pungent) — Nervous stimulant, resuscitator, mouth cleaning, anthelmintic, promotes bleeding, useful in dyspepsia, cardiac, and skin disorders. *Rookish, ushna, laghu, vata bardak;* causes salivation, lachrymation, tingling sensation in the tongue, and headache.
5. *Tikta rasa* (bitter) — Overshadows all the tastes, is appetizer, mouth cleaning, produces dryness of mouth. Anthelmintic, blood purifier, antipyretic, removes pus, toxins, serious discharge. *Vata* and *pitta* increasing. *Rooksha, sita,* and *laghu*.
6. *Kasaya* (astringent) — Produces clarity, stiffness, and traction in the tongue and the throat, dryness in the mouth, pain in the cardiac region, and heaviness. Healing, astringent, absorbent, antidiuretic. *Vata* increasing.

GUNAS

Guna is that which is located in *dravya* inherently as a causative agent and devoid of property and action. They are twenty in number, in ten pairs as follows:

1. *Guru* (heavy)	*Laghu* (light)
2. *Manda* (dull)	*Tikshna* (sharp)
3. *Shita* (cold)	*Ushna* (hot)
4. *Snigdha* (unctuous)	*Rooksha* (ununctuous)
5. *Slaksan* (smooth)	*Khara* (rough)
6. *Sandra* (dense)	*Drava* (liquid)
7. *Mridu* (soft)	*Kathina* (hard)
8. *Sthira* (stable)	*Sara* (unstable)
9. *Suksma* (subtle)	*Sthula* (grass)
10. *Visada* (nonslime)	*Piccila* (slime)

These attributes are both physical and pharmacological in nature in the context of medicine. It is the pharmacological action of the drug which determines the attributes of a substance. Effects of the *gunas* supersede the taste in drugs.

VIPAK

Vipak is the transformed state of an ingested substance after digestion. This is also called *Nisthapak*, as opposed to *Awasthapak* (stages of digestion) or *prapaka* (initial transformation). *Vipak* is said to take place at the time of division of '*rasa*' and '*mala*' after digestion is completed but, as a matter of fact, it is finalized after the next '*paka*' by *bhutagnis* in the liver where most of the drugs are metabolized. The difference between *awastha, paka,* and *vipak* may be summarized as below:

Awastha paka (pra paka)	*Nistha paka (vipaka)*
1. Transformation in stage	Final transformation
2. Production of *dosha* as *malas*	Production of *dosha* as *dhatus*
3. Perceivable	Inferable from action

Types of vipaka — As there are two final effects on the body, '*brimhen*' and '*langhan*', the '*vipak*' is said to be of two types — '*guru*' and '*laghu*'. Caraka takes three '*vipakas*', such as '*madhura*', '*amla*', and '*katu*'. Of these three, '*madhura*' is the same as *guru*, while *amla* and *katu* may be included in *laghu*. Evidently the former classification is according to effect on *dhatus*, while the latter one is based on three '*dosha*'. *Madhura, amla,* and *katu* stand for *kapha-pitta* and *vata*, respectively. These are the two views most accepted by scholars.

Effect of *Vipakas* According to Caraka

Vipak	Guna	Dosha	Dhatu	Mala
1. *Madhura*	*Snigdha guru*	*Kapha* increasing	*Semen* increasing	Laxative and diuretic
2. *Amla*	*Snigdhu laghu*	*Pitta* increasing	*Semen* increasing	Laxative and diuretic
3. *Katu*	*Ruksa laghu*	*Vata* increasing	*Semen* increasing	Constipative and antidiuretic

Effect of *Vipakas* According to Sushra

Vipak	*Dosha*	*Dhatu*	*Mala*
1. *Guru*	*Kapha* increasing, *Vata pitta*	*Semen* increasing	Laxative, Diuretic
2. *Laghu*	*Vata pitta* increasing, *Kapha* decreasing	*Semen* decreasing	Constipative, Antidiuretic

Difference between *Rasa* and *Vipaka*

Rasa	*Vipaka*
1. Taste sensation	State of metabolic transformation
2. Immediate response	Delayed response
3. Effect localized and extended to the level of digestion	Systematic effect after metabolism
4. Immediate psychological response	Delayed response of well being or otherwise
5. Perceivable	Inferable from action

VIRYA

Virya is the potency by which the drug acts. *Virya* is more dominant in drugs, while *rasa* is more dominant in dietetic substances.

Nature of Virya

Virya is interpreted differently in terms of *guna, karma*, or *dravaya* but the first view *(guna-viryavad)* is accepted in practice. According to this, *gunas* potent enough to produce action are termed *virya*. Out of twenty *gunas*, the following eight have been deemed to have potentiality to reach the state of *virya*. (1) *Laghu*, (2) *guru*, (3) *sita*, (4) *ushna*, (5) *snigdha*, (6) *ruksa*, (7) *mridu*, and (8) *tiksna*.

Again on the ancient style of generation, *viryas* have been grouped into two as *sita* and *ushna*, which represent the primordial factors of *soma* and *agni*, initiators of creation of the living world. On the level of *dosha, sheeta virya* represents *kapha varag* and *ushna virya* represents *pitta varag; vata* remains as buffer or catalytic *(yoga-vaha)*.

Effect of *Viryas*

	Effect on doshas	General effect
1. *Sita*	Pacifies *pitta*, aggravates *kapha* and *vata*	Exhilarant, moistening, cooling, life promoting; increases *semen*
2. *Ushna*	Pacifies *kapha* and *vata*, aggravates *pitta*	Heating, digestive, causes loss of consciousness, thrust, diaphoresis, emesis, purgation, solution, vertigo; decreases *semen*
3. *Snigdha*	Pacifies *vata*	Oleation, bulk increasing, sexual vigor; prevents old age
4. *Ruksa*	Aggravates *vata*, pacifies *kapha*	Astringent, roughening, healing
5. *Guru*	Pacifies *vata*	Annointing, bulk increasing; promotes union of sexual vigor and *semen* filling

	Effect on doshas	General effect
6. *Laghu*	Pacifies *kapha*	Roughening, fluid absorbing, healing; reduces fat and body weight
7. *Mridu*	Pacifies *pitta*	Saturates *rakta* and *mamsa*, softening
8. *Tiksna*	Pacifies *kapha*	Constipative; promotes secretions, tearing

PRABHAVA

The specific potency of a drug is known as *prabhava*. Thus, it is seen that two drugs, though having similar *rasa, vipaka,* and *virya,* differ in action. This difference in action is owing to the specific chemical *(bhautika)* composition of the drug, and its action can be explained by general rule on the bases of *rasa, vipaka,* and *virya*. For instance, *danti* and *citraka* are similar in *rasa (katu), vipaka (katu),* and *virya (ushna)*, but the former is purgative in action while the latter is not.

On the side of body response, a drug acting on a specific tissue, organ, or disorder is said to have *prabhava,* and action is directed to disease, not towards *dosha*. For instance, the cardiotonic action of Arjun, the anthelmintic activity of *vindanga,* and the antitoxic effect of *sirisa* are said to be *prabhava*.

Thus, two things contribute to the nature of *prabhava,* (1) specificity of chemical composition; and (2) specificity of the site of action.

In the early days, when chemistry was not so much developed, this could not be explained satisfactorily. On the other hand, such measures as god worship, the putting on of specific precious stones in rings or necklaces, the recitation of *mantras,* and various charms and amulets were said to work for *prabhava*. *Prabhava* may be called *acintya* (inexplicable or irrational) in contrast to *cintya* (explainable or rational). *Virya* is *cintya sakti,* while *prabhava* is *acintya sakti*.

GLOSSARY OF AYURVEDIC TERMS*

Aadhmaanakaree — that which causes flatulence.
Aadhmaanaghna, aadhmaanahara — antiflatulant.
Aakhuvishaghna — an antidote for rat poisoning.
Aamaateesaaraghna — antidiarrheic.
Aamadosha — broadly refers to food intoxication, usually associated with faulty digestion (and impaired metabolism).
Aamragandhiharidraa — turmeric with the flavor of mango (mango ginger).
Aanaahaghna — that which relieves constipation.
Aartawajanana — a substance which promotes ovulation/menstrual flow.
Aasyavairasyanaashana — anti-anorectic.
Aayushya — beneficial to life.
Aayushkara — a promoter of life.
Adhbobhagahar — a drug that clears the body wastes through the lower passage.
Agnisadan — a substance and drug that causes deficiency of digestive power.
Aksepa saman — anticonvulsant.
Amadoshaghnai, aamadoshahara — counteracts the effects of aamadosha.
Amadoshakara — that which causes or promotes impairment of gastrointestinal digestion.
Ama-pachan — drugs that digest *amadoshas*.
Antrika kotha-prasaman — intestinal antiseptic.
Anulomana — carminative, having the power to relieve flatulence and associated colic.
Anuvasan — a drug that causes the body to be soft, unctuous, and strong.
Apatarpana — a measure or drug that causes the body and the body element to starve.
Arshab-shatana — a drug that arrests the growth of piles or hemorrhoids, or causes them to dry up and fall.
Arshoghan — a drug that cures piles or hemorrhoids.
Asmaribhedana — urinary linthontripties; a drug that breaks a stone formed in the urinary tract.
Abha — lustrous.
Agnideepti, agnikarraka, aghnivardhaka, agnyuttejaka — that which stimulates the factors of gastrointestinal digestion.
Agnimaandyahara — antidyspeptic.
Aguru — light (not heavy).
Ajeernaghna — that which corrects indigestion.
Akshirogaghna — indicated in eye diseases.
Amritaphala — fruits with nectar (ambrosia)-like properties.
Angamardaprashamana — anodyne/analgesic.
Anilahara — alleviator of *vaata*.
Anulomana — sending or putting it the right direction (purging).
Anulomani — that which regulates the bowel movements.
Apacheeghna — that which resolves glandular enlargement.
Apasmaaraghna — anti-epileptic.
Arevatta — that which promotes excretion of body wastes.
Arshahita — beneficial in hemorrhoids.
Aruchighna, aruchihara, arushishamana — anti-anorectic.
Ashmaghna, ashmareeghna, ashmaantaka, ashmareebhedana — indicated in urolithiasis.
Asrasrutinaashana — anti-hemorrhagic.
Asthisanhaara, asthishrinkhala — that which promotes healing of bone fractures.
Ateesarashamani — anti-diarrheal.

---

* By courtesy of Dr. G. V. Satyavata et al.[26]

Atilekhana — excessive reduction of the body weight.
Atistambha — hemostatic.
Avasadan — a drug that causes depression in body and mind.
Baaddhamootraghna — that which relieves anurea.
Badakandinee — with firm/strong rhizomes.
Bajikaran — aphrodisiac.
Bala — refers to the tenacity and capacity to survive under adverse environmental conditions.
Balabuddhivivardhini — a promotor of muscle strength, resistance to disease, and intellect.
Balakara — promotor of physical strength and resistance.
Balamedhonolarvardhini — promoter of strength, intellect and digestive power.
Balaprada — that which give strength and resistance to disease.
Balva — tonic.
Balya — this term carries two meanings, viz., muscular strength, and tone, as could be judged from one's capacity to perform physical work; and resistance to disease, decay, and degeneration; resistance to disease has two aspects, viz., combatting the virulence of the disease and the capacity to inhibit or neutralize the cause of the disease.
Bhagandaraghna — useful in the treatment of fistula-in-ano.
Bhedana — purgation.
Bhootaghna, bhootaapaha — with psychoactive, and/or antimicrobial action.
Bhramahara — that which promotes clarity of mind and dispels confusion.
Bhramakara — that which produces giddiness, mental confusion.
Bhrimhana, brinhan — that which promotes the body bulk.
Bhringaara — that which blackens the hair.
Brihatkushtaghna — antileprotic.
Buddhiprada — that which promotes the intellectual faculties.
Chakshusraavahara — that which relieves ocular discharges.
Chandanagandha — that which smells like the sandalwood.
Chardi — vomiting.
Chardighna, chardishamani — anti-emetic.
Charmakantaka — a skin disease.
Chaturhikavishamajwaranaashini — antimalarial (particularly against quartan fever).
Daahahara, daahashamanee — refrigerant, relieves burning sensation.
Dadrughna — indicated in scaly affections of the skin.
Dadrupaamaahara — indicated in scaly and exudative affections of the skin.
Dantarogaghna — effective in dental diseases.
Dantya — that which promotes the health of the teeth.
Deepanapaachana — a gastric stimulant and digestant.
Deepyaka, dipana — that which improves digestion.
Dasha prasaman — refrigerant: a drug that cures the burning sensation of the body or which cools the body.
Dhaatupushtikara — that which nourishes the body tissue.
Dhaatupushtivardhaka — nutrient tissue homolog which nourishes the tissue.
Dughaanika — with milky latex.
Galagandahara — indicated in glandular swellings of the ventral part of the neck (goiter).
Gandamaalahara, gandaari — indicated in glandular swellings in the neck region.
Garbhasaya samaka — uterine sedative.
Garvasaya samkochaka — ecbolic; a drug that causes contraction of the uterus.
Garala — poison.
Garavishahara — that which is said to neutralize noxious or poisonous potions.
Garbhaantaka, garbhapaatakara — abortifacient.
Garbhaashayasamshodhani — said to cleanse the uterus.

Garbhapatana — that which induces abortion.
Garbhasraavahara — anti-abortifacient.
Garbhasthaapana — that which promotes conception (pregnancy).
Gobhee, gojee, gojhwa — having leaves which are rough like the cow's tongue.
Graahi — that which binds the bowel (astringent).
Grahaghna — useful in psychiatric involvement.
Grahanirogaraha, grahanirogaghna — indicated in malabsorption syndrome/chronic amebiasis colitis.
Grantighna — that which cures glandular enlargement.
Gudaartihara — that which relieves rectal pain.
Gudabhramshara — indicated in prolapse of the rectum.
Gudakeelakahara — indicated in hemorrhoids, rectal polyp.
Gulmaghna, gulmahara — tumor-like conditions of the abdomen.
Guru — heavy; that which is digested slowly.
Hidmanaashaka, hikkaghna, hikkaashamana — indicated in hiccough.
Hridrogaghna, hridrogahara — indicated in heart diseases.
Hridaya — cardiac tonic.
Hridayaavasaadaka — cardiac depressant.
Hridrogashamana — indicated as a palliative in heart diseases.
Jalodaranaashanee — useful in treatment of ascites.
Jantughna, jantunaashanee — anthelmintic, antimicrobial.
Jarana — digestion.
Jeernajwarahara — useful in the treatment of chronic fevers.
Jiwaniya — a drug which improves vitality of the body.
Jwaraghna, jwarahara, jwaravegav, naashaka — antipyretic and febrifuge.
Kandughan — antipyretic; a drug that cures itching sensation.
Kaamalaghna, kaamalaahara — that which cures jaundice.
Kalihaaree — that which eliminates the vitiated *doshas*.
Kandoohara, kandooghna — indicated in skin conditions associated with itching.
Kantharogavinaashani, kantharogahara — that which cures throat disorders.
Kanthashodhana — that which cleanses the throat.
Kanthashoolashodhana — that which relieves throat pain.
Kaataashobhinee — that which makes women look beautiful.
Kaasaghna, kaasahara — antitussive.
Kaasapeenasahara — indicated in chronic bronchitis and rhinitis.
Kapaalavisarpaghna — indicated in erysipelas of the scalp.
Kaphaateesaaraghna — that which cures diarrhea produced by *kapha*.
Kaphajwarahara, kaphajwaraghna — that which cures fever caused by *kapha*.
Kaphakaasahara — that which cures cough originating from *kapha*.
Kaphapittajwaraghna — that which cures fever involving *kapha* and *pitta*.
Kaphapittashamana — that which relieves the *kapha pitta dosha*.
Kaphonmaadaghna — cures insanity produced by *kapha*.
Kaphashophahara — cures edema produced by *kapha*.
Kaphodarahara — cures abdominal disease produced by *kapha*.
Karnaartihara, karnashoolaghna — that which relieves pain in the ear.
Karnanaadaghna — that which relieves tinnitus.
Karnapidakaghna — that which is indicated in the treatment of furuncles in the ear.
Karnapooyaharam — indicated in otorrhea.
Kashaaya — astringent.
Katu — pungent or bitter in taste.
Kaval dharan — gargle.

Keetamaaraka — anthelmintic, insecticidal.
Kinihee — indicated in the healing of ulcers.
Kiraatatikta — intensely bitter.
Kolam — having a sour taste.
Kotha prasaman — antiseptic.
Krimidoshara, krimidoshanaashanee — indicated in toxic conditions arising on account of insects, worms, or microbes.
Krimihara, krimighna, kriminsorka — vermicidal/anthelmintic/antimicrobial.
Krimighna — anthelmintic, antimicrobial, disinfectant.
Kshataghna — indicated in trauma.
Kshatakshayashamanee — indicated in a wasting disease associated with injury.
Kshutshamani — that which alleviates hunger.
Kukshishoolahara — indicated in abdominal colic.
Kushalee — that which promotes "health" (well-being).
Kushtaghna — indicated in skin diseases.
Laghu — light, easy to digest.
Lekhani, lekhan — that which aids in reducing corpulence.
Lochinee — soothing to the eye.
Maamsakara — that which promotes muscle growth.
Maarkava — indicated in the treatment of graying of the hair.
Mandagni — having deficient digestive power.
Madana — that which intoxicates or exhilarates.
Madhudruma — refers to sweetness of all parts of the tree.
Madhumehaghna — antidiabetic (hypoglycemic).
Mahaakleetaanika — indicated in the treatment of impotence.
Mahaakushtaghna — antileprotic.
Malamootrasamshodhaka — that which aids in the elimination of body wastes including feces and urine.
Malapaatana — aids in the elimination of body wastes, including the feces.
Medhya, medhakara, medhyakara — that which promotes memory and intellect; brain tonic.
Medohara, medoghna — causes reduction of obesity.
Mehaghna — that which relieves polyurea.
Mohanaashinee — indicated in the treatment of delusion.
Mootradoshahara — that which causes urinary disorders.
Mootrakrichrahara — indicated in dysurea.
Mootrala, mutral — diuretic.
Mootrashodhani — that which cleanses the urine.
Mootravardhana — that which promotes micturition.
Mootravirechanee — that which promotes increased micturition.
Mridurechaka, mridurechana — laxative (mild purgative).
Mukharoganaashani — indicated in the treatment of diseases of the oral cavity.
Mukharookshaghna, mukhashoshahara — that which relieves dryness of the mouth.
Mukharukhara — that which relieves pain in the oral cavity.
Mukhashodhanee — that which cleanses the mouth.
Nasya — errhine; a drug which promotes nasal discharge.
Netraogahara, netrarogaghna — indicated in the treatment of eye diseases.
Netrarukshamani — palliative in the treatment of painful eye diseases.
Netravikaarashamanee — palliative in eye disorders.
Netravranahara — indicated in ulcerative conditions of the eye.
Netrya — beneficial to the eyes.
Nidraanaashanee — that which causes insomnia.

Nidrajanan — hypnotic: a drug that brings sleep.
Nirama — free from undigested material.
Niruha vasti — a simple enema free from oil or with very little oil.
Paachana, paachani–digestive.
Paandughna, paanduhara—antianemic.
Paashaanabhdea — that which breaks stones.
Pachampacha — that which promotes digestion.
Palankasha — that which reduces the body bulk (particularly the body fat).
Parjuna — that which promotes health.
Peenasaghna, peenasanaashini — indicated in the treatment of chronic rhinitis.
Pichumarda — a cure for skin diseases.
Pipaasashamani — that which relieves polydypsia.
Pittajwaraghna — indicated in the treatment of fever due to *pitta*.
Pitta sarak — cholagogue; a drug that causes secretion of bile.
Pittaateesaaraghna — indicated in diarrhea due to pitta (enteritis).
Pittashamana — antibilious.
Pleeharogaghna, pleehodarahara — indicated in splenomegaly.
Pradaraghna — antileucorrhic.
Pramehaghna, pramehashamana — relieves polyurea (diabetes).
Prativisha — an antidote for poisons.
Pravaahikaghna — antidysenteric.
Punarnavaa — self-renewing, rejuvenating.
Puttihar — deodorants; a drug which causes purification.
Raktaateesaarashamana — palliative in hemorrhagic dysentery.
Rakta bhar samak — a vasodilator.
Rakta bhar vardhak — a vasoconstrictor.
Raktakrit — hematinic.
Raktapittaghna — indicated in the therapy of hemorrhagic diathesis.
Raktapittashamana — palliative in hemorrhagic diathesis.
Raktapittastambhana — hemostatic.
Raktashodhaka, rakata sodhak — that which purifies the blood.
Raktastambhani, rakta stambhan — that which promotes clotting.
Raktavardhaka — hematinic.
Raktavikaarashamani — indicated in hematological disorders.
Raktotklishta — blood disintegration (hemolysis?).
Rasaayana, rasaayanee — rejuvenator.
Rasaayani — lymphatics.
Rechana — purgation.
Rechani — purgative.
Rochaka — that which promotes taste and appetite improvement.
Rochana — stomachic.
Rohinee — that which promotes wound healing.
Ropana — purgative.
Ruchya — an appetizer.
Rujaapaha — analgesic.
Sama — with *anadoshas,* immature *doshas dhatus,* and *mala.*
Samagni — normal digestive power.
Samsarjana — to bring to the normal diet.
Sangraahi — that which binds the bowels (usually an astringent).
Sannipaatajwarahara — indicated in highly toxic fevers (c.f., pneumonia, typhoid, meningitis, etc.).

Sanipataja – all *"dosha"* mixed together.
Sarpadashtavishaghna — antidote for poisoning by snake bites.
Sara — laxative.
Sarnsana — a simple purgative.
Semshamana — palliative.
Semsodhan — curative.
Sharkaraanivaarana — hypoglycemic.
Sharkaraashmareenaashaka — indicated in the treatment of urinary gravel.
Sheetanut — that which lowers the body temperature and produces chills.
Shirorogaghna — indicated in the treatment of cranial diseases.
Shiro virechan — errhine, snuff.
Snodhana — radical elimination/purification.
Shodhaneeya — that which cleanses and eliminates.
Shoola prasaman — analgesic.
Shoolaghna, shoolanaashanee, shoolaapaha — said to cure colic (generally abdominal).
Shoshaghna — anticachectic.
Shothaghna, shophaghna — anti-edematous/antidropsy.
Shristavinmootrakara — that which increases the quantity of stools and urine.
Shuklaprada, shukrapravardhaka — spermatogenic.
Shukrakrit, shukrala — spermatogenic.
Shvasahar — antihistaminic, bronchial, antispasmodic.
Shwaasaghna, shwaasahara — antispasmodic, antidyspnoic.
Shwaasahita — useful in the treatment of dyspnea.
Shwaaskaasahara — antidyspnoic and antitussive.
Shwaasashamana — that which relieves dyspnea.
Shwetakushtaghna — indicated in the treatment of leucoderma (vitiligo).
Shwitraghna, shwitrahara — indicated in the treatment of vitiligo or leucoderma.
Smritiprada, smritivardhakara — that which promotes memory.
Snaayurogaghna — indicated in the treatment of diseases of the ligaments, tendons, etc.
Snehana — oleation.
Snigdha — unctuous.
Somarogaghna — indicated in the treatment of somaroga (variously interpreted as gynecological, hormonal, or metabolic disease).
Sphotajanak — vesicant.
Srotovishoshan — purifier of channels and tracts.
Stombhan — a drug that causes constipation.
Stanyadaa, stanyavardhani – that which promotes the secretion of breast milk.
Stanyajanaka, stanyakari — galactogogue.
Stanyanaashini — that which inhibits the secretion of breast milk.
Streepushpajananee — emmenagogue.
Swedahara — that which inhibits perspiration.
Swedakari, swedajanana, svedana – diaphoretic.
Swedopaga, swedya — that which promotes diaphoresis.
Teekshna — acute/pungent.
Tikshahni — excessive digestive power.
Tikta — bitter.
Timirahara, timiraghn — that which is said to cure cataracts.
Trishnaahara, trishnaghna, trishnaapaha — that which relieves thirst.
Twagrogaghna — said to cure various skin disorders.
Udaavartaghna — said to relieve retention of urine, feces, and flatus.
Udaraghna, udararogaghna — said to cure abdominal diseases.

Udarashoolahara — indicated in abdominal colic.
Unmaadahara — that which is said to cure insanity.
Unmatta — that which induces psychotropic effects (i.e., stimulates the central nervous system).
Urdhaobhag har — a drug that cleanses the upper portion of the body by throwing out waste through the upper passage, viz., the mouth and nose.
Utkleshashamana — that which relieves nausea or retching.
Vaajeekara — aphrodisiac.
Vaataghna — indicated in diseases of the nervous system.
Vaataghni, vaatapotha — anti-vaatic.
Vaataraktaghna — useful in the treatment of gout-like conditions.
Vaatavedanaahara — antineuralgic.
Valaasahara — that which relieves edema.
Vaman — emetic.
Vamanahara — that which relieves vomiting.
Vamanakara — that which induces vomiting.
Vamanashamani — anti-emetic.
Vamanopaga — emetic.
Vanhiakara — that which promotes gastrointestinal digestion.
Varnakara, varnya — useful in promoting good complexion of the skin.
Vastirogaghni — said to cure diseases of the urinary system (particularly of the bladder).
Vastirogahara, vastirogavinaashini — indicated in the treatment of diseases of the urinary system, especially relating to the bladder.
Vastishodhanee — that which cleanses the urinary bladder.
Vedanaasthaapana–analgesic, local anesthetic.
Veeryaprada — highly potent.
Vibandhashamanee — that which relieves constipation.
Virechanee, virechan — purgative.
Virekashreshtha — the best among purgatives.
Viriaya — aphrodisiac.
Visarpahara — that which is said to cure erysipelas.
Vishadoshaghna — that which is said to cure disorders caused by poisons.
Vishahara, vishaghna — antidote for poisons.
Vishalya — that which removes foreign bodies.
Vishamajwaraghna, vishamavegi — said to cure remittent and intermittent fevers (like malaria).
Vishramsana — laxative.
Vishtambhi — that which causes flatulence.
Vishtambhakaghnee — antiflatulent.
Visphotahara — indicated in eruptive conditions.
Vranaapaha, vranaghna — indicated in the treatment of wounds.
Vranapaachana — that which promotes "ripening" of abscesses.
Vranaropana — that which promotes healing of ulcers and wounds.
Vranashodhana — that which cleanses wounds.
Vriddhinaashani — indicated in the treatment of growths (benign or malignant).
Yakriduttejaka — hepatic stimulant.
Yakritodaraghna–that which is said to cure ascites involving liver pathology.
Yakritrogahara — indicated in hepatic disorders.
Yakritvriddhihara — that which is said to cure hepatomegaly.
Yonidoshaghna — that which is said to cure uterine disorders.
Yonivishodhana — that which cleanses the uterus.

Yonivyaapadvinaashani — useful in the treatment of gynecological disorders.
Yooka-likshaashamanee — indicated in the treatment of pediculosis.
Yoshinee — stated to be useful in the promotion of health in women.

REFERENCES

1. **Chopra R. N., Chopra, I. C., Handa, K. L., and Kapur, L. D.,** *Indigenous Drugs of India,* 2nd ed., Academic Publishers, Calcutta, 1958; reprint 1982.
2. **Dash, B. and Laltish, K.,** *Materia Medica of Ayurvedia,* Vol. 32, Concept Publishing, New Delhi, 1980.
3. **Nadkarni, A. K,** *Dr. K. M. Nadkarni's Indian Materia Medica,* revised ed., Popular Book Depot, Bombay, 1954.
4. *Charaka Samhita, Shree Gulabkunverba,* Vol. 1 to 6, Ayurvedic Society, Jamnagar, 1949.
5. *Sushruta Samhita, Chowkhamba,* 2nd ed., Sanskrit Series Office, Varanasi, 1968.
6. **Dr. Wise;** cited in **Badhwar, R. L. and Ghosh, S.,** *Poisonous Plants of India,* Vol. 1 and 2, Indian Council of Agricultural Research, New Delhi, 1965.
7. **Jacolliot,** cited in **Chopra, R. N., Chopra, I. C., Handa, K. L., and Kapur, L. D.,** *Indigenous Drugs of India,* 2nd ed., Academic Publishers, Calcutta, 1958; reprint, 1982.
8. **Captain Johnston Saint;** cited in **Chopra, R. N., Chopra, I. C., Handa, K. L., and Kapur, L. D.,** *Indigenous Drugs of India,* 2nd ed., Academic Publishers, Calcutta, 1958; reprint 1982.
9. **Prof. Brown;** cited in **Chopra, R. N., Chopra, I. C., Handa, K. L., and Kapur, L. D.,** Academic Publishers, Calcutta, 1958; reprint 1982.
10. **Willaman, J. J. and Li, H.-L.,** Alkaloid bearing plants and their contained alkaloids, *Lloydia, J. Nat. Prod. Suppl.,* 33(3A), 1970.
11. **Bhakuni, D. S., Dhar M. L., Dhar, M. N., Dhawan, B. N., and Mehrotra, B. N.,** Screening of Indian plants for biological activity. II, *Indian J. Exp. Biol.,* 7, 250, 1969.
12. **LeCler, D.,** *Histoire de la Medicine Arabe;* cited in **Chopra, R. N., Chopra, I. C., Handa, K. L., and Kapur, L. D.,** *Indigenous Drugs of India,* 2nd ed., Academic Publishers, Calcutta, 1958; reprint 1982.
13. **Mitra R. and Prasad S.,** Pharmacognostic study of *ulat kambal* (*Abroma augusta* Linn.), *J. Res. Indian Med.,* 6, 41, 1941.
14. **Srivastva, G. P. and Basu, N. K.;** cited in **Dasgupta, B. and Basu, N. K.,** *Experientia,* 26, 477, 1970.
15. **Ali, S., Ahsan, A. M., and Hann, G.,** 1958; cited in **Dasgupta, B. and Basu, N. K.,** *Experientia,* 26, 477, 1970.
16. **Adityachaudhury, N. and Gupta, P. K.,** 1969; cited in **Dasgupta, B. and Basu, N. K.,** *Experientia,* 26, 477, 1970.
17. **Dasgupta, B. and Basu, N. K.,** Chemical investigation of *Abroma augusta* Linn. Identity of Abromine with Betaine, *Experientia,* 26, 477, 1970.
18. **Ghosal, S. and Dutta, S. K.,** Alkaloids of *Abrus precatorius* L., *Phytochemistry,* 10, 195, 1971.
19. **Hooker, J. D.,** *Flora of British India,* Vol. 1 to 7, L. Reeve and Co. England, 1875—1897.
20. **Saldanha, C. J. and Nicolson, D. H.,** *Flora of Hassan Dist., (Karnatka),* Amerind Publishing, New Delhi, 1976.
21. **Prasad, S. and Bhattacharya, I. C.,** Pharmacognostical studies of *Achyranthes aspera* Linn., *J. Sci. Ind. Res. Sect. C,* 20c, 246, 1961.
22. **Mehra, P. N. and Karnick, C. R.,** Pharmacognostic studies of *Achyranthes aspera* Linn., *Indian J. Pharm.,* 31, 170, 1969.
23. **Kapoor, V. K. and Singh, H.,** Isolation of betaine from *Achyranthes aspera* L., *Indian J. Chem.,* 4, 461, 1966.
24. **Kapoor, V. K. and Singh, H.,** Investigations of *Achyranthes aspera, Indian J. Pharm.,* 29, 285, 1967.
25. **Hariharan, V. and Rangaswami, S.,** Structure of saponins A and B from the seeds of *Achyranthes aspera L., Phytochemistry,* 9, 409, 1970.
26. **Satyavati, G. V., Raina, M. K., and Sharma, M.,** *Medicinal Plants of India,* Vol. 1, Indian Council on Medical Research, New Delhi, 1976.
27. **Mehra, P. N. and Puri, H. S.,** Pharmacognostic investigations on *Radix aconiti heterophylli (ativisha)* and its adulterants, *Res. Bull. Punjab Univ.,* 19, 439, 1968.
28. **Chopra, R. N., Badhwar, R. L., and Ghosh, S.,** *Poisonous Plants of India,* Vol. 1. and 2, Indian Council of Agricultural Research, New Delhi, 1965.
29. **Lawson, A. and Tops, J. E. C.,** The relationship between aconitine and atisine and some degradation products of the latter, *J. Chem. Soc.,* p. 1640, 1937.
30. **Jacobs, W. A. and Craig, C. L.,** Aconite alkaloids. VIII. Atisine, *Biol. Chem.,* 143, 589, 1942.
31. **Kelkar, N. C. and Rao, B. S.,** Indian essential oils. V. Essential from the rhizomes of *Curcuma longo* Linn. VI. Essential oil from the rhizome of *Acorus calamus* Linn., *J. Indian Inst.,* 17A, 7, 1934; *Br. Chem. Abstr. A,* p. 529, 1934.
32. **Dandiya, P. C.,** Trimethoxy-benzene derivatives from indigenous drugs, *Indian J. Physiol. Pharmacol.,* 14, 87, 1970.
33. **Agarwal, S. L., Dandiya, P. C., Singh, K. P., and Arora, R. B.,** A note on the preliminary studies of certain pharmacological actions of *Acorus calamus* L., *J. Am. Pharm. Assoc.,* 45, 655, 1956.
34. **Dixit, R. S., Perti, S. L., and Rangathan, S. K.,** *J. Sci. Ind. Res. Sect. C,* 15, K, 1956.
35. **Anon.,** *Indian Pharmacopeia,* Manager of Publications, Delhi, 1955.

36. **Sen, J. N. and Ghosh, T. P.,** Vasicine — an alkaloid present in *Adhatoda vasica, Q. J. Indian Chem. Soc.,* 1, 315, 1925; *Chem. Abstr.,* 19, 1501, 1925.
37. **Gupta, K. G. and Chopra, I. C.,** Antitubercular action of *Atoda vasica* (N.O. Acanthaceae), *Indian J. Med. Res.,* 42, 355, 1954.
38. **Prakash, A. and Prasad, S.,** Pharmacognostical studies on the bark of *Aegle marmelos* Correa (Bilva), *J. Res. Indian Med.,* 4, 97, 1969.
39. **Chakarvarti, R. N. and Das Gupta, B.,** Sterol from *Aegle marmelos, Experientia*,12, 335, 1956.
40. **Chakarvarti, R. N. and Das Gupta, B.,** The structure of Aegelin, *J. Chem. Soc.,* p. 1580, 1958.
41. **Chatterjee, A., Chaudhury, R. R., and Das, B. C.,** Minor alkaloids of *Aegle marmelos* Correa, *Sci. and Cult.,* 33, 279, 1967.
42. **Chatterjee, A. , Dutta, G. P. , Bhattacharya, S., Audier, H. E., and Dass, B. C.,** The structure of marmin, *Tetrahedron Lett.,* 5, 471, 1967.
43. **Dikshit, B. B. L. and Dutt, S.,** Chemistry of *Aegle marmelos* (the Indian bel), *J. Ind. Chem. Soc.,* 7, 759, 1930; *Br. Chem. Abstr. A,* p. 1628, 1930.
44. *Bhava Prakasha Nighantu* (with commentary by K. C. Chunekar), Pandey, G. S., Ed., Chowkhamba Sanskrit Sansthan, Varanasi, 1969.
45. **Shah, C. S. and Bhattacharya, A. R.,** Pharmacognostic study of *Albizia lebbeck* Benth., Bark., *J. Sci. Ind. Res. Sect. C,* 19, 199, 1960.
46. **Varshney, I. P. and Geeta, B.,** The study of the saponins of *Albizia lebbeck* Benth., seeds from Madhya Pradesh, *J. Indian Chem. Soc.,* 48, 907, 1970.
47. **Varshney, I. P., Khan, A. A., and Asha, S.,** *Indian J. Appl. Chem.,* 34, 214, 1971.
48. **Varshney, I. P., Geeta, B., Srivastva, H. C., and Krishamoorthy, T. N.,** Partial structure of lebbekanin-A, a new saponin from the seeds of *Albizzia lebbeck* Benth., *Indian J. Chem.,* 11, 1094, 1973.
49. **Tripath, V. J. and Dasgupta, B.,** Neutral constituents of *Albizzia lebbeck, Curr. Sci.,* 43, 46, 1974.
50. **Chopra, R. N., Nayer, S. L., and Chopra, I. C.,** Glossary of Indian Medicinal Plants, Council of Scientific and Industrial Research, New Delhi, 1956.
51. **Dhar, M. L., Dhar, M. M., Dhawan, B. N., Mehrotra, B. N., and Ray, C.,** Screening of Indian plants for biological activity. I, *Indian J. Exp. Biol.,* 6, 232, 1968.
52. **Bhakuni, D. S., Dhar, M. M., and Dhar, M. L.,** Structure of Alangine-A, an alkaloid from *Alangium lamarckii* Thwaits, *J. Sci. Ind. Res. Sect. B,* 19, 8, 1960.
53. **Dasgupta, B.,** Chemical investigations of *Alangium lamarckii*. I. Isolation of new alkaloid ankorine from the leaves, *J. Pharm. Sci.,* 54, 487, 1965.
54. **Dasgupta, B.,** Chemical investigations of *Alangium lamarckii*. II. Isolation of choline from the leaves, *Experientia,* 22, 287, 1966.
55. **Dasgupta, B. and Sharma, S.,** Chemical investigation of *Alangium lamarckii*. Isolation of steroids and terpenoides from the leaves, *Experienia,* 22, 647. 1966.
56. **Battersby, A. R., Kapil, R. S., Bhukuni, D. S., Popli, S.P., Merchant, J. R., and Salgar, S. S.,** New alkaloids from *Alangium lamarckii* Thw., *Tetrahedron Lett.,* 41, 4965, 1966.
57. **Sharma, R. K. and Gupta, P. C.,** Acreyl alcohol and wax from the root bark of *Alangium lamarckii, Vijnana Parishad Anusandhan Patrika,* 9, 147, 1966.
58. **Salgar, S. S. and Merchant, J. R.,** Chemical investigation of root bark and stem bark of *Alangium lamarckii, Curr. Sci.,* 35, 281, 1966.
59. **Pakrashi, S. C. and Eshakh, Ali,** Newer alkaloids from *Alangium lamarckii* Thw., *Tetrahedron Lett.,* 23, 2143, 1967.
60. **Pakrashi, S. C., Bhattacharyya, J., Mookerjee, S., Samanta, T. B., and Vorbnuggen, H.,** Studies on Indian medicinal plants. XVIII. The non-alkaloidal constituents from the seeds of *Alangium lamarckii* Thw., *Phytochemistry,* 7, 461, 1968.
61. **Ivanova, N.,** Chemical composition of Alhagi camelorium, *J. Appl. Chem. (U.S.S.R.),* p. 9, 1936; *Chem. Abstr.,* 31, 3104, 1937.
62. **Bhandari, P. R. and Mukerjee, B.,** Garlic *(Allium sativum)* and its medicinal uses, *Nagarjun,* p. 121, 1959.
63. **Anon.,** *Health Bull.,* Delhi, 24, 36, 1942.
64. **Parry, E. J.,** *The Chemistry of Essential Oils and Artificial Perfumes,* 2nd ed., Vol. 1 and 2, Scott Greenwood & Sons, London, 1921—1922.
65. **Bordia, A. and Bansal, H. C.,** Essential oil of garlic in prevention of atheroscerosis, *Lancet,* 2, 1491, 1973.
66. **Bordia, A., Bansal, H. C., Arora, S. K., Rathore, A. S., Ranawat, R. V. S., and Singh, S. V.,** Effect of the essential oil (active principle) of garlic on serum cholesterol, plasma fibrinogen, whole blood, coagulation time and fibrinolytic activity in alimentary lipaemia, *J. Assoc. Phys. India.,* 22, 267, 1974.
67. **Jain, R. C. , Vyas, C. R., and Mahatma, O. P.,** Hypoglycemic action of onions and garlic, *Lancet,* 2, 1491, 1973.
68. **Bhandari, P. R. and Mukerji, B.,** Aloes, C.D.R.I., Lucknow, 1958.
69. **Trease, G. E. and Evans, W. C.,** *Pharmacognosy,* 12th ed., Bailliere Tindal, London, 1983.
70. **Shah, C. S. and Mody, K. D.,** Estimation of barbalion in Indian aloes, *Indian J. Pharm.,* 29, 10, 1967.

71. **Singh, M., Sharma, J. N., Arora, R. B., and Kocher, B. R.,** Beneficial effects of *Aloe vera* in the healing of thermal burns and radiation injury in albino rats, *Indian J. Pharm.*, 5, 258, 1973.
72. **Gupta, K.,** Aloes compound (a herbal drug) in functional sterility, in 16th Indian Obstet. Gynaecol. Congr., New Delhi, 1972.
73. **Sastry, M. S.,** Comparative chemical study of two varieties of Galangal, *Indian J. Pharm.*, 23, 76, 1961.
74. **Inamdar, M. C., Khorana, M. L., and Rao Rajarama, M. R.,** Expectorant activity of *Alpina galanga*, *Indian J. Physiol. Pharmacol.*, 6, 150, 1962.
75. **Chakravarti, D., Chakravarti, R. N., and Ghose, R.,** Chemical examination of Dita bark. III, *Bull. Calcutta Sch. Trop. Med.*, 4, 4, 1956.
76. **Chatterje, A., Mukherjee, B., and Ray, A. B.,** The alkaloids of leaves of *Alstonia scholaris* R. Br., *Tetrahedron Lett.*, 41, 3633, 1965.
77. **Goodson, J. A., Henry, T. A., and MacFie, J. W. S.,** Action of cinchona and certain other alkaloids in bird malaria, *Biochem. J.*, 24, 874, 1930; *Chem. Abstr.*, 25, 348, 1931.
78. **Henry, T. A.,** *The Plant Alkaloids*, 4th ed., Blakiston Company, Philadelphia, 1949.
79. **Goodson, J. A. and Henry, T. A.,** The alkaloid echitamine from the bark of *Alstonia scholaris* R. Br., *J. Chem. Soc.*, p. 1640, 1925.
80. **Siddappa, S.,** *J. Mysore Univ.*, 5, 63, 1945.
81. **Anon.,** *Indian For. Leafl.*, No. 70, 4, 1970.
82. **Anon.,** *Health Bull., (Delhi),* No. 23, 31, 1941.
83. **Garg, S. K., Saksena, S. K., and Chaudhary, R. R.,** Antifertility screening of plants. VI. Effect of five indigenous plants on early pregnancy in albino rats, *Indian J. Med. Res.*, 58, 1285, 1970.
84. **Singh, A. , Kapoor, L. D., and Chandra, V.,** Pharmaco-botanic studies of kalmegh, *J. Res. Indian Med.*, 7, 93, 1972.
85. **Prasad S. and Gupta, K. C.,** A. pharmacognostical study of *Andrographis paniculta* Nees., *Indian J. Pharm.*, 19, 163, 1957.
86. **Bhaduri, K.,** The oil of *Argemone mexicana, Am. J. Pharm.*, 86, 49, 1914; *Chem. Abstr.*, 8, 1186, 1914.
87. **Govinadachari, T. R., Pai, B. R., Srinivasan, M., and Kalyanarman, P. S.,** Chemical investigation of *Andragraphis paniculata*, *Indian J. Chem.*, 7, 306, 1969.
88. **Dwarkanath C.,** *Introduction to Kayachikitsa*, Popular Book Depot, Bombay, 1959.
89. **Sobyanin, N. P. and Saakov, S. G.,** The coloring matter of flowers of *Althea rosea, J. Chem. Ind. (Russia),* 1929; *Chem. Abstr.*, 24, 2962, 1930.
90. **Shah, N. C., Mitra, R., and Kapoor, L. D.,** Pharmacognostic studies of *Angelica glauca*, Edgew., *Indian J. Pharm.*, 34, 171, 1972.
91. **Raina, M. K.,** Pharmacognostic Investigation on Some Indian Umbelliferous Plants, Ph. D. thesis, Punjab University, Punjab, 1974.
92. **Bhakuni, D. S., Tewari, S., and Dhar, M. M.,** Aporpine alkaloids of *Annona squamosa*, *Phytochemistry*, 11, 1819, 1972.
93. **Mukerjee, T. D. and Govind Ram,** Studies on indigenous insecticidal plants. II. *Annona squamosa, J. Sci. Ind. Res. Sect. C,* 17, 9, 1958.
94. **Mishra, M. B., Tewari, J. P., and Mishra, S. S.,** cited in *Advances in Indian Medicine,* B.H.U., Varansi, 1970, 199.
95. **Bandopadhyay, M., Pardeshi, N. P., and Seshadri, J. R.,** Comparative study of *Anethum graveolens* and *A. sowa, Curr. Sci.*, 41, 50, 1972.
96. **Malivaya, B. K. and Dutt, S.,** Chemical examination of the essential oil derived from *Anthem soa* Roxb. Oil from green herb and the seeds, *Proc. Indian Acad. Sci.*, 12A, 251, 1940; *Br. Chem. Abstr. B*, 3, 56, 1941.
97. **Rao, B. S., Sudborogh, J. J., and Watson, H. E.,** Notes on some Indian essential oils, *J. Ind. Inst. Sci. Sect. A*, 8, 183, 1925.
98. **Anon.,** *Health Bull.(Delhi),* No. 23, 29, 1941.
99. Ber Schimmel u. Co., *Melitz Leipzig,* October, 105, 1909; April, 95, 1910.
100. **Finnemore, H.,** *The Essential Oils,* E. Benn, London, 1926.
101. **Kohli, R. P. , Dua, P. R. , Shankar, K., and Sakena, R. C.,** Some central effects of an essential oil of *Apium graveolens* Linn., *Indian J. Med. Res.*, 55, 1099, 1967.
102. **Farooq, M. O., Gupta, S. R., Kiamuddin, M., Rehman, W., and Seshadri, T. R.,** Chemical examination of celery seeds, *J. Sci. Ind. Res. Sect. B,* 12, 400, 1953.
103. **Kulshreshtha ,V. K., Singh, N., Saxena, R. C., and Kohli, R. P.,** A study of central pharmacological activity of alkaloid fraction of *Apium graveolens* Linn., *Indian J. Med. Res.*, 58, 99, 1970.
104. **Kinzo Kafuku and Nobutoshi Ichikawa,** Odorous principle of lignum aloe, *J. Chem. Soc. Jpn.*, 56, 1155, 1935; *Chem Abstr.*, 30, 240, 1936.
105. **Nagrajan, G. R. and Seshadri, T. R.,** Leucocyanidinis of areca nut and toon *(Cedrella toona)* wood, *J. Sci. Ind. Res. Sect B,* 20, 615, 1961.
106. **Bose, B. C., Vijavargiya, R., Safi, A. A., and Sharma, S. K.,** Chemical and pharmacological studies on *Argemone mexicana, J. Pharm. Sci.*, 52, 1172, 1963.

107. **Chakravarti, R. N., Maiti, P. C., and Saha, T. K.,** A preliminary note on fractionation of *Argemone* alkaloids, *Bull. Calcutta Sch. Trop. Med.,* 2, 41, 1954.
108. **Kirtikar, K. R. and Basu, B. D., and I. C. S. (retired),** *Indian Medicinal Plants,* Vol. 1 to 4, Plates I to IV, L. M. Basu, Publ., Allahabad, 1918; 2nd ed., 1935; reprinted., Bishensingh Mahendrapal Singh, Dehra Dun, 1975.
109. **Murray, J. A.,** *The Plants and Drugs of Sind,* Richardson, London, 1881; reprint, Indian Book Gallery, Delhi, 1983.
110. **Dutta, N. K. and Sastry, M. S.,** Pharmacological action of *Aristolochia bracteata* Retz. on the uterus, *Ind. J. Pharm.,* 20, 302, 1959.
111. **Sastry, M. S.,** Chemical investigations on *Aristolochia bracteata* Retz., *Ind. J. Pharm.,* 27, 265, 1965.
112. **Govinadachari, T. R. and Vishwanathan, N.,** The identity of aristochine, *Indian J. Chem.,* 5, 655, 1967.
113. **Ganguly, A. K., Gopinath, K. W., Govindachari, T. R., Nagrajan, K., Pai, B. R., and Parthasarthy, P. C.,** Ishwarone, a novel tetracyclic sesquiterpene, *Tetrahedron Lett.,* 3, 133, 1969.
114. **Govindachari, T. R., Mohamed, P. A., and Parthasarthy, P. C.,** Ishwarone and arostolochene, two new sesquiterpene hydrocarbons from *Aristolochia indica, Tetrahedron,* 26, 615, 1970.
115. **Govindachari, T. R. and Partasharthy, P. C.,** Ishwarol, a new tetra cycle sesquiterpene alcohol from *Aristolochia indica* Linn., *Indian J. Chem.,* 9, 1310, 1971.
116. **Tewari, P. V., Prasad, D. N., and Das, P. K.,** Preliminary studies on uterine activity of some Indian medicinal plants, *J. Res. Indian Med.,* 1, 68, 1966.
117. **Dwarkanath, C.,** *Digestion and Metabolism in Ayurveda,* Shri Baidyanath Ayurved Bhawan, Calcutta, 1967.
118. **Madanpala,** *Madanpala Nighantu,* Shree Venkatashewara Steam Press, Bombay, 1954.
119. **Chopra, R. N., Roy, D. N., and Ghosh, S. M.,** *Blumea densiflora* and *Artemesia vulgaris,* their insecticidal and larvicidal properties, *J. Malar. Inst. India,* 3, 495, 1940.
120. **Atal, C. K. and Kapur, B. M.,** Cultivation and Utilization of Aromatic Plants, Regional Research Laboratory, CSIR, Jammu-Tawl, 1982.
121. **Atal, C. K. and Kapur, B. M.,** Cultivation and Utilization of Medicinal Plants, Regional Research Laboratory, CSIR, Jammu-Tawl, 1982.
122. **Burkill, I. H.,** A Dictionary of Economic Products of the Malaya Peninsula, Vol. 1 and 2, Government of Malaysia and Singapore, Ministry of Agriculture and Cooperatives, Kuala Lumpur, Malaysia, 1966.
123. **Willis, J. C.,** *A Dictionary of Flowering Plants and Ferns,* 8th ed., Cambridge University Press, London, 1937.
124. **Steinmetz, E. F.,** *Codex Vegetabilis,* Amsterdam, 1957.
125. **Lal, S. K. and Rao, S. K. S.,** Some observations of the presence of acetylcholine in Indian Jack fruit. *Artocarpus integrifolia* Linn., *Arch. Int. Pharmacodyn.,* 148, 397, 1964.
126. **Mehra, P. N. and Bhatnager, J. K.,** A comparative study of "Shatawar" and its supposed botanical source, *Indian J. Pharm.,* 20, 33, 1958.
127. **Roy, R. N., Chavan, S. R., Bhagwager, S., Dutta, N. K., and Iyer, N. S.,** Preliminary pharmacological studies on different extracts of roots of *Asparagus racemosus* (Satawari) Willd., *Indian J. Pharm.,* 30, 289. 1968.
128. **Dange, P. S., Kanitkar, U. K., and Pendse, G. S.,** Amylase and lipase activities in the roots of *Asparagus racemosus, Planta Med.,* 17, 393, 1969.
129. **Mehrotra, B. N. and Kandu, B. C.,** Pharmacognostic studies on *Astercantha longifolia* Nees., *Planta Med.,* 10, 474, 1962.
130. **Godbole, R. N., Gunde, B. G., and Srivastva, P. D.,** *J. Oil Soap,* 206, 1941.
131. **Prashar, V. V. and Singh, Harkishan,** Investigation of *Asteracantha longifolia* Nees., *Indian J. Pharm.,* 27, 109, 1965.
132. **Parasher, V. V. and Singh, Harkishan,** A sterol from seeds of *Asteracantha longifolia* Nees., *Indian J. Pharm.,* 27, 118, 1965.
133. **Bhandari, P. R. and Mukerjei, B.,** The neem: Indian lilac, *East. Pharm.,* p. 1, 1959.
134. **Murthy, A. L. N., Rangaswami, S., and Seshadri, T. R.,** Bitter principle of neem oil, *Indian J. Pharm.,* 2, 206, 1940.
135. **Mitra, C. R.,** Neem, Indian Central Oilseeds Committee, Hyderabad, 1963.
136. **Narayanan, C. R. and Iyer, K. N.,** Isolation and characterization of desacetylnimbin, *Indian J. Chem.,* 5, 460, 1976.
137. **Barak, S. P. and Chakraborty, D. P.,** Chemical investigations of *Azadirachta indica* leaf *(M. azadirachta), J. Indian Chem. Soc.,* 45, 466, 1969.
138. **Rao, A. R., Sukumar, S., Paramasivam, T. B., Kamalakshi S., Parashuraman, R. R., and Shantha, M.,** Study of antiviral activity of tender leaves of margosa tree *(Melia azadirachta)* on vissinia and variola virus. A preliminary report, *Indian J. Med. Res.,* 57, 495, 1969.
139. **Murthy, S. P. and Sirsi, M.,** Pharmacological studies on *Melia azadirachta.* I. Antibacterial, antifungal and antitubercular activity of neem oil and its fractions, in Symp. Utilization of Indian Med. Plants, Lucknow, 1957.
140. **Sharma, V. N. and Saksena, K. P.,** Sodium nimbidinate. *In vitro* study of its spermicidal action, *Indian J. Med. Sci.,* 13, 1038, 1959.
141. **Sharma, V. N. and Saksena, K. P.,** Spermicidal activity of sodium nimbinate, *Indian J. Med. Res.,* 47, 322, 1959.

142. **Guenther E.,** *The Essential Oils,* Vol. 1 to 6, D. Van Nostrand, New York, 1948—1952.
143. **Kon, G. A. R. and Woolman, A. M.,** Sapogenins. III. The dehydrogenation products of methylsarsapogenin and methylcholestanol, *J. Chem. Soc.,* p. 794, 1939; *Chem. Abstr,* 33, 6325, 1939.
144. **Bapat, S.K., Ansari, K. U., and Chandra, V.,** Hypoglycemic effect of *Bambusa dendrocalamus. Indian J. Physiol. Pharmacol.,* 13, 189. 1969.
145. **Bapat, S. K., Ansari, K. U., Jauhari, A. C., and Chandra, V.,** Hypoglycemic effect of two indigenous plants, *Indian J. Physiol. Pharmacol.,* 14, 59, 1970.
146. **Tewari, P. V., Prasad, D. N., and Das, P.K.,** Preliminary studies on uterine activity of some Indian medicinal plants, *J. Res. Indian Med.,* 1, 68, 1966.
147. **Datta, P. C. and Biswas, C.,** Pharmacognostic study of *Barleria prionitis* Linn., *Q. J. Crude Drug Res.,* 8, 1161, 1968.
148. **Gujral, M. L., Saxena, P. N., and Misra, S. S.,** An experimental study of the comparative activity of indigenous diuretics, *J. Indian Med. Assoc.,* 25, 49, 1955.
149. **Khwaja, A. J.,** Pharmacognosy of the leaf and root of *Barringtonia acutangula* Gaertn., *Planta Med.,* 17, 338, 1969.
150. **Row, L., Ramachandra, and Sastry, C. S.,** Isolation of barringtonic acid and a new triterpene dicarboxylic acid from *B. acutangula* Gaertn., *Indian J. Chem.,* 2, 463, 1964.
151. **Barua, A. K. and Chakrabarti, P.,** Triterpenoids. XIX. The constitution of barringtogenol C, a new triterpenoid sapogenin from *Barrington acutangula* Gaertn., *Tetrahedron Lett.,* 21, 381, 1965.
152. **Viswanatham, N., Sastry, B. S., and Rao, E. V.,** Re-examination of the flower petals of *Bauhinia tomentosa* Linn. as a commercial source of Rutin, in Proc. 22nd Indian Pharmaceutical Congr., Calcutta, 1970.
153. **Prasad, S. and Prakesh, A.,** Pharmacognostical study of *Bauhinia variegata, Indian J. Pharm.,* 34, 170, 1972.
154. **Ramchandran, R. and Joshi, B. C.,** Chemical examination of *B. purpurea* flowers, *Curr. Sci.,* 36, 574, 1967.
155. **Puntambekar, S.V. and Krishna, S.,** The fatty oil from the seeds of *Bauhinia variegata* Linn., *J. Indian Chem. Soc.,* 17, 96, 1940.
156. **Bhakuni, D. S., Shoeb, A., and Popli, S. P.,** Studies in medicinal plants. I. Chemical constituents of *Berberis asiatica* Roxb., *Indian J. Chem.,* 6, 123, 1968.
157. **Sahdev, R. K., Handa, K. L., and Rao, P. R.,** A note on the alkaloids of *Berberis lycium* Royle, *Indian J. Chem.,* 9, 503, 1971.
158. **Halder, R. K., Neogi, N. C., and Rathor, R. S.,** Pharmacological investigations of berberine hydrochloride, *Indian J. Pharm.,* 2, 26, 1970.
159. **Jagraj Beharilal,** Constituents of the seeds of *Blepharis edulis* Pers., *J. Indian Chem. Soc.,* 13, 109, 1936; 17, 259, 1940.
160. **Haravey, S. K.,** A brief comparative pharmacognostic study of certain indigenous drugs, *Nat. Med.,* J-1, 1967.
161. **Ghoshal, L. M.,** *Food Drugs,* October, 80, 1910.
162. **Basu, N. K., Lal, S. B., and Sharma, S. N.,** Investigations on Indian medicinal plants, *Q. J. Pharmacol.,* 20, 38, 1947.
163. **Chopra, R. N., Gosh, S., Gosh, B. N., and De, P.,** The pharmacology and therapeutics of *Boerhaavia diffusa* (Punarnava), *Indian Med. Gaz.,* 58, 203, 1923.
164. **Misra, A. N. and Tiwari, H. P.,** Constituents of the roots of *B. diffusa, Phytochemistry,* 10, 3318, 1971.
165. **Surange, S. R. and Pendse, G. S.,** Pharmacognostical study of root of *B. repanda* Willd. (Punarnava), *J. Res. Indian Med.,* 7, 1, 1972.
166. **Srivastava, D. N., Singh, R. H., and Udupa, K. N.,** Studies on the Indian indigenous drug, punarnava (*Boehraavia diffusa* Linn.). V. Isolation and identification of steroid, *J. Res. Indian Med.,* 7, 34, 1972.
167. **Singh, R. H. and Udupa, K. N.,** Studies on the Indian indigenous drug punarnava (*B. diffussa* Linn.). III. Experimental and pharmacological studies, *J. Res. Indian Med.,* 7, 17, 1972.
168. **Bhalla, T. N., Gupta, M. B., Seth, P. K., and Bhargava, K. P.,** Antiinflammatory activity of *B. diffusa, Indian J. Physiol. Pharmacol.,* 12, 37, 1968.
169. **Bhalla, T. N., Gupta, M. B., and Bhargava, K. P.,** Anti-inflammatory and biochemical study of *B. diffusa,* *J. Res. Indian Med.,* 6, 11, 1971.
170. **Singh, R. H. and Udupa, K. N.,** Studies on the Indian indigenous drug Punarnava (*B. diffusa*). IV. Preliminary controlled clinical trial in nephrotic syndrome, *J. Res. Indian Med.,* 7, 28, 1972.
171. **Mehra, P. N. and Karnik, C. R.,** Pharmacognostic studies on *Bombax ceiba* Linn., *Indian J. Pharm.,* 30, 284, 1968.
172. **Mukkerjee, J. and Roy, B.,** Chemical examination of *Salmalia malabarica* Schott Endl. (syn. *Bombax malabaricum* DC.), *J. Indian Chem. Soc.,* 48, 769, 1971.
173. **Seshadri, V., Bhatta, A. K., and Rangaswani, S.,** Phenolic components of *Bombax malabaricum* (root bark), *Curr. Sci.,* 40, 630, 1971.
174. **Seshadri, V., Bhatta, A. K., and Rangaswarni, S.,** Phenolic components of *Bombax malabaricum, Indian J. Chem.,* 11, 825, 1973.
175. **Harish, G. and Gupta, R. K.,** Chemical constituents of *Salmalia malabaricum* Schott et Endl. flowers, *J. Pharm. Sci.,* 61, 807, 1972.

176. **Misra, M. B., Mishra, S. S., and Misra, R. K.,** Pharmacology of *Bombax malabaricum* D.C., *Indian J. Pharm.,* 30, 165, 1968.
177. **Chopra, R. N.,** Alcoholic beverages in India, *Indian Med. Gaz.,* 77, 224, 1942.
178. **Chopra, R. N., Chopra, I. C., and Verma, B. S.,** Supplement to Glossary of Indian Medicinal Plants, Publication and Information Directorate (CSIR), New Delhi, 1969.
179. **Menon, M. K. and Kar, A.,** Analgesic and psychopharmacological effects of the gum resin of *Boswellia serrata, Planta Med.,* 19, 333, 1971.
180. **Dhawan, B. N. and Saxena, P. N.,** Evaluation of some indigenous drugs for stimulant effect on the rat uterus: a preliminary report, *Indian J. Med. Res.,* 46, 808, 1968.
181. **Mitra, R. and Mehrotra, S.,** Pharmacognostical study of priyal, *Buchanania lanzan* Spreng., *Herba Hung. Tom.,* 20, N1, 1981.
182. **Tandon, S. P., Tewari, K. P., and Saxena, V. K.,** Chemical examination of the roots of *Butea monosperma, Proc. Nat. Acad. Sci. India Sect. A,* 32, 237, 1969.
183. **Gupta, S. R., Ravinderanath, B., and Seshadri, T. R.,** The glucosides of *Butea monosperma, Phytochemistry,* 9, 2231, 1970.
184. **Seshadri, T. R. and Trikha. R. K.,** Proanthocyanidins from the bark and gum of *Butea monosperma, Indian J. Chem.,* 9, 1201, 1971.
185. **Razdan, M. K., Kapila Kanti, and Bhide, N. K.,** Antifertility effect and some pharmacological actions of *Butea frondosa* seed extracts, *Indian J. Physiol. Pharmacol.,* 13, 239, 1969.
186. **Khanna, U., Handa, S., and Chaudhury, R. R.,** The effects of *Butea monosperma* (Lam.) Kuntz on the fertility of female rats, *Indian J. Pharm.,* 28, 343, 1966.
187. **Kaleysa Raj, R. and Kurup, P. A.,** Anthelmintic activity, toxity and other pharmacological properties of palasonin, the active principle of *B. frondosa* seeds and its piperazine salt, *Indian J. Med. Res.,* 56, 1818, 1968.
188. **Sharma, B. M. and Singh, P.,** Pharmacognostic study of seeds of *Caesalpinia crista* L., *J. Res. Indian Med.,* 7, 8, 1972.
189. **Tummin Katti, M. C.,** Chemical examination of the seeds of *Caesalpinia bonducella* Flem., *J. Indian Chem. Soc.,* 7, 207, 1930.
190. **Ghatak, N. G.,** Chemical examination of the kernels of the seeds of *Caesalpinia bonducella, Proc. Indian Acad. Sci.,* 4, 141, 1934.
191. **Bhide, M. B., Nikam, S. T., and Chavan, S. R.,** Effects of seeds of sagarghota *(Caesalpinia bonducella)* on some aspects of reproductive system, in 16th Annu. Conf. Assoc. Physiol. Pharmacol. India, 1970, 28.
192. **Iyengar, M. A. and Pendse, G. S.,** Anti-diarrheal activity of the nut of *Caesalpinia bonducella* Flem., *Indian J. Pharm.,* 27, 307. 1965.
193. **Chatterjee A., Desmukh, S. K., and Chandrasekharan, S.,** Diterpenoid constituents of *Callicarpa macrophylla*. The structures and stereochemistry of calliterpenone and calliterpenone monoacetate, *Tetrahedron,* 28, 4319, 1972.
194. **Chaudhuri Rai, H. N.,** Pharmacognostic studies on the stem bark of *C. inophyllum* Linn., *Bull. Bot. Soc. Bengal,* 19, 54, 1965.
195. **Salvador, P. P. and Florencio, S. A.,** Chemical analysis for possible source of oils of forty five species of oil bearing seeds, *Philipp. Agric.,* 22, 408, 1933; *Chem. Abstr.,* 28, 2207, 1934.
196. **Mitra, C.,** *Calophyllum inophyllum* Linn. I. Chemical constituents of nut oil and the stem bark, *J. Sci. Ind. Res. Sect. B,* 16, 120, 1957.
197. **Govindachari, T. R., Pai, B. R., Mithukumarswamy, N., Rao, R. U., and Rao, N. N.,** Chemical components of heartwood of *C. inophyllum*. I. Isolation of mesuaxanthone B and a new xanthone, calophyllin B, *Indian J. Chem.,* 6, 57, 1968.
198. **Govindachari, T. R., Viswanathan, N., Pai, B. R., Rao Ramdas, U., and Srinivasan, M.,** *Calophyllum inophyllum, Tetrahedron,* 23, 1901, 1967.
199. **Subramanian, S. S. and Nair, A. G. R.,** Myricetin-7-glucoside from the androecium of the flowers of *C. inophyllum, Phytochemistry,* 10, 1679, 1971.
200. **Arora, R. B., Mathur, C. N., and Seth, S. D. S.,** Anticoagulant and cardiovascular actions of caryophyllolide, an indigenous complex coumarin, *Arch. Int. Pharmacodyn.,* 139, 75, 1962.
201. **Chaudhuri Rai, H. N.,** Pharmacognostic studies on the leaf of *Calatropis giganta* R. Br. ex Ait, *Bull. Bot. Surv. India,* 3, 171, 1961.
202. **Gupta, R. C., Sharma, P. C., and Kapoor, L. D.,** Pharmacognostic study of the leaf of *Calotropis* (Ait) R.Br. (Arka) *procera* (Madar), *J. Res. Indian Med.,* 6, 167, 1971.
203. **Singh, B. and Rastogi, R. P.,** Structure of ascelpin and some observations on the NMR spectra of *Calatropis* glycosides, *Phytochemistry,* 11, 757, 1972.
204. **Aylsworth, J. W.,** In the manufacture of plastic compositions India rubber or gutta percha is mixed, *Chem. Abstr.,* 7, 1426, 1913.
205. **Gerber, C. and Flourens, P.,** Rennet of the latex of *Calatropis procera, C. R.,* 155, 408, 1913; *Chem. Abstr.,* 7, 627, 1913.

206. **Sharma, G. K.,** *Calatropis procera* and *Calatropis gigantea, Indian J. Vet. Sci.*, 4, 63, 1934; *Chem. Abstr.*, 28, 6197, 1934.
207. **Shukla, O. P. and Krishnamurti, C. R.,** Properties and partial purification of bacteriolytic enzyme from the latex of *Calatropis procera* (Madar), *J. Sci. Ind. Res. Sect. C*, 20, 109, 1961.
208. **Derasari, H. R. and Shah, G. F.,** Preliminary pharmacological investigation of the roots of *Calatropis procera*. R. Br., *Indian J. Pharm.*, 27, 278, 1965.
209. **Srivastava, G. N., Chakarvarti, R.,N., and Zaidi, S. H.,** Studies on anticoagulant therapy. III. *In vitro* screening of some Indian plant latices for fibrinolytic and anti-coagulant activity, *Indian J. Med. Sci.*, 16, 873, 1962.
210. **Mehra, P. N.,** Biology, pharmacognosy and pharmaceutical preparations of Cannabis, in Seminar on Long Term Effects of *Cannabis* Use in India, ICMR, New Delhi, 1972.
211. **Kapoor, L. D.,** General introduction on cannabis preparations including pharmacognostic aspects, in Seminar on Long Term Effects of *Cannabis* Use in India, ICMR, New Delhi, 1972.
212. **Sharma, A. K., Singh, P. P., Nath, V., and Gode, K. D.,** Recent trends in qualitative and quantitative characterizations of cannabinoids, in Seminar on Long Term Effects of *Cannabis* Use in India, ICMR, New Delhi, 1972.
213. **Bose, B. C., Vijayvargiya, R., Saifi, A. Q., and Bhagwat, A. W.,** Chemical and pharmacological investigations of *Cannabis indica* Linn. I, *Arch. Int. Pharmacodyn.*, 146, 99, 1963.
214. **Bose, B. C., Saifi, A. Q., and Bhagwat, A. W.,** Observations on the pharmacological actions of *Cannabis indica*. II, *Arch. Int. Pharmacodyn.*, 147, 285, 1964.
215. **Bose, B. C. and Mukerji, B.,** Observation on the physiologically active fraction of Indian hemp, *Cannabis sativa* Linn., *Indian J. Med. Res.*, 33, 265, 1954.
216. **Arora, R. B., Gupta, L., Sharma, R. C., and Tayal, G.,** Pharmacodynamic and Toxicological aspects of different Cannabis preparations, Seminar on Long Term Effects of *Cannabis* use in India, ICMR, New Delhi. 1972.
217. **Gaind, K. N. and Juneja ,T. R.,** Investigations on *Capparis decidua, Planta Med.*, 17, 95. 1969.
218. **Gaind, K. N. and Juneja, T. R.,** Investigations on *Capparis decidua* Edgew. Phytochemical study of flowers and fruits, *Res. Bull. Punjab Univ.*, 21, 67, 1970.
219. **Juneja, T. R., Gaind, K. N., and Panisar, A. S.,** Investigations on *Capparis decidua* Edgew., *Res. Bull. Punjab Univ.*, 21, 519, 1970.
220. **Gaind, K. N., Juneja, T. R., and Jain, P. C.,** Anthelmintic and purgative activity of *Capparis decidua* Edgew., *Indian J. Hosp. Pharm.*, 2, 153, 1969.
221. **Gaind, K. N., Juneja, T. R., and Jain, P. C.,** Investigations on *Capparis decidua* Edgew. II. Antibacterial and antifungal studies, *Indian J. Pharm.*, 31, 24, 1969.
222. **Panse, T. B. and Paranjape, A. S.,** A study of 'carpasemine' isolated from *Carica papaya* seeds, *Proc. Indian Acad. Sci.*, 18, 140, 1943.
223. **Subbarayan, C. and Cama, H. R.,** Carotenoids in *Carica papaya* (papaya fruit), *Indian J. Chem.*, 2, 451, 1964.
224. **Govindachari, T. R., Nagarajan, K., and Viswanathan, N.,** Carpaine and pseudo carpaine, *Tetrahedron Lett.*, 24, 1907, 1965.
225. **Garg, S. K. and Garg, G. P.,** Antifertility effect of *Areca catechu* Linn. and *Carica papaya* Linn. in female albino rats, *Indian J. Pharm.*, 3, 23, 1971.
226. **Saha, et al.;** cited in *Advances in Research in Indian Medicine,* Banaras Hindu University, Varansai, 1970, 202.
227. **Sareen, et al.;** cited in *Advances in Reseach in Indian Medicine,* Banaras Hindu University, Varanasi, 1970, 202.
228. **Pillai, N. C., Rao, G. S. S., and Sirsi, M.,** Plant anticoagulants, *J. Sci. Ind. Res. Sect. C,* 16, 106, 1957.
229. **Chandrasekhar, N., Vaidyanathan, C. S., and Sirsi, M.,** Some pharmacological properties of the blood anticoagulant from *Carica papaya, J. Sci. Ind. Res. Sect. C,* 20, 213, 1961.
230. **Bose, B. C., Saifi, A. Q., Vijayvargiya, R., and Bhagwat, A. W.,** Phamacological study of *C. papaya* seed with special reference to its anthelmintic action, *Indian J. Med. Sci.*, 15, 888, 1961.
231. **Dar, R. N., Garg, L. C., and Pathak, R. D.,** Anthelmintic activity of *Carica papaya* seeds, *Indian J. Pharm.*, 27, 335. 1965.
232. **Rastogi, R. C., Vohra, M. M., Rastogi, R. P., and Dhar, M. L.,** Studies on *Carissa carandas* Linn. I. Isolation of the cardiac active principles, *Indian J. Chem.*, 4, 132, 1966.
233. **Rastogi, R. C., Rastogi, R. P., and Dhar, M. L.,** Studies on *Carissa carandas* Linn. II. The polar glycosides, *Indian J. Chem.*, 5, 215, 1967.
234. **Pakrashi, S. C., Datta, S., and Ghosh Dastidar, P. P.,** Indian medicinal plants. XVII. Phytochemical examination of *Carissa* spp., *Phytochemistry,* 7, 495, 1968.
235. **Chatterjee, M. L. and Roy, A. R.,** Pharmacological action of *Carissa carandas* roots (N.O. Apocynaceae). Preliminary observations, *Bull. Calcutta Sch. Trop. Med.*, 13, 15, 1965.
236. **Vohra, M. M. and De, N. N.,** Comparative cardiotonic activity of *Carissa carandas* L. and *C. spinarum* A. DC., *Indian J. Med. Res.*, 51, 937, 1963.

237. **Joglekar, S. N. and Gaitonde, B. B.,** Histamine releasing activity of *C. carandas* roots (Apocyanaceae) *Jpn. J. Pharmacol.,* 20, 367, 1970.
238. **Atal, C. K. and Sood, N. M.,** Study of Indian caraway and its substitutes. I. Essential oil of *Carum carvi* Linn., *Indian J. Pharm.,* 29, 42. 1967.
239. **Kapur, H. R., Gaind, K. N., Naring, K. S., and Ray, J. N.,** A new formula for chaksine, the alkaloid of *Cassia absus* and some experiments of its constitution, *J. Indian Chem. Soc.,* 17, 281, 1940.
240. **Pradhan, S. N., Varadan, K. S., Ray, C., and De, N. N.,** Pharmacological investigations of chaksine, an alkaloid from *Cassia absus* Linn., *J. Sci. Ind. Res. Sect. B.,* 12, 358, 1953.
241. **Stoll, A., Becker, B., and Helfenstein, A.,** Die Constitution der Sennoside, *Helv. Chim. Acta,* 33, 313, 1950.
242. **Fairbairn, J. W. and Saleh, M. R. I.,** Vegetable purgatives containing anthracene derivatives, *J. Pharm. Pharmacol.,* 3, 918, 1951.
243. **Chaudhuri Rai, H. N. and Kayal, A. N.,** Pharmacognostic studies on the leaves of *Cassia fistula* Linn., *J. Sen Memorial Comm. Bot. Soc. Bengal (Calcutta),* p. 477, 1969.
244. **Rao, K. R. V. and Ali, Nazar,** Pharmacognostic studies on *Cassia fistula* Linn. leaves, *Indian J. Pharm.,* 32, 174, 1970.
245. **Lilly Kutty, L. and Santha Kumari, G.,** Antimicrobial activities of *Cassia fistula* Linn., *J. Res. Indian Med.,* 4, 25, 1969.
246. **Kaji, N.N. and Khorana, M.L.,** Studies in *Cassia fistula* Linn. leaves, *Curr. Sci.,* 33, 462, 1964.
247. **Kaji, N. N., Khorana, M. L., and Sanghavi, M. M.,** One more glucoside from *Cassia fistula* Linn. pods, *Indian J. Pharm.,* 27, 71, 1965.
248. **Sen, A. B. and Shukla, Y. N.,** Chemical examination of *Cassia fistula, J. Indian Chem. Soc.,* 45, 744, 1968.
249. **Narayanan, V. and Seshadri, T. R.,** Proanthocyanidins of *Cassia fistula, Indian J. Chem.,* 10, 379, 1972.
250. **Agarwal, G. D., Rizvi, S. A. I., Gupta, P. C., and Tewari, J. D.,** Structure of fistulic acid, a new coloring matter from the pods of *Cassia fistula, Planta Med.,* 21, 150, 1972.
251. **Shah, C. S. and Shinde, M. V.,** Phytochemical studies of seeds of *Cassia tora* L. and *C. occidentalis* L., *Indian J. Pharm.,* 31, 27, 1969.
252. **Ginde, B. S., Kudavna, H., Nayak, K. V., and Kulkarni, A. B.,** Chemical investigations on *Cassia occidentalis* Linn. Isolation and structure of cassiollin, a new xanthone, *J. Chem. Soc.,* 9, 1285, 1970.
253. **Prakash, D. and Prasad, S.,** Pharmacognostical studies on *Cassia tora* Linn. (Chakwad), *J. Res. Indian Med.,* 6, 270, 1971.
254. **Sastry, M. S.,** Chemical investigations on *Cassia tora* Linn., *Curr. Sci.,* 34, 481, 1965.
255. **Krishna Roa, G. S., Sukh Dev, and Guha, P. C.,** Studies in sesquiterpenes. XII. Sesquiterpenes of the essential oil from the wood of Himalayan deodar, *J. Indian Chem. Soc.,* 29, 721, 1952.
256. **Pande, B. S., Krishnappa, S., Bisarya, S. C., and Sukh Dev,** Studies in Sesquiterpenes. XLVI. Cis and trans-atlantones from *C. deodara* Loud., *Tetrahedron,* 27, 841, 1971.
257. **Godbole, N. N. and Gunde, B. G.,** Analysis of the seed oil of *Celastrus paniculatus* (DUKUDU oil), *Fette Seifen,* 43, 249, 1936.
258. **Shah, M. M., Phalnikar, N. L., and Bhide, B. V.,** A note on the chemical investigation of the fruits of *Celastrus paniculata* Willd., *Curr. Sci.,* 16, 57, 1947.
259. **Basu, N. K. and Pabrai, P. R.,** Chemical investigations of *Celastrus paniculata* Willd., *J. Am. Pharm. Assoc.,* 35, 272, 1946.
260. **Gaitonde, B. B., Raiker, K. P., Shroff, F. N., and Patel, J. R.,** Pharmacological studies with malakangni, an indigenous tranquillizing drug (preliminary report), *Curr. Med. Pract.,* 1, 619, 1957.
261. **Sheth, U. K., Vaz, A., Bellare, R. A., and Deliwala, C. C.,** Behavioral and pharmacological studies on tranquillizing fraction from the oil of *Celastrus paniculatus* (malkanguni oil), *Arch. Int. Pharmacodyn.,* 144, 34, 1963.
262. **Majumdar, D. N.,** *Indian J. Pharm.,* p. 61, 1943.
263. **Vidyarathi, N. L.,** Composition and constitution of the oil from *Vernonia anthelmintica, Patna Univ. J.,* p. 51, 1945.
264. **Oliver,** Tea, coffee, *Tr. J.,* 443, 1932.
265. **Thorpe,** *Thorpe's Dictionary of Applied Chemistry,* Vol. 1 to 10, 1945—1950.
266. **Nietzki, R.,** Über ein neues, in der Blüthen von *Cichorium intybus* enthaltenes Glucosid, *Arch. Pharm.,* 5, 327, 1876.
267. **Misra, R. N. and Dutt, S.,** Chemical examination of the seeds of *Cichorium intybus* Linn. Constituents of the oil from the seeds, *J. Ind. Chem. Soc.,* 14, 141, 1937.
268. **Bhattacharji, S. and Dhar, M. L.,** Hayatinin and other alkaloids of *Cissampelos pareira* Linn., in Symp. at Colombo, Ceylon, 1967.
269. **Pradhan, S. N., Ray, C., and Varadhan, K. S.,** Pharmacology of hayatin methiodide, *J. Sci. Ind. Res. Sect C,* 17, 107, 1958.
270. **Pradhan, S. N., Pandey, K, and Badola, R. P.,** A clinical trial of hayatin methiodide as a relaxant in 100 cases, *Br. J. Anaesth.,* 36, 604, 1964.

271. **Basu, D. K.,** Studies on curariform activity of hayatin methochloride, an alkaloid of *C. pareira, Jpn. J. Pharmacol.,* 20, 246, 1970.
272. **Madan, C. L. and Nayar, S. L.,** A pharmacognostic study of the stem of *Cissus quadrangularis, J. Sci. Ind. Res. Sect. C,* 18, 253, 1959.
273. **Sen, S. P.,** Studies on the active constituents of *Cissus quadrangularis.* II. *Curr. Sci.,* 35, 317, 1966.
274. **Subbu, V. S. V.,** Pharmacological evaluation of a glucoside obtained from the plant *Vitis quadrangularis, Indian J. Physiol. Pharmacol.,* 12, 15, 1968.
275. **Subbu, V. S. V.,** Pharmacological and toxicological evaluation of an active principle obtained from the plant *Vitis quadrangularis, Indian J. Pharm.,* 2, 91, 1970.
276. **Subbu, V. S. V.,** Mechanisms of action of *Vitis glucoside* on myocardial tissue, *Indian J. Med. Sci.,* 25, 400, 1971.
277. **Udupa, K. N. and Prasad, G. C.,** Biochemical and Ca^{45} studies on the effect of *Cissus quadrangularis* in fracture healing, *Indian J. Med. Res ,* 52, 480, 1964.
278. **Prasad, G. C. and Udupa, K. N.,** Role of *Cissus quadrangularis* on fracture healing, in *Advances in Research in Indian Medicine,* Banaras Hindu University, Varansi, 1970, 163.
279. **Banerjee, S. P. and Dandiya, P. C.,** Smooth muscle and cardiovascular pharmacology of α-eletrin-2-D-glycopyranoside, glycoside of *Citrullus colocynthis, J. Pharm. Sci.,* 56, 1965, 1967.
280. **Banerjee, H. N.,** Chemical examination of *Clerodendron infortunadum, J. Indian Chem. Soc.,* 14, 51, 1937.
281. **Banerjee, S. K. and Chakravarti, R. N.,** Taraxerone from *Clitoria ternatea* Linn., *Bull. Calcutta Sch. Trop. Med.,* 12, 23, 1964.
282. **Dymack, W., Warden, C. J. H., and Hooper, D.,** *Pharmacographia Indica,* Vol. 1 to 3, Thanker Spink, Calcutta, 1890—1899.
283. **Bisht, B. S. and Nayar, S. L.,** Pharmacognosy of root and leaf of *Coccinia indica* Wight and Arn., *J. Sci. Ind. Res. Sect. C,* 17, 46, 1958.
284. **Gupta, S. S.,** Experimental studies on pituitary diabetes. II. Effect of indigenous antidiabetic drugs against the acute hyperglycemic response of anterior pituitary extract in glucose fed albino rats, *Indian J. Med. Res.,* 51, 716, 1963.
285. **Sitholey, R. V.,** The Unani drug gaozaban, *Q. J. Crude Drug Res.,* 10, 1581, 1970.
286. **Ghosh, S.,** 1942; cited in Satyavati, G. V., Effect of an Indigenous Drug on Disorders of Lipid Metabolism with Special Reference to Atherosclerosis and Obesity (Medaroga), thesis, D. Ag. M., Banaras Hindu University, Varanasi, 1966.
287. **Bhati, A.,** Essential oil from the resin of *Commiphora mukul* Hook. ex Storks, *J. Indian Chem. Soc.,* 27, 436, 1950.
288. **Patil, V. D., Nayak, U. R., and Sukhdev,** Component of Guggulu — resin from *Commiphora mukul,* in Proc. Semin. Disorders Lipid Metabolism, New Delhi, 1971.
289. **Patil, V. D.,** Chemistry of *Commiphora mukul* Gum Resin, Ph.D. thesis, Poona University, Poona, 1972.
290. **Patil, V. D., Nayak, U. R., and Sukhdev,** Chemistry of Ayurvedic crude drugs. I. Guggulu (resin from *Commiphora mukul*). I. Steroidal constituents, *Tetrahedron,* 8, 2341, 1972.
291. **Gujral, M. L., Sareen, K., Tangri, K. K., Amma, M. K. P., and Roy, A. K.,** Anti-arthritic and anti-inflammatory activity of gum guggal (*Balsamodendron mukul* Hook), *Indian J. Physiol. Pharmacol.,* 4, 267, 1960.
292. **Khanna, D. S., Agarwal, O. P., Gupta, S. K., and Arora, R. B.,** A biochemical approach to antiatherosclerotic action of *Commiphora mukul:* an Indian indigenous drug in Indian domestic pigs *(Sus scrofa), Indian J. Med. Res.,* 57, 900, 1969.
293. **Das, D., Sharma, R. C., and Arora, R. B.,** Antihyperlipidaemic activity of fraction A of *Commiphora mukul* in monkeys, *Indian J. Pharm.,* 5, 283, 1973.
294. **Satyavati, G. V.,** Effect of an Indigenous Drug on Disorders of Lipid Metabolism with Special Reference to Atherosclerosis and Obesity (Medaroga),.Thesis, D. Ay. M., Banaras Hindu University, Varanasi, 1966.
295. **Dwarakanath, C. and Satyavati, G. V.,** Research in some of the concepts of Ayurveda and application of modern chemistry and experimental pharmacology therefore, *Ayurveda Pradeepika,* 1, 69, 1970.
296. **Sastry, V. V. S.,** Experimental and Clinical Studies on the Effect of the Oleogum Resin of *Commiphora mukul* Engl. on Thrombotic Phenomena Associated with Hyperlipaemia (Snehavyapat), Thesis, D. Ay, M., Banaras Hindu University, Varanasi, 1967.
297. **Tripathi, S. N., Shastri, V. V. S., and Satyavati, G. V.,** (Achayra) Experimental and clinical studies on the effect of Guggulu *(Commiphora mukul)* in hyperlipaemia and thrombosis, *J. Res. Indian Med.,* 2, 10, 1968.
298. **Chatterjee, R., Guha, M. P., and Dasgupta, A. K.,** Plant alkaloids. IV. *Berberis Himalaica* Ahrendt and *B. tinctoria* Leschenault, *J. Ind. Chem. Soc.,* p. 921, 1952.
299. **Tiwari, R. D., SriVastra, K. C., Shukla, S., and Bajpai, R. K.,** Chemical examination of the fixed oil from the seeds of *Cordya myxa, Planta med.,* 15, 240, 1967.
300. **Basu, N. M., Ray, G. K., and De, N. K.,** On the vitamin C and carotene content of several herbs and flowers used in Ayurvedic system of medicine, *J. Indian Chem. Soc.,* 24, 358, 1947.

301. **Prasad, S., Khosa, R. L., and Durve, R. S.,** Pharmacognostical studies on Kalihari *(Gloriosa superba)* and its adulterant *Costus speciosus* Linn.(Kustha), *J. Res. Indian Med.,* 3, 126, 1969.
302. **Pandey, V. B. and Dasgupta, B.,** Chemical investigation of *Costus speciosus*: a new source of diosgenin, *J. Inst. Chem.,* 42, 131, 1970.
303. **Dasgupta, B. and Pandey, V. B.,** A new Indian source of diosgenin: *Costus speciosus, Experientia,* 26, 475, 1970.
304. **Pandey, V. B., Dasgupta, B., Bhattacharya, S. K., Debnath, P. K., Singh, S., and Sanyal, A. K.,** Chemical and pharmacological investigations of saponins of *Costus speciosus, Indian J. Pharm.,* 34, 116, 1972.
305. **Singh, S., Sanyal, A. K., Bhattacharya, S. K., and Pandey, V. B.,** Oestrogenic activity of saponins from *Costus speciosus* (Koen.) Sm., *Indian J. Med. Res.,* 60, 287, 1972.
306. **Bhattacharya, S. K., Parikh, A. K., Neogy, N. C., Lal, R., Debnath, P. K., and Pandey, V. B.,** Investigations on the anti-inflammatory and anti-arthritic activities of saponins isolated from *Costus speciosus, Rheumatism,* 6, 1, 1971.
307. **Rangaswami, S. and Venkata, Rao, E.,** Isolation of lycorine from *C. defixum* Ker-Gawl, *Curr. Sci.,* 23, 265, 1954.
308. **Rangaswami, S. and Venkata, Rao, E.,** Chemical examination of the seeds of *C. defixum, Indian J. Pharm.,* 17, 140, 1955.
309. **Aiyar, V. N. and Seshadri, T. R.,** Chemical components of *Croton oblongifolius.* IV. Constitution of oblongifoliol and deoxyoblongifoliol, *Indian J. Chem.,* 9, 1055, 1971.
310. **Aiyar, V. N. and Seshadri, T. R.,** Isolation of acetyl aleuritolic acid from *Croton oblongifolius* Roxb., *Indian J. Chem.,* 9, 1028, 1971.
311. **Soni, P., Gupta, S. C. and Aggarwal, J. S.,** Chemical examination of the seeds of *Cucumis sativus, J. Sci. Ind. Res. Sect. B,* 8, 210, 1949.
312. **Anon.,** *Health Bull. (Delhi),* No. 23, 32, 1941.
313. **Krisha, S. and Badhwar, R. L.,** Aromatic plants of India, *J. Sci. Ind. Res.,* Suppl. 242, 1950.
314. **Bisht, B. S. and Nayer, S. L.,** Pharmacognostic study of the rhizome of *Curculigo orchioides* Gaertn., *J. Sci. Ind. Res. Sect. C,* 19, 252, 1960.
315. **Mitra, R. and Kapoor, L. D.,** Pharmacognostical study of Amargundi, *Indian J. Pharm.,* 33, 125, 1971.
316. **Dutt, S. and Tayab, J. N.,** Chemical examination of essential oil derived from the rhizome of *Curcuma amada* Roxb., *Indian Soap J.,* 7, 200, 1941; *Vide Chem. Abstr.,* 35, 6393, 1941.
317. **Pachauri, S. P. and Mukerjee, S. K.,** Effect of *Curcuma longa* (Haridra) and *C. amada* (Amargundi) on the cholesterol level in experimental hypercholesterolemia of rabbits, *J. Res. Indian Med.,* 5, 27, 1970.
318. **Mehra, P. N. and Puri, H. S.,** Study of different samples of Haridra *(Curcuma longa* L.), *Indian J. Pharm.,* 33, 132, 1971.
319. **Anon.,** *Health Bull. (Delhi),* No. 23, 37, 1941.
320. **Subramanian, A. K. and Rao, G. S. K.,** Studies in terpenoids. IV. (1) Synthesis of (+) juvabione and (+)-ar-juvabione; (2) conversion of turmerone to (+)-ar-juvabione *(Curcuma longa), Tetrahedron Lett.,* 47, 4677, 1967.
321. **Arora, R. B., Basu, N., and Jain, A. P.,** Chemical, pharmacological and toxicological studies of *Curcuma longa* (turmeric), in 2nd Indo Soviet Symp. Chem. Nat. Products including Pharmacology, 1970, 133.
322. **Arora, R. B., Basu, N., Kapoor, V., and Jain, A. P.,** Anti-inflammatory studies on *Curcuma longa* (turmeric), *Indian J. Med. Res.,* 59, 1289, 1971.
323. **Mukherji, B., Zaidi, S. H., and Singh, G. B.,** Spices and gastric function. I. Effect of *Curcuma longa* on the gastric secretion in rabbits, *J. Sci. Ind. Res. Sect. C,* 20, 25, 1961.
324. **Agarwal, R. R. and Dutt, S.,** Chemical examination of *Cuscuta reflexa, J. Indian Chem. Soc.,* 12, 384, 1935.
325. **Gopinath, K. W., Sharma, R. C., and Kidwai, A. R.,** Chemical examination of *C. reflexa* Roxb., *J. Sci. Ind. Res. Sect. B,* 21, 601, 1962.
326. **Subramanian, S. S. and Nair, A. G. R.,** Chemical components of *Cuscuta reflexa* Roxb., *Indian J. Chem.,* 1, 501, 1963.
327. **Subramanian, S. S. and Nair, A. G. R.,** Isolation of mannitol from *C. reflexa* growing on *Santalum album., Indian J. Chem.,* 2, 81, 1964.
328. **Subramanian, S. S. and Nair, A. G. R.,** Isolation of luteolin from *C. reflexa* growing on *Glycosmis triphylla, Indian J. Chem.,* 2, 378, 1964.
329. **Subramanian, S. S. and Nair, A. G. R.,** Occurrence of mangiferin in *C. reflexa* growing on *Mangifera indica, Indian J. Chem.,* 4, 335, 1966.
330. **Prasad, D. N.,** Preliminary pharmacological investigations on *C. reflexa, Indian J. Med. Res.,* 53, 465, 1965.
331. **Mukherjee and Bhattacharya,** *J. Indian Chem. Soc. Ind. News Ed.,* 4, 1945.
332. **Rao, B. S. and Simonsen, J. L.,** Occurrence of sylvestrene, *J. Chem. Soc.,* 127, 2494, 1925; *Chem. Abstr.,* 407, 1926.
333. **Kapadia, V. H., Naik, N. G., Wadia, M. S., and Sukh Dev,** Sesquiterpenoids from the essential oils of *Cyprus rotundus* L., *Tetrahedron Lett.,* 47, 4661, 1967.

334. **Gupta, M. B., Singh, N., Palit, T. K., and Bhargava, K. P.,** Antiinflammatory activity of active constituent of *C. rotundus, Indian J. Pharm.,* 2, 23, 1970.
335. **Gupta, M. B., Palit, T. K., Singh, N., and Bhargava, K. P.,** Pharmacological studies to isolate the active contituents from *Cyprus scariosus* possessing anti-inflammatory, anti-pyretic and analgesic activities, *Indian J. Med. Res.,* 59, 76, 1971.
336. **Indira, M., Sirsi, M., Radomire, S., and Sukhdev,** The occurrence of some estrogenic substances in plants. I. Estrogenic activity of *C. rotundus,.J. Sci. Ind. Res. Sect. C,* 15, 202, 1956.
337. **Singh, N., Kulshreshtha, V. K., Gupta, H. B., and Bhargava, K. P.,** Pharmacological studies on *Cyprus rotundus, Indian J. Pharm.,* 1, 19, 1969.
338. **Singh, N., Kulshreshta, V. K., Gupta, H. B., and Bhargava, K. P.,** A pharmacological study of *C. rotundus, Indian J. Med. Res.,* 58, 103, 1970.
339. **Aggarwal, S. L., Dandiya, P. C., and Sharma, V. N.,** Chemical and pharmacological investigations of seeds of *Daucus carota* Linn., *Indian Pharm.,* 8, 291, 1953.
340. **Ghambhir, S. S., Sanyal, A. K., Sen, S. P., and Das, P. K.,** Studies on *D. carota* Linn. I. Pharmacological studies with the water-soluble fraction of the alcoholic extract of the seeds: a preliminary report, *Indian J. Med. Res.,* 54, 178, 1966.
341. **Garg, S. K. and Garg, G. P.,** Antifertility screening of plants. VII. Effect of five indigenous plants on early pregnancy in albino rats, *Indian J. Med. Res.,* 59, 302, 1970.
342. **Gupta, S. P., Ghatale, N., Farooq, A., and Arora, R. B.,** Pharmacological screening of seeds of *D. carota* with special reference to their antifertility activity, *Indian J. Pharm.,* 4, 101, 1972.
343. **Ghambir, S. S., Sanyal, A. K., Sen, S. P., and Das, P. K.,** Studies on *D. Carota* Linn. II. Cholinergic activity of the quarternary base isolated from water-soluble fraction of alcoholic extract of seeds, *Indian J. Med. Res.,* 54, 1053, 1966.
344. **Ghambir, S. S., Sen, S. P, Sanyal, A. K., Raina, M. K., and Das, P. K.,** Pharmacological study of tertiary base from seeds of *D. carota, Indian J. Physiol. Pharmacol.,* 12, 35, 1968.
345. **Bhargava, A. K., Ali, S. M., and Chauhan, C. S.,** Pharmacological investigations of the essential oil of *D. carota* var. *sativa* D. C., *Indian J. Pharm.,* 29, 127, 1967.
346. **Mukherjee, B., Bhattacharya, S. S., and Sarkar, P. K.,** *Dendrobium macraei* — an effective blood pressure lowering agent, *Indian J. Pharm.,* 5, 252, 1973.
347. **Bhargava, K. P. and Gupta, K. P.,** Effect of digoxin on cardiac centers, *Indian J. Physiol. Pharmacol.,* 6, 19, 1962.
348. **Chaudhuri, Rai H. N.,** Pharmacognostic studies on the stem bark of *Dillenia indica* Linn., *Bull. Bot. Soc. Bengal,* 22, 55, 1968.
349. **Joshi, B. S. and Rane, D. F.,** The structure of disolbulbine, *Indian J. Chem.,* 7, 452, 1969.
350. **Jindal, M. N., Kelkar, V. V., and Doctor, R. B.,** The anorexient activity of Kalio-Kund *(Discorea bulbifera)* methyl phenidate and cocaine in rats; a preliminary study, *Indian J. Med. Res.,* 57, 1075, 1969.
351. **Rao, G. S. K., Sukh Dev, and Guha, P. C.,** Studies in sesquiterpenes. VIII. Sesquiterpenes of the essential oil from *Dipterocarpus indicus* Bedd., *J. Indian Chem. Soc.,* 29, 589, 1952.
352. **Banerjee B., Chakravarti, R. N., and Chaudhuri, R. N.,** Seeds of *Dolichos biflorus* as diuretic, *Bull. Calcutta Sch. Trop. Med.,* 5, 71, 1957.
353. **Mehra, P. N. and Handa, S. S.,** Pharmacognosy of Bhringraja — antihepato-toxic drug of Indian origin, *Indian J. Pharm.,* 30, 284, 1968.
354. **Bhargava, K. K., Krishnaswamy, N. R., and Seshadri, T. R.,** Isolation of desmethylwedelolactone and its glucoside from *Eclipta alba, Indian J. Chem.,* 8, 664, 1970.
355. **Govindachari, T. R, Sidhaye, A. R., and Vishwanathan, N.,** Deoxyelephantopin a new sesquiterpene from *Elephantapus scaber* Linn., *Indian J. Chem.,* 8, 762, 1970.
356. **Govindachari, T. R., Viswanathan, N., and Fuhrer, H.,** Isodeoxyelephantopin, a new germacranodiolide from *E. scaber* Linn., *Indian J. Chem.,* 10, 272, 1972.
357. **Anon.,** *Health Bull. (Delhi),* No. 23, 42, 1951.
358. **Shah, C. S. and Khanna, P. N.,** Pharmacognostic comparison of *Embelia ribes* Burm. and *Embelia robusta* C.B. Clarke fruits, *Indian J. Pharm.,* 23, 275, 1961.
359. **Shah, N. C. and Kapoor, L. D.,** Botany of *Embelia ribes* Burm. — the Vidanga from Meghalaya, *Indian J. Pharm.,* 33, 125, 1971.
360. **Rao, Ch. B. and Venkateswarlu ,V.,** Chemical examination of *Embelia ribes.* I. Isolation of a new constituent "vilangin", its constitution and synthesis, *J. Org. Chem.,* 26, 4529, 1961.
361. **Khurana, S. K., Krishnamurthy, V., and Seshadri, T. R.,** Mass spectral analysis of the pigments of *Embelia ribes* and *Connarus monocerpas* Linn., *Curr. Sci.,* 41, 331, 1972.
362. **Arora, R. B., Ghatak, N., and Gupta, S. P.,** Antifertility activity of *Embelia ribes, J. Res. Indian Med.,* 6, 107, 1971.
363. **Guru, L. V. and Mishra, D. N.,** Effect of the alcoholic and aqueous extractives of *Embelia ribes* (Burm.) in patients infested by ascarides. Certain clinical studies, *J. Res. Indian Med.,* 1, 47, 1966.

364. **Anon.,** *Health Bull. (Delhi),* No. 23, 38, 1951.
365. **Srinivasan, M.,** Indian gooseberry *(Phyllanthus embelica), Nature,* 153, 684, 1944.
366. **Krishnamurti, G. R. and Giri, K. V.,** Preparation, purification, and composition of pectins from Indian fruits and vegetables, *Indian Acad. of Sci.,* 29B, 155, 1949.
367. **Dhar, D. C., Srivastva, D. L., and Sreenivasaya, M.,** Studies on *Embelica officinalis* Gaertn. I. Chromatographic study of some constituents of *Amla, J. Sci. Ind. Res. Sect. C,* 15, 205, 1956.
368. **Bhandari, P. R., Dhar, M. L., and Sharma, V. N.,** Chemical constituents of rootbark of *Crataeva nurvala, J. Sci. Ind. Res. Sect. B,* 10, 88, 1951.
369. **Anon.,** Progress Report, CDRI, Lucknow, 1951—1952.
370. **Rao, M. R. R. and Siddiqui, H. H.,** Pharmacological studies on *E. officinalis* Gaertn., *Indian J. Exp. Biol.,* 2, 29, 1964.
371. **Khorana, M. L., Rao, M. R. R., and Siddiqui, H. H.,** Antibacterial and antifungal activity of *Phyllanthus emblica, Indian J. Pharm.,* 21, 331, 1959.
372. **Bisht, B. S. and Nayer, S. L.,** Pharmacognostical study of the bark of *Erythrina variegata* Linn. var. *orientalis* (Linn.) Merill., *J. Sci. Ind. Res. Sect. C,* 18, 224, 1959.
373. **Bhakuni, D. S. and Khanna, N. M.,** Chemical examination of the bark of *E. indica* Lam., *J. Sci. Ind. Res. Sect. B,* 18, 494, 1959.
374. **Singh, Harkishan, and Chawla, A. K.,** Investigation of *Erythrina* sp. IV. Study of chemical constituents of seeds of *E. variegata* var. *orientalis, Planta Med.,* 19, 72, 1970.
375. **Ghosal, S., Ghosh, D. K., and Dutta, S. K.,** Occurrence of erysotrine and other alkaloids in *E. variegata, Phytochemistry,* 9, 2397, 1970.
376. **Bhattacharya, S. K., Debnath, P. K, Sanyal, A. K., and Ghosal, S.,** Pharmacological studies on the alkaloids of *Erythrina variegata* (Mandav), *J. Res. Indian Med.,* 6, 235, 1971.
377. **Sengupta, P. and Das, B. P.,** Terpenoids and related compounds. V. Triterpenoids from the flowers of *Eugenia jambolana* Lam., *J. Indian Chem. Soc.,* 42, 539, 1965.
378. **Vaish, S. K.,** Therapeutic uses of Jaman seed in alloxan diabetes, *Proc. 41st Indian Sci. Congr.,* Part 3, p. 230, 1954.
379. **Aiman, R.,** Indigenous antidiabetic substances of plant origin, *Symp. Indian Drugs Bombay,* p. 3, 1961.
380. **Shrotri, D. S., Kelkar, M., Deshmukh, V. K., and Aiman, R.,** Investigations of the hypoglycemic properties of *Vinca rosea, Cassia auriculata* and *E. jambolana, Indian J. Med. Res.,* 51, 464, 1963.
381. **Chaudhury, Rai H.N. and Datta, S. C.,** Pharmacognostic studies on Ayapana, *Indian Pharm.,* 5, 331, 1950.
382. **Anon.,** *The British Codex,* Pharmaceutical Press, London, 1949.
383. **Veda, H. and Chiao, M. -H.S.U.,** A chemical study of *Euphorbia, J. Taiwan Pharm. Assoc.,* 1, 40, 1949; *Chem. Abstr.,* 45, 7306, 1951.
384. **Iqbal, N. K., Chandra, O. M., Gupta, K. P., Singhal, K. C., and Saxena, P. N.,** A preliminary report on the pharmacological actions of the latex of *Euphorbia neriifolia.* II, *Indian J. Pharm.,* 1, 8, 1969.
385. **Rizvi, S. S. A, Chandra, O. M., Singhal, K. C., Gupta, K. P., Iqbal, V. J., and Saxena, P. N.,** Toxicological and insecticidal activity of latex of *E. neriifolia* L. I, *Indian J. Pharm.,* 1, 9, 1969.
386. **Baveja, S. K. and Singla, R. D.,** Investigation of *Evolvulus alsinoides* Linn. (Shankhpushpi), *Indian J. Pharm.,* 31, 108, 1969.
387. **Mehra, P. N. and Raina, M. K.,** Pharmacognostic studies of *Ferula* species, *Indian J. Pharm.,* 34, 170, 1970.
388. **Subahrmanyan, V., Sreenivasamurthy, V., Krishnamurthy, K., and Swaminathan, M.,** Studies on the antibacterial activity of spices, *J. Sci. Ind. Res. Sect. C,* 16, 240, 1957.
389. **Chatterjee, D. and Chakraborty, D. P.,** Chemical examination of *Ficus bengalensis, Indian J. Chem. Soc.,* 45, 185, 1968.
390. **Subramanian, S. S. and Nair, A. G. R.,** Sterols and flavonols of *F. benghalensis, Phytochemistry,* 9, 2583, 1970.
391. **Deshmukh, V. K., Shrotri, D. S., and Aiman, R.,** Isolation of hypoglycemic principle from the bark of *F. bengalensis, Indian J. Physiol. Pharmacol.,* 4, 182, 1960.
392. **Brahmachari, H. D. and Augusti, K. T.,** Isolation of orally effective hypoglycemic compounds from *F. bengalensis,* Linn. *Indian J. Physiol. Pharm.,* 8, 60, 1964.
393. **Nayer, S. L. and Bisht, B. S.,** The pharmacognosy of the bark of *F. glomerata* Roxb., *J. Sci. Ind. Res. Sect. C,* 18, 15, 1959.
394. **Momin and Ray,** *Indian J. Vet. Sci.,* 13, 183, 1953.
395. **Sharma, R. C., Zaman, A., and Kidwai, A. R.,** Chemical examination of *F. racemosa* Linn., *Indian J. Chem.,* 1, 365, 1963.
396. **Shrotri, D. S. and Aiman, R.,** The relationship of post absorptive state to the hypoglycemic action studies on *F. bengalensis* and *F. glomerata, Indian J. Med. Res.,* 48, 162, 1960.
397. **Ambike, S. H. and Rao, M. R. R.,** Studies on a phytosterolin from the bark of *F. religiosa.* I, *Indian J. Pharm.,* 29, 91, 1967.
398. **Shah, C. S., Qadry, J. S., and Chauhan, M. G.,** Chemical races in fennel, *Indian J. Pharm.,* 31, 171, 1969.

399. **Shah, C. S., Qadry, J. S., and Chauhan, M. G.,** Chemical races in fennel, *Planta Med.,* 18, 285, 1970.
400. **Mehra, P. N., Jolly, S. S., and Puri, H. S.,** Pharmacognosy of *Fumaria indica* (Haussk) Pungsley and its adulterants, *Indian J. Pharm.,* 30, 284, 1968.
401. **Panday, V. B., Dasgupta, B., Bhattacharya, S. K., Lal, R., and Das, P. K.,** Chemistry and pharmacology of the major alkaloid of *F. indica* (Haussk, Pungsley), *Curr. Sci.,* 40, 455, 1971.
402. **Bhattacharya, S. K., Lal, R., Sanyal, A. K., and Das, P. K.,** Pharmacological studies on the *Furmaria parviflora* Lam., *Indian J. Pharm.,* 1, 8, 1969.
403. **Karanjgaokar, C. G., Radhkrishnan, P. V., and Venkataraman, K.,** Morelloflavone, a 3-(8) flavonyl-flavonone from the heartwood of *Garcinia morella, Tetrahedron Lett.,* 33, 3195, 1967.
404. **Deshpande, P. J., Pathak, S. N., and Gode, J. D.,** Wound healing under the influence of certain indigenous drugs, in *Advances in Research in Indian Medicine,* Banaras Hindu University, Varanasi, 1970, 280.
405. **Santhanan, K. and Rao, P. L. N.,** Antibiotic principles of *G. morella.* XII. Characterization of Baud α-l-, guttiferins as cathartic principles of gamboge and seed coat of *G. morella, Indian J. Exp. Biol.,* 6, 158, 1968.
406. **Sitholey, R. V.,** Distinguishing characters of the species known as Goazaban, *Q. J. Crude Drug Res.,* 11, 1818, 1971.
407. **Tewari, P., Prasad, D. N., Chaturvedi, C., and Das, P. K.,** Preliminary studies on uterine activity of *G. superba* Linn. and its adulterant *Costus speciousus* Sm., *J. Res. Indian Med.,* 1, 196, 1967.
408. **Tewari, P. V., Sharma, P. V., Prasad, D. N., and Pandey, V. B.,** Phytochemical and pharmacological studies (action on uterine musculature) of *Costus speciousus* (Koen) Sm. (Kevuka), *J. Res. Indian Med.,* 7, 14, 1972.
409. **Haravey, S. K.,** *Glycyrrhiza glabra* Linn. A review on the importance of the plant (roots) *Natl. Med. J.,* 1, 1959.
410. **Datta, S. C. and Mukerji, B.,** *Glycyrrhiza.* Pharmacognosy of Indian Root and Rhizome Drugs, Bull. No. 1, Manager of Publications, Pharmacognosy Laboratory Ministry of Health. Government of India, Delhi, 1950, 53.
411. **Puri, B. and Seshadri, T. R.,** Survey of anthoxanthins. V. Coloring matter of liquorice roots, *J. Sci. Ind. Res. Sect. B.,* 13, 475, 1954.
412. **Gujral, M. L., Sareen, K., Phukan, D. P., and Amma, M. K. P.,** Antiarthritic activity of glycyrrhizin in adrenalectomised rats, *Indian J. Med. Sci.,* 15, 625, 1961.
413. **Prasad, S.,** Pharmacognosy of anti-inflammatory drugs, in *Advances in Research in Indian Medicine,* Banaras Hindu University, Varanasi, 1970, 117.
414. **Krishna and Ramaswami,** *Indian Forest Bull.,* N.S., No. 79, 17, 1932.
415. **Kedare, B. S. and Tendolkar, G. S.,** Destructive distillation of some hardwood species of Bombay state, *J. Sci. Ind. Res. Sect. B,* 12, 125, 1953.
416. **Rao Venkata, D., Rao Venkata, E., and Viswanadhan, N.,** Occurrence of luteolin in the leaves of *Gmelina arborea* Linn., *Curr. Sci.,* 36, 71, 1967.
417. **Rao Venkata, D. and Rao Venkata, E.,** Chemical components of the leaves of *Gmelina arborea, Indian J. Pharm.,* 32, 40, 1970.
418. **Govindachari, T. R., Parthasarthy, P. C., Desi, H. K., and Mohamed, P. A.,** Clutyl ferulate — a new long chain ester from *Gmelina arborea* L. and *Lunnea grandis* (Dennst.), *Indian J. Chem.,* 9, 1027, 1971.
419. **Gaur, D. S. and Gupta, L. P.,** Experimental studies on the rasayanlike effect of Kashmari (*Gmelina arborea*) on serum proteins, *J. Res. Indian Med.,* 3, 43, 1968.
420. **Chander, K. and Seshadri, T. R.,** Leucoanthocyanidin in cotton seed, *J. Sci. Ind. Res. Sect. A,* 16, 319, 1957.
421. **Seshadri, T. R.,** Recent developments in our knowledge of gossypol, the coloring matter and active principle of cotton and related plants, *Proc. Indian Natl. Sci. Acad. Part A,* 37, 411, 1971.
422. **Prasad, S.,** Pharmacognosy of anti-inflammatory drugs, in *Advances in Research in Indian Medicine,* Banaras Hindu University, Varanasi, 1970, 139.
423. **Sarkar, B. and Khanna, N. M.,** Chemical examination of *Grewia populifolia* Vahl., *J. Sci. Ind. Res. Sect. C.,* 18, 20, 1959.
424. **Wahi, S. P. and Chunekar, K. C.,** Pharmacognostical studies of *Gymnema sylvestre* R. Br., *J. Sci. Res. Banaras Hindu Univ.,* 15, 205, 1964.
425. **Dymock, W., Warden, C. J. H., and Hooper, D.,** *Gymnema sylvestre, Pharmographis Indica,* Vol. 2, Thacker Spink, Calcutta, 1891, 450.
426. **Chakaravarti, R. N., Chakravarti, D., and Itty, M. I.,** Crystalline saponin from *G. sylvestre, Bull. Calcutta Sch. Trop. Med.,* 14, 126, 1966.
427. **Gupta, S. S. and Seth, C. B.,** Experimental studies on pituitary diabetes. II. Comparison of blood sugar level in normal and anterior pituitary extract induced hyperglycaemic rats treated with a few Ayurvedic remedies, *Indian J. Med. Res.,* 50, 708, 1962.
428. **Gupta, S. S., Seth, C. B., and Mathur, V. S.,** Effect of gurmar and Shilajet on bodyweight of young rats, *Indian J. Physiol. Pharmacol.,* 9, 87, 1965.
429. **Deshmukh, V. K. and Pandit, U.,** A study of fruits of *Helicteres isora* Linn., *Indian J. Pharm.,* 30, 283, 1968.
430. **Kapoor, L. D., Kapoor, S. L., Srivastava, S. N., Singh, A., and Sharma, P. C.,** Survey of Indian plants for saponins, alkaloids and flavonoids. II, *Lloydia,* 34, 94, 1971.

431. **Atal, C. K., Srivastava, J. B., Wali, B. K., Chakravarty, R. B., Dhawan, B. N., and Rastogi, R. P.,** Screening of Indian plants for biological activity. VIII, *Indian J. Exp. Biol.*, 16, 330, 1978.
432. **Barik, B. R., Dey, A. K., and Das, P. C.,** *Helicteres isora* L. A new source of diosgenin, *Indian J. Chem.*, 20B, 938, 1981.
433. **Singh, S. B., Singh, A. K., and Thakur, R. S.,** Chemical constituents of the leaves of *H. isora* L., *Indian J. Pharm. Sci.*, 46, 148, 1984.
434. **Dhawan, B. N. and Saxena, P. N.,** Evaluation of some indigenous drugs for stimulant effect on the rat uterus. A preliminary report, *Indian J. Med. Res.*, 46, 808, 1958.
435. **Jain, S. K.,** *Medicinal Plants,* National Book Trust of India, New Delhi, 1968.
436. **Prasad, S., Wahi, S. P., and Misra, L. C.,** Pharmacognostic investigations on Indian sarsaparilla, its substitutes and adulterants. I. *Hemidesmus indicus* R. Br., *Indian J. Pharm.*, 26, 81, 1964.
437. **Prasad, S. and Wahi, S. P.,** Pharmacognostical investigations on Indian Sarsaparilla. I. Root and rootstock of *Hemidesmus indicus* R. Br., *Indian J. Pharm.*, 27, 35, 1965.
438. **Krishnamurthy, K. H. and Sundaram, R.,** Foliar epidermis and pharmacognosy in some members of Aslepiadaceae, *J. Indian Bot. Soc.*, 46, 160, 1967.
439. **Subramanian, S. S. and Nair, A. G. R.,** Flavonoids of some Asclepiadacious plants, *Phytochemistry*, 7, 1703, 1968.
440. **Bose, K. C. and Bose, N. K.,** *J. Indian Med. Assoc.*, 1, October, 1931; cited in **Rastogi, R. P. and Dhar, M. L.,** *J. Sci. Ind. Res. Sect. B*, 19, 455, 1960.
441. **Basu, N. K. and Pabrai, P. R.,** *Q. J. Pharm. Pharmacol.*, 137, 1947; cited in **Rastogi, R. P. and Dhar, M. L.,** *J. Sci. Ind. Res. Sect. B*, 19, 455, 1960.
442. **Sastry, M. M., Dhalla, N. S., and Malhotra, C. L.,** 1959; cited in **Rastogi, R. P. and Dhar, M. L.,** *J. Sci. Ind. Res. Sect. B*, 19, 455, 1960.
443. **Rastogi, R. P. and Dhar, M. L.,** Chemical examination of *Bacopa monniera* Weltest, *J. Sci. Ind. Res. Sect. B*, 19, 455, 1960.
444. **Chatterji, N., Rastogi, R. P., and Dhar, M. L.,** Chemical examination of *Bacopa monniera*. I. Isolation of chemical constituents, *Indian J. Chem.*, 1, 212, 1963.
445. **Banerjee, S. K. and Chakravarti, R. N.,** Stigmasterol from *Herpestris monniera, Bull. Calcutta Sch. Trop. Med.*, 11, 57, 1963.
446. **Malhotra, C. L. and Das, P. K.,** Pharmacological studies of *H. monniera* L. (Brahmi), *Indian J. Med. Res.*, 47, 294, 1959.
447. **Das, P. K., Malhotra, C. L., and Dhalla, N. S.,** Studies on alkaloids of *H. monniera* L., *Indian J. Physiol. Pharmacol.*, 5, 136, 1961.
448. **Anon.,** *Pharmacopoeia of India,* Manager of Publications, Delhi, 1966, 390.
449. **Kapoor, L. D., Singh, A., Kapoor, S. L., and Srivastava, S. N.,** Survey of Indian plants for saponins, alkaloids and flavonoids, *Lloydia*, 32, 297, 1969.
450. **Basu, N. K. and Jayaswal, S. B.,** Amoebicidal activity of the alkaloids *in vitro, Indian J. Pharm.*, 30, 289, 1968.
451. **Malhotra, C. L., Das, P. K., Sastry, M. S., and Dhalla, N. S.,** Chemical and pharmacological studies on *Hydrocotyle asiatica* Linn., *Indian J. Pharm.*, 23, 106, 1961.
452. **Rastogi, R. P., Sarkar, B., and Dhar, M. L.,** Chemical examination of *Centella asiatica* Linn. I. Isolation of the chemical constituents, *J. Sci. Ind. Res. Sect. B*, 19, 252, 1960.
453. **Dutta, T. and Basu, U. P.,** Triterpenoids. I. Thankuniside and thankunic acid: a new triterpene glycoside and acid from *Centella asiatica* Linn (Urb.), *J. Sci. Ind. Res. Sect. B*, 21, 239, 1962.
454. **Singh, B. and Rastogi, R. P.,** A reinvestigation of the triterpenes of *C. asiatica, Phytochemistry*, 8, 917, 1969.
455. **Aithal, H. N. and Sirsi, M.,** Preliminary pharmacological studies on *C. asiatica* Linn. *(N.O. Umbelliferae), Antiseptic,* May 1, 1961.
456. **Ramaswamy, A. S., Pariyswami, S. M., and Basu, N.,** Pharmacological studies on *C. asiatica* Linn. *(Brahma manduki) (N.O. Umbelliferae), J. Res. Indian Med.*, 4, 160, 1970.
457. **Appa Rao, M. V. R., Usha, S. P., Rajagopalan, S. S., and Sarangan, R.,** Six months results of double blind trial to study the effect of Mandookaparni and Punarnava on normal adults, *J. Res. Indian Med.*, 2, 79, 1967.
458. **Appa Rao, M. V. R., Rajgopalan, S. S., Srinavasan, V. R., and Sarangan, R.,** Study of Mandookaparni and Punarnava for their Rasayana effect on normal healthy adults, *Nagarjun*, 12, 33, 1969.
459. **Appa Rao, M. V. R., Srinivasan, K., and Rao, K. T.,** Effect of Mandookaparni *(Centella asiatica)* on the general mental ability (Medhya) of mentally retarded children, *J. Res. Indian Med.*, 8, 9, 1973.
460. **Satyavati, G. V., Gupta, K., Ashok, and Tandon Neeraj,** Medicinal plants of India, Vol. 2, Indian Council of Medical Research, New Delhi, 1987.
461. **Thomas, A. F.,** Geigerone (3-isopropenyl-4-methyl-4-vinylcyclohexanone), a new naturally occurring C_{12} terpenoid, *Helv. Chim. Acta*, 55, 2429, 1972.
462. **Sharma, S. D., Sethi, A. S., and Sharm, M. L.,** Synthesis of *dl*-geigerone, *Indian J. Chem.*, 15B, 958, 1977.
463. **Sharma, G. D.,** A Study of Some Indigenous Drugs against Dermatomycosis in Animals, M.V.Sc. thesis, Agra University, Agra, 1968.

464. **Gayatri Devi and Sisodia, C. S.,** Effect of ether extract of *Juniperus communis* on sarcoptic mange in sheep, *Indian J. Anim. Sci.,* 39, 345, 1969.
465. **Srivastava, S. C. and Sisodia, C. S.,** Treatment of psoroptic mange in sheep with *J. communis* (Hipush) extract, *Indian Vet. J.,* 46, 826, 1969.
466. **Rao, V. R. and Gupta, I.,** *In vivo* studies on the antifungal activity of some indigenous drugs in experimentally infected goats and rabbits, *Indian J. Pharmacol.,* 3, 29, 1971.
467. **Rao, V. R. and Gupta, I.,** A note on the antifungal activity of some indigenous plants, *Indian J. Anim. Sci.,* 47, 226, 1977.
468. **Chauhan, N. S., Uniyal, M. R., and Sannd, B. S.,** A preliminary study of the indigenous drug used in Tibetan medical center, Dharamsala (H.P.) *Nagarjun,* 22, 190, 1979.
469. **Desai, H. K., Gawad, D. H., Govindachari, T. R., Joshi, B. S., Kamat, V. N., Modi, J. D., Parthasarathy, P. C., Patankar, S. J., Sidhaye, A. R., and Viswanathan, N.,** Chemical investigation of some Indian plants. VI, *Indian J. Chem.,* 9, 611, 1971.
470. **Mohan, H., Sudhanshu, A., and Pathak, H. D.,** Flavonoids from the leaves of *Litsea glutinosa* (Lauracea), *Natl. Appl. Sci. Bull.,* 27, 95, 1977.
471. **Agrawal, M. and Gupta, R.,** Constituents of the essential oil from the seeds of *L. glutinosa, Perfum Kosmet Aromes* (spec. ed.), October, 138, 1980.
472. **Upreti, H., Bhakuni, D. S., and Dhar, M. M.,** Aporphine alkaloids of *Litsea sebifera, L. Wightiana* and *Actinodaphne obovata, Phytochemistry,* 11, 3057, 1972.
473. **Sivakumaran, M. and Gopinath, K. W.,** Sebiferine and litseferine, two new alkaloids from *Litsea sebifera, Indian J. Chem.,* 14B, 150, 1976.
474. **Kar, A., Menon, M. K., and Chauhan, C. S.,** Effect of essential oil of *Litsea glutinosa* (Lour.) C. B. Robins on cardiovascular system and isolated tissues, *Indian J. Exp. Biol.,* 8, 61, 1970.
475. **Menon, M. K., Kar, A., and Chauhan, C. S.,** Some psychopharmacological actions of the essential oil of *L. glutinosa* (Lour.) C. B. Robins, *Indian J. Physiol. Pharmacol.,* 14, 185, 1970.
476. **Maheshwari, J. K., Singh, K. K., and Saha, S.,** Ethnomedicinal uses of plants by the Tharus of Kheri district, U.P., *Bull. Med. Ethnobot. Res.,* 1, 318, 1980.
477. **Shah, C. S. and Contractor, A. M.,** Pharmacognostic study of *Lobelia nicotianifolia* Heyne, *J. Sci. Ind. Res. Sect. C,* 20, 120, 1961.
478. **Shah, C. S., Qadry, J. S., and Bhatt, M. G.,** Campanulaceae. Lobeline from *Lobelia nicotianifolia, Phytochemistry,* 11, 2884, 1972.
479. **Handa, K. L. and Nazir, B. N.,** Nor-lobelanine from *Lobelia nicotianifolia* Heyne, *J. Inst. Chem. India,* 32, 146, 1960.
480. **Mukharya, D. and Dahia, M. S.,** Antimicrobial activity of some plant extracts, *Indian Drugs,* 14, 160, 1977.
481. **Mukerji, B. and Ghosh, S. K.,** *Lobelia nicotianifolia* Heyne as substitute for *Lobelia inflata* Linn. *B. P., Curr. Sci.,* 14, 198, 1945.
482. **Gopolan ,C., Rama Sastri, B. V., and Balasubramanian, S. C.,** Nutritive Value of Indian Foods, National Institute of Nutrition, Hyderabad, ICMR, New Delhi, 1984, 78.
483. **Barua, A. K., Chakraborti, S. K., and Ray, A. K.,** Chemical examination of the seeds of *Luffa acutangula, J. Indian Chem. Soc.,* 35, 480, 1958.
484. **Mallavarapu, G. R. and Rao, L. R.,** Chemical examination of some Cucurbitaceae plants, *Indian J. Chem.,* 17B, 417, 1979.
485. **Kaul, S. and Verma, S. L.,** Oxalate contents of foods commonly used in Kashmir, *Indian J. Med. Res.,* 55, 274, 1967.
486. **Nanda, R. S.,** Fluoride content of North Indian foods, *Indian J. Med. Res.,* 60, 1470, 1972.
487. **Singh, P. P.,** Oxalic acid content of Indian foods, *Qual. Plant. Mater. Veg.,* 22, 335, 1973.
488. **Sengupta, S. R. and Pal, B.,** Iodine and fluorine contents of food stuffs, *Indian J. Nutr. Diet.,* 8, 66, 1971.
489. **Aswal, B. S., Bhakuni, D. S., Goel, A. K., Kar, K., Mehrotra, B. N., and Mukherjee, K. C.,** Screening of Indian plants for biological activity. X, *Indian J. Exp. Biol.,* 22, 312, 1984.
490. **Singh, S. P., Singh, N., and Kohli, R. P.,** Antifertility studies of some indigenous plants, *Indian J. Pharmacol.,* 10(1), 88, 1978.
491. **Deshmukh, P. D., Rathore, Y. S., and Bhattacharya, A. K.,** Larval survival of *Diacrisia obliqua* Walker on several plant species, *Indian J. Entomol.,* 41, 5, 1979.
492. **Grewal, K. S. and Kochhar, B. D.,** *Luffa acutangula:* the chemical and pharmacological investigation of luffa seeds, *Indian J. Med. Res.,* 31, 63, 1943.
493. **Jaiswal, S., Batra, A., Verma, A., and Bokadia, M. M.,** Free amino acid of some regionally available medicinally important plant seeds, *Sci. Cult.,* 50, 24, 1984.
494. **Rangaswami, S. and Samba Murthy, K.,** Crystalline bitter principle of seeds of Indian edible sponge gourd (*L. aegyptica* Mill.), *Indian J. Pharm.,* 16, 225, 1954.
495. **Varshney, I. P., Khan, M. S. Y., and Sharma, S. C.,** Saponins and sapogenins. XXVII. On the unidentified acid and neutral sapogenins from the seeds of *Luffa aegyptica* Mill. (black variety), *Aust. J. Chem.,* 14, 1689, 1965.

496. **Varshney, I. P. and Beg, M. F. A.,** Saponins from seeds of *L. aegyptica* Mill. Isolation of aegyptinin A and aegyptinin B, *Indian J. Chem.,* 15B, 394, 1977.
497. **Pal, S., Chakraborty, S. K., Banerjee, A., and Mukerji, B.,** Search for anticancer drugs from Indian medicinal plants (Ayurvedic, Unani, etc.), *Indian J. Med. Res.,* 56, 445, 1968.
498. **Mitra, C. R. and Awasthi, Y. C.,** Constituents of *Madhuca latifolia* and *M. butyracea* nuts, *J. Sci. Ind. Res. Sect. D,* 21, 102, 1962.
499. **Awasthi, Y. C. and Mitra, C. R.,** *Madhuca latifolia.* Constituents of fruit pulp and nut shell, *Phytochemistry,* 6, 121, 1967.
500. **Hariharan, V., Rangaswami, S., and Sarangan, S.,** Saponins of the seeds of *Bassia latifolia, Phytochemistry,* 11, 1791, 1972.
501. **Singh, R. P.,** Studies on sterols of Mahua oil, *Indian J. Appl. Chem.,* 22, 61, 1959.
502. **Subramanian, S. S. and Nair, A. G. R.,** Sapotaceae. Myricetin and myricetin 3-O-L rhamnoside from the leaves of *Madhuca indica* and *Achras sapota, Phytochemistry,* 11, 3090, 1972.
503. **Daniel, M., Sabnis, S. D., and Mani, N. V.,** Estimation of tannins in some of the forest resources of Gujarat, *Indian J. For.,* 1, 223, 1978.
504. **Awasthi, Y. C. and Mitra, C. R.,** Madhuca latifolia, Triterpenoid constituents of the trunk bark, *Phytochemistry,* 7, 1433, 1968.
505. **Nanda, R. S.,** Fluoride contents of North Indian foods, *Indian J. Med. Res.,* 60, 1470, 1972.
506. **Gopalan, C., Rama Sastri, B. V., and Balasubramanian, S. C.,** Nutritive Value of Indian Foods, ICMR, New Delhi, 1984.
507. **Singh, R. and Singh, R.,** Screening of some plant extracts for antiviral properties, *Technology (Sindri),* 9, 415, 1972.
508. **Setty, B. S., Kamboj, V. P., Garg, H. S., and Khanna, N. M.,** Spermicidal potentials of saponins isolated from Indian medicinal plants, *Contraception,* 14, 571, 1976.
509. **Trivedi, V. B., Kazmi, S. M., and Kazmi, S. N.,** Comparative bactericidal activity of two angiosperms, *Bull. Bot. Soc. Univ. Saugar,* 27, 36, 1980.
510. **Kapoor, L. D. and Ram Krishan,** *Advances in Essential Oil Industry,* Today & Tomorrow's Printers & Publishers, New Delhi, 1977.
511. **Lounasmaa, M., Widen, C. J., Tuuf, C. M., and Hubtikangas, A.,** On the phloroglucinol derivates of *Mallotus philippinensis, Planta Med.,* 28, 16, 1975.
512. **Bhattacharjee, A. K. and Das, A. K.,** Phytochemical screening of some Indian plants, *Q. J. Crude Drug Res.,* 9, 1408, 1969.
513. **Kapoor, L. D., Singh, A., Kapoor, S. L., and Srivastavas, S. N.,** Survey of Indian medicinal plants for saponins, alkaloids and flavonoids. I, *Lloydia,* 32, 297, 1969.
514. **Crombie, L., Greew, C. L., Tuck, B., and Whitnig, D. A.,** Constituents of Kamala. Isolation and structure of two new components, *J. Chem. Soc.,* p. 2625, 1968.
515. **Bandopadhyay, M., Dingra, V. K., Mukerjee, S. K., Pardeshi, N. P., and Seshadri, T. R.,** Triterpenoids and other components of *Mallotus philippinensis.* Euphorbiaceae, *Phytochemistry,* 11, 1511, 1972.
516. **Gujral, M. L., Verma, D. R., and Sareen, K. N.,** Oral contraceptives. I. Preliminary observations on the antifertility effect of some indigenous drugs, *Indian J. Med. Res.,* 48, 46, 1960.
517. **Gujral, M. L., Verma, D. R., Sareen, K. N., and Roy, A. K.,** Oral contraceptives. II. Antifertility effect of *M. philippinensis, Indian J. Med. Res.,* 48, 52, 1960.
518. **Verma, P., Gupta, S. S., and Agarwal, S. K.,** Purgative and anthelmintic activity in *M. philippinensis, Indian J. Pharmacol.,* 13(1), 103, 1981.
519. **Gupta, S. S., Verma, P., and Hishikar, K.,** Purgative and anthelmintic effects of *M. philippinensis,* in rats against tapeworm, *Indian J. Physiol. Pharmacol.,* 28, 63, 1984.
520. **Sanyal, M. and Datta, P. C.,** Macroscopic, microscopic and preliminary phytochemical studies on the bark of *Melia azedarach* Linn., *Indian Drugs,* 20, 85, 1982.
521. **Patel, B. R., Bhatt, I. P., and Satakopan, S.,** Diagnostic key for Ayurvedic powdered crude drugs, *Indian Drugs,* 14, 201, 1977.
522. **Pandey, Y. N.,** Studies on the cuticular characters of some Meliaceae, *Bull. Bot. Surv. India,* 11(3,4), 377, 1969.
523. **Nair, A. G. R. and Subramanian, S. S.,** Quercetin glycosides from the leaves of *Soymida febrifuga* and *Melia azedarach, Indian J. Chem.,* 13, 527, 1975.
524. **Amirchand, Mitra C. R. and Siddiqui, S.,** Chemical examination of the bakayan fruit (*Melia azedarach* Linn.), *J. Sci. Ind. Res. Sect. B,* 7, 69, 1948.
525. **Bholanath,** Chemical examination of the heartwood of *M. azedarach, J. Sci. Ind. Res. Sect. B.,* 13, 740, 1954.
526. **Nair, M. G. and Chang, F. C.,** 6β-Hydroxy-4-stigmasten-3-one and 6β-hydroxy-4-campesten-3-one, *Phytochemistry,* 12, 903, 1973.
527. **Mishra, M. and Srivastava, S. K.,** A new flavone glycoside from M*elia azedarach* Linn., *Curr. Sci.,* 53, 694, 1984.
528. **Shrivastava, A. K., Chauhan, C. S., and Upadhyaya, R. K.,** Studies on seed protein of *Melia azedarach* L., *Indian J. Chem.,* 7, 1179, 1969.

529. **Atwal, A. S. and Pajni, H. R.**, Preliminary studies on the insecticidal properties of drupes of *M. azedarach* against caterpillars of *Pieris brassicae* (Lepidoptera: Pieridae), *Indian J. Entomol.*, 26(2), 221, 1964.
530. **Shrivastava, A. K. and Chauhan, C. S.**, Antibacterial activity of unsaponifiable matter obtained from fixed oil of seeds of *Melia azedarach, Indian Drugs Pharm. Ind.*, 12(3), 7, 1977.
531. **Neogi, N. G., Baliga, P. A. C., and Srivastava, R. K.**, *In vitro* anthelmintic activity of some indigenous drugs, *J. Indian Med. Assoc.*, 41, 435, 1963.
532. **Talwar, Y. P., Nigam, M. C., Handa, K. L., and Kapoor, L. D.**, Essential oil of *Mentha arvensis* grown in Kashmir, *Reichstoffe Aromen Koerperflegimmittel*, 14, 2, 1964.
533. **Malla, S., Shukla, V. S., Nigam, M. C., and Handa, K. L.**, Essential oil of *Mentha arvensis, Indian Oil Soap J.*, 27, 173, 1962.
534. **Arora, S. K. and Singh, M.**, Free amino acids of *Mentha arvensis, Sci. Cult.*, 34, 373, 1968.
535. **Bodhankar, S. L, Garg, S. K., and Mather, V. S.**, Effect of *Mentha arvensis* on the fertility in female albino rats, *Bull. P. G. I. Chandigarh*, 5(2), 66, 1971.
536. **Kapoor, M., Garg, S. K., and Mather, V. S.**, Antiovulatory activity of five indigenous plants in rabbits, *Indian J. Med. Res.*, 62, 1225, 1974.
537. **Khanna, P. and Staba, E. J.**, Antimicrobials from plant tissue culture, *Lloydia*, 31, 1180, 1968.
538. **Chaurasia, S. C. and Vyas, K. M.**, *In vitro* effect of some volatile oils against *Phytophthora parasitica* var. *piperina, J. Res. Indian Med. Yoga Homoeop.*, 12(3), 139, 1977.
539. **Shome, U., Mehrotra, S., and Sharma, H. P.**, Pharmacognostic studies on the flowers of *Mesua ferrea* L., *Proc. Indian Acad. Sci. (Plant Sci.)*, 91, 211, 1982.
540. **Subramanyan Raju, M., and Subba Rao, N. V.**, Chemical study of the stamen of *Mesua ferrea, J. Res. Indian Med. Yoga Homoeop.*, 12(3), 124, 1977.
541. **Subramanyan Raju, M., Srimannarayana, G. and Subba Rao, N. V.**, Two new novel biflavones, mesuaferrone A and B from *Mesua ferrea* stamens, in 8th Int.Symp.Chem.Nat.Prod., New Delhi, 1972, 116.
542. **Subramanyan Raju, M., Srimannarayana, G., and Subba Rao, N. V.**, Structure of mesuanic acid, *Indian J.Chem.*, 12, 884, 1974.
543. **Govindachari. T. R., Pai, B. R., Subramaniam, P. S., Rao, U. R., and Muthukumaraswamy, N.**, Constituents of *Mesua ferrea* L. I. Mesuaxanthone A and mesuaxanthone B, *Tetrahedron*, 23, 243, 1967.
544. **Govindachari, T. R., Pai, B. R., Subramaniam, P. S., Rao, U. R., and Muthukamaraswamy, N.**, Constituents of *Mesua ferrea* L. II. Ferruol A, a new 4-alkylcoumarin, *Tetrahedron*, 23, 4161, 1967.
545. **Gupta, B. K., Dhar, K. L., and Atal, C. K.**, Guttiferol, a new triterpenoid from *Mesua ferrea* L. *Indian J.Pharm.*, 38, 157, 1976.
546. **Sreenivasan, B.**, Components fatty acids and composition of some oils and fats, *J. Am. Oil Chem. Soc.*, 45, 259, 1968.
547. **Namboodiripad, C. P.**, Investigations of certain seed oils. V. Fixed oil from *Mesua ferrea, Indian Oil Soap J.*, 32(4), 97, 1966.
548. **Subramanyam, Raju M. and Subba Rao, N. V.**, Fatty acid composition of Nahor (*Mesua ferrea* Linn.) seed oil, *J. Res. Indian Med. Yoga Homoeop.*, 12(4), 97, 1977.
549. **Chakraborty, D. P., Purkayastha, M., and Bose, P. K.**, on the antibiotic properties of some constituents of *Mesua ferrea, Proc. Natl. Inst. Sci. India Part B*, 25, 8, 1959.
550. **Kakrani, H. K., Nair, G. V., Dennis, T. J., and Jagdale, M. H.**, Antimicrobial and anthelmintic activity of essential oil of *Mesua ferrea* Linn. (letters to editors), *Indian Drugs*, 21, 261, 1984.
551. **Bhide, M. B. and Naik, P. V.**, Antiasthmatic activity of some indigenous and synthetic drugs, in Drug potential of Indian Medicinal Plants, Institute Medical Science, Banaras Hindu University, Varanasi, March 1979, 14.
552. **Gopala Krishnan, C., Shankaranarayanan, D., Nazimudeen, S. K., Vishwanathan, S., and Kameswaran, L.**, Antiinflammatory and CNS depressant activities of xanthones from *Calophyllum inophyllum* and *Mesua ferrea, Indian J.Pharmacol.*, 12, 181., 1980.
553. **Chaudhuri Rai, H. N.**, Pharmacognostic studies on the stem bark of *Michelia champaca* Linn., *Bull. Bot. Soc. Bengal*, 17(1 and 2), 1, 1964.
554. **Sharma, M.**, Origin and development of sclereids in *Michelia champaca* Linn., *Proc. Indian Natl. Sci. Acad. Part B*, 36, 298, 1970.
555. **Nair, P. K. K.**, Pollen morphology of some Indian medicinal plants, *J. Sci. Ind. Res. Sect. C*, 20, 45, 1961.
556. **Majumdar, P. L. and Chatterji, A.**, Active principles of the trunk bark of *Michelia champaca* L., *J. Indian Chem. Soc.*, 40, 929, 1963.
557. **Banerjee, K. and Chakravarti, R. N.**, Micheline A from *Michelia champaca* L., *Bull. Calcutta Sch. Trop. Med.*, 12(3), 113, 1964.
558. **Bheemasankara Rao, C. H., Suseela, K., Subba Rao, P. V., Gopala Krishna, P., and Subba Raju, G. V.**, Chemical examination of some Indian medicinal plants, *Indian J.Chem.*, 23B, 787,.1984.
559. **Govindachari, T. R., Joshi, B. S., and Karmat, V. N.**, Structure of parthenolide, *Tetrahedron*, 21, 1509, 1965.
560. **Sethi, V. K., Dhar, K. L., and Atal, C. K.**, A new sesquiterpene lactone from *Michelia champaca* roots, *Indian J. Pharm. Sci.*, 40, 226, 1978.

561. **Sethi, V. K., Thapa, R. K., Dhar, K. L., and Atal, C. K.,** Constituents of *Michelia champaca* and Lewis acid catalyzed transformations of parthenolide into guaianolides, *Planta Med.,* 51, 364, 1984.
562. **Nigam, M. C.,** Studies in some essential oils. Essential oil from flowers of *Michelia champaca.* I. Identification of components and isolation of a new sesquiterpene — Champacene. II. Studies on the constitution of Champacene, *Agra Univ. J. Res. (Sci.),* 12(1), 299, 1963.
563. **Chopra, M. M. and Handa, K. L.,** The essential oil of the fruit rinds of *Michelia champaca, Perfum. Essent. Oil Rec.,* 54, 817, 1963.
564. **Sharma, M. L., Nigam, M. C., Handa, K. L., and Rao, P. R.,** Chemical and gas chromatographic investigation on linalool and linalyl acetate bearing plants in India, *Indian Oil Soap J.,* 31, 303, 1966.
565. **Bedi, K. L. and Atal, C. K.,** Study of Indian seed oils. IV. Hexadecenoic acid and other fatty acids in *Michelia champaca* and *Corydalis adiantifolius, Indian J. Chem.,* 8, 325, 1970.
566. **Mehra, P. N. and Raina, M. K.,** Pharmacognostic studies on the stem bark of *Mimusops elengi* Linn., *Indian J. Pharm.,* 32, 175, 1970.
567. **Misra, G. and Mitra, C. R.,** Constituents of fruit and seed of *Mimusops elengi, Phytochemistry,* 6, 453, 1967.
568. **Misra, G. and Mitra, C. R.,** Constituents of leaves, heartwood and roots of *Mimusops elengi, Phytochemistry,* 7, 501, 1968.
569. **Sinha, A.,** Chemical examination of *Mimusops elengi* Linn. I. Examination of the fatty oil from the seeds, *Proc. Natl. Acad. Sci. India Sect. A,* 32(1), 56, 1962.
570. **Misra, G. and Mitra, C. R.,** Constituents of bark of *Mimusops elengi, Phytochemistry,* 6, 1309, 1967.
571. **Bahl, C. P., Banerjee, A., and Seshadri, T. R.,** Chemical observations of some Indian plant drugs, *Curr. Sci.,* 37, 1, 1968.
572. **Chopra, I. C. and Kapoor, L. D.,** Steroid sapogenin bearing plants of India, *Indian For.,* 94, 620, 1968.
573. **Varshney, I. P. and Logani, K. M.,** Study of the saponins and sapogenins from *Mimusops elengi* Linn., *Indian J. Appl. Chem.,* 32, 173, 1969.
574. **Gupta, G. K., Dhar, K. L., and Atal, C. K.,** Chemical constituents of *Mimusops elengi, Indian J. Chem.,* 14B, 818, 1976.
575. **Vijayalakshmi, K., Mishra, S. D., and Prasad, S. K.,** Nematicidal properties of some indigenous plant materials against second stage juveniles of *Meloidogyne incognita* (koffoid and white) chitwood, *Indian J. Entomol.,* 41, 326, 1979.
576. **Sharma, M. L., Chandokhe, N., Ray Ghatak, B. J., Jamwal, K. S., Gupta, O. P., Singh, G. B., Mohd Ali, M., Thakur, R. S., Handa, K. L., Rao, P. R., Jamwal, P. S., and Sareen, Y. K.,** Pharmacological screening of Indian medicinal plants, *Indian J. Exp. Biol.,* 16, 228, 1978.
577. **Banerji, R., Srivastava, A. K., Misra, G., Nigam, S. K., Singh, S., Nigam, S. C., and Saxena, R. C.,** Steroid and triterpenoid saponins as spermicidal agents, *Indian Drugs,* 17, 6, 1978.
578. **Banerji, R., Prakash, D., Patnaik, G. K., and Nigam, S. K.,** Sapasmolytic activity of saponins, *Indian Drugs,* 20, 51, 1982.
579. **Jain, S. K. and Tarafder, C. R.,** Medicinal plant-lore of the Santals (a revival of P. O. Bodding's work), *Econ. Bot.,* 24, 241, 1970.
580. **Rao, R., Raghunanda, I., and George, M.,** Investigations on plant antibiotics, *Indian J. Med. Res.,* 49, 159, 1949.
581. **Rangaswami, S. and Sankarasubramanian, S.,** Chemical components of the flowers of *Moringa pterogosperma, Current Sci.,* 15, 316, 1946.
582. **Bhattacharjee, A. K. and Das, A. K.,** Phytochemical screening of some Indian plants, *Q. J. Crude Drug Res.,* 9, 1408, 1969.
583. **Anjaneyulu, B., Babu Rao, V., Ganguly, A. K., Govindachari, T. R., Joshi, B. S., Kamat, V. N., Manmade, A. H., Mohamed, P. A., Rahimtula, A. D., Saksena, A. K., Vadre, D. S., and Viswanathan, N.,** Chemical investigation of some Indian plants, *Indian J.Chem.,* 3, 237, 1965.
584. **Das, J. M.,** Free amino acids and carotenes in the leaves of *Moringa oleifera* Lam. syn. *Moringa pterygosperma* Gaertn., *Curr.Sci.,* 34, 374, 1965.
585. **Ramiah, N. and Nair, G. N.,** Amino acid and sugars in the flowers and fruits of *Moringa oleifera* Lam., *J. Inst. Chem.,* 49, 163, 1977.
586. **Saluja, M. P., Kapil, R. S., and Popli, S. P.,** Studies on medicinal plants. VI. Chemical constituents of *Moringa oleifera* Lamk. (hybrid variety) and isolation of 4-hydroxymellein, *Indian J.Chem.,* 16B, 1044, 1978.
587. **Sengupta, A., Sengupta, C., and Das, P. K.,** Triglycerides composition of *Moringa concanensis* seed fat, *Lloydia,* 6, 666, 1971.
588. **Verma, S. C., Banerji, R., Misra, G., and Nigam, S. K.,** Nutritional value of *Moringa* (letters to editor), *Curr. Sci.,* 45, 769, 1976.
589. **Bhatnagar, S. S., Santapau, H., Desa, J. D. H., Maniar, A. C., Ghadially, N. C., Solomon, M. J., Yellore, S., and Rao, T. N. S.,** Biological activity of Indian medicinal plants. I. Antibacterial, antitubercular and antifungal action, *Indian J. Med. Res.,* 49, 799, 1961.
590. **Bhawasar, G. C., Guru, L. V., and Chadda, A. K.,** Antibacterial activity of some indigenous medicinal plants, *Med. Surg.,* 5, 11, 1965.

591. **Dhawan, B. N., Dubey, M. P., Mehrotra, B. N., Rastogi, R. P., and Tandon, J. S.,** Screening of Indian plants for biological activity. IX, *Indian J. Exp. Biol.,* 18, 594, 1980.
592. **Singh, R. and Singh, R.,** Screening of some plant extracts for antiviral properties, *Technology (Sindri),* 9, 415, 1972.
593. **Ghosh, G.,** Pharmacognostic study of *Mecuna pruriens* Baker non DC seed, *Nagarjun,* 27, 9, 1983.
594. **Ghosh, G.,** A note on pharmacognostic and chemical identification of *Mucuna utilis* seeds, a substitute of *Mucuna pruriens.* Short notes, *Indian Drugs,* 20, 24, 1982.
595. **Damodaran, M. and Ramaswamy, R.,** Isolation of 1-3,4-dihyroxyphenylalanine from the seeds of *Mucuna pruriens, Biochem. J.,* 31, 2149, 1937.
596. **Niranjan, G. S. and Katiyar, S. K.,** Chemical composition of some legumes, *J. Indian Chem. Soc.,* 56, 822, 1979.
597. **Santra, D. K. and Majumdar, D. N.,** The *Mucuna pruriens* DC. II. Isolation of water insoluble alkaloids, *Indian J. Pharm.,* 15, 60, 1953.
598. **Majumdar, D. N. and Zalani, C. D.,** *Mucuna pruriens* DC. Alkaloidal constituents. III. Isolation of water soluble alkaloids and a study of their chemical and physiological characterization, *Indian J. Pharm.,* 5, 62, 1953.
599. **Ghosal, S. and Singh, S.,** Alkaloids of *Mucuna pruriens* DC.: chemistry and pharmacology, in 2nd Soviet Symp. Chem. Nat. Product including Pharmacology, New Delhi, February, 1970, 135.
600. **Ghosal, S., Singh, S., and Bhattacharya, S. K.,** Alkaloids of *Mucuna pruriens:* chemistry and pharmacology, *Planta Med.,* 19, 279, 1971.
601. **Mehta, J. C. and Majumdar, D. N.,** Indian medicinal plants. V. *Mucuna pruriens* bark (N.O. Papilionaceae). I, *Indian J. Pharm.,* 6, 92, 1944.
602. **Ramaswamy, S., Nazimudeen, S. K., Viswanathan, S., Kulanthivel, P., Rajeskaran, V., and Kameswaran, L.,** Some pharmacological effects of *Mucuna pruriens, Proc. M.M.C. Res. Soc.,* 2, 39, 1979.
603. **Nath, C., Gupta, G. P., Bhargava, K. P., Lakshmi, V., Singh, S., and Popli, S. P.,** Study of anti-parkinsonian activity of seed of *Mucuna prurita* Hook., in Proc. 13th Annu. Conf. Indian Pharmacol. Soc. Jammu, 1980; *Indian J. Pharmacol.,* 13, 94, 1981.
604. **Pant, M. C., Uddin, I., Bhardwaj, U. R., and Tewari, R. D.,** Blood sugar and total cholesterol lowering effect of *Glycine soja* (Sieb & Zucc.), *Mucuna pruriens* DC and *Dolichos biflorus* Linn. Seed diets in normal fasting albino rats, *Indian J. Med. Res.,* 56, 1808, 1968.
605. **Singh, K. N., Barthwal, K. C., and Misra, M. N.,** Liver and blood cholesterol lowering effect of proteins of *Dolichos biflorus, Mucuna pruriens,* and *Acacia melanoxylon* leguminous seeds in albino rats, *Med. Surg.,* 16(4), 5, 1976.
606. **Bhattacharya, S. K. and Sanyal, A. K.,** Curariform activity of some Indian medicinal plants, *Nagarjun,* 13(2), 19, 1969.
607. **Bhattachary, S. K., Banerjee, S. K., Prakash, D., and Raina, M. K.,** Pharmacological identification of an identical quaternary base isolated from eleven species of Leguminosae, *Indian Med. Gaz.,* 11, 16, 1971.
608. **Varshney, I. P. and Sharma, S. C.,** Saponins and sapogenins. XXII. Chemical investigations of seeds of *Myristica fragrans* Hout., *Indian J. Chem.,* 6, 474, 1968.
609. **Purushothaman, K. K. and Sarada, A.,** Chemical examination of the aril of *Myristica fragrans* (Jathipathri), *Bull. Med. Ethnobot. Res.,* 1, 223, 1980.
610. **Purushothaman, K. K. and Sarada, A.,** Isolation of *dl*-dehydro-di-isoeugenol from the aril of *Myristica fragrans, Indian J. Chem.,* 19B, 236, 1980.
611. **Gopalakrishnan, M. and Mathew, A. G.,** Proanthocyanidins of nutmeg. Research note, *Indian Cocoa Arecanut Spices J.,* 6(4), 105, 1983.
612. **Gopalan, C., Rama Sastri, B. V., and Balasubramanian, S. C.,** Nutritive Value of Indian Foods, National Institute of Nutrition, Hyderabad, Indian Council of Medical Research, New Delhi, 1948.
613. **Bhat, J. V. and Broker, R.,** Action of some plant extracts on pathogenic staphylococci, *J. Sci. Ind. Res. Sect. B,* 12, 540, 1953.
614. **Pathak, R. K., Chourasia, S. C., and Singh, K. V.,** Antibacterial activity of essential oil of two medicinal plants, *Indian Drugs Pharm. Ind.,* 14(2), 7, 1979.
615. **Bhagwat, A. W. and Saifi, A. Q.,** Observations on the pharmacological actions of the volatile oil of *Myristica fragrans* (Houtt.), *J. Sci. Res. (Bhopal),* 2, 183, 1980.
616. **Mehra, P. N. and Garg, L. C.,** Pharmacognostic study of *Nardostachys jatamansi* DC., *Planta Med.,* 10, 433, 1962.
617. **Govindachari, T. R., Rajadurai, S., and Pai, B. R.,** Structure of jatamansone, *Chem. Ber.,* 91, 908, 1958.
618. **Maheshwari, M. L. and Saxena, D. B.,** Seychellene and seychelane from *Nardostacys jatamansi, Indian J. Chem.,* 12, 1221, 1974.
619. **Anjaneyulu, A., Babu Rao, V., Ganguly, A. K., Govindachari, T. R, Joshi, B. S., Kamat, V. N., Manmade, A. H., Mohamed, P. A., Rahimtula, A. D., Saksena, A. K., Varde, D. S., and Viswanathan, N.,** Chemical investigation of some Indian plants, *Indian J. Chem.,* 3, 237, 1965.

620. **Sastry, S. D., Maheswari, M. L., and Bhattacharyya, S. C.,** Terpenoids. LXXXVI.The structure of nardol, *Tetrahedron Lett.,* No. 10, 1035, 1966.
621. **Sastry, S. D., Maheswari, M. L., Chakravarti, K. K., and Bhattacharyya, S. C.,** Terpenoids. CXV. Chemical constitutents of *Nardostachys jatamansi, Perfum. Essent. Oil Rec.,* 58, 154, 1967.
622. **Sastry, S. D., Maheswari, M. L., Chakravarti, K. K., and Bhattacharyya, S. C.,** Terpenoids. CVI.The structure of calarenol, *Tetrahedron,* 23, 1997, 1976.
623. **Rucker, G., Tautges,, J., Maheswari, M. L., and Saxena, D. B.,** Norseychelanone, α and β-patchoulenes and patchouli alcohol from *Nardostachys jatamansi, Phytochemistry,* 15, 224, 1976.
624. **Shanbhag, S. N., Mesta, C. K., Maheswari, M. L, Paknikar, S. K., and Bhattacharyya, S. C.,** Terpenoids. LII.Jatamanson, new terpenic coumarin from *Nardostachys jatamansi, Tetrahedron Lett.,* No. 20, 2605, 1964.
625. **Shanbhag, S. N., Mesta, C. K., Maheswari, M. L., and Bhattacharyya, S. C.,** Terpenoids. LXXV. Constituents of *Nardostachys jatamansi* and synthesis of (+) dihydrosamidin and visnadin from jatamansin, *Tetrahedron,* 21, 3591, 1965.
626. **Hamied, K. A., Bakshi, V. M., and Aghara, L. P.,** Pharmacological investigation of *Nardostachys jatamansi, J. Sci. Ind. Res. Sect. C,* 21, 100, 1962.
627. **Arora, R. B., Singh, M., and Chandra Kanta,** Tranquillizing activity of jatamansone, a sesquiterpene from *Nardostachys jatamansi, Life Sci.,* No. 6, 225, 1962.
628. **Bose, B. C., Gupta, S. S., Vijayvarigiya, R., Saifi, A. Q., and Bhatnager, J. N.,** Preliminary observation on the pharmacological actions of various fractions of *Nardostachys jatamansi* DC., *Curr. Sci.,* 26, 278, 1957.
629. **Arora, R. B., Singh, K. P., Das, P. K., and Mistry, P. N.,** Prolonged hypotensive effect of essential oil of *Nardostachys jatamansi, Arch. Int. Pharmacodyn. Ther.,* 113(3/4), 367.1958.
630. **George, M. and Pandalai, K. M.,** Investigations on plant antibiotics. IV.Further search for antibiotic substances in Indian medicinal plants, *Indian J. Med. Res.,* 37, 169, 1949.
631. **Girgune, J. B., Jain, N. K., and Garg, B. D.,** Antifungal activity of some essential oils.II, *Indian Drugs,* 16, 24, 1978.
632. **Amin, M. G., Dixit, Y. B., and Pathak, J. D.,** Reaction time studies in relation to an indigenous drug — *Nardostachys jatamansi, Antiseptic,* 58, 565, 1961.
633. **Arora, R. B.,** Cardiovascular pharmacotherapeutics of six medicinal plants indigenous to India, Award Monogr. Ser. No. 1, Hamdard National Foundation, New Delhi, 1965.
634. **Arora, R. B. and Madan, B. R.,** Antiarrhythmics. III. Antiarrthymic activity of *Nardostachys jatamansi* — an Indian indigenous drug, *Indian J. Med. Res.,* 44, 259, 1956.
635. **Mitra, R., Mehrotra, S., and Kapoor, L. D.,** Pharmacognostic study of *Nelumbo nucifera* (Kamal) leaf. II, *J. Res. Indian Med. Yoga Homoeop.,* 11(3), 67, 1976.
636. **Mitra, R., Mehrotra, S., and Kapoor, L. D.,** Pharmacognostical study of *Nelumbo nucifera* (Kamal) rhizome. I, *J. Res. Indian Med. Yoga Homoeop.,* 11(3) 45, 1976.
637. **Nagarajan, S., Nair, A. G. R., Ramakrishnan, S., and Subramanian, S. S.,** Chemical examinations of the flowers of *Nelumbo speciosum* Willd., *Curr. Sci.,* 35, 176, 1966.
638. **Gangrade, H. H. and Kaushal, R.,** The composition of the oil of *Nelumbum speciosum* Willd. seeds, *Acta Cien. Indica,* 8C(1), 38, 1982.
639. **Dhawan, B. N., Patnaik, G. K., Rastogi, R. P., Singh, K. K., and Tandon, J. S.,** Screening of Indian plants for biological activity. VI, *Indian J. Exp. Biol.,* 15, 208, 1977.
640. **Sharma, P. C. and Kapoor, L. D.,** Pharmacognostic study of *Nerium indicum* Mill (Kaner) (syn. *N. odorum* Soland.), *J. Res. Indian Med.,* 5(1), 39, 1970.
641. **Basu, D. and Chatterjee, A.,** Occurrence of plumericin in *N. indicum, Indian J. Chem.,* 11, 297, 1973.
642. **Satyanarayana, T., Parameswara Prasad, P., Vimla Devi, M., and Krishna Rao, R. V.,** Phytochemical studies on *Nerium odorum* (root bark), *Indian J. Pharm.,* 37, 126, 1975.
643. **Siddiqui, S., Farrukh Hafeez, Sabir Begum, and Bina S. Siddiqui,** Oleanderol, a new pentacyclic triterpene from the leaves of *Nerium oleander, J. Nat. Prod., (Lloydia),* 51(2), 229, 1988.
644. **Singh, N., Nath, R., Agarwal, A. K., and Kohli, R. P.,** Pharmacological investigation of some indigenous drugs of plant origin for evaluation of their antipyretic, analgesic and antiinflammatory activity, *J. Res. Indian Med. Yoga Homoeop.,* 13(2), 58, 1978.
645. **Singh, N., Kulshrestha, V. K., Kohli, R. P., and Chatterjee, A.,** A glycoside from the roots of *N. indicum* as an adaptogen producing a state of nonspecific increase in resistance in animals, 4th Indo-Soviet Symp.Chemistry of Natural Products including Pharmacology, Lucknow, 1976, 122.
646. **Bose, B., Ghosh, C., and Singh, R. P.,** Study on the chemical constituents of seeds of *Nigella sativa* (Kalajira).A preliminary note, *J. Inst. Chem. (Calcutta),* 53, 273, 1981.
647. **Chandra, G.,** Chemical composition of the flower oil of *Nyctanthes arbor-tristis* Linn., *Indian Perfum.,* 14(1), 19, 1970.
648. **Dhingra, V. K., Seshadri, T. R., and Mukerjee, S. K.,** Carotenoid glycosides of *Nyctanthes arbor-tristis* Linn., *Indian J. Chem.,* 14B, 231, 1976.
649. **Chauhan, J. S. and Saraswat, M.,** A new glycoside from the stem of *Nyctanthes arbor-tristis, J. Indian Chem. Soc.,* 55, 1049, 1978.

650. **Sen, A. B. and Singh, S. P.,** Chemical examination of *Nyctanthes arbor-tristis* Linn., *J. Indian Chem. Soc.*, 41, 192, 1964.
651. **Singh, S. P., Bhattacharji, S., and Sen, A. B.,** Flavonoids of *Nyctanthes arbor-tristis* Linn., *Bull Natl. Inst. Sci. India*, No. 31, 41, 1965.
652. **Majumdar, D. N. and Agrawal, A. K.,** Investigation on the chemical constituents of *Nycanthes arbor-tristis* Linn. abstr. of the paper presented at the 18th Meet., Indian Pharmacol. Soc., Varanasi, 1966; *Indian J. Pharm.*, 28, 340, 1966.
653. **Kapoor, L. D., Kapoor, S. L., Srivastava, S. N., Singh, A., and Sharma, P. C.,** Survey of Indian plants for saponins, alkaloids and flavonoids. II, *Lloydia*, 34, 94, 1971.
654. **Sabir, M., Raviprakash, V., Suresh, S., and Jawahar Lal,** Pharmacological actions of *Nyctanthes arbor-tristis*, in Proc. 6th Annu. Conf., Indian Pharmacol. Soc., Hissar, 1974; *Indian J. Pharmacol.*, 6, 17, 1974.
655. **Saxena, R. S.,** Some pharmacological studies of *Nyctanthes arbor-tristis* Linn., M.D. thesis, Meerut University, Meerut, 1980.
656. **Singh, R. C., Saxena, K. K., Gupta, B., Saxena, K. K., and Prasad, D. N.,** On some more pharmacological properties of *Nyctanthes arbor-tristis* Linn. (Harsingar) — the plant known for antiinflammatory actions, *Indian J. Pharmacol.*, 16, 47, 1984.
657. **Gupta, S.,** Pharmacognosy of leaves of different species of *Ocimum* (Tulsi), *J. Res. Indian Med.*, 2(1), 35, 1967.
658. **Dey, B. B. and Choudhuri, M. A.,** Essential oil of *Ocimum sanctum* L. and its antimicrobial activity, *Indian Perfum.*, 28, 82, 1984.
659. **Lal, R. N., Sen, T. K., and Nigam, M. C.,** Gas chromatography of the essential oil of *Ocimum sanctum*, *Perfum. Kosmet.*, 59, 230, 1979.
660. **Nair, A. G. R. and Gunasegaran, R.,** Chemical investigation of certain South Indian plants, *Indian J. Chem.*, 21B, 979, 1982.
661. **Singh, P. P., Sharma, N. M., and Suri, B. K.,** Value of green leaves as sources of available calcium, *Indian J. Med. Res.*, 57, 204, 1969.
662. **Vohora, S. B., Garg, S. K., and Chaudhury, R. R.,** Effects of six indigenous plants on early pregnancy in albino rats, *Indian J. Med. Res.*, 57, 893, 1969.
663. **Singh, T. J., Dasgupta, P., Khan, S. Y., and Mishra, K. C.,** Preliminary pharmacological investigations of *Ocimum sanctum* Linn., *Indian J. Pharm.*, 32, 92, 1970.
664. **Chavan, S. R., Shah, M. P., and Nikam, S. T.,** Individual and synergistic activity of some essential oils as mosquito larvicidal agents, *Bull Haffkine Inst.*, 11, 18, 1983.
665. **Sharma, S. K. and Wattal, B. L.,** Efficacy of some mucilaginous seeds as biological control agents against mosquito larvae, *J. Entomol. Res.*, 3(2), 172, 1979.
666. **Grover, G. S. and Rao, J. T.,** Investigations on the antimicrobial efficiency of essential oils from *Ocimum sanctum* and *Ocimum gratissimum*, *Perfum. Kosmet.*, 58, 326, 1977.
667. **Bhat, J. V. and Broker, R.,** Action of some plant extracts on pathogenic staphylococci, *J. Sci. Ind. Res. Sect. B*, 12, 540, 1953.
668. **Kishore, N., Dubey, N. K., Tripathi, R. D., and Singh, S. K.,** Fungitoxic activity of leaves of some higher plants, *Natl. Acad. Sci. Lett.*, 5(1), 9, 1982.
669. **Shah, C. S., Medora, R. S., and Bhavsar, G. C.,** Pharmacognostic comparison and botanical identity of white and black turpeth (Nisoth), *Indian J. Pharm.*, 23, 192, 1961.
670. **Shah, C. S., Qadry, J. S., and Krishnamurthy, T. N.,** Sugars and coumarins in black turpeth *(Ipomoea turpethum)*, *Indian J. Pharm.*, 34, 126, 1972.
671. **Khare, A. K., Srivastava, M. C., Tewari, J. P., Puri, J. N., Singh, S., and Ansari, N. A.,** A preliminary study of antiinflammatory activity of *Ipomoea turpethum* (Nisoth), *Indian Drugs*, 19, 224, 1982.
672. **Prakash, A. and Prasad, S.,** Pharmacognostical studies on the bark of *Oroxylum indicum* Vent. (Shyonaka), *J. Res. Indian Med.*, 4(1), 73, 1969.
673. **Subramanian, S. S. and Nair, A. G. R.,** Flavonoids of the leaves of *Oroxylum indicum* and *Pajanelia longifolia*, *Phytochemistry*, 11, 439, 1972.
674. **Dey, A. K., Mukherjee, A., Das, P. C., and Chatterjee, A.,** Occurrence of aloeemodin in the leaves of *Oroxylum indicum* Vent., *Indian J. Chem.*, 16B, 1042, 1978.
675. **Subramanian, S. S. and Nair, A. G. R.,** Flavonoids of the stem bark of *Oroxylum indicum*, *Curr. Sci.*, 41, 62, 1972.
676. **Joshi, K. C., Prakash, L., and Shah, R. K.,** Chemical examination of the roots of *Tabebuia rosea* and heartwood of *Oroxylum indicum*, *Planta Med.*, 31, 257, 1977.
677. **Bhattacharje, A. K. and Das, A. K.,** Phytochemical screening of some Indian plants, *Q. J. Crude Drug Res.*, 9, 1408, 1969.
678. **Nair, A. G. R. and Joshi, B. S.,** Oroxindin, a new flavone glucuronide from *Oroxylum indicum* Vent., *Proc. Indian Acad. Sci. Sect. A*, 88A, 323, 1979.
679. **Satsokar, R. S., Shah, L. G., Bhatt, K., and Sheth, U. K.,** Preliminary study of pharmacological properties of Anantmul *(Hemidesmus indicus)*, *Indian J. Physiol. Pharmacol.*, 6, 68, 1962.

680. **Gupta, R. C., Ansari, M. S., and Kapoor, L. D.,** Pharmacognostical studies on *Paederia foetida* Linn., *Q. J. Crude Drug Res.,* 11, 1697, 1971.
681. **Prasad, S., Sapru, H. N., and Goel, R. K.,** Pharmacognostical studies on *Paederia foetida,* Linn., *J. Res. Indian Med.,* 6(1), 55, 1971.
682. **Basu, N. M., Ray, G. K. and De, N. K.,** On the vitamin C and carotene content of several herbs and flowers used in Ayurvedic system of medicine, *Indian Chem. Soc.,* 24, 358, 1947.
683. **Desai, H. K., Gwad, D. H., Govindachari, T. R., Joshi, J. S., Kamat, V. N., Modi, J. D., Parthasarathy, P. S., Patankar, S. J., Sidhaye, A. R., and Viswanathan, N.,** Chemical investigations of some Indian plants, VI, *Indian J. Chem.,* 9, 611, 1971.
684. **Tripathy, V. J. and Das Gupta, B.,** Neutral constituents of *P. foetida* Linn., *J. Indian Chem. Soc.,* 51, 1057, 1974.
685. **Shukla, Y. N., Llyod, H. A., Morton, J. F., and Kapadia, G. J.,** Iridoid glycosides and other constituents of *Paederia foetida, Phytochemistry,* 15, 1989, 1976.
686. **Bose, P. K., Banerjee, A. K., and Ghosh, C.,** Chemical investigation of *Paederia foetida, Trans. Bose Res. Inst. Calcutta,* 19, 77, 1953—1955.
687. **Chaturvedi, G. N. and Singh, R. H.,** Experimental studies on the anti-arthritic effect of certain indigenous drugs, *Indian J. Med. Res.,* 53, 71, 1965.
688. **Roychoudhury, G. K., Chakrabarty, A. K., and Dutta, B.,** A preliminary observation on the effects of *Paederia foetida* on gastrointestinal helminths in bovines, *Indian Vet. J.,* 47, 767, 1970.
689. **Deshapande, S. S.,** Essential oil from flowers of kewda *(Pandanus odoratissimus), Indian J. Chem. Soc.,* 509, 1938; *Chem. Abstr.,* 15, 2282, 1938.
690. **Dingra, D. R., Gupta, G. N., Jain, C., and Shukla, V. N.,** Otto of *Pandanus odoratissimus* Linn. or Kewda, *Perfum. Essent. Oil Rec.,* 42, 114, 1951; *Chem. Abstr.,* p. 8915, 1938.
691. **Dhingra, D. R., Gupta, G. N., Jain, J. C., and Scuckla, V. N.,** Otto of *Pandanus odoratissimus* L. or Kewada, *Perfum. Essential Rec.,* 42. 114, 1951; **Dhingra, S. N., Dhingra, D. R., and Gupta, G. N.,** Essential oil of kewada, *Perfum Essential Rec.,* 45, 219, 1954.
692. **Dwarkanath, C.,** Use of opium and cannabis in the traditional systems of medicine in India, *Bull. Narcotics,* 17(1), 1965.
693. **Saha, J. C. and Krishnamurthy, K. H.,** Pharmacognostic study of *Phyllanthus niruri* Linn., *Bull. Bot. Soc. Bengal,* 13(1 and 2), 53, 1959.
694. **Krishnamurthy, G. V. and Seshadri, T. R.,** The bitter principle of *Phyllanthus niruri, Proc. Indian Acad. Sci. Sect. A,* 24(4), 357, 1946.
695. **Row, L. R., Srinivasulu, C., Smith, M., and Subba Rao, G. S. R.,** New lignans from *Phyllanthus niruri* L., *Tetrahedron Lett.,* No. 24, 1557, 1964.
696. **Anjaneyulu, A. S. R., Jaganmohan Rao, K., Row, L. R., and Subrahmanyam, C.,** Crystalline constituents of Euphorbiaceae. XII. Isolation and structural elucidation of three new lignans from the leaves of *Phyllanthus niruri* L., *Tetrahedron,* 29, 1291, 1973.
697. **Mulchandani, N. B. and Hassarajani, S. A.,** 4-Methoxy-nor-securinine, a new alkaloid from *Phyllanthus niruri, Planta Med.,* 50, 104, 1984.
698. **Chauhan, J. S., Sultan, M., and Srivastava, S. K.,** Chemical investigation of the roots of *Phyllanthus niruri, J. Indian Chem. Soc.,* 56, 326, 1979.
699. **Ward, R. S., Satyanarayana, P., Row, L. R., and Rao, B. V. G.,** The case for a revised structure for hypophyllanthin.An analysis of the ^{13}C NMR spectra of aryltetralins, *Tetrahedron Lett.,* No. 32, 3043, 1979.
700. **Bhatnagar, S. S., Santapau, H., Desa, J. D. H., Maniar, A. C., Ghadially, N. C., Soloman, M. J., Yellore, S., and Rao, T. N. S.,** Biological activity of Indian medicinal plants. I. Antibacterial, antitubercular and antifungal action, *Indian J. Med. Res.,* 49, 799, 1961.
701. **Bhowmick, B. N. and Chaudhary, B. K.,** Antifungal activity of leaf extracts of medicinal plants on *Alternaria alternata* (Fr.) Keissler, *Indian Bot. Rep.,* 1, 164, 1982.
702. **Ramakrishnan, P. N., Murgesan, R., Palanichamy, S., and Murugesh, N.,** Oral hypoglycaemic effect of *Phyllanthus niruri* L. leaves, *Indian J. Pharm. Sci.,* 44, 10, 1982.
703. **Dixit, S. P. and Achar, M. P.,** Bhumyamalaki *(Phyllantus niruri)* and jaundice in children, *J. Natl. Integ. Med. Assoc.,* 25(8), 269, 1983.
704. **Mitra, R. and Prasad, S.,** Further studies on the pharmacognosy of *Picrorhiza kurroa* Linn., (Kutki), *J. Res. Indian Med.,* 6(3), 300, 1972.
705. **Mehra, P. N. and Jolly, S. S.,** Pharmacognosy of Indian bitters. I. *Gentiana kurrao* Royle and *Picrorhiza kurroa* Royal ex. Benth., *Res. Bull. (N.S.) Punjab Univ.,* 19(1 to 11), 141, 1968.
706. **Sharma, A. P.,** Pharmacognostic Studies in the Identity of Certain Plants and Plant Parts Used as Drugs in the Ayurvedic and Unani Systems of Medicine, found in Saharnpur, Dehradun, Tehri and Garhwal Districts, Ph.D. thesis, Agra University, Agra, 1972.
707. **Uniyal, M. R. and Issar, R. K.,** Commercially and traditionally important medicinal plants in Mandakini Valley of Uttarakhand, Himalayas, *J. Res. Indian Med.,* 4(1) 83, 1969.

708. **Kapoor, L. D., Srivastva, S. N., Singh, A., Kapoor, S. L., and Shah, N. C.,** Survey of Indian plants for saponins, alkaloids and flavonoids. III, *Lloydia,* 35, 288, 1972.
709. **Rastogi, R. P., Sharma, V. N., and Siddiqui, S.,** Chemical examination of *Picrorhiza kurroa* Benth. I, *J. Sci. Ind. Res. Sect. B,* 8, 173, 1949.
710. **Singh, B. and Rastogi, R. P.,** Chemical examination of *Picrorhiza kurroa* Benth. VI. Re-investigation of Kutkin, *Indian J. Chem.,* 10, 29, 1972.
711. **Rastogi, R. P. and Dhar, M. L.,** Chemical examination of *Pricrorhiza kurroa* Benth. IV. Identification of Kurrin as D-mannitol, *J. Sci. Ind. Res. Sect. B,* 18, 219, 1959.
712. **Basu, K., Das Gupta, B., Bhattacharya, S. K. and Debnath, P. K.,** Chemistry and pharmacology of apocynin, isolated from *Picrorhiza kurroa* Royle ex Benth., *Curr. Sci.,* 40, 603, 1971.
713. **Chaturvedi, G. N. and Singh, R. H.,** Treatment of jaundice with an indigenous drug *Picrorhiza kurroa* (a clinical and experimental study), *Curr. Med. Pract.,* 9, 451, 1965.
714. **Rajaram, D.,** A preliminary clinical trial of *P. kurroa* Benth. on bronchial asthma, *Bombay Hosp. J.,* 18(2), 14/66, 1976.
715. **Tewari, N. S. and Jain, P. C.,** A clinical evaluation of Arogyawardhani as hypocholesteremic agent with special reference to obesity/corpulency, *J. Res. Ayur Siddha,* 1(1), 121, 1980.
716. **Bisht, B. S.,** Pharmacognosy of Piplamul — the root and stem of *Piper longum* L., *Planta Med.,* 11, 410, 1963.
717. **Mehra, P. N. and Puri, H. S.,** Pharmacognostic studies on Pipalmul. XXII, *Indian J. Pharm.,* 32, 184, 1970.
718. **Dasgupta, A. and Datta, P. C.,** Medicinal species of Piper.Pharmacognostic delimitation, *Q. J. Crude Drug Res.,* 18(1), 17, 1980.
719. **Chatterjee, A. and Dutta, C. P.,** The structure of piper longumine, a new alkaloid isolated from the roots of *Piper longum* Linn. (Piperaceae), *Sci. Cult.,* 29, 568, 1963.
720. **Chatterjee, A. and Dutta, C. P.,** The alkaloids of *P. longum* Linn. I. Structure and synthesis of piperlongumine and piperlonguminine, *Tetrahedron,* 23, 1769, 1967.
721. **Dutta, C. P., Banerjee, N., and Sil, A. K.,** Studies on the genus *Piper:* studies on the roots of *Piper longum* Linn., *Indian J. Chem.,* 15B, 583, 1977.
722. **Atal, C. K., Girotra, R. N., and Dhar, K. L.,** Occurrence of sesamin in *Piper longum* L., *Indian J. Chem.,* 4, 252, 1966.
723. **Dhar, K. L. and Atal, C. K.,** Occurrence of *N*-isobutyl-deca-trans 2-4-dienamide in *Piper longum* Linn. and *Piper peepuloides* Royle, *Indian J. Chem.,* 5, 588, 1967.
724. **Govindachari, T. R., Jadha, V. S. J., Joshi, B. S., Kamat, V. N., Mohamed, P. R., Parthasarathy, P. C., Patankar, S. J., Prakash, D., Rane, D. F., and Viswanathan, N.,** Chemical investigations of some Indian plants. IV, *Indian J. Chem.,* 7, 308, 1969.
725. **Handa, K. L., Sharma, M. L., and Nigam, M. C.,** Essential oil of *Piper longum.* Properties of the components and isolation of two monocyclic sesquitepenes, *Perfum. Kosmet.,* 44, 233, 1963.
726. **Handa, K. L., Nigam, M. C., and Sharma, M. L.,** Decomposition studies on the sesquiterpenes from the oil of *Piper longum, Perfum. Kosmet.,* 44, 261, 1963.
727. **Sharma, R. K., Rathore, Y. K. S., and Sanjeev Kumar,** Chemical examination of dried fruit of Pippli *(Piper longum), J. Sci. Res. Plant Med.,* 4(4), 63, 1983.
728. **Dutta, C. P., Banerjee, N., and Roy, D. N.,** Studies on the genus *Piper lignan* in the seeds of *Piper longum, Phytochemistry,* 14, 2090, 1975.
729. **Bedi, K. L, Atal, C. K,. and Achaya, K. T.,** Study of Indian seed oils. VIII. Component-fatty acids of some seed fats of Piperaceae, *Lloydia,* 34, 256, 1971.
730. **Manavalan, G. and Singh, J.,** Chemical and some pharmacological studies on leaves of *P. longum* Linn., *Indian J. Pharm. Sci.,* 41, 190, 1979.
731. **Gopalan, C., Rama Sastri, B. V., and Balasubramanian, S. C.,** Nutritive Value of Indian Foods, National Institute of Nutrition, Hyderabad, Indian Council of Medical Research, New Delhi, 1984, 85.
732. **Anon.,** Report of the Composite Drug Research Scheme, ICMR, New Delhi, 1967—1968.
733. **Gupta, O. P., Nath, A., Gupta, S. C., and Srivastava, T. N.,** Preparation of semi-synthetic analogues of *Piper amides* and their antitubercular activity, *Bull. Med. Ethnobot. Res.,* 1(1), 99, 1980.
734. **Bhargava, A. K. and Chauhan, C. S.,** Antibacterial activity of essential oils, *Indian J. Pharm.,* 30, 150, 1968.
735. **Rao, C. S. S. and Nigam, S. S.,** Antimicrobial activity of some Indian essential oils, *Indian Drugs,* 14, 62, 1976.
736. **D'Cruz, J. L., Nimbkar, A. Y., and Kokate, C. K.,** Evaluation of fruits of *Piper longum* L. and leaves of *Adhatoda vasica* Nees for anthelmintic activity, *Indian Drugs,* 17, 99, 1980.
737. **Kokate, C. K., Tipnis, H. P., Gonsalaves, L. X., and D'Cruz, J. L.,** Anti-insect and juvenile hormone mimicking activities of essential oils of *Adhatoda vasica, Piper longum* and *Cyperus rotundus,* in 4th Asian Symp. Med. Plants, Spices, Bangkok, Thailand, September 15 to 19, 1980, 154.
738. **Sharma, A. K. and Singh, R. H.,** Screening of antiinflammatory activity of certain indigenous drugs on carrageenin induced hind paw oedema in rats, *Bull. Med. Ethnobot. Res.,* 2, 262, 1980.
739. **Kholkute, S. D., Kekare, M. B., and Munshi, S. R.,** Antifertility effects of the fruits of *Piper longum* in female rats, *Indian J. Exp. Biol.,* 17, 289, 1979.

740. **Kulshrestha, V. K., Srivastava, R. K., Singh, N., and Kohli, R. P.,** A study of central stimulant effect of *Piper longum*, *Indian J. Pharmacol.*, 1(2), 8, 1969.
741. **Kulshrestha, V. K., Singh, N., Srivastava, R. K., Rastogi, S. K., and Kohli, R. P.,** Analysis of central stimulant activity of *P. longum*, *J. Res. Indian Med.*, 6(1), 17, 1971.
742. **Banga, S. S., Garg, L. C., and Atal, C. K.,** Effect of Piplartine and crude extracts of *Piper longum* on the ciliary movements, *Indian J. Pharm.*, 26, 139, 1964.
743. **Dasgupta, A. and Datta, P. C.,** Medicinal species of *Piper*. Pharmacognostic delimitation, *Q. J. Crude Drug Res.*, 18(1), 17, 1980.
744. **Raina, M. L, Dhar, K. L., and Atal, C. K.,** Occurrence of *N*-isobutyl eicosa-trans-2-*trans*-4-dienamide in *Piper nigrum*, *Planta Med.*, 30, 198, 1976.
745. **Singh, J., Vasudevan, K., and Rai, R. S.,** Chemical constituents of stems of *Piper nigrum* L., *J. Indian Chem. Soc.*, 53, 1162, 1976.
746. **Bedi, K. L., Atal, C. K., and Achaya, K. T.,** Study of Indian seed oils. VIII. Component fatty acids of some seed fats of Piperaceae, *Lloydia*, 34, 256, 1971.
747. **Qadry, S. M. J. S.,** A note on plantago major seeds. A substitute for ispaghula, *J. Pharm. Pharmacol.*, 15, 552, 1963.
748. **Atal, C. K., Kapur, K. K., and Siddiqui, H. H.,** Studies on Indian seed oils. I. Preliminary screening for linoleic acid rich oils, *Indian J. Pharm.*, 26, 163, 1964.
749. **Patel, R. B., Rana, N. G., Patel, M. R., Dhayani, H. K., and Chauvan, H. V.,** Chromatographic screening of proteins of *Plantago ovata* Forsk., *Indian J. Pharm. Sci.*, 41, 249, 1979.
750. **Khorana, M. L., Prabhu, V. G., and Rao, M. R. R.,** Pharmacology of an alcoholic extract of *Plantago ovata*, *Indian J. Pharm.*, 20, 3, 1958.
751. **Siddiqui, H. H., Kapur K. K., and Atal C. K.,** Studies on Indian seed oil. II. Effect of *P. ovata* embryo oil on serum cholesterol levels in rabbits, *Indian J. Pharm.*, 26, 266, 1964.
752. **Bijlani, R. L. and Thomas, S.,** Effect of Isabgol on the digestion and absorption of sucrose in everted sacs of hamster ileum, *Indian J. Physiol. Pharmacol.*, 24, 400, 1980.
753. **Jain, S. K. and Tarafder, C. R.,** Medicinal plant lore of the Santals (a revival of P.O. Bodding's work), *Econ. Bot.*, 24, 241, 1970.
754. **Prasad, S., Wahi, A. K., and Khosa, R. L.,** *Pharmacognostical Studies on Roots of* Plumbago zeylanica *and P. rosea. Recent Advances in the Anatomy of Tropical Seed Plants*, Hindustan Publishing, Delhi, India, 1969, 196.
755. **Iyenger, M. A. and Pendse, G. S.,** Pharmacognosy of the root of *Plumbago zeylanica* L., *Indian J. Pharm.*, 24, 290, 1962.
756. **Sidhu, G. S. and Sankaran, A. V. B.,** A new biplumbagin and 3-chloroplumbagin from *P. zeylanica*, *Tetrahedron Lett.*, No. 26, 2385, 1971.
757. **Padhye, S. B. and Kulkarni, B. A.,** Studies on the roots of constituents of *P. zeylanica*, *J. Univ. Poona*, 44, 27, 1973.
758. **Sankaram, A. V. B., Srinivasarao, A., and Sidhu, G. S.,** Chitranone, a new binaphthoquinone from *P. zeylanica*, *Phytochemistry*, 15, 237, 1976.
759. **Gupta, M. L., Gupta, T. K., and Bhargava, K. P.,** A study of antifertility effects of some indigenous drugs, *J. Res. Indian Med.*, 6(2), 112, 1971.
760. **Aswal, B. S., Bhakuni, D. S., Goel, A. K., Kar, K., Mehrotra, B. N., and Mukerjee, K. C.,** Screening of Indian plants for biological activity. X, *Indian J. Exp. Biol.*, 22, 312, 1984.
761. **Premakumari, P., Rathunam, K., and Santhakumari, G.,** Antifertility activity of plumbagin, *Indian J. Med. Res.*, 65, 829, 1977.
762. **Bhargava, S. K.,** Effects of plumbagin on reproductive functions of male dog, *Indian J. Exp. Biol.*, 22, 153, 1984.
763. **Kundu, N. and Mondal, S.,** Studies on tumour damaging capacity of indigenous medicinal plants, *Bull. Reg. Res. Lab. Jammu*, 1, 19, 1962—1963.
764. **Mukharya, D. and Dahia, M. S.,** Antimicrobial activity of some plant extracts, *Indian Drugs*, 14, 160, 1977.
765. **Krishnaswamy, M. and Purushothaman, K. K.,** Plumbagin: a study of its anticancer, antibacterial and antifungal properties, *Indian J. Exp. Biol.*, 18, 876, 1980.
766. **Pillai, N. G. K., Menon, R. V., Pillai, G. B., Rajeskharan, S., and Nair, C. P. R.,** Effect of plumbagin in charmakeela (common warts). A case report, *J. Res. Ayur. Siddha*, 2(2), 122, 1981.
767. **Vaidya, B. G.,** Some controversial drugs of Indian medicine. VIII, *J. Res. Indian Med.*, 7(2), 45, 1972.
768. **Datta, S. C. and Mukerji, B.,** Pharmacognosy of Indian Root and Rhizome Drugs, Bull. No. 2, Manager of Publications, Pharmacognosy Laboratory Ministry of Health, Government of India, Delhi, 1950.
769. **Prasad, S.,** *Premna integrifolia.* Pharmacognosy of antiinflammatory drugs, in *Advances in Research in Indian Medicine*, Banaras Hindu University, Varanasi, 1970.
770. **Basu, N. K. and Dandiya, P. C.,** Chemical investigations of *Premna integrifolia* Linn., *J. Am. Pharm. Assoc.*, 36, 389, 1947.

771. **Deb, I. N. C.,** Chemical examination of the oil from the flowers of *Mesua ferrea, Indian Soap J.,* 8, 15, 1941; *Vide Chem. Abstr.,* p. 3535, 1948.
772. **Ramaiah, T. S., Laxman Rao, K., and Ramraj, S. K.,** Chemical examination of *Premna integrifolia* Linn., *J. Indian Chem. Soc.,* 55, 102, 1978.
773. **Bheemasankara Rao, Ch., Suseela, K., and Subba Raju, G. V.,** Chemical examination of *Premna* species. IX. Pimaradienols from *Premna latifolia* var. *mollissma, Indian J. Chem.,* 23B, 177, 1984.
774. **Dasgupta, B., Sinha, N. K., Pandey, V. B., and Ray, A. B.,** Major alkaloid and flavonoid of *P. integrifolia, Planta Med.,* 50, 281, 1984.
775. **Kurup, K. K. and Kurup, P. A.,** Antibiotic substances from the root bark of *Premna integrifolia, Naturwissenschaften,* 51, 484, 1964.
776. **Rathore, R. S., Anand Prakash, and Singh, P. P.,** *Premna integrifolia* L. A preliminary study of the inflammatory and antiarthritic activity, *Rheumatism,* 12, 130, 1977.
777. **Datta, P. C. and Das, M.,** Morphology and microscopy of *Psoralea corylifolia, Q. J. Crude Drug Res.,* 10, 1643, 1970.
778. **Chopra, R. N. and Chatterjee, N. R.,** *Psoralea corylifolia* (Babchi). Its constituents, their pharmacological action and therapeutic properties, *Indian J. Med. Res.,* 15, 49, 1927.
779. **Seshadri, T. R. and Venkata Rao, C.,** A new separation of the components of *P. corylifolia* L., *Proc. Indian Acad. Sci. Sect. A,* 5, 351, 1937.
780. **Bhattacharji, S.,** Isolation of raffinose from *P. corylifolia* L., *J. Sci. Ind. Res. Sect. B.,* 20, 135, 1961.
781. **Jois, H. S., Manjunath, B. L., and Venkata Rao,** Chemical examination of the seeds *P. corylifolia* L. I, *J. Indian Chem. Soc.,* 10, 41, 1933.
782. **Jois, H. S. and Manjunath, B. L.,** Isopsoralen from the seeds of *Psorolea corylifolia* L., *Proc. Indian Sci. Congr.,* 21(111), 243, 1934.
783. **Chakravarti, K. K., Bose, A. K., and Siddiqui, S.,** Chemical examination of the seeds of *P. corylifolia* L. (Babchi), *J. Sci. Indian Res. Sect. B,* 7, 24, 1948.
784. **Siddappa, S. and Sathyabhama Dev, Y. V.,** Chemical examination of the seeds of *P. corylifolia* L. I. On the isolation and constitution of corylifolean., *Proc. 43rd Indian Sci. Congr.,* Part 3, 126, 1956.
785. **Siddappa, S. and Sathyabhama Dev, Y.V.,** Chemical examination of seeds of *P. corylifolia* L. II. On the isolation and constitution of isopsoralidin, *Proc. 44th Indian Sci. Congr.,* Part 3, 130, 1957.
786. **Siddappa, S. and Sathyabhama Dev, Y. V.,** Chemical examination of seeds of *P. corylifolia* L. II. On the isolation and constitution of isopsoralidin, *J. Mysore Univ.,* 17B(1), 31, 1957.
787. **Gupta, B. K., Gupta, G. K., Dhar, K. L., and Atal, C. K.,** The essential oil from the seeds of *P. corylifolia, Indian Perfum.,* 23, 174, 1979.
788. **Gupta, B. K., Gupta, G. K., Dhar, K. L., and Atal, C. K.,** A C-formylated chalcone from *P. corylifolia, Phytochemistry,* 19, 2034, 1980.
789. **Rashid, Ali and Agarwala, S. C.,** Mode of action of psoralen in pigment production. III. Photooxidation of dihydroxyphenylalanine in the presence of psoralen, *Indian J. Biochem.,* 2, 271, 1965.
790. **George, M. and Pandalai, K. M.,** Investigations on plant antibiotics. IV. Further research for antibiotic substances in Indian medicinal plants, *Indian J. Med. Res.,* 37, 169, 1949.
791. **Kurup, P. A.,** Studies on plant antibiotics: screening of some Indian medicinal plants, *J. Sci. Ind. Res. Sect. C,* 15, 153, 1956.
792. **Gupta, K. C., Bhatia, M. C., Chopra, C. L., Amarnath, and Chopra, I. C.,** Antistaphylococcal activity of *Psoralea corylifolia* seed extract, *Bull. Reg. Res. Lab. Jammu,* 1, 59, 1962.
793. **Sharma, S. K. and Singh, V. P.,** The antifungal activity of some essential oils, *Indian Drugs Pharm. Ind.,* 14(1), 3, 1979.
794. **Grover, G. S. and Rao, J. T.,** *In vitro* antimicrobial studies of the essential oils of *P. corylifolia, Indian Perfum.,* 23, 135, 1979.
795. **Gaind, K. N., Dar, R. N., Chopra, B. M., and Kaul, R. N.,** Anthelmintic properties of the seeds of *P. corylifolia, Indian J. Pharm.,* 27, 198, 1965.
796. **Chandhoke, N. and Ray Ghatak, B. J.,** Pharmacological investigations of angelicin — a tranquillo-sedative and anticonvulsant agent, *Indian J. Med. Res.,* 63, 833, 1975.
797. **Anand, K. K., Sharma, M. L., Singh, B., and Ray Ghatak B. J.,** Antiinflammatory, antipyretic and analgesic properties of bavachinin — A flavanone isolated from seeds of *P. corylifolia* L. (Babchi), *Indian J. Exp. Biol.,* 16, 1216, 1978.
798. **Mukerjee, B.,** Psoralea and other indigenous drugs used in leucoderma, *J. Sci. Ind. Res. Sect. A,* 15(5)(Suppl. 1), 1956.
799. **Nayer, R. C., Pattan Shetty, J. K., Yoganarasimhan, S. N., Mary, Z., and Venkataram, B. S.,** Pharmacognostical studies on the heartwood of *Pterocarpus marsupium* Roxb., *J. Res. Indian Med. Yoga Homoeop.,* 14(1), 81, 1979.
800. **Vaidya, B. G.,** Some controversial drugs of Indian medicine. V, *J. Res. Indian Med.,* 7(2), 45, 1972.
801. **Adinarayana, D. and Syamasunder, K. V.,** A new sesquiterpene alcohol from *P. marsupium, Phytochemistry,* 21, 1083, 1982.

802. **Adinarayana, D., Syamasunder, K. V., Saligmann, A., and Wagner, H.,** Structure elucidation of pterosupin from *P. marsupium*. The first naturally occurring C-glycosyl-β-hydroxydihydrochalcone, *Z. Naturforsch. Teil C*, 37(3 and 4), 145, 1982.
803. **Sawhney, P. L. and Seshadri, T. R.,** Special chemical components of commercial woods and related plant material. IV. Phenolic components of some Pterocarpus species, *J. Sci. Ind. Res. Sect. C*, 15, 154, 1956.
804. **Subba Rao, A. V., Mathew, J., and Sankaram, A. V. B.,** Propterol A, 1,3-diarylpropan-2-ol from *P. marsupium*, *Phytochemistry*, 23, 897, 1984.
805. **Mathew, J. and Subba Rao, A. V.,** Carpusin: a novel 2-hydroxy-2-benzylcoumaranone from *P. marsupium*, *Phytochemistry*, 22, 794, 1983.
806. **Mathew, J. and Subba Rao, A. V.,** Propetrol B, a further 1,3-diaryl propan-2-ol from *P. marsupium*, *Phytochemistry*, 23, 1814, 1984.
807. **Pandey, M. C. and Sharma, P.V.,** Hypoglycaemic effect of bark of *P. marsupium* Rox. (Bijaka) on alloxan induced diabetes, *Med. Surg.*, 16(7) 9, 1976.
808. **Trivedi, C. P.,** Observations on the effect of some indigenous drug on blood sugar level of normal and diabetic rabbits, *Indian J. Physiol. Pharmacol.*, 7,11, 1963.
809. **Pandey, M. C. and Sharma, P. V.,** Hypocholestrolaemic effect of bark of *P. marsupium* Roxb. (Bijaka). An experimental study, *J. Res. Indian Med. Yoga Homoeop.*, 13(1), 137, 1978.
810. **Trivedi, J. J.,** Studies on heartwood of *P. marsupium*, *Indian J. Physiol. Pharmacol.*, 15, 51, 1971.
811. **Sepaha, G. C. and Bose, S. N.,** Clinical observations on the antidiabetic properties of *P. marsupium* and *Eugenia jambolana, J. Indian Med. Assoc.*, 27, 388, 1956.
812. **Dhir, G. G., Mohan, G., Verma, R. B., and Mishra, S. S.,** Studies on the antifungal activity of *P. marsupium*. A clinical evaluation, *Indian J. Dermatol. Venerol. Leprol.*, 48, 154, 1982.
813. **Ravindernath, B. and Seshadri, T. R.,** Chemistry of the santalin pigment. I. Structure of santalin permethyle ether, *Tetrahedron Lett.*, No. 13, 1201, 1972.
814. **Ravindernath, B. and Seshadri, T. R.,** Structural studies on santalin permethyle ether, *Phytochemistry*, 12, 2781, 1973.
815. **Kumar, N., Ravinderanath, B., and Seshadri, T. R.,** Terpenoids of *Pterocarpus santalinus* heartwood, *Phytochemistry*, 13, 633, 1974.
816. **Kumar, N. and Seshadri, T. R.,** Chemical components of *P. santalinus* sapwood, *Curr. Sci.*, 43, 611, 1974.
817. **Kumar, N. and Seshadri, T. R.,** A new triterpene from *P. santalinus* bark, *Phytochemistry*, 15, 1417, 1976.
818. **Kumar, N. and Seshadri, T. R.,** Terpenoids of *P. santalinus* leaves (letters to editor), *Curr. Sci.*, 45, 516, 1976.
819. **Dhawan, B. N., Dubey, M. P., Mehrotra, B. N., Rastogi, R. P., and Tandon, J. S.,** Screening of Indian plants for biological activity. IX, *Indian J. Exp. Biol.*, 18, 594, 1980.
820. **Chaudhuri Rai, H. N.,** Comparative pharmacognostic studies on the root bark and stem bark of *Punica granatum* Linn., *Bull. Bot. Surv. Indian*, 5(3 and 4), 329, 1963.
821. **Dey, D. and Das, M. N.,** Pharmacognostic study of the leaf and pericarp of *P. granatum* L., *Indian J. Pharm. Sci.*, 43, 66, 1981.
822. **Ponniah L. and Seshadri, T. R.,** Survey of anthocyanins from Indian sources. I, *J. Sci. Ind. Res. Sect. B*, 12, 605, 1953.
823. **Sharma, J. N. and Seshadri, T. R.,** Survey of anthocyanins from Indian sources. II, *J. Sci. Ind. Res. Sect. B*, 14, 211, 1955.
824. **Batta, A. K. and Rangaswami, S.,** Crystalline chemical components of some vegetable drugs, *Phytochemistry*, 12, 214, 1973.
825. **Nair, A. G.,R. and Gunasegaran, R.,** Chemical investigations of certain South Indian plants, *Indian J. Chem.*, 21, 979, 1982.
826. **Trivedi, V. B. and Kazmi, S. M.,** *Kachnar* and *Anar* as antibacterial drugs, *Indian Drugs*, 16, 295, 1979.
827. **Janardhanan, K. K., Ganguly, D., Baruah, J. N., and Rao, P. R.,** Fungitoxicity of extracts from tannin-bearing plants, *Curr. Sci.*, 32, 226, 1963.
828. **Charya, M. A. S., Reddy, S. M., Kumar, B. P., and Reddy, S. R.,** Laboratory evaluation of some medicinal plants against two pathogenic fungi, *New Bot.*, 6, 171, 1979.
829. **Vishwa Prakash, Singhal, K. C., and Gupta, R. R.,** Anthelmintic activity of *Punica granatum* and *Artemisia siversiana, Indian J. Pharmacol.*, 12, 62, 1980.
830. **Youngken, H. W., Sr.,** A pharmacognostical study of *Rauwolfia, J. Am. Pharm. Assoc. (Sci. Ed.)*, 43, 70, 1954.
831. **Woodson, R. E., Younken, H. W., and Schlittler, E.,** *Rauwolfia: Botany, Pharmacognosy, Chemistry and Pharmacology*, Little, Brown, Boston, 1957.
832. **Mukerji, B.,** Indian Pharmaceutical Codex, Part I, Council of Scientific Industrial Research, New Delhi, 1953.
833. **Datta, S. C. and Mukerji, B.,** Pharmacognosy of Indian roots and rhizome drugs, Bull. No. 1, Manager of Publications, Pharmacognosy Laboratory, Ministry of Health, Government of India, Delhi, 1950.
834. **Sharma, A. P. and Sharma, R. K.,** Pharmacognostic studies on the rhizome of *Rheum emodi* Wall., *J. Sci. Res. Plant Med.*, 1(2), 32, 1980.
835. **Narayana Aiyer, K. and Kolammal, M.,** Erandah: *Ricinus communis* Linn., *Pharmacognosy of Ayurvedic Drugs, Trivandrum*, 1(9), 9, 1966.

836. **Shah, C. S. and Shah, N.**, Pharmacognostic study of the root bark of *Ricinus communis* Linn., *J. Res. Indian Med.*, 7(4), 50, 1972.
837. **Watt, J. M. and Breyer-Brandwijk, M. G.**, *The Medicinal and Poisonous Plants of Southern Africa*, E & S Livingstone, Edinburgh, 1932.
838. **Mitra, R. and Kapoor, L. D.**, Ashoka. A pharmacognostic Monograph. Pharmacognosy Research Unit, Central Council for Research in Indian Medicine and Homeoepathy, National Botanic Gardens, Lucknow, 1973—1974, unpublished.
839. **Sen, S. P.**, Chemical study of the indigenous plants (Asoka), *Curr. Sci.*, 32, 502, 1963.
840. **Prasad, S.**, A histological study of the roots of *Saussurea lappa*, *Indian J. Pharm.*, 7, 81, 1945.
841. **Ghosh, S., Chatterjee, N. R., and Dutta, A.**, Chemical examination of the plant, its medicinal uses, with chemical analysis of roots and leaves of *Saussurea lappa* Clark. I, *J. Indian Chem. Soc.*, 6(4), 517, 1929.
842. **Mehra, P. N. and Raina, M. K.**, Pharmacognosy of Pashanbheda, *Indian J. Pharm.*, 33, 126, 1971.
843. **Bahl, C. P. and Seshadri, T. R.**, "Pashanbhedi" — drugs for urinary calculus, *Advances in Research in Indian Medicine*, Banaras Hindu University, Varanasi, 1970, 77.
844. **Bahl, C. P., Murari, R., Parthasarathy, M. R., and Seshadri, T. R.**, Components of *Bergenia stracheyi* and *B. ligulata.*, *Indian J. Chem.*, 12, 1038. 1974.
845. **Sharma, V. N.**, A study of pharmacological actions of *Bergenia ciliata* (Pashanbedha), in 16th Annu. Conf. Assn. Physiol. Pharmacol., India, 1970, 15.
846. **Maurya, D. P. S., Prabhakar, M. C., Seth, S. D. S., and Arora, R. B.**, Pharmacological studies on Phashanbheda *(Bergenia ligulata)* and gokhru *(Tribulus terrestris)* with special reference to antilithiatic activity, *Indian J. Pharm.*, 4, 112, 1972.
847. **Siddiqui, S.**, Industrial utilization of "bhilawan", *J. Sci. Ind. Res.*, 17, 1942.
848. **Ghosh, S. and Dutt, A.**, Chemical examination of *Sida cordifolia* Linn., *J. Indian Chem. Soc.*, 7, 825, 1930.
849. **Chopra, R. N. and De, P.**, The action of sympathomimetic alkaloids in *Sida cordifolia* (Brela), *Indian J. Med. Res.*, 18, 467, 1930.
850. **Varshney, I. P. and Khan, A. A.**, Chemical examination of fruits and stem of *Solanum indicum* L., *Indian J. Pharm.*, 23(3), 49, 1971.
851. **Ansari, M. S., Gupta, R. C., and Prasad, S.**, Pharmacognostical studies on *Solanum surattense* Burm. (Kantkari), *J. Res. Indian Med.*, 6(2), 143, 1971.
852. **Manjunath, B. L. and Shadakshara Swamy, M.**, Reexamination of the alkaloids of the fruits of *S. xanthocarpum*, *J. Mysore Univ. Sec.*, 3B(16), 117, 1942.
853. **Sate, Y. and Latham, H. G.**, The isolation of diosgenin from *Solanum xanthocarpum*, *J. Am. Chem. Soc.*, 75, 6067, 1953.
854. **Kusano, G., Beisler, J., and Sato, Y.**, Steroidal constituents of *S. xanthocarpum*, *Phytochemistry*, 12(2), 397, 1973.
855. **Shah, C. S., Quadry, J. S., and Mody, K. D.**, A contribution to Chirata, *Indian J. Pharm.*, 32.1, 15, 1970.
856. **Mishra, S. S., Bapa, S. K., and Tiwari, J. P.**, Preliminary phytochemical and pharmacological study of *Symplocos racemosa*, *Indian J. Physiol. Pharmacol.*, 8, 3, 1964.
857. **Mitra, R.**, Lodhra — an important drug, *Nagarjun*, 23, 9, 1980.
858. **Späth, E.**, Loturin is identical with aribin and harmin, *Monatsh Chem.*, 41, 401, 1920; Chopra, R. N. et al., 1982.
859. **Clark, E.**, A relation between rotenone, deguelin, and tephrosin, *Science*, 73, 17, 1931; *Chem. Abstr.*, p. 1509, 1931.
860. **Clark, E. P. and Claborn, H. V.**, Tephrosia. II. Isolepresin, *J. Am. Chem. Soc.*, 54, 4454, 1932; *Chem. Abstr.*, 27, 86, 1933.
861. **Clark, J. R. and Bannerjee, S. C.**, Glucoside from *Tephrosia purpurea*, *Proc. Chem. Soc. London*, 25, 16, 1910; *Chem. Abstr.*, 440, 1910.
862. **Rangaswami, S. and Sastry, B. V. Rama**, Chemical examination of plant insecticides, VIII. Pods of *Tephros lanceolata*, *Proc. Indian Acad. Sci. Sect. A*, 38, 257, 1953.
863. **Rangaswami, S. and Rao, G. R.**, Occurrence of rutin in the leaves of *Tephrosia lanceolata* Grab., *Indian J. Pharm.*, 17, 136, 1955.
864. **Gupta, R. C.**, Pharmacognostic study of Sharpunkha-*Tephrosia purpurea* Pers., *Indian J. Pharm.*, 34(6), 170, 1970.
865. **Pandey, Y. N.**, A study on an important drug plant, *Tephrosia purpurea* Pers., *Q. J. Crude Drug Res.*, 13, 2, 1975.
866. **Gupta, L. P., Sen, S. P., and Udupa, K. N.**, Pharmacognostical and pharmacological studies on *Terminalia arjuna*, *J. Res. Indian Med. Yoga Homoeop.*, 11, 4, 1976.
867. **Asthana, J. G. and Chauhan, C. S.**, Preliminary chemical investigation of the fruits of *Terminalia belerica*, *Pharmakon. Sagar.*, 7, 26, 1963—1964.
868. **Chatterjee, R., Guha, M. P., and Chatterjee, A.**, Plant alkaloids. III. *Thalictrum foliosum* D. C., *J. Indian Chem. Soc.*, p. 371, 1952.
869. **Gopinath, K. W., Govindachari, T. R., Rajappa, S., and Ramdas, C. V.**, Identity of thalictrine, *J. Sci. Ind. Res. Sect. B*, 18, 444, 1959.

870. **Kidwai, A., Salooja, K. C., Sharma, V. N., and Siddiqui, S.**, Chemical examination of *Tinospora cordifolia* Miers, *J. Sci. Ind. Res. Sect. B*, 8, 115, 1949.
871. **Chatterjee, A. and Ghosh, S.**, Tinosporin, the furanoid bitter principle of *Tinospore cordifolia, Sci. Cult.*, 26, 140, 1960.
872. **Hanuman, J.B., Bhatt, R.K., and Sabata Balkrishna**, A clerodane furano-diterpene from *Tinospora cordifolia, J. Nat. Prod.*, 51(2), 197, 1988.
873. **Fazli, F. R. Y. and Hardman, R.**, The spice fenugreek (*Trigonella foenum-graecum* L.), *Trop. Sci.*, 10, 66, 1968.
874. **Seshadri, T. R., Varshney ,I. P., and Sood, A. R.**,Study of glycosides from *Trigonella foenum-graecum* Linn. seeds, *Curr. Sci.*, 42, 421, 1973.
875. **Sood, A. R.**, Chemical components from the leaves of *Trigonella foenum-graecum* L., *Indian J. Pharm.*, 37, 100, 1975.
876. **Datta, S. C. and Mukerji, B.**, Urginea: Squill. Pharmacognosy of Indian Leaf Drugs, Bull. No. 2, Ministry of Health, Government of India, Calcutta, 1952, 96.
877. **Beri, R. M. and Pharasi, S. C.**, Studies on the bulbs of *Urginea indica, Indian For.*, 98, 408, 1971.
878. **Seshadri, T. R. and Subramanium, S. S.**, *J. Sci. Ind. Res.*, 9, 114, 1950.
879. **Mitra, R.**, Pharmacognostical Studies on Some Important Indian Medicinal Plants, Ph.D. thesis, Department of Pharmaceutics, Banaras Hindu University, Varansai, 1969.
880. **Prasad, S., Mitra, R., and Ansari, M. S.**, Pharmacognostical studies on rasna (*Vanda roxburghii* R. Br.), *J. Res. Indian Med.*, 2, 208, 1968.
881. **Mehrotra, B. N. and Aswal, B. S.**, A study of the Banafsha of Indian market, *Indian J. Pharm.*, 38(Abstr.), 153, 1976.
882. **Singh, P.**, Pharmacognosy of *Viola odorata* L. and its adulterants, *Q. J. Crude Drug Res.*, 5, 712, 1965.
883. **Prasad, S. and Luthera, S. P.**, Pharmacognostical studies on *Withania somnifera* Dunal, *Indian J. Pharm.*, 21, 189, 1959.
884. **Shah, C. B., Sukkawala, V. M., and Vyas, L. S.**, Pharmacognosy of *Withania somnifera* Dunal, *Indian J. Pharm.*, 21, 195, 1959.
885. **Srivastava, S. K., Iyer, S. S., and Rey, G. K.**, Estimation of the total alkaloids of *Withania somnifera* Dunal, *Indian J. Pharm.*, 22, 94, 1961.
886. **Chandra, V., Singh, A., and Kapoor, L. D.**, Studies on alkaloid-bearing plants. I. *Withania somnifera* Dunal, *Indian Drugs Pharm. Ind.*, September-October, 1, 1970.
887. **Majumdar, D. N.**, *Withania somnifera* Dunal. II. Alkaloidal constituents and their chemical characterization, *Indian J. Pharm.*, 17, 158, 1955.
888. **Das, B. and Mukherjee, R.**, Chemical examination of the stem bark of *Leuconotis aegenifolia* DC, *J. Sci. Ind. Res. Sect. B*, 21, 506, 1962.
889. **Khanna, K.L., Schwarting, A. E., and Bobbitt, J. M.**, The occurrence of isopelletierine in *Withania somnifera, J. Pharm. Sci.*, 51, 1194, 1962.
890. **Schwarting, A. E., Bobbitt, J. M., Rother, A., Atal, C. K., Khanna, K. L., Leary, J. D., and Walter, W. G.**, The alkaloids of *Withania somnifera, Lloydia*, 26, 258, 1963.
891. **Dhalla, N. S., Sastry, M. S., and Malhotra, C. L.**, Chemical studies of the leaves of *Withania somnifera, J. Pharm. Sci.*, 50, 876, 1961.
892. **Atal, C. K. and Schwarting, A. E.**, Investigation of amino acids in the berries of *Withania somnifera* Dunal, *Curr. Sci.*, 29, 22, 1960.
893. **Qadry, S. M. J. S. and Hamid, H. A.**, A pharmacognostical study of *Zanthoxylum alatum* Roxb., *Planta Med.*, 10, 275, 1962.
894. **Narayana Aiyer, K. and Kolammal, M.**, Ardarakah: *Zingiber officinale* Rosc., Pharmacognosy of Ayurvedic Drugs, *Trivandrum*, 1(9), 102, 1966.
895. **Taran, E. N.**, Anesthetics from leaves of *Zizyphus vulgaris* Lam, *Formatsiya*, 4(11/12), 20, 1941; *Chem. Abstr.*, 38, 2792, 1944.
896. **Mitra, R.**, Bibliography on Pharmacognosy of Medicinal Plants, Economic Botany Information Service, National Botanical Research Institute, Lucknow, 1985.
897. **Kurup, P. N. V., Ramdas, V. N. K., and Joshi, P.**, Handbook of Medicinal Plants, revised and enlarged, Central Council for Research in Ayurveda and Siddha, New Delhi, 1979.
898. **Anon.**, *Hamdard Pharmacopoeia of Eastern Medicine*, Hamdard National Foundation, Pakistan, 1970.
899. **King, G. and Dutt, U. C.**, *Materia Medica of the Hindus with a Glossary of Indian Plants*, Thaker, Spink, Calcutta, 1877.
900. **Fluckiger, F. A.**, *Pharmacographia. A History of the Principal Drugs of Vegetable Origin Met with in Great Britain and British India*, Macmillan, London, 1879.

LIST OF FIGURES AND CREDITS*

With the exception of *Rauwolfia serpentina* (L.) Benth., the drawings in this volume are reproduced by permission of Bishen Singh Mahendra Pal Singh and the Indian Council of Medical Research from: Kirtikar, K. R., Basu, B. D., and I.C.S., *Indian Medicinal Plants* (4 vols. text, 4 vols. plates), L. M. Basu, Publ., Allahabad, 1918; 2nd ed., 1935; reprint ed., Singh, Dehra-Dun, India, 1975. *Rauwolfia serpentina* is reproduced by permission from Woodson, R. E., Younken, H. W., and Schlittler, E., *Rauwolfia: Botany, Pharmacognosy, Chemistry, and Pharmacology*, copyright 1957 by Little, Brown and Company, Boston.

Acacia catechu Willd. 8
Acacia farnesiana Willd. 10
Achyranthes aspera Linn. 13
Aconitum ferox Wall. 15
Alpinia galanga Sw. 29
Amomum subulatum Roxb. 34
Anona squamosa Linn. (*Annona squamosa* Linn.) 41
Argyreia speciosa Sweet. 48
Azadirachta indica A. Juss. 59
Balanites roxburghii Planch. *(B. aegyptiaca)* 62
Bambusa arundinacea Willd. *(B. bambos)* 65
Barringtonia acutangula Gaertn. 68
Bombax malabaricum DC. *(B. ceiba)* 80
Boswellia serrata Roxb. 83
Caesalpinia bonducella Flem. *(C. crista)* 87
Calophyllum inophyllum Linn. 90
Calotropis gigantea R. Br. 92
Calotropis procera Ait. 92
Carica papaya Linn. 99
Cassia absus Linn. 103
Cassia fistula Linn. 105
Cassia tora Linn. 108
Cicorium intybus Linn. 114
Cissampelos pareira Linn. 119
Citrullus colocynthis Schrad. 122
Clerodendron infortunatum Gaertn. (*Clerodendrum infortunatum* Linn.) 124
Clitoria ternatea Linn. 126
Coptis teeta Wall. 133
Cordia myxa Linn. 135
Costus speciosus Smith 138
Cynodon dactylon Pers. 153
Cyperus rotundus Linn. 155
Dendrobium macraei Lindl. 160
Dioscorea bulbifera Linn. 164
Dolichos biflorus Linn. 167
Elephantopus scaber Linn. 170
Embelia ribes Burm. f. 173
Erythrina indica Lam. *(E. variegata)* 177

* Where two names are listed, the first is the name on the original plate, the second, the currently recognized binomial used in this volume.

Eugenia jambolana Lam. 179
Gmelina arborea Roxb. 196
Gymnema sylvestre Br. 200
Mesua ferrea Linn. 228
Nyctanthes arbor-tristis Linn. 246
Ocimum sanctum Linn. 249
Papaver somniferum Linn. 256
Peganum harmala Linn. 258
Phyllanthus niruri Linn. (*P. fraternus* Linn.) 261
Premna integrifolia Linn. *(P. obtusifolia)* 270
Ptercarpus marsupium Roxb. 276
Pterocarpus santalinus Linn. 278
Punica granatum Linn. 280
Rauwolfia serpentina (L.) Benth. 284
Rhus succedanea Linn. 288
Ruta graveolens Linn. 293
Salvadora persica Linn. 295
Saxifraga ligulata Wall. *(Bergenia ligulata)* 74
Solanum indicum Linn. 304
Sphaeranthus indicus Linn. 308
Tamarindus indica Linn. 314
Terminalia arjuna W. & A. 319
Tribulus terrestris Linn. 326
Urginea indica Kunth. 328
Vanda roxburghii R. Br. 331
Vernonia cinerea Less. 333
Zanthoxylum alatum Roxb. 339
Zizyphus jujuba Lam. *(Z. martiana* Lam.) 343

INDEX

A

Abelmoschus moschatus Medic, see *Hibiscus abelmoschus* Linn.
Abies spectabilis (D. Don) G. Don, 5
Abies webbiana Lindl., see *Abies spectabilis* (D. Don) G. Don
Abortifacients, 42, 50, 66, 198, 220, 243, 245, 259, 268, 269, 286, 294, 301
Abortion preventative, 53, 197
Abroma augusta (Linn.) L. f., 6
Abromine alkaloid (betaine), 6
Abrus precatorius Linn. 7
Abscesses, 186, 260, 327, 344
Acacatechin, 9
Acacia arabica, see *Acacia nilotica*
Acacia catechu (Linn.f.) Willd., 8—9
Acacia farnesiana (L.) Willd., 10, 11
Acacia nilotica (Linn.) Willd. Delile ssp. *Indica* (Benth.) Brenan, 12
Acanthaceae
 Andrographis paniculata, 39
 Barleria prionitis, 67
 Blepharis edulis, 77
 Hygrophila auriculata, 210
 Justicia adhatoda, 216
Acetaldehyde, 115
Acetate, 86
Acetic acid, 111, 115, 143, 162, 314
Acetone, 115
Acetoxyl/methoxyl proton ratio, 86
Acetyl aleuritolic acid, 142
3-*O*-Acetyl-calotropin, 93
Acetylcholine-like action, of *Cissus quadrangularis,* 121
Acetylcholine-like substance, *Artocarpus integrifolia* seeds, 54
Acetyl oleanolic acid, 279
Acetyl oleanolic aldehyde, 279
Achrosine, 107
Achyranthes aspera Linn., 13—14
Achyranthine, 14
Acids, see specific acid
Acne, 30
Aconine alkaloid, 16
Aconitine alkaloid, 16
Aconitum ferox, Wall ex Ser., 15
Aconitum heterophyllum Wall, 17—18
Acorin, 18
Acorus calamus Linn., 18—19
Acronarcotic poison, 219, 268
Actinodaphnine, 218
Adhatoda vasica Nees, see *Justicia adhatoda* Linn.
Adhatoda zeylanica Medic, see *Justicia adhatoda* Linn.
Adiantum lunulatum Burm f., see *Adiantum philippense* Linn.
Adiantum philippense Linn., 20

Adrenergic blocking agent, 286
Aegelin, 21
Aegelinin, 21
Aegle marmelos (L.) corr. ex Roxb., 21
Aegyptinin A, 221
Afzelin, 75
Agada tantra, toxicology, 1
Aglucone, 194
Aglycone, 44, 61
Aglycoside, 86
Ahimsa, Buddhist doctrine, see *Himsa*
Ajmalicine, 286
AL60, 22
Alangiaceae
 Alangium salviifolium, 22
Alangicine, 22
Alangimarckine, 22
Alangine, 22
Alangium, see *Alangium salviifolium,* 22
Alangium lamarckii (Thwaits), see *Alangium salviifolium* (Linn. f.) Wang
Alangium salviifolium (Linn. f.) Wang, 22
Alanine, 127, 129, 221, 227, 267
α-Alanine, 208, 234
Albigenic acid, 23
Albizzia lebbek (Willd.) Benth., 23
Albrine, 313
Albuminoids, 85, 113
Alcohols, see specific compound
Aldehyde, 117, 255
Aleprestic acid, 207
Alepric acid, 207
Aleprolic acid, 207
Aleprytic acid, 207
Alexander the Great, 2
Alexandrian laurel, see *Ochrocarpus longifolius* Benth. & Hook f.
Alexandrian pods, *Cassia angustifolia,* 104
Alexipharmic, 334
Alexiteric, 332
Alhagi camelorum Fisch., see *Alhagi pseudalhagi* (Beib.) Desv.
Alhagi maurorium Baker, see *Alhagi pseudalhagi* (Beib.) Desv.
Alhagi pseudalhagi (Beib.) Desv., 24
Aliphatic triols, 131
Alizarin, 292
dl-Allantoin, 77
Alliin (active principle of garlic), 26
Allium cepa Linn., 25
Allium sativum Linn., 26
α-Allocryptopine, 47
Alloimperatorin, 21
iso-Allorottlerin, 224
Allyl cyanide, 84
5-Allyl cysteine sulfoxide, 26
Allyl mustard oil, 84
Allyl-propyle disulfide, 25, 26

Almond, cuddapa, see *Buchanania lanzan*
Almond, Jangli, see *Hydnocarpus laurifolia*
Almond oil, substitute for, 85
Alocasia macrorrhiza (L.) Schott, 27
Aloe barbadensis Mill., 28
Aloe-emodin, 104, 107, 109, 252
Aloe vera Linn., see *Aloe barbadensis* Mill.
Aloe wood, see *Aquilaria agallocha*
Alpinia calcarata Rosc., see *Alpinia galanga* Willd.
Alpinia galanga (Willd.), 29—30
Alstonia scholaris (Linn.) R.Br., 31
Alterative substances, 11, 14, 20, 21, 23, 26, 31, 38, 39, 47, 49, 50, 67, 71, 72, 73, 82, 83, 85, 91, 93, 109, 111, 121, 128, 130, 142, 146, 153, 165, 166, 169, 171, 184, 188, 193, 203, 207, 208, 216, 222, 229, 259, 264, 268, 271, 272, 277, 300, 303, 306, 307, 309, 312, 318, 322, 324, 334, 337
Althaea officinalis Linn., 32
Altingia excelsa Noronha, 33
Amaranthaceae
   Achyranthes aspera, 13
Amarin, 221
Amaryllidaceae
   Curculigo orchiodes, 146
Amaryllideae
   Crinum defixum, 140
Amaryllis zeylanicum Linn., see *Crinum defixum* Ker-Gawl.
Amebic dysentery, 113, 206
Amebicide, 100
Amenorrhea, see Menstrual disorders
α-Aminobutyric acid, 221
3α-Aminocan-5-ene, 205
3α-Aminoconan-5-ene, 205
Aminotransferase activity, 79
Amomum subulatum Roxb., 34—35
Amorphophallus campanulatus (Roxb.) Blume, 36
Amphetamine toxicity (in mice), 30
Amulets, as repository for medicinal drugs, 1
Amygdalin, 272, 273
Amygdalus communis Linn., see *Prunus amygdalus* Batsch.
Amylase and lipase activities, see *Asparagus racemosus*
Amylodextrin, 238
α-Amyrin, 31, 199, 203, 228, 243
β-Amyrin, 199, 201, 203
β-Amyrin acetate, 128, 218, 222, 334
β-Amyrone, 334
Anacardiaceae
   Buchanania lanzan, 85
   Rhus succedanea, 288
   Semecarpus anacardium, 301
Anacyclus pyrethrum DC., 37
Anaferine, 337
Analeptic medicinal property, *Cinnamomum camphora*, 116
Analgesics, 257
Ananas comosus (Stickm.) Merr., 38
Ananas sativus, see *Ananas comosus* (Stickm.) Merr.
Anasarca, treatment for, 14, 64, 79, 183, 210

Andrographis paniculata (Burm.f.) Nees, 39
Anemia, 79
Anetheum sowa Kurz., see *Peucedanum graveolens* Linn.
Anethine, 260
Anethole or anise camphor, 189
Anethum foeniculum Linn., see *Foeniculum vulgare* Mill.
Angelica glauca Edgew., 40
Angelicin, 44, 239, 274
Angina, treatment of, 320
Anhygrine, 337
Anisaldehyde, 247, 260
Ankorine, phenolic alkaloid, 22
Annona squamosa Linn., 41—42
Annonaceae
   Annona squamosa, 41
Anodyne antispasmodic: jatiphaladi churna, 238
Anodyne substances in action, 16, 33, 47, 57, 159, 211, 332
Anodynes, 252, 259, 306
Anorexia, treatment of, 117
Anthelmintics 20, 22, 23, 27, 28, 39, 42, 50, 52, 53, 60, 63, 66, 70, 72, 79, 86, 88, 100, 109, 112, 125, 127, 147, 149, 152, 156, 157, 174, 182, 184, 185, 187, 189, 190, 192, 193, 206, 213, 214, 224, 225, 232, 236, 245, 253, 265, 274, 281, 300, 309, 312, 315, 318, 334, 337
Antheole, 260
Anthocephalus cadamba Miq., see *Anthocephalus indicus* A. Rich
Anthocephalus indicus A. Rich, 43
Anthocyanidin, 279
Anthoxanthins, 44
Anthraquinone, 28, 104, 287, 322
Antibacterial action, 224, 234, 250, 265, 266, 268, 274, see also Bacterial growth, inhibition
   Escherichia coli, 188, 245
   Micrococcus pyogenes var. aureus, 245, 271
   Staphylococcus aureus, 188, 271
Anticancer action, 335
   against human epidermal carcinoma of the nasopharynx, 23, 42, 71, 73, 83, 93, 110
   in Walker carcinosarcoma 256, 75
Anticatarrhal medicine for children, 67
Anticonvulsants, 44, 111
Antidote to opium, 75
Antidysenteric, 9
Antiemetic activity, of ginger, 342
Antiestrogenic activity, in mice and rabbits, 88, 158
Antifat remedy, 71
Antifertility activity, in rodents, 38, 86, 88
Antifungal activity, see Fungitoxic effects
Anti-inflammatory activity, 73, 79, 139, 150, see also Inflammatory conditions
Antilithic activity, 75, 317
Antiperiodic substances, activity, 14, 16, 17, 31, 50, 52, 73, 88, 107, 109, 125
Antiprogestatin, 158
Antiprotozoal activity, 22, 23, 69, 128, 150
Antipyretics, 16, 39, 73, 89, 91, 199, 206, 318, 332
Antiscorbutics, 38, 191, 283, 296

Antiseptics, 76
Antispasmodic substances, activity, 5, 18, 56, 57, 66, 78, 88, 93, 96, 102, 116, 128, 157, 204, 216, 232, 274, 294, 300, 317
Antiviral activity, 154
Aparajitin, 127
Aperients and laxatives, 24, 52, 61, 69, 83, 84, 109, 127, 130, 131, 163, 190, 327
Aphelandrine, 270
Aphonia, 61
Aphrodisiacs, 7, 12, 17, 18, 25, 31, 49, 56, 66, 70, 77, 81, 86, 96, 132, 137, 141, 146, 185, 186, 198, 210, 237, 294, 297, 300
Aphthous condition of mouth, treatment for, 27
Apiin, 44
Apigenin, 197, 250, 289
Apiol, 260
Apiose glucosides, 44
Apium graveolens Linn., 44
Aplotaxene, 300
Apocynaceae
  Alstonia scholaris, 31
  Hollarrhena antidysenterica, 205
  Nerium indicum, 243
  Rauwolfia serpentina, 284
Apocynin, 263
Aporphone alkaloids, 42
Appetite, treatment for loss of, 203, 289
Aquilaria agallocha Roxb., 45
Arabian jasmine, see *Jasminum sambac*
Arabic medicine, influenced by Hindu, 1, 3—4
Arabinose, 61, 185
L-Arabinose, 60
Araceae
  Acorus calamus, 18
  Alocasia macrorrhiza, 27
  Amorphophallus campanulatus, 36
Arachidic acid, 143, 236, 264, 302
Arbutin derivates, 75
Areca catechu Linn., 46
Areca nut, see *Areca catechu*
Arecaidine alkaloid, 46
Arecolidine alkaloid, 46
Arecoline alkaloid, 46
Argemone mexicana Linn., 47
Argentine, 129
Arginine, 46, 127, 220, 225, 234, 253
Argyreia speciosa Sweet., 48—49
Aristochine alkaloid, 51
Aristo red, 50
Aristolochene, 51
Aristolochia bracteata Retz., 50
Aristolochia indica Linn., 51
Aristolochiaceae
  Aristolochia bracteata, 50
  Aristolochia indica, 51
Aristolochic acid alkaloid, 50
Arjunalic acid, 320
Arjuna myrobalan, see *Terminalia arjuna* Wight & Arn.
Arjunetin, 320
Arjunine, 320

Armepavine, 241
Aromatic additive in medications, 240
Arsenic, 38, 336
Artemisia brevifolia Wall., see *Artemisia maritima* Linn.,
Artemisia maritima Linn., 52
Artemisia vulgaris Linn., 53
Artemisin, crystalline alkaloid, 52
Arthritis in experimental rats, 131
Artocarpus heterophyllus Lam., see *Artocarpus integrifolia* Linn.
Artocarpus integrifolia Linn., 54
Artostenone, 54
Artosterone, androgenic properties, 54
Asafetida, see *Ferula foetida*
Asaresinol feruate, 185
α-Asarone, *trans*-isomer, 18
Ascardi galli, 157
Ascarides, anthelmintic action of *Ebelia ribes,* 147
Ascaridole, 113
Ascaris lumbricoides, anthelmintic action on, 100
Ascarone, 18
Ascites, 127, 143
Asclepiadaceae
  Calotropis gigantea, 93
  Calotropis procera, 95
  Gymnema sylvestre, 200
Asclepias geminata Roxb., see *Gymnema sylvestre* R.Br.
Asclepias gigantea Linn., see *Calotropis gigantea* Linn. R.Br.
Asclepin, 95
Ash-colored fleabane, see *Vernonia cinerea* Less.
Asiatic acid, 208, 281
Asiaticoside, 208
Asoka, Buddhist ruler, 2
Asoka tree, see *Saraca indica* Linn.
Asparagin, 55
Asparagus, see *Asparagus racemosus*
Asparagus adscendens Roxb., 55
Asparagus racemosus Willd., 56
Aspargin, 194
Aspartic acid, 127, 208, 221, 227, 234
L-Aspartic acid, 264
Asteracantha longifolia Nees, see *Hygrophila auriculata*
Asthenia, 61
Asthma, 44, 46, 64, 72, 79, 82, 86, 88, 146, 184, 211, 212, 216, 238, 301, 312, 317, 318
Astragalin, 71
Astringents, 5, 9, 11, 12, 14, 17, 21, 27, 31, 32, 42, 43, 46, 66, 71, 73, 75, 78, 81, 86, 91, 103, 110, 136, 147, 211, 232—233, 300, 248, 305
Atharvaveda, 1
"Atisine", 17
Atlantone, 110
Atonic dyspepsia, see Dyspepsia
Atreya, 3
Atropa acuminata Royle ex Lindley, 57
Atropamine, 211
Atropine, 157
Attar, 214

Aucubin, 267
Aurone glucoside, 84
Averrhoa carambola Linn., 58
Ayapanin, 181
Ayurveda, 1
 basic concepts, 347—348
 pharmacodynamic aspects, 347—348
 pharmacology, principles of, 349—350
 terminology, 353—359
 tridoshas, 347
Azadirachta indica A. Juss., 59—60

B

Babul tree, see *Acacia nilotica*
Babula, 226—227
Bacillus megatherium, 106
Bacopa monniera (L.) Pennell, 61
Bacoside A, saponin, 61
Bacoside B, saponin, 61
Bacterial growth, inhibited, 106, 281, 300
Bacteriolytic agent, in *Calotropis procera*, 95
Bael fruit, see *Aegle marmelos*
Baicalein, 252
Bakalactone, 225
Bakayanin, 225
Bakuchial, 274
Bala tantra, pediatrics, 1
Balanites aegyptiaca (Linn.) Delile, 62—63
Balanites roxburghii Planch., see *Balanites aegyptiaca*
Baliospermum axillare Blume, see *Baliospermum montanum* Muell. Arg.
Baliospermum montanum Muell. Arg., 64
Balsamodendron mukul Hook., see *Commiphora mukul* (Hooker, Stedor) Engl.
Bamboo, see *Bambusa bambos*
Bambusa arundinacea Willd., see *Bambusa bambos* (Linn.) Voss.
Bambusa bambos (Linn.) Voss., 65—66
Banyan tree, see *Ficus benghalensis*
β-Barbaloin, 28
Barbaloin, 28
Barleria prionitis Linn., 67
Barrington, 69
Barringtonia acutangula (Linn.) Gaertn., 68—69
Barringtonic acid (triterpene dicarboxylic acid), 69
Bassia longifolia Linn., see *Madhuca longifolia* (Linn.) Macbride
Bassia oil, 222
Bastard teak, see *Butea monosperma*
Bauhinia tomentosa Linn., 70
Bauhinia variegata Linn., 71
Behenic acid, 234, 264
Belladonna, Indian, see *Atropa acuminata*
Belladonna-like action, 157
Belleric myrobalan, see *Terminalia belerica* Gaertn. Roxb.
Benincasa cerifera Savi, see *Benincasa hispida* (Thunb.) Cogn.
Benincasa hispida (Thunb.) Cogn., 72
Benzaldehyde, 11, 33, 118, 145
Benzaldehyde cyanhydrin, 272
Benzaldehyde hydrocyanic acid, 272
Benzoic acid, 111, 162
Benzopate, 136
Benzoylheteratisine alkaloid, 17
Benzyl acetate, 214, 255
Benzyl aconine, 16
Benzyl alcohol, 11, 255
Benzyl benzoate, 255
Benzyl salicylate, 255
Benzythiourea, anthelmintic, 100
Berbamine, 73
Berberidaceae
 Berberis aristata DC, 73
Berberine, 47, 73, 134, 323
Berberis aristata DC, 73
Berberis aristata var. *floribunda* Hook. f. and Thomas, see *Berberis aristata* DC
Berberis coriaria Royle ex Lindl, see *Berberis aristata* DC
Bergapten, 44
Bergenia ligulata (Wall.) Engl., 74—75
Bergenin, 75
Beriberi, treatment for, 100, 111
Betaine, 14, 115, 184, 201
Betula utilis D. Don, 76
Betulic acid, 61, 297
Betulin, 22, 76, 243, 270, 279
Betulinaldehyde, 22
Betulinic acid, 22, 243
Bhang, see *Cannabis indica*
Bhilawanol, 301
Bhoja prabandha, treatise mentioning anaesthetics, 1
Bhringaraj, see *Eclipta alba*
Bhuta vidya, management of seizures, 1
Bicyclic sesquiterpene hydrocarbaon, 166
Bignoniaceae
 Oroxylum indicum, 252
Bilious affections, 56, 82, 152, 197, 287, 309, 335
Binaphthoquinone, 268
3-Biplumbagin, 268
3', 6'-Biplumbagin, 268
Birch, see *Betula utilis*
Bird-cherry, see *Prunus puddum*
Bitter apple, see *Citrullus colocynthis*
Bitter tonic substances, 17, 18, 30, 31, 38, 60, 69, 79, 84, 88, 103, 120, 134, 286, 287, 312
Blackberry, see *Eugenia jambolana*
Black cumin, see *Nigella sativa*
Black musalie, see *Curculigo orchiodes*
Black nightshade, see *Solanum nigrum* Linn.
Black pepper, see *Piper nigrum*
Bladder disorders, 81, 302, 334
Bleeding, internal, 298
Bleeding piles, treatment for, 58, 100, 303
Blenorrhagia, treatment of, 129
Blepharin (optically active), 77
Blepharis edulis Pers.,77
Blistering agent, 93
Blocking action, cardiotonic, vasoconstrictor, and neuromuscular, in frogs, 61
Blood, 89

coagulation, 101, 162
composition, 131, 208
disorders, 60
Blood pressure
  arterial, 158
  hypertensive effects, 103, 161
  hypotensive effects, 190
Blood sugar level reduction, 66, 186
Blood vessel musculature, effect of chaksine on, 103
Blume, see *Blumea lacera*
Blumea camphor, 78
Blumea lacera DC., 78
Boerhaavia diffusa Linn., 79
Boerhaavia repens Linn., see *Boerhaavia diffusa* Linn.
Boils, treatment of, 36, 217, 260, 277, 279, 291, 327, 344
Boldine, 218
Bombacaceae
  Bombax ceiba, 80
Bombax ceiba Linn., 80—81
Bonducella nut, see *Caesalpinia arista*
Bonducin glycoside, 88
Bone fracture, 218
Bone setter, see *Cissus quadrangularis*
Borassus flabellifer Linn., 82
Borassus flabelliformis Murr. see *Borassus flabellifer* Linn.
Borneol, 53, 149, 341
Boswellia serrata Roxb., 83
Bowel disorders, see Stomach and bowel disorders
Bradycardia in rabbit heart, 91
Brahmi ghrita, 61
Brahmine alkaloid, 61
Brain, "heating of", 287
Brassica juncea (L.) Czern. and Coss, 84
Brassica juncea Hook f. and Thoms, see *Brassica juncea* (L.) Czern. and Coss
Brassicasterol, 112
Breath and mouth, treatment for, 46
Bromeliaceae
  Ananas comosus, 38
Bromelin, 38
Bromostyrene, 255
Bronchial catarrh, 297
Bronchitis, treatment of, 212, 281, 318, 324, 329, 336, 345
Brown mustard, see *Brassica juncea* (L.) Czern. and Coss
Brucine, 310
Buchanania lanzan Spreng., 85
Buchanania latifolia Roxb., see *Buchanania lanzan* Spreng.
Buddhism, advent of, 2
Buriyat Republic, 2
Burns, 222
Burseraceae
  Boswellia serrata, 83
  Commiphora mukul, 131
Butea frondosa Koen. ex Roxb., see *Butea monosperma* (Lam.) Kuntze
Butea monosperma (Lam.) Kuntze, 86

Butein, 86
Butrin, 86
Butterfly pea, see *Clitoria terneata*
Butter tree, Indian, see *Madhuca longifolia*

C

Cadalene, 166
d-α-Cadinene, 40, 215
Caesalpinia bonduc (Linn.) Roxb., see *Caesalpinia crista* Linn.
Caesalpinia bonducella Flem., see *Caesalpinia crista* Linn.
Caesalpinia crista Linn., 87—88
Caesalpiniaceae
  Bauhinia tomentosa, 70
  Bauhinia variegata, 71
  Caesalpinia crista, 87
  Cassia fistula, 105
  Tamarindus indica, 314
Calarene, 239
Calarenol, 239
Calcium oxalate, 6
Calculous affections, 81, 143
Callicarpa macrophylla Vahl., 89
Calliterpenone, 89
Calomel, 109
Calophyllic acid, 91
Calophyllolide, 91
Calophyllonide, 91
Calophyllum inophyllum Linn., 90—91
Calotoxin, 93, 95
Calotropagenin, 95
Calotropin, 93, 95
Calotropis gigantea (Linn.) R.Br., 92—94
Calotropis procera (Ait.) R.Br., 93, 95
Campesterol, 253
Camphene, 18, 52, 215, 238, 300, 341
Camphor, 30, 250, see also *Cinnamomum camphora*
Camphoric acid, 116
Cancer, see Anticancer action; Tumors; specific condition
Cannabidiol, 96
Cannabidiolic acid, 96
Cannabigerol, 96
Cannabinol-6-9-trimethyl-3-pentyl 6H-dibenzene(b,d)-pyran-1-ol, 96
Cannabis indica Linn., 96
Cannabis sativa Linn., see *Cannabis indica* Linn.
Cannabistrin, 204
Capillaries, dilation of, 103
Capsaicin, 98
Capsicin, 98
Capsicum annum Linn., 98
Caraway, see *Carum Carvi*
Carbachol, 250
Carbon dioxide, 197
Carbon monoxide, 197
Carbuncles, 217, 327, 337, 344
Cardamom, lesser, see *Elettaria cardamomum*
Cardiac contraction, 277
Cardiac edema, treatment of, 162

Cardiac failure, 320
Cardiac stimulant substances, 35, 52, 61, 162, 172, 181, 271, 301, 303, 305
Cardiotonic, vasoconstrictor, and neuromuscular blocking action in frogs, 61
Cardiovascular system, effects of *Digitalis purpurea* on, 162
Cardomon, greater, see *Amomum subulatum*
Carica papaya Linn., 99—100
Caricaceae
  Carica papaya, 99
Caricin, 100
Carminative substances, 5, 18, 25, 26, 30, 31, 33, 35, 36, 44, 45, 69, 71, 101, 102, 110, 116, 117, 118, 130, 131, 132, 137, 145, 147, 149, 150, 151, 152, 156, 158, 174, 179, 185, 187, 189, 204, 212, 215, 227, 229, 230, 238, 239, 245, 260, 264, 265, 266, 270, 271, 280, 296, 300, 305, 307, 314, 317, 327, 330, 332, 340, 342
Carotene, 181, 220
  α-Carotene, 234
  β-Carotene, 217, 232, 234
Carotenoid pigments, 141
Carpaine, 100
Carpasemine, 100
Carpesteral, 307
Carpine, 100
Carposide, 100
Carpusin, 277
Carrot, see *Daucus carota*
Carum carvi Linn., 101
Carum copticum Benth., 102
Carvacol, 101
Carvene, 101
Carvol (*d*-carvone), 260
Carvomenthone, 227
Carvone, 101, 218, 260
β-Carophylenoxide, 274
Caryophyllene, 118, 166, 218, 250
Cassia, fetid, see *Cassia tora*
Cassia absus Linn., 103
Cassia angustifolia Wahl., 104
Cassia cinnamon, see *Cinnamomum tamala*
Cassia fistula Linn., 105—106
Cassia foetida Pers., see *Cassia tora* Linn.
Cassia flower, see *Acacia farnesiana*
Cassia obtusifolia Linn., see *Cassia tora* Linn.
Cassia occidentalis Linn., 107
Cassia tora Linn., 108—109
Cassia toroides Roxb., see *Cassia tora* Linn.
Cassiollin 1,7-dihydroxy-5-methoxy carbonyl-3-methyl xanthone, 107
Castor oil plant, see *Ricinus communis* Linn.
Catarrhal affections, 30, 78
Catechin, 9
α, β, γ-Catechin, 9, 298
(+) Catechin, 46
(+) Catechin-3-gallate, 75
Catechol, 25
Catecholamines, pharmacological interaction with *cannabis*, 96

Catechu, see *Acacia catechu*
Catechutannic acid, 9
Cathartic substances, 127, 192, 251, 263, 287, 329
  hydragogues, and 28, 178, 221, 251
Cathartin, 107, 136
Cattle, purgative and anthelmintic of, 63
Cedrol, 110
Cedrus deodara Loud., 110
Cedrus libani Barrel var. *deodara* Hook. f., see *Cedrus deodara* Loud.
Celastraceae
  Celastrus paniculatus, 111
Celastrine, 111
Celastrol, 111
Celastrus montana Roth, see *Celastrus paniculatus* Willd.
Celastrus multiflora Roxb., see *Celastrus paniculatus* Willd.
Celastrus nutans Roxb., see *Celastrus paniculatus* Willd.
Celastrus paniculatus Willd., 111
Celery, see *Apium graveolens*
Cellulose, 55
Centella asiatica (Linn.) Urban, see *Hydrocotyle asiatica* Linn.
Centratherum anthelminticum (Willd.) Kuntze, 112
Cephaeline, 22
Cephalalgia, 231
Cerotic acid, 162
Ceryl alcohol, 50, 106, 253
Chaksine, 103
Chalcone, 224
Chalta, see *Dillenia indica*
Champa, yellow, see *Michelia champaca*
Champacene, 230
Chancres, 243
Chanoclavine, 49
Charaka (later book of Ayurveda), 1
Charas, see *Cannabis indica*
Chaulmoogric acid, 207
Chavicine, 266
Chebulic myrobalan, see *Terminalia chebula* Retz.
Chebulininic acid, 332
Chelerythrine, 47
Chelidonic acid, 193
Chenopodiaceae
  Chenopodium ambrosioides, 113
Chenopodium ambrosioides Linn., 113
Chicory, see *Cichorium intybus*
Chilblains, 106
China, Ayurvedic teachings, 2
Chinese flower, see *Paederia foetida*
Chinese gooseberry, see *Averrhoa carambola*
Chiratin, 312
Chiratogenin, 312
Chireta, see *Swertia chirata* Buch-Ham
Chlorophyll, 109
3-Choroplumbagin, 268
Cholagogue substances, 24, 39, 45, 191, 242, 247, 263, 289, 318
Cholera, 302, 311, 340

Cholesterol, 131, 236, 263
Choline, 115, 201, 236, 327, 337
Cholinergic activity, 66, 158
Chromogen, 198
Chrysin, 252
Chrysoeriol, 44
Chrysophanic acid, 104, 107, 109, 287
Chrysophenol, 107, 109
Chyluria, 313
Cicorin, 115
Cichorium intybus Linn., 114—115
n-Cicosane, 264
Cinchio-tannic acid, 43
Cinchona and quinine substitute, *Alstonia scholaris*, 31
1:8 Cineola, 230
Cineole, 30, 52, 149, 172, 341
Cinnamaldehyde, 118
Cinnamic acid, 33, 194
Cinnamic aldehyde, 33
Cinnamomum camphora Nees and Ebern., 116
Cinnamomum cassia Blume, see *Cinnamomum zeylanicum* Breyn.
Cinnamomum tamala Nees and Ebern., 117
Cinnamomum zeylanicum Breyn., 118
Cinnamon, see *Cinnamomum zeylanicum*
Cirrhosis of the liver, 320
Cissampelos pareira Linn., 119—120
Cissus quadrangularis Wall., 121
"Citbittol", 122
Citral, 255
Citric acid, 100, 162, 314, 330
Citrulluin, 122
Citrulluric acid, 122
Citrullus colocynthis Schrad., 122—123
Clerodane furano-diterpene 2, 324
Clerodendrum infortunatum, 124—125
Clerodin, 125
Climbing lily, see *Gloriosa superba*
Clitoria terneata Linn., 126—127
Clutyl ferulate, 197
CNS depressant action, 12, 69, 71, 83, 147
Coagulant activity in semen, 279
Cobra bite, 292
Cobra's saffron, see *Mesua ferrea*
Cocarcinogens, long and short acid compounds of *Croton tiglium*, 143
Coccinia cordifolia Cogn., see *Coccinia indica* Wight and Arun
Coccinia grandis [Linn.] Voight, see *Coccinia indica* Wight and Arun
Coccinia indica Wight and Arun, 128
Cocos nucifera Linn., 129
Codeine, 257
Colchicine, 130, 193
Colchicum luteum Baker, 130
Colic, 27, 33, 88, 294, 305, 311
Colloturine, 313
Collyrium, 112, 134
Colocynthin, 122
α- and β-Colubrine, 310

Columbamine, 73
Combretaceae
   Terminalia arjuna, 319
   Terminalia belerica, 321
   Terminalia chebula, 322
Commiphora mukul (Hooker, Stedor) Engl., 131
Common cucumber, see *Cucumis sativis*
Common fumitory, see *Fumaria indica*
Common milk hedge, see *Euphorbia neriifolia*
Compositae
   Anacyclus pyrethrum, 37
   Artemisia maritima Linn., 52
   Artemisia vulgaris, 53
   Blumea lacera, 78
   Centratherum anthelminticum, 112
   Cichorium intybus, 114
   Eclipta alba, 169
   Elephantopus scaber, 170
   Eupatorium triplinerve Vahl., 181
   Saussurea lappa, 299
   Sphaeranthus indicus, 308
   Taraxacum officinale, 316
   Vernonia cinerea, 333
Conception, 307
Concuressine, 205
Condelphine, 159
Conessimine, 205
Conessine, 205
Conessi tree, see *Hollarrhena antidysenterica*
Coniferae
   Abies spectabilis, 5
   Taxus baccata, 317
Conjunctivitis, 103, 281, 291, 309, 315, 334, see also Eye, inflammatory conditions
Convolvulaceae
   Argyreia speciosa, 49
   Cuscuta reflexa, 152
   Evolvulus alsinoides, 184
   Ipomoea nil, 213
   Operculina turpethum, 251
Convolvulus turpethum Linn., see *Operculina turpethum* (Linn.) Silva, Manso
Coptine, 134
Coptisine, alkaloid, 47
Coptis teeta Wall., 133—134
Coptizine, 134
Coral tree, Indian, see *Erythrina variegata*
Cordia dichotoma Forst. f. [*C.obliqua* Willd.; *C. myxa*] C.B. Clark. non Linn., see *Cordia myxa* Roxb.
Cordia myxa Roxb., 135—136
Coreopsin, 86
Coriander, see *Coriandrum sativum*
Coriandrum sativum Linn., 137
Corn mint, see *Mentha arvensis*
Cornutia corymbosa Burm. f., see *Premna obtustifolia* R.Br.
Corylifolean, 274
Corynebacterium diphtheriae, 106
Coryza, treatment for, 16, 117
Cosmetics and skin, 45

β-Costen, 300
d-Costen, 300
Costic acid, 300
Costol, 300
Costos lactone, 300
Costunolide, 230
Costus, see *Costus speciosus; Saussurea lappa* C.B. Clarke
Costus speciosus Sm., 138—139
Cotton, Indian, see *Gossypium herbaceum*
Cough and colic medication, 32, 63
Coumarin, 194, 251
Coumasterol, 274
Counterirritants, 84, 217
Country fig tree, see *Ficus racemosa*
Country mallow, see *Sida cordifolia* Linn.
Cowitch plant, see *Mucuna pruriens*
Crategelic acid, 180
Crinum asiaticum Linn., see *Crinum defixum* Ker-Gawl.
Crinum bracteatum Willd., see *Crinum defixum* Ker-Gawl.
Crinum defixum Ker-Gawl., 140
Crinum latifolium Linn., see *Crinum defixum* Ker-Gawl.
Crocin, 141
Crocus sativus Linn., 141
Croton albumin,143
Croton globulin, 143
Croton oblongifolius Roxb., 142
Croton philippense Lam., see *Mallotus philippensis* (Lam.) Muell.-Arg.
Croton tiglium Linn., 143
Crotonyl mustard oil, 84
Cruciferae
  Brassica juncea, 84
  Raphanus sativus, 283
Cryptomeridiol, 279
Cryptoxanthene, 100
Crystalline steroketon, 54
Cucumber, common, see *Cucumis sativus*
Cucumus sativus Linn., 144
Cucurbitaceae (Cucurbitae)
  Benincasa hispida, 72
  Citrullus colocynthis, 122
  Coccinia indica, 128
  Cucumis sativus, 144
  Luffa acutangula, 220
  Luffa cylindrica, 221
Cucurbitacin B, 220
Cucurbitacin β-glycoside, 128
Cuddapa almond, see *Buchanania lanzan*
Cumaldehyde, 145
Cumene, 102
Cumin, black see *Nigella sativa*
Cuminic acid, 145
Cuminic alcohol, 145
Cuminic aldehyde, 145
Cumin seed, see *Cuminum cyminum*
Cuminum cyminum Linn., 145
Cupressaceae

Juniperus communis, 215
Cupuliferae
  Betula utilis, 76
Curariform activity, 103
Curculigo orchiodes Gaertn., 146
Curcuma amada Roxb., 147
Curcuma angustifolia Roxb., 148
Curcuma domestica Valeton, see *Curcuma longa* Linn.
Curcuma longa Linn., 149—150
Curcuma starch, see *Curcuma angustifolia*
Curcuma zedoaria Rosc., 151
Curcumin, 149, 151
Curcuminoides, 149
Cuscohygrine, 337
Cuscuta reflexa Roxb., 152
Cuscutin, 152
Cyanidin, 238
Cyanin, 283
Cyanomaclurin, 54
Cycleanine alkaloid, 120
Cymene, 116, 118, 245
p-Cymene, 145, 247
Cynodon dactylon (Linn.) Pers., 153—154
Cyperaceae
  Cyperus rotundus, 155
Cyperus rotundus Linn., 155—156
L-Cysteine hydrochloride, 264
Cystine, 221, 225, 234, 253, 267
Cystitis, chronic, 81

D

Daidzein, 274
Dandelion, see *Taraxacum officinale* Weber.
Dasamula, Ayurvedic medication, 197, 271, 305, 325
Dashmul asava, 307
Datura alba Nees, see *Datura metal* Linn.
Datura metal Linn., 157
Datura, to treat asthma, 2
Daucus carota Linn., 158
Deafness, 66
Decylaldehyde, 250
Deguelin, 318
Dehydrocarveol, 264
Delphininene, 159
Delphinium denudatum Wall. ex Hook. f. & Thoms., 159
Demethylpsychotrine alkaloid, 22
Demethyltublosine, 22
Demulcent substances 6, 11, 12, 32, 54, 55, 75, 81 82, 83, 85, 127, 136, 139, 144, 148, 154, 161, 191, 194, 197, 202, 210, 218, 220, 267, 272, 303, 325, 327, 335, 336
Dendrobium macraei Lindl., 160—161
Dentifrice, 129
Denudatine, 159
Deobstruant substances, 45, 73, 88, 117, 169, 239, 292, 296, 318, 320, 329, 337
Deodar, see *Cedrus deodara*
Deodorizer, 30

Deoxyelephantopin, 171
5-Deoxykaempferol, 277
Deoxyoblongifoliol, 142
Deoxytubulosine alkaloid, 22
Dermatitis, produces effect of, 6
Desacetylnimbin, 60
Devil's cotton, see *Abroma augusta*
Dextrin, 100
Diabetes, treatment of, 128, 180, 201, 202, 272, 277
Diacetyl, 115
Diaphoretics, 110, 297, 305, 318
Diaphoretic substances in action in, 16, 73, 79, 116, 117, 140, 143, 181, 190, 203, 274, 279, 282, 306, 334, 335
Diarrhea, treatment of, 54, 55, 75, 81, 83, 85, 110, 156, 186, 191, 199, 252, 254, 257, 281, 302, 320, 324, 327, see also Dysentery
  chronic, 318, 322, 336
  infantile, 209, 289
Dictamine, 21
Dieudesmin, 264
Diferuloyl methane, 149
Digestive and bowel complaints, 40, 42, see also Stomach and bowel disorders
Digestive substances, activity, 21, 27, 100, 115
Digitalis purpurea Linn., 162
Digitomide, 136
Digitoxin, 162
5-Diglucoside, 281
Dihydroconcuressine, 205
Dihydrofurano-dihydroiso-coumarin, 77
Dihydrofuro coumarin, 21
Dihydroprotoemetine alkaloid, 22
Dihydrostimasterol, 264
Dihydroxybenzoic acid, 198
5-Dihydroxy-7-methoxyisoflavone, 252
Dihydroxystearic acid, 290
Diketopentane, 115
Dilleniaceae
  Dillenia indica, 163
Dillenia indica Linn., 163
Dillenia speciosa Griff., see *Dillenia indica* Linn., 163
Dill, Indian, see *Peucedanum graveolens*
Dillo oil, 91
Dimethyl colchicine, 193
Dimethyl sulfide, 84
Dimyrcene, 131
Diosbulbine, 165
Dioscoreaceae
  Dioscorea bulbifera, 164
Dioscorea bulbifera Linn., 164—165
Dioscorea crispata Roxb., see *Dioscorea bulbifera* Linn.
Dioscorea pulchella Rev., see *Dioscorea bulbifera* Linn.
Dioscorea sativa Thunb., see *Dioscorea bulbifera* Linn.
Dioscorea versicolor Buch. Ham ex Wall., see *Dioscorea bulbifera* Linn.
Dioscorides, 2

Diosgenin, 139, 327
Dipentene, 145, 255
Dipterocarpaceae
  Dipterocarpus indicus, 166
Dipterocarpus indicus Bedd., 166
Dipterocarpus turbinatus Dyer, see *Dipterocarpus indicus* Bedd.
Dita bark, see *Alstonia scholaris*
Ditamine alkaloid, 31
Diterpenoids, 131
Diuretic substances, 14, 16, 18, 20, 24, 25, 27, 32, 42, 44, 47, 56, 57, 60, 61, 67, 70, 72, 75, 77, 78, 79, 81, 82, 83, 85, 86, 107, 112, 117, 120, 123, 127, 129, 137, 144, 154, 162, 166, 167, 172, 176, 180, 199, 201, 208, 210, 215, 217, 220, 234, 242, 243, 253, 260, 262, 265, 267, 274, 283, 287, 292, 294, 296, 297, 300, 302, 303, 305, 306, 307, 316, 318, 323, 324, 325, 327, 329, 335, 336, 337, 345
Dodder, see *Cuscuta reflexa*
Dogbite, see *Alangium salviifolium*
Dog grass, see *Cynodon dactylon*
Dolichos biflorus Linn., 167—168
Dolichos uniflorus Lam., see *Dolichos biflorus* Linn.
Dropsy, treatment of, 64, 67, 143, 183, 237, 262, 305, 320, see also Anasarca; Ascites; Elephantiasis
  renal, 14, 329
Droserone, 268
Dulcitol, 152
Duodenal ulcer, 26
"Dvarapala", 3
Dysentery, treatment of, 55, 70, 81, 83, 93—94, 95, 110, 156, 158, 171, 182, 186, 191, 199, 248, 250, 252, 257, 260, 262, 279, 281, 282, 286, 287, 297, 302, 320, 324, 327, see also Diarrhea
  amebic, 113, 206
  chronic, 202, 303, 311, 322
  hemorrhagic, 298
Dysmenorrhea, see Menstrual disorders
Dyspepsia, treatment of, 25, 36, 71, 100, 107, 139, 248, 250, 294, 300, 301, 311, 318, 323, 324, 332, 336, 340
  atonic, 98, 102
Dysuria, treatment of, 97, 305, 336

E

Ear, 66, 202, see also Otitis; Otorrhea
Earache, 183, 255
East Indian globe thistle, see *Sphaeranthus indicus*
East Indian screw tree, see *Helicteres isora*
Echinocystic acid, 23
Echitamidine alkaloid, 31
Echitenine alkaloid, 31
Eclipta alba Hassk., 169
Ecliptine, 169
Eczema, 318
Egypt, Hindu medicine in, 1
α-Elatrine, 122—123
Elephant creeper, see *Argyreia speciosa*

Elephantiasis, treatment for, 49, 95, 169, 237, 303, 313
Elephantopus scaber Linn., 170—171
Elephants foot, see *Elephantopus scaber*
Elettaria cardamomum Maton, 172
Elixir of life, 4
Ellagic acid, 175, 281, 321
Elliptinone, 268
Embelia glandulifera Wight, see *Embelia ribes* Burm. f.
Embelia ribes Burm. f., 173—174
Embelin, 174
Emblic myrobalan, see *Emblica officinalis*
Emblica officinalis Gaertn., 175—176
Emblicol, 175
Emetic nut, see *Randia dumetorum* Lam.
Emetic substances, activity, 7, 18, 69, 79, 84, 93, 123, 128, 221, 254, 329, 335
Emetine alkaloid, 22
Emmenagogue substances, activity, 6, 18, 26, 28, 44, 50, 51, 60, 66, 78, 83, 88, 100, 115, 131, 156, 185, 194, 214, 227, 236, 239, 245, 259, 260, 265, 287, 292, 294, 302, 317, 327, 329, see also Menstrual disorders
Emodin glucoside, 104, 287
Emollient substances, 12, 49, 84, 106, 217, 252, 267, 272, 302, 309, 327, 335, 345, see also Counterirritants
Endocarditis, treatment of, 320
English, in India, 4
Enlarged spleen, treatment of, 5
Entamoeba histolytica, antiprotozoal properties against, 22, 23, 69, 128, 150
Ephedrine, 303
Ephemerantha macraei (Lindl.) Hunt et Sunmeh, see *Dendrobium macraei* Lindl.
l-Epicatechin, 9, 277
Epidermal carcinoma, 83
Epifriedelinol acetate, 253
Epilepsy, treatment of, 37, 61, 72, 116, 311, 317, 330, 335
Epilupiol, 279
Ergine, 49
Ergonovine, 49
Ergot substitute, 50
Erysipelas, 297
Erysodine, 178
Erysovine, 178
Erystorine, 178
Erythema, treatment of, 16
Erythraline, 178
Erythrina corallodendron Linn., see *Erythrina variegata* Linn.
Erythrina indica Lam., see *Erythrina variegata* Linn.
Erythrina stricta Roxb., see *Erythrina variegata* Linn.
Erythrina variegata Linn., 177—178
Erythrine, antidote to strychnine poisoning, 178
Erythrodiol, 279
Erythrodiol-3-monoacetate, 277
Escharotic activity, 301
Escherichia coli, 106

Ether extract, 113
Ethyl gallate, 175
β-Eudesmol, 277
Eugenia jambolana Lam., 179—180
Eugenol, 18, 117, 118, 249, 250, 260, 287
Eugenol methyl ether, 250
Eupatorium ayapana Vent., see *Eupatorium triplinerve* Vahl.
Eupatorium triplinerve Vahl.,181
Euphorbiaceae
  Baliospermum montanum, 64
  Croton oblongifolius, 142
  Croton tiglium, 143
  Emblica officinalis, 175
  Euphorbia hirta, 182
  Euphorbia neriifolia, 183
  Mallotus philippensis, 224
  Phyllanthus fraternus, 261
  Ricinus communis, 290
Euphorbia hirta Linn., 182
Euphorbia neriifolia Linn., 183
Euphorbia pilulifera Linn., see *Euphorbia hirta* Linn.
Euphosterol, 182
Evolvine, 184
Evolvulus alsinoides Linn., 184
Expectorant substances, 5, 12, 18, 20, 21, 24, 25, 31, 33, 35, 53, 69, 77, 79, 93, 116, 128, 151, 178, 182, 202, 212, 283, 289, 294, 297, 300, 305, 306, 317, 332
Eye, 57, 60, 73, 78, 262, 287, 313, 321
  conjunctivitis, 103, 281, 291, 309, 315, 334
  inflammatory conditions, 35, 41, 46, 176, 315, 344
  nearsightedness, 342
  surgery, 57

F

γ-Fagarine, 340
Febrifuge substances, 17, 20, 21, 31, 32, 38, 39, 43, 60, 66, 67, 73, 79, 88, 106, 107, 110, 134, 171, 178, 190, 206, 250, 286, 303, 307, 318, 323, 334, 335, see also Fevers
Febrile delirium, treatment of, 116
Feet, burning, 329
Fenchone, 189
Fennel, see *Foeniculum vulgare*
Fenugreek, see *Trigonella foenum-graecum* Linn.
Ferroquinoline alkaloid, 21
Ferula foetida Regel., 185
Ferulic acid, 131, 185
Fetid cassia, see *Cassia tora*
Fevers, treatment of, 139, 163, 282, 289, 303, 307, 311, 313, 322, see also Refrigerants
  chronic, 287, 305, 324
  intermittent, 110, 259, 262, 312
  remittent, 259, 300
Ficus benghalensis Linn., 186
Ficus glomerata Roxb., see *Ficus racemosa* Linn.
Ficus indica Linn., see *Ficus benghalensis* Linn.
Ficus racemosa Linn., 187
Ficus religiosa Linn., 188

Fig sacred, see *Ficus religiosa*
Fig tree, country, see *Ficus racemosa*
Fir, Himalayan silver, see *Abies spectabilis*
Fish poisons, 183
Fistucacidin, 106
Fistulic acid, 106
Fistulin, 106
Flacourtiaceae
   Hydnocarpus laurifolia, 207
Flatulence, 19, 86, 260, 291, 322, 323
Flavonoids, see individual compounds
Flower, Chinese, see *Paederia foetida*
Fluoride, 266, 281
Foeniculum capillaceum Gilib., see *Foeniculum vulgare* Mill.
Foeniculum vulgare Mill., 189
Foods, see Nutritive substances
Formic acid, 143, 162, 330
N-Formyldeacetyl-colchicine, 193
Foxglove, see *Digitalis purpurea*
Fragrant screwpine, see *Pandanus tectorius*
French, in India, 4
Friedelin, 6, 23, 91, 186, 253
Fructose, 115
Fucose, 60
L-Fucose, 60
Fumariaceae
   Fumaria indica, 160
Fumaria indica (Haussk.) Pungsley, 190
Fumaria parviflora Lamk., see *Fumaria indica* (Haussk.) Pungsley
Fumarilicine, 190
Fumariline, 190
Fumitory, common, see *Fumaria indica*
Fungitoxic effects, 215, 250, 265, 266, 281
Furan alcohol, 115
Furano-diterpenes, 324
Furfuraldehyde, 115
Fufuryl alcohol, 115

G

Galactogogue substances, 9, 27, 55, 56, 100, 101, 117, 156, 198, 260, 302, 327, 338
Galactomannan, 82, 129
Galactose, 185
D-Galactose, 60
Gallic acid, 162, 175, 182, 281, 287, 321, 322
Galloylated leukoanthocyanidin-4-(2-*O*-galloyl)glucoside, 75
Galls, 289
Gamboge, Indian, see *Garcinia morella*
Ganiarine, 270
Ganja, see *Cannabis indica*
Garancin, 292
Garbanzol, 277
Garcinia indica Choisy, 191
Garcinia morella Desr., 192
Garcinia purpurea Roxb., see *Garcinia indica* Choisy
Garden rue, see *Ruta graveolens*
Gargle for teeth and gums, 35, 222, 289

Garjanic acid, 166
Garlic, see *Allium sativum*
Gastric catarrh, 82
Gastric mucosa, affected by *Alpinia galanga*, 30
Gastrointestinal irritant, 94
Gaurjan balsam, in treatment of gonorrhea, 166
Genitourinary disorders, 262, 266, 267, 271, 283, 292, 325
Gentianaceae
   Swertia chirata, 312
Geraniol, 255
Geranylacetate, 274
Ghrita, see *Brahmi ghrita*
Gilenin, 324
Giloin, 324
Gilosterol, 324
Gingediacetates, 342
Gingediols, 342
Ginger, see *Zingiber officinale*
"Gingerin", 341
Ginger mango, see *Curcuma amada*
Gingerol, 341
Gitalin, 162
Gitoxin, 162
Glandular swelling, 183, 268—269
Gleet, see Gonorrhea and gleet
Gloriosa superba Linn., 193
Glossary of Ayurvedic terms, 353—359
4-*O*(*d*-Glucopyranosol uronic acid)-D-galactopyranose, 60
Glucose, 61, 139, 329
3-Glucoside, 71
C-Glucoside, 28
Glucosides, see specfiic compounds
β-D-Glucoside, 81
Glucuronic acid, 185
D-Glucuronic acid, 60
Glutamic acid, 127, 199, 267
Glyceride of lenolic acid, 125
Glyceride of lingnoceric acid, 125
Glyceride of oleic acid, 125
Glyceride of stearic acid, 125
Glycerol, 103
Glycine, 86, 127, 267
Glycoflavones, 262
Glycolic acid, 25
Glycosides, see specific compounds
Glycosuria, treatment of, 86, 180
Glycosyl-β-hydroxy-dihydrochalkone pterosupin, 277
Glycyrrhetinic acid, 194
Glycyrrhiza glabra Linn., 194
Glycyrrhizin, 194
Gmelina arborea Linn., 196
Golden collyrium, see *Colchicum luteum*
Gold thread, see *Coptis teeta, Thalictrum foliolosum*
Gonorrhea and gleet, treatment for, 6, 9, 35, 55, 56, 81, 128, 129, 151, 166, 296, 297, 303, 306, 307, 315, 324, 344
Gossypium herbaceum Linn., 198
Gossypium indicum Lam., see *Gossypium herbaceum* Linn.

Gossypetin, 198
Gossypol, 198
Gout, treatment for, 45, 104, 251, 265, 267, 325, 336, 337
Gram-negative bacteria, 26
Gramineae
   Bambusa bambos, 65
   Cynodon dactylon, 153
Grapes, see *Vitis vinifera* Linn.
Grass, dog, see *Cynodon dactylon*
Grass, nut, see *Cyperus rotundus*
Graveobioside A, 44
Graveobioside B, 44
Great-leaved caladium, see *Alocasia macrorrhiza*
Greece, influence of Hindu medicine in, 1
Greek invasions of India, 3
Grewia hirsuta Vahl. 198
Grewia pilosa Roxb., see *Grewia hirsuta* Vahl.
Grewia polygama Mast., see *Grewia hirsuta* Vahl.
Guaiacol, 255
Gugal, gum, see *Commiphora mukul*
Guineaworms, 334
Gulsakri, see *Grewia hirsuta*
Gum gugal, see *Commiphora mukul*
Gums, treatment of, 222, 289, 322
Gunas, defined, 350
Guttiferae
   Calophyllum inophyllum, 91
   Garcinia indica, 191
   Garcinia morella, 192
   Mesua ferrea, 228
   Ochrocarpus longifolius, 248
Guvacine, 46
Guvaoline 46
Gymnema sylvestre R.Br., 200

H

Hair growth, stimulant, 28
Hamamelidaceae
   Altingia excelsa, 33
Harmaline, 259
Harman, 313, 325
Harmatol, 259
Harmine, 259, 325
Harmolal, 259
Hayatidin, 120
Hayatin, 120
Hayatinin, 120
Headache, 272, 279, 296
Head lice, 193, 259, 272
Heart, see also Blood pressure
   disease, 162, 320, 336
      ischemic, 131
      palpitations, 22
   physiology, effects on, 54, 95, 100, 184, 277
   stimulants, see Cardiac stimulants
Helicteres isora Linn., 202
Heliotropum indicum Linn., see *Cissus quadrangularis* Wall.
Hellenic civilization, 2

Hematuria, treatment of, 154, 281, 303
Hemicrania, treatment of, 127
Hemidesmus indicus (L.) Schult., 203
Hemoptysis, 281, 317
   from internal organs, 72
   pulmonary TB, 81
Hemorrhage, 265, 289, 298, 320, 336
   of gums, 313, see also Gums, treatment of
   internal, 88, 91
   prevention of, 294
   uterine, 9
   urethral, 297
Hemorrhoids, 9, 36, 243, 248, 281, 287, 296, 298, 301, 302, 303, 306, 309, 315, 322, 334
Hemotoxyline, 298
Hemp, Indian, see *Cannabis indica*
Henbane, see *Hyoscyamus niger*
n-Hencosane, 264
Hentriacontane, 79, 81, 123, 253, 264
Hentriacontane-16-one, 264
Hentriacontanol, 81, 197, 253, 266
Hentriacontanone, 266
Hepatic dropsy, 215
Hepatitis, 263, 316
n-Heptadecane, 264
Hermaphrodite flowers, *Boerhaavia diffusa*, 79
Herpestine, 61
Herpestis monniera (L.) H.B.K., see *Bacopa monniera* (L.) Pennell
Heteratisine, 17
Heterophyllidine, 17
Heterophylline, 17
Hetisine, 17
Hexacosanol, 106
n-Hexadecane, 264
Hexadecanoic acid, 264
Hexanoate, 342
Hexobarbitone-induced hypnosis in mice, 250
Hibiscus abelmoschus Linn., 204
Hiccup, treatment of, 82, 211, 300, 312, 317, 332
α- and β-Himachalene, 110
Himachalol alcohol, 110
Himalayan fir, see *Taxus baccata* Linn.
Himalayan silver fir, see *Abies spectabilis*
Himsa or "Ahimsa", Buddhist teaching against violence, 2, 3
Hippocrates, visited India, 2
Histamine, 250
Histidine, 129, 253
Hogweed, spreading, see *Boerhaavia diffusa*
Holadysamine, 205
Holarrhena antidysenterica (Linn.) Wall., 205—206
Holy basil, see *Ocimum sanctum*
Homonapelline, 16
Homotaraxasterol, 316
Homotetrahydrocannabinol, 96
Hookworm, treatment of, 113, 189, 225, see also Anthelmintics
Horse gram, see *Dolichos biflorus*
Horseradish tree, see *Moringa oleifera*

Humuelene, 166
Huns, invasions of India, 3
Hydnocarpus laurifolia (Dennst.) Sleumer, 207
Hydnocarpus wightiana Blume, see *Hydnocarpus laurifolia* (Dennst.) Sleumer
Hydragogues, 28, 178, 221, 251, see also Cathartics
Hydrocarbon, 340
Hydrocyanic acid, 89, 272, 273
Hydrocele, 33
Hydrocinnamic acid, 45
Hydrocortisone acetate, 150
di-Hydrocostus lactone, 300
Hydrocotyle asiatica Linn., 208—209
Hydrophilic mucilage, 327
Hydrophobia, 183, 306, 311
Hydroxy acids, 103
p-Hydroxybenzaldehyde, 277
p-Hydroxybenzyl methyl ester, 122
5-Hydroxy-8-methoxy-7-*O*-β-D-glucopyranuronosyl flavone, 252
5-Hydroxymethyl furfuraldehyde maltol, 115
Hydroxypeganine, 216
Hygrophila auriculata (Schumach.) Heine., 210
Hyoscine, 157
Hyoscyamine, 157, 211
Hyoscyamus niger Linn., 211
Hypaphorine, 178
Hyperglycemic action, to control, 26
Hypertension, 320
Hypnotics, 330, 337
Hypocholesterolemic action, 267, 277
Hypochondria, see Hypochondriasis
Hypochondriasis, treatment of, 21, 185, 286
Hypoglycemic action, 23, 66, 67, 73, 83, 128, 180, 250, 262, 265, 271, 277, 279
Hypolipemic action, 131
Hypophyllanthum, 262
Hypothermia in mice, 30, 67, 69, 71
Hypotensive action, in experimental animals, 75, 271, 286
Hyssop, see *Hyssopus officinalis*
Hyssopus officinalis Linn., 212
Hysteria, treatment of, 61, 76, 107, 294, 330

I

Igasurine, 310
Impotence, treatment for, 77, 324, 325
Indian aconite, see *Aconitum ferox*
Indian atees, see *Aconitum heterophyllum*
Indian belladonna, see *Atropa acuminata*
Indian birthwort, see *Aristolochia indica*
Indian butter tree, see *Madhuca longifolia*
Indian cotton, see *Gossypium herbaceum*
Indian coral tree, see *Erythrina variegata*
Indian dill, see *Peucedanum graveolens*
Indian gamboge, see *Garcinia morella*
Indian hemp, see *Cannabis indica*
Indian Jack tree, see *Artocarpus integrifolia*
Indian jalap, see *Operculina turpethum*
Indian madder, see *Rubia cordifolia*

Indian nightshade, see *Solanum indicum*
Indian olibaum, see *Boswellia serrata*
Indian pennywort, see *Hydrocotyle asiatica*
Indian rhubarb, see *Rheum emodi*
Indian sarsaparilla, see *Hemidesmus indicus*
Indian senna, see *Cassia angustifolia*
Indian squill, see *Urginea indica*
Indian valerian, see *Valeriana jatamansi*
Indian worm-wood, see *Artemisia vulgaris*
Indigestion, 316
Indole-3-alkylamines, 236
Inflammatory conditions, 35, 116, 139, 194, 318, 332
  bladder, 306
  bowels, 195, 287
  bronchial tubes, 195
  chest, 292
  external, 52
  of eye, 35, 46, 176, 315, 344
  genitourinary passage, 195
  gums, 344
  internal, 66
  joints, 66, 89, see also Rheumatism
  kidneys, 306
  of liver, 70
  lungs, 176
  skin, 266, 267, 279, 292, 297
  of stomach, 43, 289
Inophyllic acid, 91
Inositol, 162
L-Inositol, 182
Inotropic and chronotropic effects, 54, 250
Insanity, 72
Insecticides, 18, 265
Insomnia, 286
Insulin secretion, 201, see also Diabetes
Intermittent fevers, 110, 259, 262, 312
Intestinal problems, see also Diarrhea; Dysentary
  disorders, 109
  parasites, 113, see also Anthelmintics; Vermifuges
  and urinary problems, 53
Inulide, 115
Inulin, 115, 300, 316
Inulocoagulase, 115
Invert sugar, 24
Ipecacuanha, substitute for, 140
Ipomoea hederacea auct. non Jacq., see *Ipomoea nil* (L.) Roth
Ipomoea nil (L.) Roth, 213
Ipomoea turpethum Linn., see *Operculina turpethum* (Linn.) Silva, Manso
Iridaceae
  Crocus sativus, 141
Ischemic heart disease, 131
Ishwarane, 51
Ishwarol, 51
Ishwarone, 51
Isoajmaline, 286
Isobarbaloin, 28
Isobutrin, 86
N-Isobutyl-deca-*trans*-2-*trans*-4-dienamide, 264
N-Isobutyl eicosa-*trans*-2-*trans*-4-dienamide, 266

Isobutylamide, 264
Isobutyric acid, 118
Isochaksine, 103
Isoconessimine, 205
Isocoreoptin, 86
Isodeoxyelephantopin, 171
Isoergine, 49
Isoflavonoid glycol marsupol, 277
Isoleucine, 127, 199
Isoliquirtigenin, 277
Isoliquiritin, 194
Isomonospermoside, 86
Isomorelin, 192
Isopelletierine, 281, 337
Isopsoralen, 274
Isopsoralidin, 274
Isopterocarpone, 279
Isopterocarpolone, 279
Isoquercitrin, 71
Isorhamnetin, 104
Isoreinoleic acid, 290
Isotalatizidine, 159
Isotephrosin, 318
Isovalerianic acid, 330
Isovaleric aldehyde, 297
Isotubulosine, 22
Isozeylinone, 268

J

Jalap, Indian, see *Operculina turpethum*
Jalap, substitute, 213
Jamboline (glucoside), 180
Jambulon, 182
Jangli almond, see *Hydnocarpus laurifolia*
Japan, Ayurvedic teachings, 2
Jasmine, see *Jasminum sambac*
Jasmine, night, see *Nyctanthes arbor-tristis*
Jasminum sambac Ait., 214
Jatamansinol, 239
Jatamansone, 239
"Jatiphaladi churna", 238
Jatrorrhizine, 73, 134, 323
Jaundice, treatment of, 64, 79, 158, 210, 262, 287, 292, 306, 316, 323, 336
Java glangal, see *Alpinia galanga*
Jequirity, see *Abrus precatorius*
Jibantine, 161
Jivaka, crowned King of Physicians, 3
Jivantic acids, 161
Joints, inflamed, 66, 89, see also Rheumatism
Jujube fruit, see *Zizyphus martiana*
Juncea Prain var. *oleira* Prain, see *Brassica juncea* (L.) Czern. and Coss
Juniper berry, see *Juniperus communis*
Juniperus communis Linn., 215
Justicia adhatoda Linn., 216

K

Kaempferol, 60, 75, 81, 104, 106, 152, 243,
Kalmeghin, 39

Kamala powder, for syphilitic ulcers, 224
Kamalin I, 224
Kantakari, see *Solanum surattense*
Kaolin, 109
Kaya cikitsa, internal medicine, 1
Kempferol, see Kaempferol
Ketone, 101, 294
3-Ketosteroid, 121
Ketosterol, 298
Kidney disorders, 56, 211, 316, see also Urinary disorders
  kidney stones, 35, 75, 162, 167—168
King of bitters, see *Andrographis paniculata*
Kinovin, 313
Kokum oil, for diarrhea and dysentery, 191
Kidney and liver complaints
Korea, Ayurvedic teachings, 2
Kurchaline, 205
Kurchitree, see *Holarrhena antidysenterica*, 205
Kutkin, 263
Kutkiol, 263
Kutkisterol, 263
Kutkoside, 263

L

Labiateae
  Hyssopus officinalis, 212
  Mentha arvensis, 227
  Ocimum sanctum, 249
Laccol, 289
Lachrymation, to induce, 244
Lactagogue, 291
Lactic acid, 115, 162
Lactone, 40, 52, 109
Lamarckinine , 22
"Lanceolatine A", 318
Larvicidal action, 250
Laryngeal tuberculosis, 26
Lauraceae
  Cinnamomum camphora, 116
  Cinnamomum tamale, 117
  Cinnamomum zeylanicum, 118
  Litsea glutinosa, 218
Lauric acid, 129
Laxative substances, 18, 21, 24, 27, 54, 58, 71, 72, 79, 81, 82, 85, 86, 106, 109, 129, 163, 176, 188, 197, 262, 263, 267, 272, 274, 275, 283, 296, 302, 312, 314, 318, 321, 335
Leadwort, white, see *Plumbago zeylanica*
Lecithin, 217
Leclerc, observations on Arabian medicine, 4
Leguminosae
  Abrus precatorius, 7
  Alhagi pseudalhagi, 24
  Cassia absus, 103
  Cassia angustifolia, 104
  Cassia occidentalis, 107
  Cassia tora, 108
  Saraca indica, 298
  Tephrosia purpurea, 318

Trigonella foenum-graecum, 327
Leprosy, treatment of, 60, 91, 93, 111, 207, 208, 274, 296, 300, 301, 329
Lesser cardamom, see *Elettaria cardamomum*
Lettsomia nervosa Roxb., see *Argyreia speciosa*
Leucine, 127, 129, 267
Leukoanthocyanidin, 198
Leukoanthocyanin, 279
Leukocyanidin, 46, 86
(±)Leukocynidin, 91
Leukodelphinidin, 175
(+)-Leukodelphinidin, 320
Leukoderma, treatment of, 66, 77, 274
Leukopelargonidin, 106
Leukorrhea, 9, 55, 86, 199, 303, 327
Levulin, 316
Licorice, see *Glycyrrhiza glabra*
Licorice extracts, 194
Lignan sesamin, 302
Lignin, 202
Lignoceric acid, 88, 103, 234
  glyceride of, 125
Lignum pavanae, 143
Liliaceae
  Allium cepa, 25
  Allium sativum, 26
  Aloe barbadensis, 28
  Asparagus adscendens, 55
  Asparagus racemosus, 56
  Colchicum luteum, 130
  Gloriosa superba, 193
  Urginea indica, 328
Lily, climbing, see *Gloriosa superba*
Lily, superb, see *Gloriosa superba*
Limbs, diseases affecting, 111, see also Rheumatism
Limonene, 101, 172, 274
d-Limonene, 21, 44, 260
Linaceae
  Linum usitatissimum, 217
(±) Linalol, 137
Linalool, 147, 214, 255, 274
Linalool cumic aldehyde, 118
Linalool oxide, 218
Linalyl acetate, 147, 255, 340
Linoleic acid, 22, 88, 103, 112, 162, 175, 182, 217, 245, 264, 302
Linolenic acid, 103, 162, 175, 264, 302
Linoxyn, 217
Linseed, see *Linum usitatissimum*
Linum usitatissimum Linn., 217
Liposes, 291
Lintetraline, 262
Liquiritigenin, 277
Liquiritin, 194
Liquiritigenine chalcone, 194
Lithontriptic substances, 167, 283, 320
Litsea chinensis Lam., see *Litsea glutinosa* (Lour.) C.B. Robins
Litsea glutinosa (Lour.) C.B. Robins, 218
Litsea sebifera Pers., see *Litsea glutinosa* (Lour.) C.B. Robins

Liver disorders, 44, 88, 130, 142, 236—237, 321, see also Hepatitis
  cholesterol in, 236
  cirrhosis, 320
  enlargement, 93, 306
  inflamed, 70
  obstrucions
Lixorbutic medicinal property, 75
Lobeliaceae
  Lobelia nicotianaefolia, 219
Lobelia nicotianaefolia Heyne. ex Roth, 219
Lochia, to induce, 66
Lodh tree, see *Symplocos racemosa* Roxb.
Loganiaceae
  Strychnos nux vomica, 310
Loganin, 310
Long chain acids, see specific compounds
Long pepper, see *Piper longum*
Loturidine, 313
Loturine, 313
Lotus, sacred, see *Nelumbo nucifera*
Luffa aegyptiaca Mill. ex Hook. f., see *Luffa cylindrica* (Linn.) M. Roem
Luffa acutangula (Linn.) Roxb., 220
Luffa cylindrica (Linn.) M. Roem, 221
Lumbago, treatment of, 98, 287
Lumicolchicine, 193
Lung diseases, see Pulmonary disorders
Lupene-diol, 279
Lupeol, 22, 31, 81, 106, 125, 187, 279, 334
Lupeol acetate, 128, 222, 324
Lupeonone, 279
Luteolin, 44, 152, 162, 197, 250, 270
Lycorine, 140
Lymphangitis, treatment of, 151
Lysine, 129, 199, 253, 267

M

Macene, 238
Madhuca longifolia (Linn.) Macbride, 222—223
Magnoflorine, 340
Magnoliaceae
  Michelia champaca, 230
Maiden-hair fern, see *Adiantum philippense*
Malabar nut, see *Justicia adhatoda*
Malabarkino, see *Pterocarpus marsupium* Roxb.
Malaria, 250, 259, 300
Malarial fevers, treatment of, 73
Malic acid, 12, 100, 107, 194, 314, 330, 336
Maliol, 330
Mallotus philippensis (Lam.) Muell.-Arg., 224
Malonate, 342
Malvaceae
  Althaea officinalis, 32
  Gossypium herbaceum, 198
  Hibiscus abelmoschus, 204
  Sida cordifolia, 305
Malvidin, 281
Malvidine pentose glycoside, 281
Malvin chloride, 283

Mandelonitrile, 272
Mangiferin, 152
Mango ginger, see *Carcuma amada*
Mango, red, see *Garcinia indica*
Manjistin, 292
"Manna" (sugary secretion), 24
Mannite, 281
Mannitol, 152
D-Mannitol, 61, 263, 281
Marking nut tree, see *Semecarpus anacardium*
Marmasin, 21
Marsh mallow, see *Althaea officinalis*
Maslinic acid, 281
"Maxima substances" A, B, and C, in *Tephrosia purpurea*, 318
Medicinal plants
  Abies spectabilis, 5
  Abroma augusta, 6
  Abrus precatorius, 7
  Acacia catechu, 8
  Acacia farnesiana, 10
  Acacia nilotica, 12
  Achyranthes aspera, 13
  Aconitum ferox, 15
  Aconitum heterophyllum, 17
  Acorus calamus, 18
  Adiantum philippense, 20
  Aegle marmelos, 21
  Alangium salviifolium, 22
  Albizzia lebbek, 23
  Alhagi pseudoalhagi, 24
  Allium cepa, 25
  Allium sativum, 26
  Alocasia macrorrhiza, 27
  Aloe barbadensis, 28
  Alpinia galanga, 29
  Alstonia scholaris, 31
  Althaea officinalis, 32
  Altingia excelsa, 33
  Amomum subulatum, 34
  Amorphophallus campanulatus, 36
  Anacyclus pyrethrum, 37
  Ananas comosus, 38
  Andrographis paniculata, 39
  Angelica glauca, 40
  Annona squamosa, 41
  Anthocephalus indicus, 43
  Apium graveolens, 44
  Aquilaria agallocha, 45
  Areca catechu, 46
  Argemone mexicana, 47
  Argyreia speciosa, 49
  Aristolochia bracteata, 50
  Aristolochia indica, 51
  Artemisia maritima, 52
  Artemisia vulgaris, 53
  Artocarpus integrifolia, 54
  Asparagus adscendens, 55
  Asparagus racemosus, 56
  Atropa acuminata, 57
  Averrhoa carambola, 58
  Azadirachta indica, 59
  Bacopa monniera, 61
  Balanites aegyptiaca, 62
  Baliospermum montanum, 64
  Bambusa bambos, 65
  Barleria prionitis, 67
  Barringtonia acutangula, 68
  Bauhinia tomentosa, 70
  Bauhinia variegata, 71
  Benincasa hispida, 72
  Berberis aristata, 73
  Bergenia ligulata, 74
  Betula utilis, 76
  Blepharis edulis, 77
  Blumea lacera, 78
  Boerhaavia diffusa, 79
  Bombax ceiba, 80
  Borassus flabellifer, 82
  Boswellia serrata, 83
  Brassica juncea, 84
  Buchanania lanzan, 85
  Butea monosperma, 86
  Caesalpinia crista, 87
  Callicarpa macrophylla, 89
  Calophyllum inophyllum, 91
  Calotropis gigantea, 93
  Calotropis procera, 95
  Cannabis indica, 96
  Capsicum annuum, 98
  Carica papaya, 99
  Carum carvi, 101
  Carum copticum, 102
  Cassia absus, 103
  Cassia angustifolia, 104
  Cassia fistula, 105
  Cassia occidentalis, 107
  Cassia tora, 108
  Cedrus deodara, 110
  Celastrus paniculatus, 111
  Centratherum anthelminticum, 112
  Chenopodium ambrosioides, 113
  Cicorium intybus, 114
  Cinnamomum camphora, 116
  Cinnamomum tamala, 117
  Cinnamomum zeylanicum, 118
  Cissampelos pareira, 119
  Cissus quadrangularis, 121
  Citrullus colocynthis, 122
  Clerodendrum infortunatum, 124
  Clitoria ternatea, 126
  Coccinia indica, 128
  Cocos nucifera, 129
  Colchicum luteum, 130
  Commiphora mukul, 131
  Coptis teeta, 133
  Cordia myxa, 135
  Coriandrum sativum, 137
  Costus speciosus, 138
  Crinum defixum, 140
  Crocus sativus, 141
  Croton oblongifolius, 142

Croton tiglium, 143
Cucumis sativus, 144
Cuminum cyminum, 145
Curculigo orchioides, 146
Curcuma amada, 147
Curcuma angustifolia, 148
Curcuma longa, 149
Curcuma zedoaria, 151
Cuscuta reflexa, 152
Cynodon dactylon, 153
Cyperus rotundus, 155
Datura metal, 157
Daucus carota, 158
Delphinium denudatum, 159
Dendrobium macraei, 160
Digitalis purpurea, 162
Dillenia indica, 163
Dioscorea bulbifera, 164
Dipterocarpus indicus, 166
Dolichos biflorus, 167
Eclipta alba, 169
Elephantopus scaber, 170
Elettaria cardamomum, 172
Embelia ribes, 173
Emblica officinalis, 175
Erythrina variegata, 177
Eugenia jambolana, 179
Eupatorium triplinerve, 181
Euphorbia hirta, 182
Euphorbia neriifolia, 183
Evolvulus alsinoides, 184
Ferula foetida, 185
Ficus benghalensis, 186
Ficus racemosa, 187
Ficus religiosa, 188
Foeniculum vulgare, 189
Fumaria indica, 190
Garcinia indica, 191
Garcinia morella, 192
Gloriosa superba, 193
Glycyrrhiza glabra, 194
Gmelina arborea, 196
Gossypium herbaceum, 198
Grewia hirsuta, 199
Gymnema sylvestre, 200
Helicteres isora, 202
Hemidesmus indicus, 203
Hibiscus abelmoschus, 204
Holarrhena antidysenterica, 205
Hydnocarpus laurifolia, 207
Hydrocotyle asiatica, 208
Hygrophila auriculata, 210
Hyoscyamus niger, 211
Hyssopus officinalis, 212
Ipomoea nil, 213
Jasminum sambac, 214
Juniperus communis, 215
Justicia adhatoda, 216
Linum usitatissimum, 217
Litsea glutinosa, 218
Lobelia nicotianaefolia, 219
Luffa acutangula, 220
Luffa cylindrica, 221
Madhuca longifolia, 222
Mallotus philippinensis, 224
Melia azedarach, 225
Mentha arvensis, 227
Mesua ferrea, 228
Michelia champaca, 229
Mimusops elengi, 232
Moringa oleifera, 234
Mucuna pruriens, 236
Myristica frangrans, 238
Nardostachys jatamansi, 239
Nelumbo nucifera, 241
Nerium indicum, 243
Nigella sativa, 245
Nyctanthes arbor-tristis, 246
Ochrocarpus longifolius, 248
Ocimum sanctum, 249
Operculina turpethum, 251
Oroxylum indicum, 252
Paederia foetida, 253
Pandanus tectorius, 255
Papaver somniferum, 256
Paganum harmala, 258
Peucedanum graveolens, 260
Phyllanthus fraternus, 261
Picrorhiza kurroa, 263
Piper longum, 264
Piper nigrum, 266
Plantago ovata, 267
Plumbago zeylanica, 268
Premna obtusifolia, 270
Prunus amygdalus, 272
Prunus puddum, 273
Psoralea corylifolia, 274
Pterocarpus marsupium, 276
Pterocarpus santalinus, 278
Punica granatum, 280
Randia dumetorum, 282
Raphanus sativus, 283
Rauwolfia serpentia, 284
Rheum emodi, 287
Rhus succedanea, 288
Ricinus communis, 290
Rubia cordifolia, 292
Ruta graveolens, 293
Salvadora persica, 295
Santalum album, 297
Saraca indica, 298
Saussurea lappa, 299
Semecarpus anacardium, 301
Sesamum indicum, 302
Sida cordifolia, 303
Solanum indicum, 304
Solanum nigrum, 306
Solanum surattense, 307
Sphaeranthus indicus, 308
Strychnos nux-vomica, 310
Swertia chirata, 312
Symplocos racemosa, 313

Tamarindus indica, 314
Taraxacum officinale, 316
Taxus baccata, 317
Tephrosia purpurea, 318
Terminalia arjuna, 319
Terminalia bellerica, 321
Terminalia chebula, 322
Thalictrum foliolosum, 323
Tinospora cordifolia, 324
Tribulus terrestris, 325
Trigonella foenum-graecum, 327
Urginea indica, 328
Valeriana jatamansi, 330
Vanda roxburghii, 331
Vernonia cinerea, 333
Viola odorata, 335
Vitis vinifera, 336
Withania somnifera, 337
Zanthoxylum alatum, 339
Zingiber officinale, 341
Zizyphus martiana, 343
Zizyphus vulgaris, 345
Melanthin, 245
Melancholia, 21
Melia azadirachta Linn., see *Azadirachta indica* A. Juss.
Melia azedarach Linn., 225—226
Meliaceae
  Azadirachta indica, 59
  Melia azedarach, 225
Melissic acid, 182
Melissyl alcohol, 162
Melizitose, 24
Memory, improved, 111
Menispermaceae
  Cissampelos pareira, 119
  Tinospora cordifolia, 324
Menorrhagia, see Menstrual disorders
Menstrual disorders, 6, 50, 53, 71, 81, 83, 97, 100, 104, 111, 115, 116, 118, 131, 132, 141, 157, 178, 187, 192, 198, 237, 242, 245, 262, 291, 298, 302, 313
Mental derangement, 77
Mentha arvensis Linn., 227
α-Menthol, 227
Mercurial salivation, 9
Mesua ferrea Linn., 228—229
Mesuanic acid, 229
Mesuaxanthone B, 91
Methionin, 46, 253
p-Methylacetophenone, 110, 264
4-Methoxy-norsecurinine, 262
4-Methoxy-securinine (phylanthine), 262
Methyl alcohol, 115
Methyl anthanilate, 214
Methyl amyl ketone, 118
Methyl cinnamate, 30
Methyl ether, 86
Methyl ether acetate, 86
Methyl ether of β-phenylethyl alcohol, 255
Methylgingediacetates, 342

Methylgingediols, 342
Methylheptyl ketone, 287
Methyl-*n*-heptyl-ketone, 294
N_2-Methylholarrhimine, 206
2-Methyl isoflavones, 194
Methyl mercaptan, 253, 283
Methyl-*n*-nonyl ketone, 294, 340
Methyl pelletieriene, 281
Methyl salicylate, 11, 335
Methyl-3,4,5-trimethoxycinnamate, 264
Mexican tea, see *Chenopodium ambrosioides*
Michelia champaca Linn., 230—231
Micrococcus spp., 95, 106
Migraine, 91, 302
Milk hedge, common, see *Euphorbia neriifolia*
Mimosaceae
  Acacia catechu, 8
  Acacia farnesiana, 10
  Acacia nilotica, 12
  Albizzia lebbek, 23
Mimusops elengi Linn., 232—233
Mint, corn, see *Mentha arvensis*
Mocharas, 81
Modar, see *Calotropis procera*
Moghuls, effect of fall of, 4
Mohammedan invasion of India, 3
Molludistin, 250
Mongolia, Ayurvedic teachings, 2
Monniera cuneifolia Mich. see *Bacopa monniera*(L.) Pennell
Monohydroxyoleic acid, 112
Monospermoside, 86
Moraceae
  Artocarpus integrifolia, 54
  Ficus benghalensis, 186
  Ficus racemosa, 187
Morello flavone, 3-(8) flavonoyl-flavonone, 192
Morin, 54
Moringaceae
  Moringa oleifera, 234
Moringa oleifera Lam., 234—235
Moringa pterygosperma C. F. Gaertn., see *Moringa oleifera* Lam.
Morphine, 257
Mountain ebony, see *Bauhinia tomentosa*
Mouth, ulcers of, 9
Mucic acid, 175
Mucilage, 32, 55
Mucuna pruriens (Stickm.) DC, 236—237
Mucunine, 236
Mudar, swallowwort, see *Calotropis gigantea*
Mudarine, 93
Mugwort, see *Artemisia vulgaris* Linn.
Muscular rheumatism, 84
Muscular strain, treatment of, 116
Musk mallow, see *Hibisucs abelmoschus*
Musk root, see *Nardostachys jatamansi*
Myocardial contraction, stimulation, 95, 162
Myocarditis, treatment of, 116
Myrcene, 131
Myricetin, 60, 91, 204

Myricetin-7, 91
Myristicaceae
   Myristica fragrans, 238
Myristic acid, 22, 112, 129, 143, 162, 175
Myristica fragrans Houtt., 238
Myrsinaceae
   Embelia ribes, 173
Myrtaceae
   Barringtonia acutangula, 68
   Eugenia jambolana, 179

N

Nalanda University, 3
Naphthalene derivatives, 81
Narceimine, 190
Narceine, 257
Narcotic poisoning, 291, 294, 329
Narcotic substances, 16, 57, 211, 337
Narcotine, 257
Nardostachys jatamansi DC., 239—240
Nardostechone, 239
Naringenin-4-O-β-glucopyranosyl-α-xylopyranoside, 247
Negro coffee, see *Cassia occidentalis*
Nelumbium nelumbo (Linn.) Druce, see *Nelumbo nucifera* Gaertn.
Nelumbium speciosum Willd., see *Nelumbo nucifera* Gaertn.
Nelumbo nucifera Gaertn., 241—242
Neoajmaline, 286
Nepal, Ayurvedic teachings, 2
Neriodorein, 243
Nerium indicum Mill., 243—244
Nerium odorum Soland., see *Nerium indicum* Mill.
Nerol, 250
Nervine substances, 61, 111, 272, 282, 310
Nervous disorders, 286, 301, 303, 311, 330, 332, 335, 337
Neuralgia, treatment of, 35, 107, 301
Neurasthenia, 61
Neuromuscular blocking action, 120
Nicotiflorin, 247
Nicotinamide, 291
Nicotine, 337
Nicotinic acid, 327
Nigella sativa Linn., 245
Night jasmine, see *Nyctanthes arbor-tristis*
Nimatone, 60
Nimbin, 60
Nimbinin, 60
Nimbosterol, 60
Niranthin, 262
Nirtetralin, 262
Nitrogen, 113
Nitrogenin, 63
Nonacosane, 60
n-Nonadecane, 264
Nondrying oil, 81
Nonyl aldehyde, 118
Norhyoscyamine, 157
Norlobelanine, 219
Nornuciferine, 241
Nose
   bleeding, 289
   discharges, 214
Nucleoprotein, 127
Nut grass, see *Cyperus rotundus*
Nutmeg, see *Myristica fragrans*
Nutmeg butter, 238
Nutritive substances, 12, 21, 54—56, 72, 82, 144, 272, 327
Nux vomica, 2, see also *Strychnos nux vomica* Linn.
Nyctaginaceae
   Boerhaavia diffusa, 79
Nyctanthes arbor-tristis Linn., 246
Nymphaea nelumbo Linn., see *Nelumbo nucifera* Gaertn.
Nymphaeaceae
   Nelumbo nucifera, 241

O

Obesity, 49
Oblongifoliol, 142
Ochrocarpus longifolius Benth. & Hook f., 248
Ocimene, 147
Ocimum sanctum Linn., 184, 249—250
n-Octadecane, 264
12,13-Epoxy-9-octadecenoic acid, 112
Odoratine, 335
Oleaceae
   Jasminum sambac Ait., 214
   Nyctanthes arbor-ristis, 246
Oleanolic acid, 23, 180, 220
Oleic acid, 22, 88, 112, 125, 143, 162, 175, 182, 264, 302
Oleic oil, 103
Olein, 222
Oil, drying, 32
Omum, see *Carum Copticum*
Onion, see *Allium cepa*
Operculina turpethum (Linn.) Silva, Manso, 251
Ophelic acid, 312
Ophthalmia, 36, 103, 212, 323, see also Eye, inflammatory conditions
Ophthalmic surgery, 57
Opium, poisoning antidotes, 57, 111
Opium poppy, see *Papaver somniferum*
Orchidaceae
   Dendrobium macraei, 160
   Vanda roxburghii, 331
Orchiditis, see Orchitis
Orchitis, 33, 91
Orientin, 250
Oroxinden, 252
Oroxylin A, 252
Oroxylum indicum (Linn.) Vent., 252
Osyritin, 318
Otitis, 49, 67, 332
Otorrhea, treatment of, 121, 202
Oxalidaceae

Averrhoa carambola, 58
Oxalic acid, 336
Oxyacanthine, 73
Oxyanthrequinone, 106
Oxyberberine, 73
Oxytocic substances, see Uterus, stimulation

P

Paederia foetida Linn., 253—254
α-Paederine, 253
β-Paederine, 253
Paederoside, 253
Palasonin, anthelmintic action, 86
Palmae
  Areca catechu, 46
  Borassus flabellifer, 82
  Cocos nucifera, 129
Palmitic acid, 22, 88, 103, 112, 115, 143, 162, 175, 182, 253, 264, 302
Palmitin, 73, 134, 323
Palmyra palm, see *Borassus flabellifer*
Palpitation of the heart, 21
Pandanaceae
  Pandanus tectorius, 255
Pandanus odoratissimus Roxb., see *Pandanus tectorius* Soland. ex Parkinson
Pandanus tectorius Soland. ex Parkinson, 255
Paniculatine, 111
Pannay tree, see *Calophyllum inophyllum*
Papain (papayotinn), 100
Papaveraceae
  Argemone mexicana, 47
  Papaver somniferum, 256
Papaverine, 257, 286
Papaver somniferum Linn., 256—257
Pawpaw, see *Carica papaya*
Papilionaceae
  Butea monosperma, 86
  Clitoria terneata, 126
  Dolichos biflorus, 167
  Erythrina variegata, 177
  Glycyrrhiza glabra, 194
  Mucuna pruriens, 236
  Psoralea corylifolia, 274
  Pterocarpus marsupium, 276
  Pterocarpus santalinus, 278
Paracelsus, Greek philosopher in India, 2
Paralysis, treatment of, 16, 66, 269, 292, 294, 303, 311, 315, 329
Parasympathetic nerve endings, action of chaksine on, 103
Parturition, difficult, 305
Patchouli, odor resembling, 18
Pea, butterfly, see *Clitoria ternatea*
Pectin, 100
Pedaliaceae
  Sesamum indicum Linn.
Peganine, 259
Peganum harmala Linn., 258—259
Pelargonidin 3, 281

Pelargonin, 283
Pelletierine, 281
Pellitorine, 37
Pellitory, see *Anacyclus pyrethrum*
Pemphigus, 226
Pennywort, Indian, see *Hydrocotyle asiatica*
Pentose glycosides, 281
Pentunidin, 281
Pepper, black, see *Piper nigrum*
Pepper, long, see *Piper longum*
Peptic, 283
Peptidase, 125
Peptide, 169
Pericarditis, 320
Periplocaceae
  Hemidesmus indicus, 203
Peristaltic movement, to assist, 22
Persian lilac, see *Azadirachta indica*; *Melia azedarach*
Petroaselinic acid, 260
Peucedanum graveolens Linn., 260
Pharbitin, 213
Pharbitis nil Choisy, see *Ipomoea nil* (L.) Roth
Pharbitis seeds, see *Ipomoea nil*
Pharyngitis, 222
Phellandrene, 118, 260, 300, 341
  d-α-Phellandrene, 21, 117, 149, 340
  β-Phellandrene, 145, 227
Phenethylamine, 272
Phenol, 115
Phenylacetaldehyde, 247
Phenylalanine, 46, 199, 208, 253, 342
Phenylalanine E, 221
Phenylbutazone, 150
4-Phenylcoumarin, 229
Phenylethyl acetate, 255
Phenylethyl alcohol, 255, 264
"Philosophers stone", of chemists, 4
Phlegm, 221, 296, 300
Phlobaphenes, 30, 106
Phlobotannin, 202
Phosphaturia, 325
Phosphohydrolase activity, 79
Phosphoric anhydride, 296
Photooxidation, 274
Photosterin, 238
Phthalic acid, 255
Phthisis, 66, 72, 82, 86
Phyllanthum, 262
Phyllanthus emblica Linn., see *Emblica officinalis* Gaert.
Phyllanthus fraternus Webster, 261—262
Phyllanthus niruri auct. non Linn. see *Phyllanthus fraternus* Webster
Phyllemblin, 175
Phyltetralin, 262
Physcion(3-menthyl-6-methoxy-1,8-dihydroxy anthraquinone), 107
Phytosteria, 32
Phytosterol, 83, 87, 123, 129, 202, 316
Phytosterolin, 123

Picatorine alkaloid, 7
Picro-aconine, 16
Picrocrocin, 141
Picrorhiza kurroa Royle ex Benth., 263
Picroside-1, 263
Piles, see Hemorrhoids
Pinaceae
   Cedrus deodara, 110
Pinene, 18, 118, 137, 145
   α-Pinene, 246, 250
   β-Pinene, 250
   d-Pinene, 30
   d-α-Pinene, 147
α-Pinosterol, 122
l-Pinocamphene, 212
Pinocampherol, 230
Pinus deodara Roxb. S.P.B., see *Cedrus deodara* Loud.
Piperaceae
   Piper longum, 264
   Piper nigrum, 266
Piperidine, 266
Piperine, 264, 266
Piperlin, 266
Piper longum Linn., 264—265
Piperlongumine, 264
Piper nigrum Linn., 266
Piperolein A, 266
Piperolein B, 266
Plantaginaceae
   Plantago ovata, 267
Plantago ovata Forsk., 267
Plasma fibrinogen, 26
Plumbaginaceae
   Plumbago zeylanica, 268
Plumbago zeylanica Linn., 238, 268—269
Plumericin, 243
Pneumonia, 301
Poisons, see Toxic substances
Poison bulb, see *Crinum defixum*
Polygonaceae
   Rheum emodi, 287
Polyisoprenoid, 230, 270
Polypodiaceae
   Adiantum philippense, 20
Polysaccharides, 314
Pomegranate, see *Punica granatum*
Poppy, opium, see *Papaver somniferum*
Portuguese in India, 4
Potassium nitrate, 50, 79
Prabhava, discussed, 352
Premna corymbosa auct. non Rottl. and Willd., see *Premna obtusifolia* R. Br.
Premna integrifolia Linn. see *Premna obtusifolia* R. Br.
Premna obtusifolia R. Br., 270—271
Premnine, 270
Prickly heat, 297
Prickly leaves, see *Elephantopus scaber*
Proanthocyanidin, 106
Procyanidin, 46, 86

Prolamin, 327
Proline, 46, 129, 199, 225
Prophylactic, 300
Propterol, 277
Proteids, 266
Proteinase, 125
Proteolytic enzymes, 175
Protocatechic acid, 25
Protocrocin, 141
Protoemetinol, 22
Protopine, 47, 190
Prunetin-4′, 252
Prunus amygdalus Batsch., 272
Prunus puddum Roxb. ex Wall., 273
Prurienine, 236
Prurigo, 297
Pseudoaconitine, 16
Pseudobaptigenin, 277
Pseudocarpaine, dimeric, 100
Pseudo-pelletieriene, 281
Pseudostrychnine, 310
Pseudotropine, 337
Pseudowithanine, 337
Psoralea corylifolia Linn., 274—275
Psoralen, 44, 274
Psoralidin, 274
Psoriasis, 208, 289
Psychomimetic effect, 96
Psychotrine, 22
Ptergospermin, 234
Pterocarpdiolone, 279
Pterocarptriol, 279
Pterocarpol, 279
Pterocarpus marsupium Roxb., 276—277
Pterocarpus santalinus Linn., 278—279
Pterostilbine, 277, 279
Ptychotis ajowan DC., see *Carum copticum* Benth.
Pulmonary disorders, 5, 26, 47, 72, 75, 83, 95, 97, 110
Punarnavine, 79
Punicaceae
   Punica granatum, 280
Punica granatum Linn., 280—281
Punico tannic acid, 281
Purgative croton, see *Croton tiglium*
Purgative substances, 7, 14, 22, 28, 42, 50, 63, 64, 73, 79, 91, 94, 104, 106, 107, 140, 142, 183, 189, 192, 193, 213, 220, 229, 230, 247, 251, 255, 260, 263, 287, 291, 314, 322, 323, 335, 345
Purging cassia, see *Cassia fistula*
Purple tephrosia see *Tephrosia purpurea* (Linn) Pers.
Purpurins, 292, 318
Pyrexia, 73
Pyroligneous acid, 197
Pyromucic acid, 115
Pyruvic acid, 115
Pythagoras, visited India, 2

Q

Quercetin, 25, 75, 60, 81, 91, 197, 241

Quercetin-3 galactoside, 186
Quercitol, 232
Quercitrin, 225
Quercitrin (3-rhamnoside), 75
Quinazoline, 259
Quinovin, 313

R

Racemic acid, 336
Radish, see *Raphanus sativus* Linn.
Raffinose, 274
Randia dumetorum Lam., 282
Ranunculaceae
   Aconitum ferox, 15
   Aconitum heterophyllum, 17
   Coptis teeta, 133
   Delphinium denudatum, 159
   Nigella sativa, 245
   Thalictrum foliolosum, 323
Ranikhat disease virus, 47, 111
Rapanone, 174
Raphanus sativus Linn., 283
Rasa sastra, iatro-chemistry, 2
"Rasant", 73
Rasayana tantra, geriatrics, 1
Raubasine, 286
Rauhimbine, 286
Rauwolfia serpentina (L.) Benth. ex Kurz., 284—286
Rauwolfinine, 286
Red mango, see *Garcinia indica*
Red sandalwood, see *Pterocarpus santalinus* Linn.
Refrigerant and cooling substances, 21, 43, 56, 58, 66, 82, 83, 219, 279, 281, 297, 303, 313, 314, 325, 326
Reidin B, 287
Reidin C, 287
Renal dropsy, 14, 329
Rennin, 95
Rescinnamine, 286
Reserpine, 286
Reserpinine, 286
Resolvent substances, 77, 260, 294, 305
Respiratory diseases, 257, 263, 265, 271, 282, 283, 287, 289, 296, 300, 302, 305, 307, 321, 322, 329
Rhamnaceae
   Zizyphus martiana, 232
   Zizyphus vulgaris, 345
Rhamnose, 139, 185, 251, 329
Rhaponticin, 287
Rhein, 104, 106, 107, 109, 287
Rheinglucoside, 106
Rheo-tannic acid, 287
Rheumatism, 30, 36, 37, 45, 89, 104, 106, 203, 251, 252, 254, 255, 265, 267, 269, 271, 282, 291, 296, 300, 303, 307, 311, 324, 327, 329, 332, 337
Rheumatoid arthritis, treatment of, 44, 194
Rheum emodi Wall., 287
Rhus succedanea Linn., 288—289

Ribbed gourd, see *Luffa acutangula* (Linn.) Roxb.
Ricin, 290
Ricinine, 291
Ricinoleic acid, 290
Ricinus communis Linn., 290—291
Rigveda, 1
Ringworm, 83, 103, 106, 136
Rome, influence of Hindu medicine in, 1
Rosaceae
   Prunus amygdalus, 272
   Prunus puddum, 273
Rotenone, 318
Rottlera, see *Mallotus philippensis*
Rough chaff tree, see *Achyranthes aspera*
Round zedoary, see *Curcuma zedoaria*
Roundworm, 86, 113, 206
Rubefacients, action, 25, 64, 111, 129, 140, 143, 151, 183, 266, 294, 342
   local stimulants, and 26, 98, 100
Rubiaceae
   Anthocephalus indicus, 43
   Paederia foetida, 253
   Randia dumetorum, 282
   Rubia cordifolia, 292
Rubia cordifolia Linn., 292
Rue, Syrian, see *Peganum harmala*
Rutaceae
   Aegle marmelos, 21
   Peganum harmala, 258
   Ruta graveolens, 293
   Zanthoxylum alatum, 339
Ruta graveolens Linn., 293—294
Rutin, 186, 225, 294, 318

S

Sabinene, 149, 172
Sacred fig, see *Ficus religiosa*
Sacred lotus, see *Nelumbo nucifera*
Saffron, see *Crocus sativus*
Saffron, cobra's, see *Mesua ferrea*
Safrole, 147
Safronal, 141
Sage leaves, see *Alangium salviifolium*
Salicylic acid, 11
Salmalia malabarica Schott and Endl., see *Bombax ceiba*
Salvadoraceae
   Salvadora persica, 295
Salvadora persica Linn., 295—296
Salya toutra, surgery, 1
Sammohini, anaesthetic, 1
Sanguinarine, 47
Santalaceae
   Santalum album, 297
Santalin A, 279
Santalin B, 279
Santalols, 297
Santalone, 297
Santalum album Linn., 297
Santonone, 297

Saponin, 14, 56, 61, 63, 85, 88, 123
Sapotaceae
    Madhuca longifolia, 222
    Mimusops elengi, 232
Saraca indica Linn., 298
Sarcoma 180, 23
Sarpagine, 286
Sarsaparilla, Indian, see *Hemidesmus indicus*
Saussurea lappa C.B. Clarke, 299—300
Saussurine, 300
Saxifraga ligulata Wall., see *Bergenia ligulata* (Wall.) Engl.
Saxifragaceae
    Bergenia ligulata, 74
Scabies, treatment of, 91, 101, 297
Scandoside, 253
Scillaren A, 329
Scillaren B, 329
Scitaminiaceae
    Alpinia galanga, 29
    Costus speciosus, 138
Sclerenchyma, 55
Scopolamine, 211
Scopoletin, 251, 282
Scorpion sting, 35, 112, 137, 156, 248, 263, 265, 282, 289, 292, 305, 312, 315, 320, 332, 334
Screwpine, fragrant, see *Pandanus tectorius*
Screwtree, East Indian, see *Helicteres isora*
Scrofula, treatment of, 60, 71, 190, 301
Scrophulariaceae
    Bacopa monniera, 61
    Digitalis purpurea, 162
    Picrorhiza kurroa, 263
Scurvy, 296, see also Antiscorbutics
Scutellarein-7-rutinoside, 252
Scythian invasion of India, 3
Sebiferine, 218
Sedanolide, 44
Sedanomic acid anhydride, 44
Sedative substances, 16, 44, 57, 83, 96, 97, 116, 141, 240, 242, 282, 286, 297, 298, 300, 301, 306, 317, 330, 337
Seeds, babchi, see *Psoralea corylifolia*
Seeds, spogel, see *Plantago ovata*
Selin-4 (15) en-1b, 277
d-Selinene, 44
γ-Selinene, 250
Semecarpol, 301
Semecarpus anacardium Linn. f., 301
Senna, Indian, see *Cassia angustifolia*
Sennoside A, 104, 106
Sennoside B, 104, 106
Serine, 199
DL-Serine, 264
Serpentina, see *Rauwolfia serpentina* (L.) Benth. ex Kurz.
Serpentine, 286
Serpentinine, 286
Serum cholesterol studies, 131—132
Sesamin, 131, 264
Sesamum, see *Sesamum indicum* Linn.

Sesamum indicum Linn., 302
Sesamum orientale Linn., see *Sesamum indicum* Linn.
Sesomolin, 302
Sesquiterpene alcohol, 18, 40, 44, 45, 60, 166
Sesquiterpene hydrocarbons, 156
Sesquiterpenes, 149, 340
Sexual impotence, 311
Shell shock, 330
Shigella somnei, 185
Sialogogue substances, 37, 342
Sida cordifolia Linn., 303
Silajit, 11
Silicon dioxide, 131
Silk cotton tree, see *Bombax ceiba*
Simaroubaceae
    Balanites aegyptiaca, 62
Siris, see *Albizzia lebbek*
Sitosterol, 22, 162, 253, 274, 279, 281
    α-3-Sitosterol, 107
    β-Sitosterol, 6, 21, 22, 46, 60, 61, 75, 79, 81, 103, 106, 125, 128, 136, 139, 167, 186, 199, 252, 264, 270, 281, 297, 334
    γ-Sitosterol, 21, 23
β-Sitosterol glucoside, 188, 260
Skimmianine, 340
Skin diseases, treatment of, 26, 33, 71, 109, 112, 125, 128, 139, 146, 147, 150, 152, 188, 203, 250, 262, 269, 272, 277, 279, 287, 292, 297, 300, 301, 306, 309, 312, 313, 324, 329, 334
Small caltrops, see *Tribulus terrestris*
Small pox, 327
Snake bite, treatment for, 32, 35, 43, 45, 54, 64, 70, 79, 109, 110, 136, 137, 142, 265, 289, 324
Solamargine, 306
Solanaceae
    Atropa acuminata, 57
    Capsicum annuum, 98
    Datura metal, 157
    Hyoscyamus niger, 211
    Solanum indicum, 304
    Solanum nigrum, 306
    Solanum surattense, 307
    Withania somnifera, 337
Solanidine, 98, 304, 307
Solanine, 304, 306, 307
Solanocarpine, 307
Solanum indicum Linn., 304
Solanum nigrum Linn., 306
Solanum surattense Burm f., 307
Solanum xanthocarpum Schrad. and Wendl., see *Solanum surattense* Burm f.
Solasodine, 98
Soma plant, 1
Somniferiene, 337
Somniferinine, 337
Spanish pepper, see *Capsicum annuum*
Spasmodic conditions, treatment for, 116, 311
Spasmolytic action, 152
Spastic disorders, 330
Spermatorrhea, treatment of, 55, 67, 199, 237, 303, 324, 325, 337

Spermicides, 222
Spermine, 270
Sphaeranthine, 308
Sphaeranthus indicus Linn., 308
α-Spinasterol, 222, 232, 334
Spleen, 44, 130
　enlargement, 5, 93, 115
　inflamed, 254
Spogel seeds, see *Plantago ovata*
Sponge gourd, see *Luffa cylindrica*
Spreading hogweed, see *Boerhaavia diffusa*
Sprue, 320
Squalene, 129
Sri Lanka, Ayurvedic teachings, 2
Staff tree, see *Celastrus paniculatus* Willd.
Staphisagrine, 159
Staphyllococcus aureus, growth inhibited, 185, 188, 271
Starch, curcuma, see *Curcuma angustifolia*
Starch as valuable article of diet, 148
Stearic acid, 88, 103, 112, 115, 143, 175, 264, 290, 302
Stearic acid, glyceride of, 125
Sterculiaceae
　Abroma augusta, 6
　Helicteres isora, 202
Steroidal ketones, 131
Sterols, 31, 50, 85, 111, 112
Steroptene, 203, 255
Steroptin, 102
Stigmasterol, 22, 122, 253, 334
Stimulant substances, 18, 25, 26, 30, 31, 33, 37, 40, 44—46, 51, 64, 66, 81, 82, 84, 96, 98, 101, 102, 111, 117, 161, 234, 238, 259, 260, 264—266, 272, 286, 289, 294, 297, 300, 301, 303, 309, 310, 316, 329, 330, 342
Stomach and bowel disorders, 39, 89, 102, 120, 137, 167, 181, 184, 185, 189, 252, 262, 263, 265, 266, 267, 271, 272, 283, 289, 291, 297, 313, 321, 327, 342
Stomachic substances, 5, 17, 21, 28, 30, 31, 33, 35, 36, 39, 52, 73, 78, 79, 98, 100, 101, 112, 115, 118, 120, 121, 145, 147, 159, 172, 176, 180, 197, 201, 204, 206, 227, 245, 248, 250, 252, 260, 263, 271, 281, 287, 303, 309, 310, 312, 317, 330, 336, 340, 342, 344
Storax, see *Altingia excelsa*
Strangury and urinary disorders, 66, 242
Streptogenin, 167
Strychnicine, 310—311
Strychnine, 310
Strychnos nux vomica Linn., 310
Styptics, 15
Styracin, 33
Styrol, 33
Succinic acid, 162, 330
Sucrose, 24, 82
Sugar apple, see *Annona squamosa*
Sulfaguanidine, compared to Berberine, 73
Sulfurein, 86
Superb lily, see *Gloriosa superba*

Susruta, later book of Ayurveda, 1
Sweet flag, see *Acoris calamus*
Sweet wood, see *Glycyrrhiza glabra*
Swertia chirata Buch-Ham, 312
Sylvatine, 264
Symplocaceae
　Symplocos racemosa, 313
Symplocos racemosa Roxb., 313
Syphilis, 82, 203, 224, 324, 332
Syrian rue, see *Peganum harmala*
Syzygium jambolanum DC., see *Eugenia jambolana* Lam.

T

Tabashir (bamboo manna), 66
Taeniafuge substances, 46, 47, 266, 281, see also Anthelmintics
Tamarind, see *Tamarindus indica*
Tamarindus indica Linn., 314—315
Tannic acid, 107, 322
Tannin, 11, 12, 30, 49, 54, 85, 106, 162
Tapeworms, treatment for, see Taeniafuge substances
Taraxacerin, 316
Taraxacin, 316
Taraxacum officinale Weber, 316
Taraxasterol, 316
γ-Taraxasterol, 186
Taraxerol, 6, 127
Taraxerone, 127
Taraxeryl acetate, 6
Tartaric acid, 100, 314, 330, 336
Tautra and *Siddhas*, 1, 3
Taxicatin, 317
Taxila, university, 3
Taxin, 5, 317
Taxines, 317
Taxus baccata Linn., 317
Teeth, carious, 37
Telgupotato, see *Amorphophallus campanulatus* (Roxb.)Blume
Tephrosia lanceolata R. Grah., see *Tephrosia purpurea* (Linn.) Pers.
Tephrosia maxima Pers., see *Tephrosia purpurea* (Linn.) Pers.
Tephrosia purpurea (Linn.) Pers., 318
Tephrosin, 318
Terminalia arjuna Wight & Arn., 319—320
Terminalia bellerica (Gaertn.) Roxb., 321
Terminalia chebula Retz., 322
Terpene alcohol, 300
Terpenic ketone, 215
α-Terpenol, 145
Terpinene, 172
Terpinene-4-ol, 250
Terpineol, 172, 274
Terpinolene, 264
α-Terthienylmethanol, 169
Tetracasanol, 111
Tetrahydrocannabinol, 96
Tetrahydrocoptisine, 190

Tetrahydropalmitine, 73
Tetratriacontanyl-tetratriacontanoate, 202
Thalictrine, 323
Thalictrum foliolosum DC., 323
Thebaine, 257, 286
Thermal burns, 28
Thiophene derivatives, 169
Thirst, quenching, 58, 248, 297
Thornapple, see *Datura metal*
Thread worms, treatment for, 66, 112
Threonine, 225, 253
α-Thujene, 264
α-Thujone, 218
Thujone oil, 52, 53
Thymelaeaceae
  Aguilaria agallocha, 45
Thymene, 102
Thymol, 102, 145, 260
Tibet, Ayurvedic teachings learned, 2
3α-Tigloyloxytropane, 337
Tigogenin, 139
Tiliaceae
  Grewia hirsuta, 199
Tinnevelly, see *Cassia angustifolia*
Tinosporine, 324
Tinospora cordifolia (Willd.) Miers, 324
Tonic and restorative substances, 5, 7, 17, 18, 20, 24, 28, 36, 39, 42—45, 49, 51, 52, 55, 56, 66, 67, 70—73, 81, 82, 84, 86, 89, 93, 98, 102, 106, 111, 112, 146, 161, 163, 167, 222, 225, 257, 262—264, 271, 289, 292, 300, 301, 303, 306, 309, 310, 312, 314, 316—318, 320, 321, 323—325, 327, 332, 337, 390
  cardiac, 303, 320
  fever, 340
Tonsillitis, treatment of, 37
Tonsils, hypertrophy of, 9
Toothache, 37, 116, 248, 266, 277, 307, 323, 340
Tooth brush tree, see *Salvadora persica*
Toxic substances, 16, 243
Toxins
  Aconitum ferox, 15—16, 159
  Cinnamomum zeylanicum, 118
  Gloriosa superba, 193
  Lobelia nicotianaefolia, 219
Trachyspermum ammi [Linn.] Sprag., see *Carum copticum* Benth.
Tree turmeric, see *Berberis aristata* DC.
Triacontane, 162, 182
Triacontanol, 264
Tribulus terrestris Linn., 325—326
Trichomes, 14
Tricosane, 340
Tridosh fever, 77
Tridoshas, defined, 347
Triglycerides, 131, 260
Trigonella foenum-graecum Linn., 327
Trigonelline, 327
Trilaurin, 274
Trimethoxy-l-propenyl benzane, 18
Trimethylamine, 296

Trimyristicin, 238
Triphala, 176, 321, 322
Triterpenoids, 69, 222
Tropine, 306, 337
Trypsin, 38
Tryptophan, 46, 129, 253
Tubulosine, 22
Tumors, 42, 49, 52, 181, 165, 296, 300, 309
Turmeric, see *Curcuma longa*
Turmeric oil, 149
Turpethins, 251
Tuscan jasmine, see *Jasminum sambac*
Tympanitis, 318
Tyrosine, 46, 127, 129, 253, 267
L-Tyrosine, 264

U

Ulcerations, 292, 300
Ulcers, 313
Ulcers and wounds, 42, 52, 60, 71, 72, 181
Umbelliferae
  Angelica glauca, 40
  Apium graveolens, 44
  Carum carvi, 101
  Carum copticum, 102
  Coriandrum sativum, 137
  Cuminum cyminum, 145
  Daucus carota, 158
  Ferula foetida, 185
  Foeniculum vulgare, 189
  Hydrocotyle asiatica, 208
  Peucedanum graveolens, 260
Umbelliferone, 185
Umbelliprenin, 40
Upaveda, 1
Urease, 167
Urethral discharges, 309
Urginea indica Kunth, 328—329
Urinary disorders, 32, 46, 61, 110, 120, 136, 146, 148, 150, 154, 170, 172, 217, 262, 273, 283, 303, 315, 322, 324, 325
Ursolic acid, 79, 232, 243, 250, 253, 281
Urso saponin, 282
Urticaceae
  Cannabis indica, 96
Urticaria, treatment of, 191
Urushiol, 289
Uscharin, 93, 95
Uterus
  bleeding, 81
  pain, 116, 157
  stimulation, 93, 109

V

Vaginal discharges, 315
Vajikarana tantra, science of aphrodisiacs, 1
Valeriana jatamansi Jones, 330
Valeriana wallichii DC, see *Valeriana jatamansi* Jones

Valerianaceae
  Nardostachys jatamansi, 239
  Valeriana jatamansi, 330
Valerianic oil, 330
Valeric acid, 300
Valine, 127, 253, 267
Vanda roxburghii R. Br., 331—332
Vanda tessellata Hook ex G. Don, 331
Vanillin, 33, 234
Vasicine, 216
Vasicinone, 216
Vedic period, of Hindu medicine, 1
Vegetable sponge, see Luffa cylindrica
Velvet leaf, see Cissampelos pareira
Venereal buboes, treatment of, 178
Venereal diseases, 301, see also specific diseases
Verbenaceae
  Callicarpa marcophylla, 89
  Clerodendrum infortunatum, 124
  Gmelina arborea, 196
  Premna obtusifolia, 270
Vermicides and vermifuges, 26, 70, 95, 102, 259, 265, 279, 281, 294, 296, 301, 304, 312, 322, 334
Vernolic acid, 112
Vernonia anthelminticum Willd., see Centratherum anthelminticum (Willd.) Kuntz
Vernonia cinerea Less., 333—334
Vertigo, 91, 231, 302
Vilangin, 174
Violaceae
  Viola odorata, 335
Viola odorata Linn., 335
Violaquercitrin, 335
"Violine", 335
Vipat, discussed, 350—351
Virya, discussed, 351—352
Visceral obstructions, 134
Vitaceae
  Cissus quadrangularis, 121
  Vitis vinifera, 336
Vitamin B, 72, 82
Vitis quadrangularis Wall., see Cissus quadrangularis Wall.
Vitis vinifera Linn., 336
Vomicine, 310
Vomiting, 305, 312, 315, 322, 335

W

Walker carcinosarcoma, 61
Warts, treatment of, 182
Wedelolactone, 169
White leadwort, see Plumbago zeylanica
White pumpkin, see Benincasa hispida
White sandalwood, see Santalum album Linn.
Wild cinchona, see Anthocephalus indicus
Wild violet, see Viola odorata Linn
Wind, 296, 300
Winter cherry, see Withania somnifera Dunal
Withananine, 337
Withania somnifera Dunal, 337—338
Withanine, 337
Withanone, 337
Wood oil tree, see Dipterocarpus indicus
Wounds, treatment of, 186, 345
Wormkiller, see Aristolochia bracteata, 50
Worm seed, see Artemisia maritia Linn.
Worms, treatment for, 78, 93, see Anthelmintics; Vermifuges

X

Xanthine, 292
Xanthoplanine, 340
Xanthorhamnin, 182
Xanthotoxin, 44

Y

Yam, see Dioscorea bulbifera
Yellow champa, see Michelia champaca
Yellow thistle, see Argemone mexicana
Yohimbine, 286

Z

Zanthoxylum alatum Roxb., 339—340
Zedoary, round, see Curcuma zedoaria
Zeylinone, 268
Zingiberaceae
  Amomum subulatum, 34
  Curcuma amada, 147
  Curcuma angustifolia, 148
  Curcuma longa, 149
  Curcuma zedoaria, 151
  Elettaria cardamomum, 172
  Zingiber officinale, 341
Zingiber officinale Rosc., 341—342
Zingiberene, 149, 264, 341
Zizyphic acid, 344, 345
Zizyphus jujuba Lam. non Mill., see Zizyphus martiana Lam.
Zizyphus martiana Lam., 343—344
Zizyphus vulgaris Lam., 345
Zygophyllaceae
  Tribulus terrestris, 325